Algebraic Methods in Unstable Homotopy Theory

This is a comprehensive up-to-date treatment of unstable homotopy. The focus is on those methods from algebraic topology which are needed in the presentation of results, proven by Cohen, Moore, and the author, on the exponents of homotopy groups.

The author introduces various aspects of unstable homotopy theory, including: homotopy groups with coefficients; localization and completion; the Hopf invariants of Hilton, James, and Toda; Samelson products; homotopy Bockstein spectral sequences; graded Lie algebras; differential homological algebra; and the exponent theorems concerning the homotopy groups of spheres and Moore spaces.

This book is suitable for a course in unstable homotopy theory, following a first course in homotopy theory. It is also a valuable reference for both experts and graduate students wishing to enter the field.

JOSEPH NEISENDORFER is Professor Emeritus in the Department of Mathematics at the University of Rochester, New York.

NEW MATHEMATICAL MONOGRAPHS

Editorial Board

Béla Bollobás
William Fulton
Anatole Katok
Frances Kirwan
Peter Sarnak
Barry Simon
Burt Totaro

All the titles listed below can be obtained from good booksellers or from Cambridge University Press. For a complete series listing visit
http://www.cambridge.org/uk/series/sSeries.asp?code=NMM

1 M. Cabanes and M. Enguehard *Representation Theory of Finite Reductive Groups*
2 J. B. Garnett and D. E. Marshall *Harmonic Measure*
3 P. Cohn *Free Ideal Rings and Localization in General Rings*
4 E. Bombieri and W. Gubler *Heights in Diophantine Geometry*
5 Y. J. Ionin and M. S. Shrikhande *Combinatorics of Symmetric Designs*
6 S. Berhanu, P. D. Cordaro and J. Hounie *An Introduction to Involutive Structures*
7 A. Shlapentokh *Hilbert's Tenth Problem*
8 G. O. Michler *Theory of Finite Simple Groups I*
9 A. Baker and G. Wüstholz *Logarithmic Forms and Diophantine Geometry*
10 P. Kronheimer and T. Mrowka *Monopoles and Three-Manifolds*
11 B. Bekka, P. de la Harpe and A. Valette *Kazhdan's Property (T)*
12 J. Neisendorfer *Algebraic Methods in Unstable Homotopy Theory*
13 M. Grandis *Directed Algebraic Topology*
14 G. Michler *Theory of Finite Simple Groups II*

Algebraic Methods in Unstable Homotopy Theory

JOSEPH NEISENDORFER
University of Rochester, New York

CAMBRIDGE UNIVERSITY PRESS
Cambridge, New York, Melbourne, Madrid, Cape Town, Singapore, São Paulo, Delhi

Cambridge University Press
The Edinburgh Building, Cambridge CB2 8RU, UK

Published in the United States of America by Cambridge University Press, New York

www.cambridge.org
Information on this title: www.cambridge.org/9780521760379

© J. Neisendorfer 2010

This publication is in copyright. Subject to statutory exception
and to the provisions of relevant collective licensing agreements,
no reproduction of any part may take place without the written
permission of Cambridge University Press.

First published 2010

Printed in the United Kingdom at the University Press, Cambridge

A catalogue record for this publication is available from the British Library

Library of Congress Cataloguing in Publication data
Neisendorfer, Joseph, 1945–
Algebraic methods in unstable homotopy theory / Joseph Neisendorfer.
p. cm. – (New mathematical monographs ; 12)
Includes bibliographical references and index.
ISBN 978-0-521-76037-9 (hardback)
1. Homotopy theory. 2. Algebraic topology. I. Title.
QA612.7.N45   2009
514$'$.24 – dc22   2009028748

ISBN 978-0-521-76037-9 Hardback

Cambridge University Press has no responsibility for the persistence or
accuracy of URLs for external or third-party internet websites referred to
in this publication, and does not guarantee that any content on such
websites is, or will remain, accurate or appropriate.

To the memory of my algebra teacher,
Clare DuBrock,
Tilden Technical High School,
Chicago, Illinois

Contents

Preface		*page* xiii
Acknowledgments		xix
	Introduction to unstable homotopy theory	1
1.	**Homotopy groups with coefficients**	11
	1.1. Basic definitions	12
	1.2. Long exact sequences of pairs and fibrations	15
	1.3. Universal coefficient exact sequences	16
	1.4. Functor properties	18
	1.5. The Bockstein long exact sequence	20
	1.6. Nonfinitely generated coefficient groups	23
	1.7. The mod k Hurewicz homomorphism	25
	1.8. The mod k Hurewicz isomorphism theorem	27
	1.9. The mod k Hurewicz isomorphism theorem for pairs	32
	1.10. The third homotopy group with odd coefficients is abelian	34
2.	**A general theory of localization**	35
	2.1. Dror Farjoun–Bousfield localization	37
	2.2. Localization of abelian groups	46
	2.3. Classical localization of spaces: inverting primes	47
	2.4. Limits and derived functors	53
	2.5. Hom and Ext	55
	2.6. p-completion of abelian groups	58
	2.7. p-completion of simply connected spaces	63
	2.8. Completion implies the mod k Hurewicz isomorphism	68
	2.9. Fracture lemmas	70
	2.10. Killing Eilenberg–MacLane spaces: Miller's theorem	74
	2.11. Zabrodsky mixing: the Hilton–Roitberg examples	83
	2.12. Loop structures on p-completions of spheres	88
	2.13. Serre's C-theory and finite generation	91

3. Fibre extensions of squares and the Peterson–Stein formula   94

- 3.1. Homotopy theoretic fibres   95
- 3.2. Fibre extensions of squares   96
- 3.3. The Peterson–Stein formula   99
- 3.4. Totally fibred cubes   101
- 3.5. Spaces of the homotopy type of a CW complex   104

4. Hilton–Hopf invariants and the EHP sequence   107

- 4.1. The Bott–Samelson theorem   108
- 4.2. The James construction   111
- 4.3. The Hilton–Milnor theorem   113
- 4.4. The James fibrations and the EHP sequence   118
- 4.5. James's 2-primary exponent theorem   121
- 4.6. The 3-connected cover of S^3 and its loop space   124
- 4.7. The first odd primary homotopy class   126
- 4.8. Elements of order 4   128
- 4.9. Computations with the EHP sequence   132

5. James–Hopf invariants and Toda–Hopf invariants   135

- 5.1. Divided power algebras   136
- 5.2. James–Hopf invariants   141
- 5.3. p-th Hilton–Hopf invariants   145
- 5.4. Loops on filtrations of the James construction   148
- 5.5. Toda–Hopf invariants   151
- 5.6. Toda's odd primary exponent theorem   155

6. Samelson products   158

- 6.1. The fibre of the pinch map and self maps of Moore spaces   160
- 6.2. Existence of the smash decomposition   166
- 6.3. Samelson and Whitehead products   167
- 6.4. Uniqueness of the smash decomposition   171
- 6.5. Lie identities in groups   177
- 6.6. External Samelson products   179
- 6.7. Internal Samelson products   186
- 6.8. Group models for loop spaces   190
- 6.9. Relative Samelson products   198
- 6.10. Universal models for relative Samelson products   202
- 6.11. Samelson products over the loops on an H-space   210

7. Bockstein spectral sequences — 221

- 7.1. Exact couples — 222
- 7.2. Mod p homotopy Bockstein spectral sequences — 225
- 7.3. Reduction maps and extensions — 229
- 7.4. Convergence — 230
- 7.5. Samelson products in the Bockstein spectral sequence — 232
- 7.6. Mod p homology Bockstein spectral sequences — 235
- 7.7. Mod p cohomology Bockstein spectral sequences — 238
- 7.8. Torsion in H-spaces — 241

8. Lie algebras and universal enveloping algebras — 251

- 8.1. Universal enveloping algebras of graded Lie algebras — 252
- 8.2. The graded Poincare–Birkhoff–Witt theorem — 257
- 8.3. Consequences of the graded Poincare–Birkhoff–Witt theorem — 264
- 8.4. Nakayama's lemma — 267
- 8.5. Free graded Lie algebras — 270
- 8.6. The change of rings isomorphism — 274
- 8.7. Subalgebras of free graded Lie algebras — 278

9. Applications of graded Lie algebras — 283

- 9.1. Serre's product decomposition — 284
- 9.2. Loops of odd primary even dimensional Moore spaces — 286
- 9.3. The Hilton–Milnor theorem — 290
- 9.4. Elements of mod p Hopf invariant one — 294
- 9.5. Cycles in differential graded Lie algebras — 299
- 9.6. Higher order torsion in odd primary Moore spaces — 303
- 9.7. The homology of acyclic free differential graded Lie algebras — 306

10. Differential homological algebra — 313

- 10.1. Augmented algebras and supplemented coalgebras — 315
- 10.2. Universal algebras and coalgebras — 323
- 10.3. Bar constructions and cobar constructions — 326
- 10.4. Twisted tensor products — 329
- 10.5. Universal twisting morphisms — 332
- 10.6. Acyclic twisted tensor products — 335
- 10.7. Modules over augmented algebras — 337
- 10.8. Tensor products and derived functors — 340
- 10.9. Comodules over supplemented coalgebras — 345

10.10.	Injective classes	349
10.11.	Cotensor products and derived functors	356
10.12.	Injective resolutions, total complexes, and differential Cotor	363
10.13.	Cartan's constructions	369
10.14.	Homological invariance of differential Cotor	374
10.15.	Alexander–Whitney and Eilenberg–Zilber maps	378
10.16.	Eilenberg–Moore models	383
10.17.	The Eilenberg–Moore spectral sequence	387
10.18.	The Eilenberg–Zilber theorem and the Künneth formula	390
10.19.	Coalgebra structures on differential Cotor	393
10.20.	Homotopy pullbacks and differential Cotor of several variables	395
10.21.	Eilenberg–Moore models of several variables	400
10.22.	Algebra structures and loop multiplication	403
10.23.	Commutative multiplications and coalgebra structures	407
10.24.	Fibrations which are totally nonhomologous to zero	409
10.25.	Suspension in the Eilenberg–Moore models	413
10.26.	The Bott–Samelson theorem and double loops of spheres	416
10.27.	Special unitary groups and their loop spaces	425
10.28.	Special orthogonal groups	430

11. Odd primary exponent theorems   437

11.1.	Homotopies, NDR pairs, and H-spaces	438
11.2.	Spheres, double suspensions, and power maps	444
11.3.	The fibre of the pinch map	447
11.4.	The homology exponent of the loop space	453
11.5.	The Bockstein spectral sequence of the loop space	456
11.6.	The decomposition of the homology of the loop space	460
11.7.	The weak product decomposition of the loop space	465
11.8.	The odd primary exponent theorem for spheres	473
11.9.	H-space exponents	478
11.10.	Homotopy exponents of odd primary Moore spaces	480
11.11.	Nonexistence of H-space exponents	485

12. Differential homological algebra of classifying spaces   489

12.1.	Projective classes	490
12.2.	Differential graded Hopf algebras	494
12.3.	Differential Tor	495
12.4.	Classifying spaces	502
12.5.	The Serre filtration	504
12.6.	Eilenberg–Moore models for Borel constructions	505
12.7.	Differential Tor of several variables	508

12.8.	Eilenberg–Moore models for several variables	512
12.9.	Coproducts in differential Tor	515
12.10.	Künneth theorem	517
12.11.	Products in differential Tor	518
12.12.	Coproducts and the geometric diagonal	520
12.13.	Suspension and transgression	525
12.14.	Eilenberg–Moore spectral sequence	528
12.15.	Euler class of a vector bundle	530
12.16.	Grassmann models for classifying spaces	534
12.17.	Homology and cohomology of classifying spaces	537
12.18.	Axioms for Stiefel–Whitney and Chern classes	539
12.19.	Applications of Stiefel–Whitney classes	542

Bibliography   545

Index   550

Preface

What is in this book and what is not

The purpose of this book is to present those techniques of algebraic topology which are needed in the presentation of the results on the exponents of homotopy groups which were proven by Cohen, Moore, and the author. It was decided that all of the details of those techniques would be completely and honestly presented.

Homotopy groups with coefficients are fundamental to the whole enterprise and have and will be useful in other things. The 2-primary theory was not excluded but the fact that certain things are just not true for the 2-primary case reinforces the eventual restriction, more and more, to the odd primary case and finally to the case of primes greater than 3. The argument could have been made that the exact sequences of these groups related to pairs and to fibrations are all a consequence of the fundamental work of Barratt and of Puppe on cofibration sequences and can be found as a special case in the books of G. Whitehead or of E. H. Spanier. But the general theory does not handle the low dimensional cases which correspond to the fundamental group and the only way to provide an honest uniform treatment was to present the whole theory in detail. So that is what is done.

Localization has undergone a revolution in the hands of Dror Farjoun and of Bousfield. This new theory is incredibly general. It includes both the classical theory of inverting primes and of completion. It also includes exotic forms of localization related to a theorem of Haynes Miller. Some simplifications can be made if one restricts localizations to simply connected spaces or to H-spaces. It seemed to the author that not much is lost in terms of potential applications by so doing. The same is true if no appeal is made to the arcane theory of very large sets and if we restrict Dror Farjouns fundamental existence proof so that the largest thing we have to refer to is the least uncountable ordinal.

It seemed that localization should be presented in this new incarnation and that application should be made to the construction of the Hilton–Roitberg examples of H-spaces, to the loop space structures on completed spheres, and to Serre's questions about nonvanishing of infinitely many homotopy groups of a finite complex. The last application is not traditionally thought to have anything to do with localization.

The author has been told that the theory of fibrations in cubical diagrams is out of date and should be superseded by the more general theory of limits and colimits of diagrams. Spiritually the author agrees with this. But practically he does not. The cubical theory is quite useful and specific and easier to present.

The theory of Hopf invariants due to James, Hilton, and Toda was central to the proofs of the first exponent results for spheres. Since the new methods give new exponent results, why do we include these? There are several reasons. First, the 2-primary results of James have been substantially improved by Selick but they have not been superseded, and the best possible 2-primary bounds have not been found. In order to have any 2-primary bounds on the exponents for the homotopy groups of spheres, we still need James; and James and Hilton both give us the EHP sequences which are still fundamental in the computations of unstable homotopy theory. This latter reason is also applicable to Toda's work on odd primary components. He produces a useful factorization of the double suspension in the odd primary case which leads to the odd primary EHP sequences.

Samelson products in homotopy theory with coefficients are the main tool in the proofs of the exponent results of Cohen, Moore, and the author. These products give the homotopy theory of loop spaces the structure of graded Lie algebras with the exception of some unfortunate failure of the Jacobi identity at the prime 3. This theory is included here together with an important improvement on a theory of Samelson products over the loops on an H-space. This improvement makes possible a simplification of the main exponent proofs.

The homotopy and homology Bockstein spectral sequences are presented in detail with particular attention paid to products in the spectral sequence and to the convergence of the spectral sequence in the nonfinitely generated case. It occurred to the author that no book on Bockstein spectral sequences should be written without presenting Browder's results on the unboundedness of the order of the torsion in the homology of finite H-spaces. Even though these results are well presented in the paper of Browder and in the book of McCleary, the inclusion of these results is amply justified by their beauty and by the remarkable fact that this growth in the order of the torsion in homology is precisely opposite to the bounds on the order of the torsion that we find in homotopy.

The consideration of Samelson products in homotopy and their Hurewicz representation as commutators in homology makes it vital to present a general theory of graded Lie algebras and their universal enveloping algebras. Even though it is not necessary for our applications, for the first time we make no restrictions on the ground ring. It need not contain $\frac{1}{2}$. We prove the graded versions of the Poincaré–Birkhoff–Witt theorem and the related tensor product decompositions of universal enveloping algebras related to exact sequences. There is a similarity between free groups and free Lie algebras. Subalgebras of free Lie algebras are free but they may not be finitely generated even if the ambient Lie algebra is. Nonetheless, the generators of the kernels of homomorphisms can often be determined.

The actual Eilenberg–Moore spectral sequence plays almost no role in this book. But the chain model approximations that underlie this theory play an essential role and are fully presented here. We restrict our treatment to the case when the base is simply connected. This includes most applications and avoids delicate problems related to the convergence of the approximations. Particular attention is paid to products and coproducts in these models. A new innovation is the connection to the geometry of loop multiplication via an idea which is dual to an idea presented in a Cartan seminar by John Moore.

In the chapter on exponents of the homotopy of spheres and Moore spaces, most of the above finds application.

Finally, the major omission in this book on unstable homotopy theory is that there is no systematic treatment of simplicial sets even though they are used once in a while in this book. They are used to study Eilenberg–Zilber maps, the Alexander–Whitney maps, the Serre filtration, and Kan's construction of group models for loop spaces. Too bad, you can't include everything.

Prerequisites

The reader should be familiar with homology and homotopy groups. Homology groups can be found in the classic book by S. Eilenberg and N. Steenrod [44] or in many more recent books such as those of M. Greenberg and J. Harper [49], A. Dold [33], E. Spanier [123], and A. Hatcher [51]. Homotopy groups can be found in these books and also in the highly recommended books by G. Whitehead [134] and P. Selick [114].

Some introduction to homology and homotopy is essential before beginning to read this book. All of the subsequent suggestions are not essential but some knowledge of them would be useful and historically enlightening.

The books by G. Whitehead and P. Selick provide comprehensive introductions to homotopy theory and thus to the material in this book. Whitehead's book has an excellent treatment of Samelson products. Many of the properties of Samelson products were originally proved by him. But all the properties of Samelson products that we need are proved here.

Spectral sequences are much used in this book and we assume familiarity with them when we need them. The exposition of spectral sequences by Serre [116] remains a classic but there are alternative treatments in many places such as the books of S. MacLane [77], E. Spanier, G. Whitehead, and P. Selick. We regard the Serre spectral sequence as a basic tool and use it to prove many things. The survey by J. McCleary [82] provides an excellent overview of many spectral sequences, including the Eilenberg–Moore spectral sequence to which we devote much of this book.

Obstruction theory to extending maps and homotopies is a frequent tool. It is presented in the book of Whitehead. An important generalization to sections of fibre bundles is in the book of N. Steenrod [125].

Postnikov systems are used in the treatment of the Hurewicz theorem for homotopy groups with coefficients. Postnikov systems appear in the works of Serre [117, 118]. The standard references are the books of G. Whitehead and E. Spanier. The treatment in R. Mosher and M. Tangora [98] is brief and very clear.

The main books on homological algebra are two, that of H. Cartan and S. Eilenberg [23] and that of S. MacLane [77]. Cartan–Eilenberg's treatment of spectral sequences is used in this book in order to introduce products in the mod p homotopy Bockstein spectral sequence. MacLane's book is more concrete and provides an introduction to the the details of the Eilenberg–Zilber map and to the differential bar construction.

Ways to use this book

A book this long should be read in shorter segments. Many of the chapters are self-contained and can be read independently. Here are some ideas as to how the book can be broken up. Each of the paragraphs below is meant to indicate that that material can be read independently with minimal reference to the other chapters of the book.

The first chapter on homotopy groups with coefficients introduces these groups which are the homotopy group analog of homology groups with coefficients. The essence of it is captured in the Sections 1.1 through 1.7 which start with the definition and end with the mod k Hurewicz theorem. It is basic material. When combined with Sections 6.7 through 6.9 on Samelson products and with some of the material on Bockstein spectral sequences in Sections 7.1 to 7.6 it leads via Sections 9.6 and 9.7 on the cycles in differential graded Lie algebras to a proof of the existence of higher order torsion in the integral homotopy groups of an odd primary Moore space.

The second chapter on localization is completely self-contained. Sections 2.1 through 2.7 cover the most important parts of the classical localizations and completions of topological spaces. After that, the reader can choose from applications of Miller's theorem to the nonvanishing of the homotopy groups of a finite complex in Section 2.10, applications to the Hilton–Roitberg examples, or to loop structures on completions of spheres. This chapter is one of the most accessible in the book.

The short third chapter on Peterson–Stein formulas is a self-contained introduction to these formulas and also to the theory of fibred cubes which should be better known in homotopy theory. It is a quick treatment of fundamental facts about fibrations.

The fourth chapter on Hilton–Hopf invariants and the EHP sequence introduces many of the classical methods of unstable homotopy theory, for example, the James construction, the Hilton–Milnor theorem, and the James fibrations which underlie the EHP sequence. It contains a proof of James 2-primary exponent theorem for the spheres and some elementary computations of low dimensional homotopy groups. It is an introduction to some geometric ideas which are often used in the study of homotopy groups of spheres, especially of the 2-primary components.

The fifth chapter on James–Hopf and Toda–Hopf can serve as an odd primary continuation of the fourth chapter. It contains Toda's odd primary fibrations which give the odd primary EHP sequence and it contains the proof of Toda's odd primary exponent theorem for spheres. To study the odd primary components of the homotopy of spheres, Toda realized that it could be advantageous to decompose the double suspension into a composition which is different from the obvious one of composition of single suspensions.

The sixth chapter on Samelson products contains a complete treatment of Samelson products in odd primary homotopy groups. As mentioned above, it can be combined with the Bockstein spectral sequence and material on cycles in differential graded Lie algebras to prove the existence of higher order torsion in the homotopy groups of odd primary Moore spaces. Of course, the reader will need some knowledge of Chapter 1 here.

The seventh chapter on Bockstein spectral sequences contains a presentation of Browder's results on torsion in H-spaces which is completely independent of the rest of the book. It is included because of the beauty of the results and because it was the first deep use of the Bockstein spectral sequence.

Chapters 8 and 9 present the theory and applications of graded Lie algebras and their universal enveloping algebras. Particular attention is paid to free Lie algebras and their subalgebras. Although this section contains many results of purely algebraic interest, it also has geometric applications via the Lie algebras of Samelson products and via the study of loop spaces whose homology is a universal enveloping algebra. One of the applications of differential graded Lie algebras is the previously mentioned higher order torsion in the homotopy groups of odd primary Moore spaces.

Chapter 10 on differential homological algebra is the longest in the book and only does half of the theory, albeit it is the harder half. This half deals with the cobar construction of Adams and the so-called second quadrant Eilenberg–Moore spectral sequence. The presentation here emphasizes the chain models that underlie the spectral sequences and which are often more important and useful than the spectral sequences. Special emphasis is placed on the not so obvious relation of the loop multiplication to the homological algebra. It is the detailed foundation chapter for the next section on odd primary exponent theorems and the loop space decompositions which lead to them.

Chapter 11 on odd primary exponent theorems is the chapter which guides the book in its selection of topics. It defines the central current. It uses almost the whole book as background material. Nonetheless, it can be read independently if the reader is willing to use isolated parts of the book as background material. The necessary background material includes homotopy groups with coefficients, their Bockstein spectral sequences, and Samelson products in them. These are used to construct the product decomposition theorems which are the basis for the applications to exponent theorems. Localization is necessary because, without it, these decomposition theorems would not be valid. Free graded Lie algebras, their subalgebras and universal enveloping algebras are the algebraic models for the loop spaces we study and for their decomposition into topological products. It is difficult to do but it can be read without reading all of the rest of the book.

Finally, Chapter 12 is included because it is the other half of differential homological algebra, that which is used to study classifying spaces, and it is arguably the more important and useful half of the theory. It is also often the easier half and it contains the beautiful applications of Stiefel–Whitney classes to non-immersion and non-parallelizability results for real projective spaces. It would have been a shame not to include it.

Acknowledgments

The creation of this book involved many midwives.

First and foremost is John C. Moore who was my teacher and colleague and who was present at the creation of many of the ideas in this book.

Emmanuel Dror Farjoun introduced me to his theory of localization which is prominent in this book.

The interest of an audience is vital. My audiences were excellent and this is a better book because of them.

John Harper and Inga Johnson attended a semester's course on an early version of this book. There was always a bowl of candy on Inga's desk to energize the lecturer. John dutifully read every page of text, found many typographical errors, and shared his profound knowledge of homotopy theory to suggest many clarifications.

Lucía Fernández-Suárez arrranged for lectures in Braga, Portugal. Braga is a beautiful and historic city and Lucía provided an audience of her, Thomas Kahl, Lucille Vandembroucq, and Gustavo Granja to ask clarifying questions and to make intelligent comments about localization.

Lausanne, Switzerland is another beautiful and historic city. Kathryn Hess arranged for lectures there and her understanding and appreciation of the issues of differential homological algebra is unusual and very helpful.

Brayton Gray gave inspiration by his penetrating questions and insights into unstable homotopy theory.

I owe great thanks to the ones who stood by me during the event of a dissection of the aorta. Doris and John Harper, Michelle and Doug Ravenel, Fran Crawford and Joan Robinson, and Jacques Lewin were all present and accounted for.

Last but not least, I thank Joan Robinson, Jan Pearce, and Hoss Firooznia for their help in mastering the art of LaTeX.

Introduction to unstable homotopy theory

Computation of the homotopy groups $\pi_n(X)$ of a topological space X has played a central role in homotopy theory. And knowledge of these homotopy groups has inherent use and interest. Furthermore, the development of techniques to compute these groups has proven useful in many other contexts.

The study of homotopy groups falls into three parts.

First, there is the computation of specific homotopy groups $\pi_n(X)$ of spaces. This may be traced back to Poincaré [106] in the case $n = 1$:

Poincaré: $\pi_1(X)/[\pi_1(X), \pi_1(X)]$ *is isomorphic to* $H_1(X)$.

Hurewicz [62] showed that, in the simply connected case, the Hurewicz homomorphism provides an isomorphism of the first nonzero $\pi_n(X)$ with the homology group $H_n(X)$ with $n \geq 1$:

Hurewicz: *If X is an $n - 1$ connected space with $n \geq 2$, then $\pi_n(X)$ is isomorphic to $H_n(X)$.*

Hopf [58] discovered the remarkable fact that homotopy groups could be nonzero in dimensions higher than those of nonvanishing homology groups. He did this by using linking numbers but the modern way is to use the long exact sequence of the Hopf fibration sequence $S^1 \to S^3 \to S^2$.

Hopf: $\pi_3(S^2)$ *is isomorphic to the additive group of integers* $\mathbb{Z}$.

Computation enters the modern era with the work of Serre [116, 118] on the low dimensional homotopy groups of spheres. To this end, he introduced a localization technique which he called "classes of abelian groups." A first application was:

Serre: *If $n \geq 1$ and p is an odd prime, then the group $\pi_{2n+2p-2}(S^{2n+1})$ contains a summand isomorphic to $\mathbb{Z}/p\mathbb{Z}$.*

Second, there are results which relate the homotopy groups of some spaces to those of others.

Examples are product decomposition theorems such as the result of Serre which expresses the odd primary components of the homotopy groups of an even-dimensional sphere in terms of those of odd-dimensional spheres, that is:

Serre: *Localized away from 2, there is a homotopy equivalence*
$$\Omega S^{2n} \simeq S^{2n-1} \times \Omega S^{4n-1}.$$

Localization is necessary for some results but not for all. A product decomposition which requires no localization is the Hilton–Milnor theorem [54, 89, 134] which expresses the homotopy groups of a bouquet of two suspension spaces $\pi_k(\Sigma X \vee \Sigma Y)$ in terms of the homotopy groups of the constituents of the bouquet $\Sigma X, \Sigma Y$, and of the homotopy groups of various smash products:

Hilton–Milnor: *There is a homotopy equivalence*
$$\Omega(\Sigma X \vee \Sigma Y) \simeq \Omega \Sigma X \times \Omega \Sigma (\bigvee_{j=0}^{\infty} X^{\wedge j} \wedge Y).$$

Third, Serre used his localization technique to study global properties of the homotopy groups of various spaces. What is meant by this is best made clear by giving various examples:

Serre: *For a simply connected complex with finitely many cells in each dimension, the homotopy groups are finitely generated.*

Serre: *Odd dimensional spheres have only one nonfinite homotopy group,* $\pi_{2n+1}(S^{2n+1}) = \mathbb{Z}$.

Serre: *Simply connected finite complexes with nonzero reduced homology have infinitely many nonzero homotopy groups.*

Serre [117] proved the last result by using the cohomology of Eilenberg–MacLane spaces. There is now a modern proof which uses Dror-Farjoun localization and Miller's Sullivan conjecture [83, 84].

The study of the global properties of homotopy groups was continued by James [66, 67] who introduced what are called the James–Hopf invariant maps. Using fibration sequences associated to these, James proved the following upper bound on the exponent of the 2-primary components of the homotopy groups of spheres:

James: 4^n *annihilates the 2-primary component of the homotopy groups of the sphere* S^{2n+1}.

James' result is a consequence of a more geometric result which was first formulated as a theorem about loop spaces by John Moore. For a homotopy associative H-space X and a positive integer k, let $k : X \to X$ denote the k-th power map defined by $k(x) = x^k$.

James: *Localized at 2, there is a factorization of the 4-th power map*
$$4 : \Omega^3 S^{2n+1} \to \Omega S^{2n-1} \to \Omega^3 S^{2n+1}.$$

Introduction to unstable homotopy theory 3

Toda [130, 131] defined new "secondary" Hopf invariants and used these to extend James' result to odd primes p, that is:

Toda: *For an odd prime p, p^{2n} annihilates the p-primary component of the homotopy groups of the sphere S^{2n+1}.*

Or in Moore's reformulation:

Toda: *Localized at an odd prime p, there is a factorization of the p^2-d power map*
$$p^2 : \Omega^3 S^{2n+1} \to \Omega S^{2n-1} \to \Omega^3 S^{2n+1}.$$

No progress was made in the exponents of the primary components of homotopy groups until Selick's thesis [112].

Selick: *For p an odd prime, p annihilates the p-primary component of the homotopy groups of S^3.*

Selick's result is a consequence of the following geometric result. Let $S^3\langle 3\rangle$ denote the 3-connected cover of the 3-sphere S^3 and let $S^{2p+1}\{p\}$ denote the homotopy theoretic fibre of the degree p map $p : S^{2p+1} \to S^{2p+1}$.

Selick: *Localized at an odd prime p, $\Omega^2(S^3\langle 3\rangle)$ is a retract of $\Omega^2 S^{2p+1}\{p\}$.*

Selick's work was followed almost immediately by the work of Cohen–Moore–Neisendorfer [27, 26]. They proved that, if p is a prime greater than 3, then p^n annihilates the p-primary component of the homotopy groups of S^{2n+1}. A little later, Neisendorfer [100] overcame technical difficulties and extended this result to all odd primes.

Cohen–Moore–Neisendorfer: *Localized at an odd prime there is a factorization of the p-th power map*
$$p : \Omega^2 S^{2n+1} \to S^{2n-1} \to \Omega^2 S^{2n+1}.$$

Let $C(n)$ be the homotopy theoretic fibre of the double suspension map $\Sigma^2 : S^{2n-1} \to \Omega^2 S^{2n+1}$.

Exponent corollary: *If p is an odd prime, then p annihilates the p primary components of the homotopy groups $\pi_*(C(n))$ and p^n annihilates the p primary components of the homotopy groups $\pi_*(S^{2n+1})$.*

For odd primes, Brayton Gray [46] showed that the results of Selick and Cohen–Moore–Neisendorfer are the best possible. At the prime 2, the result of James is not the best possible but the definitive bound has not yet been found.

The main point of this book is to present the proof of the result of Cohen–Moore–Neisendorfer. We present the necessary techniques from homotopy theory, graded Lie algebras, and homological algebra. To this end, we need to develop homotopy groups with coefficients and the differential homological algebra associated to fibrations. These are applied to produce loop space decompositions which yield the above theorems.

It is useful to consider two cases of homotopy groups with coefficients, the case where the coefficients are a finitely generated abelian group and the case where the coefficients are a subgroup of the additive group of the rational numbers.

For a space X and finitely generated abelian group G, $\pi_n(X;G)$ is defined as the set of pointed homotopy classes of maps $[P^n(G), X]_*$ from a space $P^n(G)$ to X where $P^n(G)$ is a space with exactly one nonzero reduced cohomology group isomorphic to G in dimension n. This definition first occurs in the thesis of Peterson [104, 99]. These homotopy groups with coefficients are related to the classical homotopy groups by a universal coefficient sequence.

Peterson: *There is a short exact sequence*

$$0 \to \pi_n(X) \otimes G \to \pi_n(X;G) \to \mathrm{Tor}^1_{\mathbb{Z}}(\pi_{n-1}(X), G) \to 0.$$

There is a Hurewicz homomorphism to homology with coefficients

$$\phi : \pi_n(X;G) \to H_n(X;G),$$

the image of which lies in the primitive elements, and a Hurewicz theorem is true.

From this point of view, the usual or classical homotopy groups are those with coefficients $\mathbb{Z}$.

In the finitely generated case, nothing is lost by considering only the case of cyclic coefficients. If 2-torsion is avoided, Samelson products were introduced into these groups for a homotopy associative H-space X in the thesis of Neisendorfer [99]:

$$[\ ,\ ] : \pi_n(X; \mathbb{Z}/k\mathbb{Z}) \otimes \pi_m(X; \mathbb{Z}/k\mathbb{Z}) \to \pi_{m+n}(X; \mathbb{Z}/k\mathbb{Z}).$$

To construct these Samelson products, it is necessary to produce decompositions of smash products into bouquets:

$$P^n(\mathbb{Z}/p^r\mathbb{Z}) \wedge P^m(\mathbb{Z}/p^r\mathbb{Z}) \simeq P^{n+m}(\mathbb{Z}/p^r\mathbb{Z}) \vee P^{n+m-1}(\mathbb{Z}/p^r\mathbb{Z})$$

when p is an odd prime. If $p = 2$, these decompositions do not always exist and therefore there is no theory of Samelson products in homotopy groups with coefficients $\mathbb{Z}/2\mathbb{Z}$. If $p = 3$, the decompositions exist but the decompositions are not "associative" and this leads to the failure of the Jacobi identity for Samelson products in homotopy with $\mathbb{Z}/3\mathbb{Z}$ coefficients.

The Hurewicz homomorphism carries these Samelson products into graded commutators in the Pontrjagin ring,

$$\phi[\alpha, \beta] = [\phi\alpha, \phi\beta] = (\phi\alpha)(\phi\beta) - (-1)^{nm}(\phi\beta)(\phi\alpha)$$

where $n = \deg(\alpha)$ and $m = \deg(\beta)$.

Neisendorfer also introduced a homotopy Bockstein spectral sequence to study the order of torsion elements in the classical homotopy groups.

With few exceptions, the first applications of homotopy groups with coefficients will be to the simple situation where the the Hurewicz homomorphism is an isomorphism through a range. In a few cases, we will need to consider situations where the Hurewicz map is merely an epimorphism but with a kernel consisting only of Whitehead products in a range. This is all we will need to develop the theory of Samelson products in homotopy groups with coefficients, where we avoid the prime 2 and sometimes the prime 3.

For a space X and a subgroup G of the rationals, $\pi_n(X; G)$ is defined as the tensor product $\pi_n(X) \otimes G$. where, if $n = 1$, we require $\pi_n(X)$ to be abelian. Once again, these homotopy groups with coefficients are related to the classical homotopy groups by a universal coefficient sequence, there is a Hurewicz homomorphism to homology with coefficients, and a Hurewicz theorem is true. Futhermore, there are Samelson products for a homotopy associative H-space X and the Hurewicz map carries these Samelson products into graded commutators in the Pontrjagin ring.

In the special case of rational coefficients Q, the Hurewicz homomorphism satisfies a strong result of Milnor–Moore [90]:

Milnor–Moore: *If X is a connected homotopy associative H-space, then the Hurewicz map $\varphi : \pi_*(X; Q) \to H_*(X; Q)$ is an isomorphism onto the primitives of the Pontrjagin ring and there is an isomorphism*

$$H_*(X; Q) \cong U(\pi_*(X; Q))$$

where UL denotes the universal enveloping algebra of a Lie algebra L.

In practice this means that the rational homotopy groups can often be completely determined and this is one of things that makes rational homotopy groups useful.

In contrast, homotopy groups with cyclic coefficients have not been much used since they are usually as difficult to completely determine as the usual homotopy groups are. Nonetheless, some applications exist. The Hurewicz map still transforms the Samelson product into graded commutators of primitive elements in the Pontrjagin ring. This representation is far from faithful but is still nontrivial. The homotopy Bockstein spectral sequence combines with the above to give information on the order of torsion homotopy elements related to Samelson products.

Many theorems in homotopy theory depend on the computation of homology. For example, in order to prove that two spaces X and Y are homotopy equivalent, one constructs a map $f : X \to Y$ and checks that the induced map in homology is an isomorphism. If X and Y are simply connected and the isomorphism is in homology with integral coefficients, then the map f is a homotopy equivalence. In general, when the isomorphism is in homology with coefficients, then the map f is some sort of local equivalence. For example, with rational coefficients, we get rational equivalences, with coefficients integers $\mathbb{Z}_{(p)}$ localized at a prime p, we get equivalences localized at p, and with $\mathbb{Z}/p\mathbb{Z}$ coeffients, we get equivalences of completions at p. The theorem of Serre, $\Omega S^{2n} \simeq S^{2n-1} \times \Omega S^{4n-1}$ localized away from 2, and the Hilton–Milnor theorem,

$$\Omega(\Sigma X \vee \Sigma Y) \simeq \Omega \Sigma X \times \Omega \Sigma \left(\bigvee_{j=0}^{\infty} X^{\wedge j} \wedge Y \right),$$

are proved in this way. A central theme of this book will be such decompositions of loop spaces.

For us, the most basic homological computation is the homology of the loops on the suspension of a connected space:

Bott–Samelson [13]: *If X is connected and the reduced homology of $\overline{H}_*(X; R)$ is free over a coefficient ring R, then there is an isomorphism of algebras*

$$T(\overline{H}_*(X; R)) \to H_*(\Omega \Sigma X; R)$$

where $T(V)$ denotes the tensor algebra generated by a module V.

Let $L(V)$ be the free graded Lie algebra generated by V. The observation that $T(V)$ is isomorphic to the universal enveloping algebra $UL(V)$ has topological consequences based on the following simple fact:

Tensor decomposition: *If $0 \to L_1 \to L_2 \to L_3 \to 0$ is a short exact sequence of graded Lie algebras which are free as R modules, then there is an isomorphism*

$$UL_2 \cong UL_1 \otimes UL_3.$$

Suppose we want to construct a homotopy equivalence of H-spaces $X \times Y \to Z$ and suppose that we compute

$$H_*(X; R) = UL_1, \quad H_*(Y; R) = UL_3, \quad \text{and} \quad H_*(Z; R) = UL_2.$$

Suppose also that we can construct maps $g : X \to Z$ and $h : Y \to Z$ such that the product $f = \mu \circ (g \times h) : X \times Y \to Z \times Z \to Z$ induces a homology isomorphism (where $\mu : Z \times Z \to Z$ is the multiplication of Z). Then we have an equivalence localized in the sense that is appropriate to the coefficients.

Here is an example. Let $L(x_\alpha)$ denote the free graded Lie algebra generated by the set $\{x_\alpha\}$). Let $\langle x_\alpha \rangle$ denote the abelianization, that is, the free module generated by the set with all Lie brackets zero. If we localize away from 2 and x is an odd degree element, then we have a short exact sequence

$$0 \to \langle [x,x] \rangle \to L(x) \to \langle x \rangle \to 0$$

and isomorphisms
$$H_*(\Omega S^{4n-1}) \cong U(\langle [x,x] \rangle), \ H_*(S^{2n-1}) \cong U(\langle x \rangle),$$
$$H_*(\Omega S^{2n}) \cong U(L(x))).$$

This leads to the result of Serre: $\Omega S^{2n} \simeq S^{2n-1} \times \Omega S^{4n-1}$ localized away from 2. Thus, Serre's result is essentially a consequence of just the Bott–Samelson theorem and the tensor decomposition of universal enveloping algebras.

Consider the following additional facts concerning Lie algebras [27]:

Free subalgebras: *If L is a free graded Lie algebra and K is a subalgebra which is a split summand as an R-module, then K is a free graded Lie algebra.*

Kernel theorem: *If K is the kernel of the natural map $L(V \oplus W) \to L(V)$ of free graded Lie algebras, then K is isomorphic to the free graded Lie algebra*

$$L\left(\bigoplus_{j=0}^{\infty} V^{\otimes j} \otimes W \right)$$

where $V^{\otimes j} = V \otimes V \otimes \cdots \otimes V$, with j factors.

A direct consequence is the Hilton–Milnor theorem,

$$\Omega(\Sigma X \vee \Sigma Y) \simeq \Omega \Sigma X \times \Omega \Sigma \left(\bigvee_{j=0}^{\infty} X^{\wedge j} \wedge Y \right).$$

In order to study torsion at a prime p, it is useful to consider the Bockstein differentials in homology with mod p coefficients. This leads to consideration of differential graded Lie algebras.

For example, let $P^n(p^r) = S^{n-1} \cup_{p^r} e^n$ be the space obtained by attaching an n-cell to an $n-1$-sphere by a map of degree p^r. Then $H_*(P^n(p^r); \mathbb{Z}/p\mathbb{Z}) = \langle u, v \rangle$ with $\deg(v) = n$ and $\deg(u) = n-1$. The r-th Bockstein differential is given by $\beta^r(v) = u, \beta^r(u) = 0$. Thus, the Bott–Samelson theorem gives isomorphisms of differential Hopf algebras

$$H_*(\Omega \Sigma P^n(p^r); \mathbb{Z}/p\mathbb{Z}) \cong T(u,v) \cong UL(u,v)$$

where $L = L(u,v)$ is a differential Lie algebra which is a free Lie algebra. Any algebraic constructions with topological implications must be compatible with these Bockstein differentials. For example, the abelianization of L is $\langle u, v \rangle$.

This is compatible with differentials, leads to the short exact sequence of differential Lie algebras

$$0 \to [L, L] \to L \to \langle u, v \rangle \to 0,$$

8   Introduction to unstable homotopy theory

and the tensor decomposition of universal enveloping algebras

$$H_*(\Omega\Sigma P^n(p^r); \mathbb{Z}/p\mathbb{Z}) \cong UL \cong U(\langle u, v \rangle) \otimes U([L, L]).$$

But this tensor decomposition can only be realized by a product decomposition of $\Omega\Sigma P^n(p^r)$ when p and n are odd. If we set $n - 1 = 2m$, then we can prove [27]:

Cohen–Moore–Neisendorfer: *If p is an odd prime and $m \geq 1$, then there is a homotopy equivalence*

$$\Omega P^{2m+2}(p^r) \simeq S^{2m+1}\{p^r\} \times \Omega\left(\bigvee_{j=0}^{\infty} P^{2m+2mj+1}(p^r)\right)$$

where $S^{2m+1}\{p^r\}$ is the homotopy theoretic fibre of the degree p^r map $p^r : S^{2m+1} \to S^{2m+1}$.

The restriction to odd primes in the above is the result of the nonexistence of a suitable theory of Samelson products in homotopy groups with 2-primary coefficients.

One reason for the above parity restriction is as follows: Suppose the coefficient ring is $\mathbb{Z}/p\mathbb{Z}$ with p an odd prime. Only when n is odd (so that μ has even dimension and ν has odd dimension) can we write that

$$[L, L] = L(\mathrm{ad}^j(u)([v, v]), \mathrm{ad}^j(u)([u, v]))_{j \geq 0} =$$

the free Lie algebra on infinitely many generators with r-th Bockstein differential given by $\beta^r(\mathrm{ad}^j(u)([v, v])) = 2\mathrm{ad}^j(u)[u, v]$ for $j \geq 0$. In this case, the module of generators of $[L, L]$ is acylic with respect to the Bockstein differential and it is possible that the universal enveloping algebra U([L,L]) represents the homology of the loop space on a bouquet of Moore spaces. In fact, the isomorphisms of differential algebras

$$H_*(S^{2m+1}\{p^r\}; \mathbb{Z}/p\mathbb{Z}) \cong U(\langle u, v \rangle),$$

$$H_*(\Omega\left(\bigvee_{j=0}^{\infty} P^{2m+2mj+1}(p^r)\right); \mathbb{Z}/p\mathbb{Z}) \cong U([L, L]),$$

$$H_*(\Omega P^{2m+2}(p^r); \mathbb{Z}/p\mathbb{Z}) \cong UL$$

then lead to the above product decomposition for $\Omega P^{2m+2}(p^r)$.

There is no analogous product decomposition for $\Omega P^{2m+1}(p^r)$. The situation is much more complicated because of the fact that $[L, L]$ does not have an acyclic module of generators when $L = L(u, v)$ with $\deg(u)$ odd and $\deg(v)$ even. To go further we need to study the homology $H(L, \beta^r)$.

Let x be an even degree element in a differential graded Lie algebra over the ring $\mathbb{Z}/p\mathbb{Z}$ with p an odd prime, let d denote the differential, and for $k \geq 1$ define new elements

$$\tau_k(x) = \mathrm{ad}^{p^k-1}(x)(dx)$$

$$\sigma_k(x) = \frac{1}{2}\sum_{j=1}^{p^k-1} p^{-1}(j, p^k - j)[\mathrm{ad}^{j-1}(x)(dx), \mathrm{ad}^{p^k-j-1}(x)(dx)]$$

where $(a,b) = \frac{(a+b)!}{(a!)(b!)}$ is the binomial coefficient. These elements are cycles, $d(\tau_k(x)) = 0, d(\sigma_k(x)) = 0$, and they determine the homology of the above L via the following proposition.

Homology of free Lie algebras with acyclic generators: *Let $L(V)$ be a free graded Lie algebra over the ring $\mathbb{Z}/p\mathbb{Z}$ with p an odd prime and with a differential d such that $d(V) \subseteq V$ and $H(V,d) = 0$. Write*

$$L(V) = H(L(V), d) \oplus K$$

where K is acyclic. If K has a basis $x_\alpha, dx_\alpha, y_\beta, dy_\beta$ with $\deg(x_\alpha)$ even and $\deg(y_\beta)$ odd, then $H(L(V), d)$ has a basis represented by the cycles $\tau_k(x_\alpha)$, $\sigma_k(x_\alpha)$ with $k \geq 1$.

This proposition has two main applications. The first application is to a decomposition theorem which leads to the determination of the odd primary exponents of the homotopy groups of spheres.

Decomposition theorem: *Let p be an odd prime and let $F^{2n+1}\{p^r\}$ be the homotopy theoretic fibre of the natural map $P^{2n+1}(p^r) \to S^{2n+1}$ which pinches the bottom $2n$-cell to a point. Localized at p, there is a homotopy equivalence*

$$\Omega F^{2n+1}\{p^r\} \simeq S^{2n-1} \times \prod_{k=1}^{\infty} S^{2p^k n - 1}\{p^{r+1}\} \times \Omega\Sigma \bigvee_\alpha P^{n_\alpha}(p^r)$$

where

$$\bigvee_\alpha P^{n_\alpha}(p^r)$$

is an infinite bouquet of mod p^r Moore spaces.

The second application is to the existence of higher order torsion in the homotopy groups of odd primary Moore spaces:

Higher order torsion: *If p is an odd prime and $n \geq 1$, then for all $k \geq 1$ the homotopy groups $\pi_{2p^k n-1}(P^{2n+1})$ contain a summand isomorphic to $\mathbb{Z}/p^{r+1}\mathbb{Z}$.*

The following decomposition theorem is valid:

Cohen–Moore–Neisendorfer: *If p is an odd prime and $m \geq 1$, then there is a homotopy equivalence*

$$\Omega P^{2m+1}(p^r) \simeq T^{2m+1}\{p^r\} \times \Omega \Sigma \bigvee_\alpha P^{n_\alpha}(p^r)$$

where there is a fibration sequence

$$C(n) \times \prod_{k=1}^{\infty} S^{2p^k n - 1}\{p^{r+1}\} \to T^{2m+1}\{p^r\} \to S^{2n+1}\{p^r\}.$$

A corollary of these decomposition theorems is [28]:

Cohen–Moore–Neisendorfer: *If p is an odd prime and $n \geq 3$, then p^{2r+1} annihilates the homotopy groups $\pi_*(P^n(p^r))$.*

In fact the best possible result is [102]:

Neisendorfer: *If p is an odd prime and $n \geq 3$, then p^{r+1} annihilates the homotopy groups $\pi_*(P^n(p^r))$.*

1 Homotopy groups with coefficients

In this chapter, we define homotopy groups with coefficients $\pi_*(X;G)$ for a pointed space X and an abelian group G. With some small restrictions, these homotopy groups are covariant functors of X and G and, most important, satisfy a universal coefficient exact sequence

$$0 \to \pi_n(X) \otimes G \to \pi_n(X;G) \to \mathrm{Tor}^{\mathbb{Z}}(\pi_{n-1}(X), G) \to 0.$$

First, $\pi_*(X;G)$ is defined when G is a finitely generated abelian group and then the definition is extended to arbitrary abelian G by using the fact that G is a direct limit of its finitely generated subgroups.

The guiding principle is that the groups $\pi_*(X;G)$ are related to the groups $\pi_*(X)$ in much the same way that the groups $H_*(X;G)$ are related to the groups $H_*(X;\mathbb{Z})$.

The definitions originated in the thesis of Frank Peterson [104] written under the direction of Norman Steenrod. Further development occurred in the thesis of the author [99] written under the direction of John Moore. Moore also influenced Peterson.

These homotopy groups with coefficients satisfy the usual long exact sequences associated to pairs and to fibration sequences. They also satisfy long exact Bockstein sequences associated to short exact sequences of coefficient groups.

In the case when $G = \mathbb{Z}/k\mathbb{Z}$ is a cyclic group, we define a mod k Hurewicz homomorphism $\varphi : \pi_*(X;G) \to H_*(X;G)$ and prove a mod k Hurewicz isomorphism theorem. The proof of the mod k Hurewicz theorem is a consequence of the fact that it is true when X is an Eilenberg–MacLane space and of the fact that any space X has a Postnikov system.

We use the usual argument to show that the mod k Hurewicz isomorphism theorem for spaces implies a similar mod k isomorphism theorem for pairs of spaces.

12 Homotopy groups with coefficients

1.1 Basic definitions

In order to relate integral homology and integral cohomology, it is convenient to introduce the following two distinct notions of duality.

Definition 1.1.1.

(A) If F is a finitely generated torsion free abelian group, let $F^* = \text{Hom}(F, \mathbb{Z})$.

(B) If T is a finite abelian group, let $T^* = \text{Hom}(T, \mathbb{Q}/\mathbb{Z})$.

Thus, $\mathbb{Z}^* \cong \mathbb{Z}$ generated by the identity map $1_{\mathbb{Z}} : \mathbb{Z} \to \mathbb{Z}$ and $(\mathbb{Z}/k\mathbb{Z})^* \cong \mathbb{Z}/k\mathbb{Z}$ generated by the map which sends 1 to $1/k$. It follows that there are unnatural isomorphisms $F^* \cong F$ and $T^* \cong T$.

The following lemma is easily verified in the cyclic case and hence in all cases.

Lemma 1.1.2. *For finitely generated free F and finite T, the natural maps $F \to (F^*)^*$ and $T \to (T^*)^*$ are isomorphisms.*

Corollary 1.1.3. *For finitely generated generated free F_1 and F_2 and finite T_1 and T_2, the natural maps*

$$\text{Hom}(F_1, F_2) \to \text{Hom}(F_2^*, F_1^*)$$

and

$$\text{Hom}(T_1, T_2) \to \text{Hom}(T_2^*, T_1^*)$$

sending a homomorphism f to its dual f^* are isomorphisms.

Since $\mathbb{Q}$ is a divisible, therefore injective, abelian group, the long exact sequence associated to the short exact sequence $0 \to \mathbb{Z} \to \mathbb{Q} \to \mathbb{Q}/\mathbb{Z} \to 0$ gives:

Lemma 1.1.4. *For finite abelian T, there is a natural isomorphism $T^* \cong \text{Ext}(T, \mathbb{Z})$.*

Let G be a finitely generated abelian group and let $P^n(G)$ be a finite complex with exactly one nonzero reduced integral cohomology group,

$$\overline{H}^k(P^n(G); \mathbb{Z}) = \begin{cases} G & \text{for } k = n \\ 0 & \text{for } k \neq n. \end{cases}$$

The universal coefficient theorem

$$0 \to \text{Ext}(H_n(X; \mathbb{Z}), G) \to H^n(X; G) \to \text{Hom}(H_n(X; \mathbb{Z}), \mathbb{Z}) \to 0$$

combines with the above lemmas to yield:

Proposition 1.1.5. *If $G = T \oplus F$ where T is finite abelian and F is finitely generated free abelian, then the reduced integral homology of $P^n(G)$ is*

$$\overline{H}_k(P^n(G); \mathbb{Z}) \cong \begin{cases} T^* & \text{if } k = n-1, \\ F^* & \text{if } k = n, \text{ and} \\ 0 & \text{if } k \neq n, n-1. \end{cases}$$

We will leave the question of the uniqueness of the homotopy type of $P^n(G)$ to the exercises.

Let $M_n(G)$ denote the Moore space with exactly one nonzero reduced integral homology group in dimension n. It follows that there is a homology equivalence

$$M_n(F^*) \vee M_{n-1}(T^*) \xrightarrow{\simeq} P^n(T \oplus F).$$

Definition 1.1.6. *If X is a pointed topological space, then the n-th homotopy group of X with G coefficients is*

$$\pi_n(X; G) = [P^n(G); X]_*$$

= the pointed homotopy classes of maps from $P^n(G)$ to X.

The two most useful examples of $P^n(G)$ are:

If $G = \mathbb{Z}$ = the additive group of integers, then $P^n(\mathbb{Z}) = S^n$ and $\pi_n(X; \mathbb{Z}) = \pi_n(X)$ = the usual homotopy groups for all $n \geq 1$.

If $G = \mathbb{Z}/k\mathbb{Z}$ = the integers mod k, then $P^n(\mathbb{Z}/k\mathbb{Z}) = P^n(k) = S^{n-1} \cup_k e^n$ = the space obtained by attaching an n-cell to an $(n-1)$-sphere by a map of degree k. Thus, $\pi_n(X; \mathbb{Z}/k\mathbb{Z})$ is defined for all $n \geq 2$.

Since $P^n(G \oplus H) \simeq P^n(G) \vee P^n(H)$, it follows that $\pi_n(X; G \oplus H) = \pi_n(X; G) \oplus \pi_n(X; H)$. Hence, the cyclic case is sufficient to define $\pi_n(X; G)$ for any finitely generated abelian group G and all $n \geq 2$ or, if G is finitely generated free abelian, $n \geq 1$.

But we can also construct $P^n(G)$ by free resolutions. Let G be any finitely generated abelian group which is free if $n = 1$. Since the case $n = 1$ is trivial ($P^1(G)$ is just a wedge of circles), we shall assume $n \geq 2$. Let

$$0 \to \bigoplus_\beta \mathbb{Z} \xrightarrow{F} \bigoplus_\alpha \mathbb{Z} \to G \to 0$$

be a finitely generated free resolution and let

$$f: \bigvee_\alpha S^{n-1} \to \bigvee_\beta S^{n-1}$$

be a map such that the induced map $f^* = F: \bigoplus_\beta \mathbb{Z} \to \bigoplus_\alpha \mathbb{Z}$ in dimension $n-1$ integral cohomology. If C_f = the mapping cone of f, the long exact cohomology sequence associated to the cofibration sequence

$$\bigvee_\alpha S^{n-1} \xrightarrow{f} \bigvee_\beta S^{n-1} \to C_f$$

shows that C_f is a $P^n(G)$.

If Y is any homotopy associative co-H-space and X is any pointed space, the comultiplication $\nu : Y \to Y \vee Y$ defines a group structure on $[Y, X]_*$. The standard example of a homotopy associative co-H-space is a suspension, $\Sigma W = S^1 \wedge W$, and the double suspension $\Sigma^2 W$ is homotopy commutative. Since $\Sigma P^n(G) = P^{n+1}(G)$:

Proposition 1.1.7. *The set $\pi_n(X; G)$ is a group if $n \geq 3$ and an abelian group if $n \geq 4$.*

On the other hand, if Y is any pointed space and X is any homotopy associative H-space, the multiplication $\mu : X \times X \to X$ defines a group structure on $[Y, X]_*$. The standard lemma is:

Proposition 1.1.8. *If Y is any co-H-space and X is any H-space, then the two structures on $[Y, X]_*$ are the same and they are both commutative and associative.*

Thus:

Proposition 1.1.9. *If X is an associative H-space, the set $\pi_n(X; G)$ is a group if $n \geq 2$ and an abelian group if $n \geq 3$.*

Exercises

(1) Let G be a finitely generated abelian group and write $G \cong T \oplus F$ where T is a torsion group and F is torsion free. Let X be any finite complex with exactly one nonzero reduced integral cohomology group which is isomorphic to G in dimension n. Thus, if $n = 1$, $G = F$ must be torsion free, $T = 0$. Assume X has an abelian fundamental group. Show that there exists an integral homology equivalence $A \vee B \to X$ where A is a Moore space with exactly one nonzero reduced homology group isomorphic to T in dimension $n - 1$ and B is a Moore space with exactly one nonzero reduced homology group isomorphic to F in dimension n, that is, B is a bouquet of spheres. Thus, if X is simply connected, it is unique up to homotopy type.

(2) (a) By considering the universal cover and the action of the fundamental group, show that $\pi_2(S^1 \vee S^2)$ is isomorphic to the group ring $\mathbb{Z}[\pi]$ where $\pi = \langle T \rangle = \{I, T^{\pm 1}, T^{\pm 2}, \dots\} =$ the infinite cyclic group generated by T.

(b) **Constructing a fake circle:** Let $\alpha = (I - 2T) \ \varepsilon \ \pi_2(S^1 \vee S^2)$ and let

$$X = (S^1 \vee S^2) \cup_\alpha e^3$$

be the result obtained by attaching a 3-cell to the bouquet by the map α. Show that X has the same integral homology as the circle S^1 but that $\pi_2(X) = \mathbb{Z}[\frac{1}{2}]$.

(3) (a) Let $P^2(\mathbb{Z}/k\mathbb{Z})$ be the standard example given above. Show that the universal cover of $P^2(\mathbb{Z}/k\mathbb{Z})$ has the homotopy type of a bouquet of $k-1$ copies of S^2 and hence that

$$\pi_2(P^2(\mathbb{Z}/k\mathbb{Z})) = \bigoplus_{i=1}^{k-1} \mathbb{Z} =$$

a direct sum of $k-1$ copies of $\mathbb{Z}$.

(b) Show that

$$\pi_2(P^2(\mathbb{Z}/k\mathbb{Z})) \vee S^2) = \pi_2(P^2(\mathbb{Z}/k\mathbb{Z})) \oplus \mathbb{Z}[\pi]$$

where $\pi = \langle T \rangle = \{I, T, T^2, \ldots, T^{k-1}\} = $ is the cyclic group generated by a generator T of order k.

(c) **Constructing a fake Moore space:** Let $\alpha = (I - 2T) \ \varepsilon \ \pi_2(P^2(\mathbb{Z}/k\mathbb{Z})) \vee S^2)$ and let

$$X = (P^2(\mathbb{Z}/k\mathbb{Z}) \vee S^2) \cup_\alpha e^3$$

be the result obtained by attaching a 3-cell to the bouquet by the map α. Show that X has the same integral homology as the Moore space $P^2(\mathbb{Z}/k\mathbb{Z})$ but that

$$\pi_2(X) = \pi_2(P^2(\mathbb{Z}/k\mathbb{Z})) \oplus \mathbb{Z}/(2^k - 1)\mathbb{Z}.$$

1.2 Long exact sequences of pairs and fibrations

Let CY denote the cone on a space Y. Suppose (X, A) is a pointed pair and G is an abelian group such that $P^{n-1}(G)$ exists. Define

$$\pi_n(X, A; G) = [(CP^{n-1}(G), P^{n-1}(G)), (X, A)]_*.$$

This is clearly a functor on pairs.

In general, $\pi_n(X, A; G)$ is a set for $n \geq 3$ and, if $n \geq 4$, the comultiplication

$$(CP^{n-1}(G), P^{n-1}(G)) \to (CP^{n-1}(G) \vee CP^{n-1}(G), P^{n-1}(G) \vee P^{n-1}(G))$$

makes $\pi_n(X, A; G)$ a group.

Of course, $\pi_n(X, *; G) = \pi_n(X; G)$.

16 Homotopy groups with coefficients

The restriction map $\partial : \pi_n(X, A) \to \pi_{n-1}(A)$ fits into the long exact sequence of a pair:

$$\ldots \pi_4(A; G) \to \pi_4(X; G) \to \pi_4(X, A; G) \xrightarrow{\partial} \pi_3(A; G) \to$$
$$\pi_3(X; G) \to \pi_3(X, A; G) \xrightarrow{\partial} \pi_2(A; G) \to \pi_2(X; G).$$

Let $F \to E \to B$ be a fibration sequence. The homotopy lifting property yields:

Lemma 1.2.1. *The projection induces an isomorphism*

$$\pi_n(E, F; G) \xrightarrow{\cong} \pi_n(B; G).$$

The long exact sequence of the pair (E, F) becomes the long exact homotopy sequence of a fibration:

$$\ldots \pi_4(F; G) \to \pi_4(E; G) \to \pi_4(B; G) \xrightarrow{\partial} \pi_3(F; G) \to$$
$$\pi_3(E; G) \to \pi_3(B; G) \xrightarrow{\partial} \pi_2(F; G) \to \pi_2(E; G) \to \pi_2(B; G).$$

The extension of the long exact sequence to $\pi_2(B; G)$ is an elementary consequence of the homotopy lifting property.

If F is a topological group and $F \xrightarrow{i} E \xrightarrow{\pi} B$ is a principal bundle with action $F \times E \to E$, then for all $n \geq 2$ there is an action $\pi_n(F; G) \times \pi_n(E; G) \to \pi_n(E; G)$, $([h], [f]) \mapsto [h] * [f]$. We have $\pi_*([f]) = \pi_*([g])$ for $[f]$ and $[g]$ in $\pi_n(E; G)$ if and only if there exists $[h]$ in $\pi_n(F; G)$ such that $[h] * [f] = [g]$.

Exercises

(1) Show that the long exact homotopy sequence of a fibration terminates in an epimorphism at $\pi_2(B; G)$ if F is simply connected.

(2) Suppose that $F \to E \to B$ is a fibration sequence of H-spaces and H-maps with $\pi_1(E) \to \pi_1(B)$ and $\pi_2(E) \otimes G \to \pi_2(B) \otimes G$ both epimorphisms. Show that the long exact homotopy sequence with coefficients can be extended to terminate in the exact sequence

$$\cdots \to \pi_2(B; G) \to \pi_1(F) \otimes G \to \pi_1(E) \otimes G \to \pi_1(B) \otimes G \to 0.$$

1.3 Universal coefficient exact sequences

Suppose $n \geq 2$. Since $P^n(\mathbb{Z}/k\mathbb{Z})$ is the mapping cone of the degree k map $k : S^{n-1} \to S^{n-1}$, the resulting cofibration sequence

$$\cdots \to S^{n-1} \xrightarrow{k} S^{n-1} \xrightarrow{\beta} P^n(\mathbb{Z}/k\mathbb{Z}) \xrightarrow{\rho} S^n \xrightarrow{k} S^n \cdots$$

1.3 Universal coefficient exact sequences

yields for every pointed space X a long exact sequence

$$\cdots \to \pi_n(X) \xrightarrow{k} \pi_n(X) \xrightarrow{\rho} \pi_n(X; \mathbb{Z}/k\mathbb{Z}) \xrightarrow{\beta} \pi_{n-1}(X) \xrightarrow{k} \pi_{n-1}(X) \cdots.$$

Of course, the map $k: S^n \to S^n$ induces multiplication by k on the abelian homotopy group $\pi_n(X)$ (or the k-th power on the fundamental group $\pi_1(X)$). The map ρ is called a mod k reduction map and the map β is called a Bockstein.

The above exact sequence is always an exact sequence of sets and an exact sequence of groups and homomorphisms except possibly at

$$\pi_2(X) \xrightarrow{\rho} \pi_2(X; \mathbb{Z}/k\mathbb{Z}) \xrightarrow{\beta} \pi_1(X)$$

when $\pi_2(X; \mathbb{Z}/k\mathbb{Z})$ is not a group. Of course, if X is a homotopy associative H-space it is always an exact sequence of groups and homomorphisms. In the general case, we have a substitute which is adequate for many purposes: The natural pinch map $P^2(\mathbb{Z}/k\mathbb{Z}) \to P^2(\mathbb{Z}/k\mathbb{Z}) \vee S^2$ yields an action $\pi_2(X) \times \pi_2(X; \mathbb{Z}/k\mathbb{Z}) \to \pi_2(X; \mathbb{Z}/k\mathbb{Z}), (h, f) \mapsto h * f$. If $f, g \varepsilon \pi_2(X; \mathbb{Z}/k\mathbb{Z})$, then $\beta(f) = \beta(g)$ if and only if there exists $h \varepsilon \pi_2(X)$ such that $h * f = g$.

If $n \geq 2$, there are short exact sequences

$$0 \to \frac{\pi_n(X)}{k\pi_n(X)} \to \pi_n(X; \mathbb{Z}/k\mathbb{Z}) \to \text{kernel}\{k: \pi_{n-1}(X) \to \pi_{n-1}(X)\} \to 0.$$

Since

$$\frac{\pi_n(X)}{k\pi_n(X)} \cong \pi_n(X) \otimes \mathbb{Z}/k\mathbb{Z},$$

$$\text{kernel}\{k: \pi_{n-1}(X) \to \pi_{n-1}(X)\} \cong \text{Tor}^{\mathbb{Z}}(\pi_{n-1}(X), \mathbb{Z}/k\mathbb{Z}),$$

we can write the universal coefficient sequence in the form in which it generalizes.

Universal coefficient exact sequence 1.3.1. For a pointed space X and $n \geq 2$ there is a natural exact sequence

$$0 \to \pi_n(X) \otimes \mathbb{Z}/k\mathbb{Z} \to \pi_n(X; \mathbb{Z}/k\mathbb{Z}) \to \text{Tor}^{\mathbb{Z}}(\pi_{n-1}(X), \mathbb{Z}/k\mathbb{Z}) \to 0.$$

If X is connected and the fundamental group is abelian, this suggests a consistent way to extend the definition of homotopy groups with coefficients to dimension 1. Set

$$\pi_1(X; \mathbb{Z}/k\mathbb{Z}) = \pi_1(X) \otimes \mathbb{Z}/k\mathbb{Z}.$$

18   Homotopy groups with coefficients

Suppose X is a nilpotent space with abelian fundamental group. It is a fundamental result of localization theory that the following are equivalent:

(a) $\pi_n(X;\mathbb{Z})$ is a $\mathbb{Z}[\frac{1}{k}]$ module for all $1 \leq n \leq \infty$.

(b) $\overline{H}_n(X;\mathbb{Z})$ is a $\mathbb{Z}[\frac{1}{k}]$ module for all $1 \leq n \leq \infty$.

The universal coefficient theorem for homotopy and homology imply that these are also equivalent to:

(c) $\pi_n(X;\mathbb{Z}/k\mathbb{Z}) = 0$ for all $1 \leq n \leq \infty$.

(d) $\overline{H}_n(X;\mathbb{Z}/k\mathbb{Z}) = 0$ for all $1 \leq n \leq \infty$.

Exercises

(1) Let G be a finitely generated abelian group and $n \geq 2$. Use the definition of $P^n(G)$ by free resolutions to show that there is a short exact universal coefficient sequence

$$0 \to \pi_n(X) \otimes G \to \pi_n(X;G) \to \mathrm{Tor}^{\mathbb{Z}}(\pi_{n-1}(X), G) \to 0.$$

(If $n = 2$, assume $\pi_1(X)$ is abelian.)

(2) Let p be a prime.

(a) Suppose there is a positive integer r such that

$$p^s \pi_m(X; \mathbb{Z}/p^s\mathbb{Z}) = 0 \quad \text{for all} \quad s \leq r.$$

If $\alpha \varepsilon \mathrm{Tor}^{\mathbb{Z}}(\pi_{m-1}(X), \mathbb{Z}/p^r z)$ has order p^s with $s \leq r$, then there is an element $\gamma \varepsilon \pi_m(X; \mathbb{Z}/p^r\mathbb{Z})$ which has order p^s and such that γ maps to α in the universal coefficient sequence

$$0 \to \pi_m(X) \otimes \mathbb{Z}/p^r\mathbb{Z} \to \pi_m(X; \mathbb{Z}/p^r\mathbb{Z})$$
$$\to \mathrm{Tor}^{\mathbb{Z}}(\pi_{m-1}(X), \mathbb{Z}/p^r\mathbb{Z}) \to 0.$$

(b) If $\pi_{m-1}(X)$ is finitely generated, together with the hypotheses in (a), show that the above universal coefficient sequence for $\mathbb{Z}/p^r\mathbb{Z}$ coefficients is split.

1.4 Functor properties

The definition by free resolutions leads immediately to the following proposition:

Proposition 1.4.1. *If $f : G \to H$ is a homomorphism of finitely generated abelian groups and $n \geq 2$, then there exists a map $F : P^n(H) \to P^n(G)$ such that the induced cohomology map $F^* = f$. We shall sometimes write $F = f^*$.*

1.4 Functor properties

Unfortunately, the homotopy class of the map F is not uniquely determined in all cases. But we do have:

Proposition 1.4.2. *The natural map $\theta : [P^n(H), P^n(G)]_* \to Hom(G, H)$ given by $\theta(F) = F^*$ is a bijection in the following cases:*

(a) *if H and G are finitely generated free abelian and $n \geq 2$.*

(b) *if H if finite abelian and G is finitely generated free abelian and $n \geq 2$.*

(c) *if H if finitely generated free abelian, and G is finite abelian, G has odd order, and $n \geq 4$.*

(d) *if H and G are finite abelian, G has odd order, and $n \geq 4$.*

Proof: The preceding proposition says that θ is always a surjection. Suppose that $H = \oplus H_\alpha$ and $G = \oplus G_\beta$. Then

$$[P^n(H), P^n(G)]_* \cong \oplus [P^n(H_\alpha), P^n(G_\beta)]_*$$

in all of the above cases since:

(1) $P^n(H) = \vee P^n(H_\alpha)$ implies

$$[P^n(H), P^n(G)]_* \cong \oplus [P^n(H_\alpha), P^n(G)]_*$$

and

(2) $P^n(G) = \vee P^n(G_\beta)$, dimension $P^n(H_\alpha) = n$, and the fact that the pair $(\prod P^n(G_\beta), \vee P^n(G_\beta))$ is $2n - 1$ connected in cases (a) and (b), $2n - 3$ connected in cases (c) and (d), implies

$$[P^n(H_\alpha), P^n(G)]_* \cong \oplus [P^n(H_\alpha), P^n(G_\beta)]_*$$

Therefore it suffices to consider the cyclic cases:

(a) $[S^n, S^n]_* = Hom(\mathbb{Z}, \mathbb{Z}) = \mathbb{Z}$, $n \geq 2$, which is a classical result true even for $n = 1$.

(b) $[P^n(\mathbb{Z}/k\mathbb{Z}), S^n]_* = Hom(\mathbb{Z}, \mathbb{Z}/k\mathbb{Z}) = \mathbb{Z}/k\mathbb{Z}$, $n \geq 2$ which is an immediate consequence of the universal coefficient theorem.

(c) $[S^n, P^n(\mathbb{Z}/k\mathbb{Z})]_* = Hom(\mathbb{Z}/k\mathbb{Z}, \mathbb{Z}) = 0$, k odd and $n \geq 4$: to see this, it is sufficient to observe that there is a fibration sequence

$$F \to P^n(\mathbb{Z}/k\mathbb{Z}) \to K(\mathbb{Z}/k\mathbb{Z}, n-1)$$

with F ℓ-connected, $\ell = \min(2n - 4, n + 2p - 5)$, where p is the smallest prime dividing k. Since dimension $S^n = n \leq \ell$, $[S^n, P^n(\mathbb{Z}/k\mathbb{Z})]_* = [S^n, K(\mathbb{Z}/k\mathbb{Z}, n-1)]_* = 0$.

(d) $[P^n(\mathbb{Z}/\ell\mathbb{Z}), P^n(\mathbb{Z}/k\mathbb{Z})]_* = \text{Hom}(\mathbb{Z}/k\mathbb{Z}, \mathbb{Z}/\ell\mathbb{Z})$, k odd and $n \geq 4$: Let $F, G : P^n(\mathbb{Z}/\ell\mathbb{Z}) \to P^n(\mathbb{Z}/k\mathbb{Z})$ be two maps. The first obstruction to homotopy of F and G is in

$$H^{n-1}(P^n(\mathbb{Z}/\ell\mathbb{Z}); \pi_{n-1}P^n(\mathbb{Z}/k\mathbb{Z})) = \text{Hom}((\mathbb{Z}/\ell\mathbb{Z})^*, (\mathbb{Z}/k\mathbb{Z})^*)$$
$$= \text{Hom}(\mathbb{Z}/k\mathbb{Z}, \mathbb{Z}/\ell\mathbb{Z}).$$

The obstruction is just $\theta(F) - \theta(G) = F^* - G^*$. All higher obstructions vanish by part (c). □

Corollary 1.4.3. *If H is a finite group of odd exponent k and $n \geq 4$, then $[P^n(H), X]_* = \pi_n(X; H)$ has exponent k for all spaces X.*

Proof: Apply part (d) of the above to the identity map of $P^n(H)$. Then use naturality. □

Corollary 1.4.4. *If $0 \to H \to G \to G/H \to 0$ is a short exact sequence of finitely generated abelian groups and $n \geq 2$, then there is a cofibration sequence $P^n(G/H) \to P^n(G) \to P^n(H)$.*

Proof: Let $f : P^n(G/H) \to P^n(G)$ be a map which induces the projection $G \to H$ in integral cohomology. The mapping cone C_f is then a $P^n(H)$. □

The maps in the above corollary are not always unique up to homotopy. But the space $P^n(H)$ is unique up to homotopy type in case $n \geq 3$. In the next section we will restrict to a short exact sequence of cyclic groups

$$0 \to \mathbb{Z}/\ell\mathbb{Z} \xrightarrow{\eta} \mathbb{Z}/k\ell\mathbb{Z} \xrightarrow{\rho} \mathbb{Z}/k\mathbb{Z} \to 0$$

and produce a more specific construction of this cofibration sequence.

1.5 The Bockstein long exact sequence

Given any continuous map $f : A \to B$, it is homotopy equivalent to a cofibration $\overline{f} : A \to \mathbb{Z}_f$ where $\mathbb{Z}_f =$ the mapping cylinder $B \cup_f (A \times I)$ obtained by identifying $(a, 1) \equiv f(a)$. The map $\overline{f}$ is the inclusion, $\overline{f}(a) = (f(a), 0)$. This leads to:

Lemma 1.5.1. *Any homotopy commutative diagram*

$$\begin{array}{ccc} A & \to & X \\ \downarrow & & \downarrow \\ Y & \to & Z \end{array}$$

1.5 The Bockstein long exact sequence

is homotopy equivalent to a strictly commutative diagram

$$\begin{array}{ccc} A & \to & X_1 \\ \downarrow & & \downarrow \\ Y_1 & \to & Z_1 \end{array}$$

where all the maps are cofibrations and it embeds in a commutative diagram

$$\begin{array}{ccccc} A & \to & X_1 & \to & X_1/A \\ \downarrow & & \downarrow & & \downarrow \\ Y_1 & \to & Z_1 & \to & Z_1/Y_1 \\ \downarrow & & \downarrow & & \downarrow \\ Y_1/A & \to & Z_1/X_1 & \to & Z_1/X_1 \cup_A Y_1 \end{array}$$

where all the rows and columns are cofibration sequences. In addition, note that

$$\begin{array}{ccc} A & \to & X_1 \\ \downarrow & & \downarrow \\ Y_1 & \to & X_1 \cup_A Y_1 \end{array}$$

is a pushout diagram and there is a cofibration sequence

$$X_1 \cup_A Y_1 \to Z_1 \to Z_1/X_1 \cup_A Y_1.$$

Proof: First replace $A \to X$ and $A \to Y$ by cofibrations $A \to X_1$ and $A \to Y_1$. Then use the homotopy extension property of the cofibration $A \to X_1$ to make the diagram strictly commutative. The inclusions $X_1 \to X_1 \cup_A Y_1$ and $Y_1 \to X_1 \cup_A Y_1$ are cofibrations. Replace the map $X_1 \cup_A Y_1 \to Z$ by a cofibration $X_1 \cup_A Y_1 \to Z_1$.

The rest follows by collapsing subspaces. □

For example, the homotopy commutative diagram

$$\begin{array}{ccc} S^{n-1} & \xrightarrow{1} & S^{n-1} \\ \downarrow k & & \downarrow k\ell \\ S^{n-1} & \xrightarrow{\ell} & S^{n-1} \end{array}$$

yields the homotopy commutative diagram below in which all rows and columns are cofibration sequences

$$\begin{array}{ccccccccc} S^{n-1} & \xrightarrow{1} & S^{n-1} & \to & * & \to & S^n & \to \\ \downarrow k & & \downarrow k\ell & & \downarrow & & \downarrow k & \\ S^{n-1} & \xrightarrow{\ell} & S^{n-1} & \to & P^n(\mathbb{Z}/\ell\mathbb{Z}) & \to & S^n & \to \\ \downarrow & & \downarrow & & \downarrow 1 & & \downarrow & \\ P^n(\mathbb{Z}/k\mathbb{Z}) & \xrightarrow{\rho} & P^n(\mathbb{Z}/k\ell\mathbb{Z}) & \xrightarrow{\eta} & P^n(\mathbb{Z}/\ell\mathbb{Z}) & \xrightarrow{\beta} & P^{n+1}(\mathbb{Z}/k\mathbb{Z}) & \to. \end{array}$$

The bottom row extends to a long sequence of cofibrations called the geometric Bockstein sequence

$$P^2(\mathbb{Z}/k\mathbb{Z}) \xrightarrow{\rho} P^2(\mathbb{Z}/k\ell\mathbb{Z}) \xrightarrow{\eta} P^2(\mathbb{Z}/\ell\mathbb{Z}) \xrightarrow{\beta}$$
$$P^3(\mathbb{Z}/k\mathbb{Z}) \xrightarrow{\rho} P^3(\mathbb{Z}/k\ell\mathbb{Z}) \xrightarrow{\eta} P^3(\mathbb{Z}/\ell\mathbb{Z}) \to \cdots.$$

Mapping this sequence to a space X yields the long exact homotopy Bockstein sequence

$$\pi_2(X;\mathbb{Z}/k\mathbb{Z}) \xleftarrow{\rho} \pi_2(X;\mathbb{Z}/k\ell\mathbb{Z}) \xleftarrow{\eta} \pi_2(X;\mathbb{Z}/\ell\mathbb{Z})$$
$$\xleftarrow{\beta} \pi_3(X;\mathbb{Z}/k\mathbb{Z}) \xleftarrow{\rho} \pi_3(X;\mathbb{Z}/k\ell\mathbb{Z}) \xleftarrow{\eta} \pi_3(X;\mathbb{Z}/\ell\mathbb{Z}) \leftarrow \cdots.$$

Remark. The homotopy commutative diagram of cofibration sequences is a good way to see the effect of $\rho, \eta,$ and β on integral chains. For example, $P^n(\mathbb{Z}/k\mathbb{Z})$ has a basis of integral chains: 1 in dimension 0, s_{n-1} in dimension $n-1$, e_n in dimension n. If we look at

$$\begin{array}{ccc} S^{n-1} & \xrightarrow{1} & S^{n-1} \\ \downarrow k & & \downarrow k\ell \\ S^{n-1} & \xrightarrow{\ell} & S^{n-1} \\ \downarrow & & \downarrow \\ P^n(\mathbb{Z}/k\mathbb{Z}) & \xrightarrow{\rho} & P^n(\mathbb{Z}/k\ell\mathbb{Z}) \end{array}$$

we see immediately that $\underline{\rho}_*(s_{n-1}) = \ell s_{n-1}, \underline{\rho}_*(e_n) = e_n$. Similarly, it is not hard to verify the commutative diagram

$$\begin{array}{ccc} S^{n-1} & \xrightarrow{k} & S^{n-1} \\ \downarrow k\ell & & \downarrow \ell \\ S^{n-1} & \xrightarrow{1} & S^{n-1} \\ \downarrow & & \downarrow \\ P^n(\mathbb{Z}/k\ell\mathbb{Z}) & \xrightarrow{\eta} & P^n(\mathbb{Z}/\ell\mathbb{Z}) \end{array}$$

and thus $\underline{\eta}_*(s_{n-1}) = s_{n-1}, \underline{\eta}_*(e_n) = ke_n$.

It is clear that $\underline{\beta}_*(e_n) = s_n, \underline{\beta}_*(s_{n-1}) = 0$.

Warning. If $k = \ell$, consider the null composition $\eta \circ \rho$. In the defining cofibration sequence, it looks like the map $k : P^n(\mathbb{Z}/k\mathbb{Z}) \to P^n(\mathbb{Z}/k\mathbb{Z})$ which is k times the identity, but, unless k is odd, it might not be. For example, the map $2 : P^n(\mathbb{Z}/2\mathbb{Z}) \to P^n(\mathbb{Z}/2\mathbb{Z})$ is not null homotopic.

Exercises

(1) If X is a simply connected space, show that the long exact homotopy Bockstein sequence terminates in a sequence of groups and homomorphisms ending

in an epimorphism
$$0 \leftarrow \pi_2(X;\mathbb{Z}/k\mathbb{Z}) \xleftarrow{\rho} \pi_2(X;\mathbb{Z}/k\ell\mathbb{Z}) \xleftarrow{\eta}$$
$$\pi_2(X;\mathbb{Z}/\ell\mathbb{Z}) \xleftarrow{\beta} \pi_3(X;\mathbb{Z}/k\mathbb{Z}) \leftarrow \cdots.$$

(2) If X is a homotopy associative H-space, show that the long exact homotopy Bockstein sequence may be extended to a long exact sequence of groups and homomorphisms
$$0 \leftarrow \pi_1(X) \otimes \mathbb{Z}/k\mathbb{Z} \xleftarrow{\rho} \pi_1(X) \otimes \mathbb{Z}/k\ell\mathbb{Z} \xleftarrow{\eta} \pi_1(X) \otimes \mathbb{Z}/\ell\mathbb{Z}$$
$$\xleftarrow{\beta} \pi_2(X;\mathbb{Z}/k\mathbb{Z}) \xleftarrow{\rho} \pi_2(X;\mathbb{Z}/k\ell\mathbb{Z}) \xleftarrow{\eta}$$
$$\pi_2(X;\mathbb{Z}/\ell\mathbb{Z}) \xleftarrow{\beta} \pi_3(X;\mathbb{Z}/k\mathbb{Z}) \leftarrow \cdots.$$

1.6 Nonfinitely generated coefficient groups

If $n \geq 2$ we can attempt to extend the definition of $\pi_n(X;G)$ to the case where G is an abelian group which is not finitely generated. Any such G can be written as a direct limit $G = \lim_\rightarrow H_\alpha$ of finitely generated subgroups H_α. Any inclusion map $H_\alpha \xrightarrow{\iota} H_\beta$ can be "realized" by a map $P^n(H_\beta) \xrightarrow{\iota} P^n(H_\alpha)$ which induces ι in integral cohomology. These maps may not be unique up to homotopy. In Section 1.4 we gave some conditions which guarantee uniqueness of these maps up to homotopy. On the other hand, it may be the case that G is a sequential limit of finitely generated subgroups and we may just make a choice of the realization of one stage into the next. We then realize the compositions to be consistent with these choices and the question of uniqueness vanishes. In any case, we get maps $\pi_n(X;H_\alpha) \to \pi_n(X;H_\beta)$ and as long as we have sufficient uniqueness we can take the direct limit and we define $\pi_n(X;G) = \lim_\rightarrow \pi_n(X;H_\alpha)$. Since direct limits commute with tensor and torsion products and since direct limits preserve exact sequences, we still have the universal coefficient exact sequence
$$0 \to \pi_n(X) \otimes G \to \pi_n(X;G) \to \operatorname{Tor}^{\mathbb{Z}}(\pi_{n-1}(X),G) \to 0.$$
For example, the rationals Q are the sequential direct limit of the subgroups $\frac{1}{k!}\mathbb{Z}$. That is, Q is the direct limit of the two isomorphic sequences

$$\begin{array}{ccccccccc}
\mathbb{Z} & \subset & \frac{1}{2!}\mathbb{Z} & \subset & \frac{1}{3!}\mathbb{Z} & \subset & \cdots & \subset & \frac{1}{k!}\mathbb{Z} & \subset \cdots \\
\downarrow 1! & & \downarrow 2! & & \downarrow 3! & & & & \downarrow k! & \\
\mathbb{Z} & \xrightarrow{2} & \mathbb{Z} & \xrightarrow{3} & \mathbb{Z} & \xrightarrow{4} & \cdots & \xrightarrow{k} & \mathbb{Z} & \xrightarrow{k+1} \cdots
\end{array}$$

Thus,
$$\pi_n(X:Q) = \lim_\rightarrow \pi_n(X;\mathbb{Z}) = \lim_\rightarrow \pi_n(X) \otimes \left(\frac{1}{k!}\mathbb{Z}\right) = \pi_n(X) \otimes Q.$$

Similarly, if p is a prime, then $\mathbb{Z}[1/p]$ is the sequential direct limit of $(1/p^\ell)\mathbb{Z}$ and
$$\pi_n(X;\mathbb{Z}[1/p]) = \pi_n(X) \otimes \mathbb{Z}[1/p].$$
We can also consider $Q/\mathbb{Z}$ to be the sequential direct limit of the two isomorphic sequences

$$\begin{array}{ccccccccc}
\frac{1}{2!}\mathbb{Z} & \subset & \frac{1}{3!}\mathbb{Z} & \subset & \frac{1}{4!}\mathbb{Z} & \subset & \cdots & \subset & \frac{1}{k!}\mathbb{Z} & \subset \cdots \\
\mathbb{Z} & & \mathbb{Z} & & \mathbb{Z} & & & & \mathbb{Z} & \\
\downarrow & & \downarrow & & \downarrow & & & & \downarrow & \\
\mathbb{Z}/2!\mathbb{Z} & \subset & \mathbb{Z}/3!\mathbb{Z} & \subset & \mathbb{Z}/4!\mathbb{Z} & \subset & \cdots & \subset & \mathbb{Z}/k!\mathbb{Z} & \subset \cdots
\end{array}$$

Thus,
$$\pi_n(X; Q/\mathbb{Z}) = \varinjlim \pi_n(X; \mathbb{Z}/k!\mathbb{Z}).$$

Finally, if p is a prime, recall that $\mathbb{Z}(p^\infty) = \mathbb{Z}[1/p]/\mathbb{Z}$ is the sequential direct limit of $\mathbb{Z}/p\mathbb{Z} \subset \mathbb{Z}/p^2\mathbb{Z} \subset \mathbb{Z}/p^3\mathbb{Z} \subset \ldots$ and thus
$$\pi_n(X; \mathbb{Z}(p^\infty)) = \varinjlim \pi_n(X; \mathbb{Z}/p^\ell \mathbb{Z}).$$

Exercises

(1) Let $p^\infty G = \{x \in G : \forall r \geq 0, \exists y \in G \text{ such that } p^r y = x\}$ and let $_{p^\infty} G = \{x \in G : \exists r \geq 0 \text{ such that } p^r x = 0\}$ = the p-torsion subgroup of G. Show that
$$G \otimes \mathbb{Z}(p^\infty) = G/p^\infty G, \qquad \operatorname{Tor}^\mathbb{Z}(G, \mathbb{Z}(p^\infty)) =_{p^\infty} G.$$
and
$$(\mathbb{Z}/p^r\mathbb{Z}) \otimes \mathbb{Z}(p^\infty) = \mathbb{Z}/p^r\mathbb{Z}, \qquad \operatorname{Tor}^\mathbb{Z}(\mathbb{Z}/p^r\mathbb{Z}, \mathbb{Z}(p^\infty)) = \mathbb{Z}/p^r\mathbb{Z},$$
$$(\mathbb{Z}/q\mathbb{Z}) \otimes \mathbb{Z}(p^\infty) = 0, \qquad \operatorname{Tor}^\mathbb{Z}(\mathbb{Z}/q\mathbb{Z}, \mathbb{Z}(p^\infty)) = 0$$
if q and p are relatively prime.

(2) Show that
$$G \otimes \mathbb{Z}[1/p] = \begin{cases} 0 & \text{if } G = \mathbb{Z}/p^r\mathbb{Z} \\ \mathbb{Z}/q\mathbb{Z} & \text{if } G = \mathbb{Z}/q\mathbb{Z} \end{cases} \text{ with } q \text{ and } p \text{ relatively prime.}$$

(3) Let X be a simply connected CW complex with
$$\pi_n(X; \mathbb{Z}[1/p]) = 0, \qquad \pi_n(X; \mathbb{Z}(p^\infty)) = 0$$
for all $n \geq 2$. Show that X is contractible.

(4) Suppose X is a simply connected space. Show that
$$\pi_n(X; Q/\mathbb{Z}) = 0 \qquad \text{for all } n \geq 2$$
if and only if $\pi_n(X)$ is a rational vector space for all $n \geq 2$.

(5) Suppose X is a simply connected space. Show that
$$\pi_n(X;Q) = 0 \text{ for all } n \geq 2$$
if and only if $\pi_n(X)$ is a torsion group for all $n \geq 2$.

1.7 The mod k Hurewicz homomorphism

The reduced homology of $P^n(\mathbb{Z}/k\mathbb{Z})$ is:

$$\overline{H}_\ell(P^n(\mathbb{Z}/k\mathbb{Z}); \mathbb{Z}/k\mathbb{Z}) = \begin{cases} (\mathbb{Z}/k\mathbb{Z})e_n & \text{if } \ell = n, \\ (\mathbb{Z}/k\mathbb{Z})s_{n-1} & \text{if } \ell = n-1, \\ 0 & \text{if } \ell \neq n, n-1, \end{cases}$$

where e_n and s_{n-1} denote generators of respective dimensions n and $n-1$.

Definition 1.7.1. For $n \geq 2$ the mod k Hurewicz homomorphism is the map

$$\varphi : \pi_n(X; \mathbb{Z}/k\mathbb{Z}) \to H_n(X; \mathbb{Z}/k\mathbb{Z})$$

defined by $\varphi(\alpha) = f_*(e_n)$ where $\alpha = [f] : P^n(\mathbb{Z}/k\mathbb{Z}) \to X$. Clearly, φ is a natural transformation.

Lemma 1.7.2. *If $n \geq 3$, the Hurewicz map φ is a homomorphism.*

Proof: Given maps $f : P^n(\mathbb{Z}/k\mathbb{Z}) \to X$ and $g : P^n(\mathbb{Z}/k\mathbb{Z}) \to X$, the sum $[f] + [g]$ is represented by the composition

$$P^n(\mathbb{Z}/k\mathbb{Z}) \xrightarrow{\nu} P^n(\mathbb{Z}/k\mathbb{Z}) \vee P^n(\mathbb{Z}/k\mathbb{Z}) \xrightarrow{f \vee g} X$$

where ν is the comultiplication and $f \vee g$ is f on the first summand and is g on the second summand. Therefore,

$$\varphi([f]+[g]) = (f \vee g)_* \circ \nu_*(e_n) = (f \vee g)_*(e_n, e_n)$$
$$= f_*(e_n) + g_*(e_n) = \varphi([f]) + \varphi([g]). \quad \square$$

Lemma 1.7.3. *Suppose X is a homotopy associative H-space. Then the Hurewicz map $\varphi : \pi_2(X; \mathbb{Z}/k\mathbb{Z}) \to H_2(X; \mathbb{Z}/k\mathbb{Z})$ is a homomorphism if k is odd or if X is simply connected.*

Proof: Consider the diagonal map $\Delta : P^2(\mathbb{Z}/k\mathbb{Z}) \to P^2(\mathbb{Z}/k\mathbb{Z}) \times P^2(\mathbb{Z}/k\mathbb{Z})$. Write $\Delta_*(e_2) = e_2 \otimes 1 + \lambda s_1 \otimes s_1 + 1 \otimes e_2$. If $\mu : X \times X \to X$ is the multiplication of X and $[f], [g] \varepsilon \pi_2(X : \mathbb{Z}/k\mathbb{Z})$, then

$$\varphi([f]+[g]) = \mu_* \circ (f_* \otimes g_*) \circ \Delta_*(e_2) = f_*(e_2) + g_*(e_2) + \lambda(f_*(s_1) \cdot g_*(s_1))$$
$$= \varphi([f]) + \varphi([g]) + \lambda(f_*(s_1) \cdot g_*(s_1)).$$

If X is simply connected the last term is 0. Otherwise, consider the twist map $T: P^2(\mathbb{Z}/k\mathbb{Z}) \times P^2(\mathbb{Z}/k\mathbb{Z}) \to P^2(\mathbb{Z}/k\mathbb{Z}) \times P^2(\mathbb{Z}/k\mathbb{Z}), T(x,y) = (y,x)$. Since $T \circ \Delta = \Delta$, it follows that $\lambda = -\lambda$, or $2\lambda = 0$. If k is odd, then $\lambda = 0$. □

Remark. Since $P^2(\mathbb{Z}/2\mathbb{Z})$ is just the two-dimensional projective space, the well known computation of the mod 2 cup product shows that $\lambda = 1$ and $\varphi([f] + [g]) = \varphi([f]) + \varphi([g]) + f_*(s_1) \cdot g_*(s_1)$ in the case: $\varphi: \pi_2(X; \mathbb{Z}/2\mathbb{Z}) \to H_2(X; \mathbb{Z}/2\mathbb{Z})$ with X a homotopy associative H-space.

The Hurewicz map is compatible with the universal coefficient sequences, the Bockstein sequences, and the action of $\pi_2(X)$ on $\pi_2(X; X/k\mathbb{Z})$. In other words, the following are commutative for $n \geq 2$:

$$\begin{array}{ccccccc}
\pi_n(X) & \xrightarrow{\rho} & \pi_n(X;\mathbb{Z}/k\mathbb{Z}) & \xrightarrow{\beta} & \pi_{n-1}(X) & \xrightarrow{k} & \pi_{n-1}(X) \\
\downarrow \varphi & & \downarrow \varphi & & \downarrow \varphi & & \downarrow \varphi \\
H_n(X) & \xrightarrow{\rho} & H_n(X;\mathbb{Z}/k\mathbb{Z}) & \xrightarrow{\beta} & H_{n-1}(X) & \xrightarrow{k} & H_{n-1}(X)
\end{array}$$

$$\begin{array}{ccccccc}
\pi_{n+1}(X;\mathbb{Z}/k\mathbb{Z}) & \xrightarrow{\beta} & \pi_n(X;\mathbb{Z}/\ell\mathbb{Z}) & \xrightarrow{\eta} & \pi_n(X;\mathbb{Z}/k\ell\mathbb{Z}) & \xrightarrow{\rho} & \pi_n(X;\mathbb{Z}/k\mathbb{Z}) \\
\downarrow \varphi & & \downarrow \varphi & & \downarrow \varphi & & \downarrow \varphi \\
H_{n+1}(X;\mathbb{Z}/k\mathbb{Z}) & \xrightarrow{\beta} & H_n(X;\mathbb{Z}/\ell\mathbb{Z}) & \xrightarrow{\eta} & H_n(X;\mathbb{Z}/k\ell\mathbb{Z}) & \xrightarrow{\rho} & H_n(X;\mathbb{Z}/k\mathbb{Z})
\end{array}$$

$$\begin{array}{ccc}
\pi_2(X) \times \pi_2(X;\mathbb{Z}/k\mathbb{Z}) & \xrightarrow{(a,b) \mapsto a*b} & \pi_2(X;\mathbb{Z}/k\mathbb{Z}) \\
\downarrow \varphi \times \varphi & & \downarrow \varphi \\
H_2(X) \times H_2(X;\mathbb{Z}/k\mathbb{Z}) & \xrightarrow{(c,d) \mapsto \rho(c)+d} & H_2(X;\mathbb{Z}/k\mathbb{Z}).
\end{array}$$

The mod k Hurewicz homomorphism φ for pairs is defined similarly. The homology $H_*(CP^{n-1}(\mathbb{Z}/k\mathbb{Z}), P^{n-1}(\mathbb{Z}/k\mathbb{Z}); \mathbb{Z}/k\mathbb{Z})$ is a free $\mathbb{Z}/k\mathbb{Z}$ module with generators e_n and s_{n-1} of respective dimensions n and $n-1$. Given $[f]$ in $\pi_n(X, A; \mathbb{Z}/k\mathbb{Z})$, define $\varphi([f]) = f_*(e_n)$. The maps φ are again natural transformations and

$$\begin{array}{ccccccc}
\pi_n(X;\mathbb{Z}/k\mathbb{Z}) & \to & \pi_n(X,*;\mathbb{Z}/k\mathbb{Z}) & & \pi_n(X,A;\mathbb{Z}/k\mathbb{Z}) & \xrightarrow{\partial} & \pi_{n-1}(A;\mathbb{Z}/k\mathbb{Z}) \\
\downarrow \varphi & & \downarrow \varphi & & \downarrow \varphi & & \downarrow \varphi \\
H_n(X;\mathbb{Z}/k\mathbb{Z}) & \to & H_n(X,*;\mathbb{Z}/k\mathbb{Z}) & & H_n(X,A;\mathbb{Z}/k\mathbb{Z}) & \xrightarrow{\partial} & H_{n-1}(A;\mathbb{Z}/k\mathbb{Z})
\end{array}$$

commute.

Thus, if $F \to E \to B$ is a fibration sequence, the following commutes:

$$\begin{array}{ccccc}
\pi_n(B,*;\mathbb{Z}/k\mathbb{Z}) & \xleftarrow{\cong} & \pi_n(E,F;\mathbb{Z}/k\mathbb{Z}) & \xrightarrow{\partial} & \pi_{n-1}(F;\mathbb{Z}/k\mathbb{Z}) \\
\downarrow \varphi & & \downarrow \varphi & & \downarrow \varphi \\
H_n(B,*;\mathbb{Z}/k\mathbb{Z}) & \leftarrow & H_n(E,F;\mathbb{Z}/k\mathbb{Z}) & \xrightarrow{\partial} & H_{n-1}(F;\mathbb{Z}/k\mathbb{Z}).
\end{array}$$

Exercises

(1) Check that the diagrams in this section commute.

(2) If X is a homotopy associative H-space, check that the Hurewicz map is compatible with the extensions of the long exact Bockstein sequences to dimension 1.

(3) If $n \geq 2$ show that $\varphi : \pi_n(X; \mathbb{Z}/k\mathbb{Z}) \to H_n(X; \mathbb{Z}/k\mathbb{Z})$ is an isomorphism if $\varphi \otimes 1 : \pi_n(X) \otimes \mathbb{Z}/k\mathbb{Z} \to H_n(X) \otimes \mathbb{Z}/k\mathbb{Z}$ and $\mathrm{Tor}^{\mathbb{Z}}(\varphi, 1) : \mathrm{Tor}^{\mathbb{Z}}(\pi_{n-1}(X), \mathbb{Z}/k\mathbb{Z}) \to \mathrm{Tor}^{\mathbb{Z}}(H_n(X), \mathbb{Z}/k\mathbb{Z})$ are isomorphisms. (The only point of this exercise is to check it when $n = 2$ and $\pi_2(X; \mathbb{Z}/k\mathbb{Z})$ may not be a group.)

1.8 The mod k Hurewicz isomorphism theorem

Recall that a connected pointed space X is called nilpotent if the fundamental group $\pi_1(X)$ acts nilpotently on all the homotopy groups $\pi_n(X)$ for $n \geq 1$. In particular, the fundamental group must be nilpotent. In the next theorem, $\pi_1(X)$ will be abelian and $\pi_1(X; \mathbb{Z}/k\mathbb{Z})$ is understood to be $\pi_1(X) \otimes \mathbb{Z}/k\mathbb{Z}$.

Mod k Hurewicz theorem 1.8.1. *Let X be a nilpotent space with abelian fundamental group and let $n \geq 1$. Suppose $\pi_i(X; \mathbb{Z}/k\mathbb{Z}) = 0$ for all $1 \leq i \leq n - 1$. Then the mod k Hurewicz homomorphism $\varphi : \pi_i(X; \mathbb{Z}/k\mathbb{Z}) \to H_i(X; \mathbb{Z}/k\mathbb{Z})$ is:*

(a) *an isomorphism for all $1 \leq i \leq n$.*

(b) *an epimorphism for $i = n + 1$ if $n \geq 2$.*

(c) *an isomorphism for $i = n + 1$ and an epimorphism for $i = n + 2$ if $n \geq 3$ and k is odd.*

Proof: The strategy of this proof is as follows:

(1) First, for all $n \geq 1$, show that it is true for Eilenberg–MacLane spaces.

(2) Second, for all $n \geq 1$, show that it is true for a general space by considering its Postnikov system.

Part (1) The mod k Hurewicz theorem for Eilenberg–MacLane spaces: For an integer k and an abelian group A, we shall write $A_k = A \otimes \mathbb{Z}/k\mathbb{Z}$ and $_k A = \mathrm{Tor}^{\mathbb{Z}}(A, \mathbb{Z}/k\mathbb{Z})$. First of all, note that the universal coefficient theorem implies:

Lemma 1.8.2.
$$\pi_i(K(A, n); \mathbb{Z}/k\mathbb{Z}) \cong \begin{cases} A_k & \text{if } i = n, \\ _k A & \text{if } i = n + 1, \\ 0 & \text{otherwise.} \end{cases}$$

If p is a prime, the following computation due to Cartan [22] expresses the homology of a $K(A, 1)$ in terms of exterior algebras $E(V, r)$ generated in odd degree

r and divided power algebras $\Gamma(W, s)$ generated in even degree s. In the cyclic case it is an immediate consequence of the collapse of the the homology Serre spectral sequence of the fibration $S^1 \to K(\mathbb{Z}/n\mathbb{Z}, 1) \to CP^\infty$. The Künneth theorem extends it to all finitely generated abelian groups. The general result then follows from direct limits, but something is missing, namely, a construction of divided powers in the homology of $K(A, 1)$. This can be found in the 1956 Cartan Seminar [22] or in the book of Brown [21].

Cartan 1.8.3. For all abelian groups A, there is an isomorphism
$$H_*(K(A,1); \mathbb{Z}/p\mathbb{Z}) \cong E(A_p, 1) \otimes \Gamma({}_pA, 2).$$

We first observe that the Hurewicz theorem is true for $K(A, 1)$ with mod p coefficients.

In dimension $n = 1$, the mod p Hurewicz map φ is an isomorphism for $K(A, 1)$. Hence, the mod p Hurewicz theorem is true for $K(A, 1)$ and $n = 1$.

On the other hand, if $\pi_1(K(A, 1); \mathbb{Z}/p\mathbb{Z}) = A_p = 0$, then the mod p Hurewicz map φ is an isomorphism in dimensions 1 and 2 and an epimorphism in dimension 3. Hence, the mod p Hurewicz theorem is true for $K(A, 1)$ and $n = 2$.

If $\pi_1(K(A, 1); \mathbb{Z}/p\mathbb{Z}) = \pi_2(K(A, 1); \mathbb{Z}/p\mathbb{Z}) = 0$, then $\pi_k(K(A, 1); \mathbb{Z}/p\mathbb{Z}) = H_k(K(A, 1); \mathbb{Z}/p\mathbb{Z}) = 0$ for all $k \geq 1$. We conclude that φ is an isomorphism in all dimensions. The mod p Hurewicz theorem is true for $K(A, 1)$ and all $n \geq 1$.

Lemma 1.8.4. *If p is a prime, then*

$$H_\ell(K(A,2); \mathbb{Z}/p\mathbb{Z}) = \begin{cases} A_p & \text{if } \ell = 2, \\ {}_pA & \text{if } \ell = 3, \\ \Gamma_2(A_p) & \text{if } \ell = 4 \\ 0 & \text{if } A_p = 0 \text{ and } \ell = 5. \end{cases}$$

Lemma 1.8.5. *If p is a prime and $m \geq 3$, then*

$$H_\ell(K(A,m); \mathbb{Z}/p\mathbb{Z}) = \begin{cases} A_p & \text{if } \ell = m, \\ {}_pA & \text{if } \ell = m+1, \\ 0 & \text{if } \ell = m+2 \text{ and } p \text{ is odd}. \\ 0 & \text{if } A_p = 0 \text{ and } \ell = m+3. \end{cases}$$

The above lemmas are a small piece of the complete computation due to Cartan. They are an elementary consequence of Cartan's calculation of the homology of a $K(A, 1)$. One uses the path space fibration $K(A, m-1) \to PK(A, n) \to K(A, m)$ and the fact that the Serre spectral sequence is a spectral sequence of algebras.

1.8 The mod k Hurewicz isomorphism theorem

Anyway, the above lemmas assert that the mod p Hurewicz theorem is true for $K(A, 1)$ and all $n \geq 1$.

Now, let k be any integer. Since $H_1(K(A, 1); \mathbb{Z}/k\mathbb{Z}) \cong \pi_1(K(A, 1); \mathbb{Z}/k\mathbb{Z}) \cong A_k$, the mod p version implies that the mod k Hurewicz theorem is true for $K(A, 1)$ and $n = 1$.

Suppose that $\pi_1(K(A, 1); \mathbb{Z}/k\mathbb{Z}) = 0$. The long exact Bockstein sequence shows that $\pi_1(K(A, 1); \mathbb{Z}/d\mathbb{Z}) = 0$ for any integer d dividing k. Since k can be factored into primes and the modular Hurewicz theorem is true for primes, $K(A, 1)$, and $n = 2$, we can use induction on the number of factors of k, the strong form of the five lemma, and long exact Bockstein sequences to show that, for any integer k, the mod k Hurewicz theorem is true for $K(A, 1)$ and $n = 2$:

Similarly, if $n \geq 3$ and if $\pi_\ell(K(A, 1); \mathbb{Z}/k\mathbb{Z}) = 0$ for all $1 \leq \ell \leq n - 1$, then $\pi_\ell(K(A, 1); \mathbb{Z}/d\mathbb{Z}) = 0$ for all d dividing k and $\ell = 1, 2$. Thus, $\pi_\ell(K(A, 1); \mathbb{Z}/p\mathbb{Z}) = H_\ell(K(A, 1); \mathbb{Z}/p\mathbb{Z}) = 0$ for all primes p dividing k and all $\ell \geq 1$. As before, induction on the number of factors of k, the strong form of the five lemma, and long exact Bockstein sequences show that $\pi_\ell(K(A, 1); \mathbb{Z}/k\mathbb{Z}) = H_\ell(K(A, 1); \mathbb{Z}/k\mathbb{Z}) = 0$ for all $\ell \geq 1$. The mod k Hurewicz theorem is true for $K(A, 1)$ and all $n \geq 1$.

Induction on the number of factors of k combines with the strong form of the five lemma and long exact Bockstein sequences to show that the mod k Hurewicz theorem is true for $K(A, 1)$ and all $n \geq 1$.

Finally, the path fibration $K(A, m-1) \to PK(A, m) \to K(A, m)$ and the Serre spectral sequence show that the mod k Hurewicz theorem is true for $K(A, m)$ for all $m \geq 1$ and all $n \geq 1$.

Part (2) The mod k Hurewicz theorem via Postnikov systems: Let A be an abelian group on which a group π acts. In other words, A is a module over the group ring $\mathbb{Z}[\pi]$. Let $\varepsilon : \mathbb{Z}[\pi] \to \mathbb{Z}$ be the augmentation epimorphism defined by $\varepsilon(g) = 1$ for all g in π. If $I = \mathrm{kernel}(\varepsilon) = $ the augmentation ideal, then π acts trivially on A if and only if $I \cdot A = 0$. The action is called nilpotent if $I^n \cdot A = 0$ for some power I^n of the augmentation ideal.

We shall say that A is mod k trivial if $A_k = 0$ and $_k A = 0$.

Lemma 1.8.6. *Let $0 \to A \to B \to C \to 0$ be a short exact sequence of abelian groups. Then:*

(a) $B_k = 0$ *implies* $C_k = 0$.

(b) *if two of the three groups are mod k trivial, then so is the third.*

Lemma 1.8.7.

(a) $A_k = 0$ *implies* $(I^n \cdot A)_k = 0$ *for all* $n \geq 1$.

(b) $_k A = 0$ *implies* $_k(I^n \cdot A) = 0$ *for all* $n \geq 1$.

The first of the two lemmas follows from the long exact sequence of the Tor functor. For the second, it is sufficient to consider the case $n = 1$. Assume $A_k = 0$. Note that $k(I \cdot A) = (I \cdot kA) = I \cdot A = 0$. Thus $(I \cdot A)_k = 0$, and, if $_kA = 0$, then $_k(I \cdot A) \subseteq_k A = 0$.

In particular, if $A_k = 0$ then $(I^n \cdot A/I^{n+1} \cdot A)_k = 0$ for all $n \geq 1$, and, if A is mod k trivial, then $(I^n \cdot A/I^{n+1} \cdot A)$ is mod k trivial for all $n \geq 1$.

Recall that a space X is called nilpotent if X is path connected, the fundamental group $\pi_1(X)$ is nilpotent, and the action of $\pi_1(X)$ is nilpotent on $\pi_m(X)$ for all $m \geq 2$. In this case, each homotopy group $\pi_m = \pi_m(X)$ has a decreasing filtration

$$\pi_m = F_1(\pi_m) \supseteq F_2(\pi_m) \supseteq F_3(\pi_m) \supseteq F_4(\pi_m) \supseteq \cdots$$

with each $F_\ell(\pi_m)/F_{\ell+1}(\pi_m)$ having a trivial $\pi_1(X)$ action and with each decreasing sequence terminating in a finite number of steps, $F_{\alpha_m+1}(\pi_m(X)) = 0$. This leads to a sequence of principal bundles, a refinement of the Postnikov system,

$$K(F_\ell(\pi_m)/F_{\ell+1}(\pi_m), m) \to E_{m,\ell} \to E_{m,\ell-1}$$

with $n \geq 1$ and $1 \leq \ell \leq \alpha_n$. It begins with

$$E_{1,0} = * \text{ and } E_{m,0} = E_{m-1,\alpha_{m-1}}$$

for $m \geq 2$.

We have

$$\pi_s(E_{m,\ell}) = \begin{cases} \pi_s(X) & \text{if } 1 \leq s \leq m-1 \\ \pi_m(X)/F_{\ell+1}(\pi_m(X)) & \text{if } s = m \\ 0 & \text{if } s > m. \end{cases}$$

Furthermore,

$$X = \lim_{m \to \infty} E_{m,\ell}$$

and this inverse limit is "finite in each degree."

Now, suppose that X satisfies the hypotheses of the mod k Hurewicz theorem for some $n \geq 1$, that is, X is nilpotent with abelian fundamental group and $\pi_i(X; \mathbb{Z}/k\mathbb{Z}) = 0$ for all $0 \leq i \leq n-1$. Then we know that all the Eilenberg–MacLane spaces $K(F_\ell(\pi_m)/F_{\ell+1}(\pi_m), m)$ which appear above also satisfy the hypotheses of the mod k Hurewicz theorem for this $n \geq 1$.

In order to perform the inductive step to prove the mod k Hurewicz theorem, we need to recall the Serre long exact homology sequence of a fibration. Suppose that

$F \to E \to B$ is an orientable fibration sequence of connected spaces with

$$\overline{H}_i(F) = \overline{H}_i(B) = 0 \text{ for } 1 \leq i \leq n-1$$

for some coefficient ring R.

The E^2 term of the homology Serre spectral sequence is

$$E^2_{s,t} = H_s(B; H_t(F)):$$

$H_{2n-1}(F)$	$\cdots$ 0 $\cdots$						
$\vdots$	$\cdots$ 0 $\cdots$						
$H_{n+1}(F)$	$\cdots$ 0 $\cdots$						
$H_n(F)$	$\cdots$ 0 $\cdots$	$H_n(B; H_n(F))$					
$\vdots$ 0 $\vdots$	$\cdots$ 0 $\cdots$	0	0	0	$\cdots$ 0 $\cdots$	0	
R	$\cdots$ 0 $\cdots$	$H_n(B)$	$H_{n+1}(B)$	$H_{n+2}(B)$	$\cdots$	$H_{2n+1}(B)$	

The first nonzero differentials are:

$$d^{n-1} : H_n(B; H_n(F)) \to H_{2n-1}(F)$$

and the transgressions

$$\tau = d^{n+j+1} : H_{n+j+1}(B) \to H_{n+j}(F)$$

with $0 \leq j \leq n-2$.

It follows that we have the Serre long exact homology sequence

$$H_{2n-1}(F) \to H_{2n-1}(E) \to H_{2n-1}(B) \xrightarrow{\tau}$$
$$H_{2n-2}(F) \to H_{2n-2}(E) \to H_{2n-2}(B) \xrightarrow{\tau}$$
$$\cdots$$
$$H_{n+2}(F) \to H_{n+2}(E) \to H_{n+2}(B) \xrightarrow{\tau}$$
$$H_{n+1}(F) \to H_{n+1}(E) \to H_{n+1}(B) \xrightarrow{\tau}$$
$$H_n(F) \to H_n(E) \to H_n(B) \to 0.$$

Suppose now that the coefficients are $\mathbb{Z}/k\mathbb{Z}$. Since the transgression is defined by

$$\tau : H_{n+j+1}(B, *) \leftarrow H_{n+j+1}(E, F) \xrightarrow{\partial} H_{n+j}(F),$$

it follows that the transgression is compatible with the connecting homomorphism of the long exact homotopy sequence of the fibration:

$$\begin{array}{ccc} \pi_{n+j+1}(B; \mathbb{Z}/k\mathbb{Z}) & \xrightarrow{\partial} & \pi_{n+j}(F; \mathbb{Z}/k\mathbb{Z}) \\ \downarrow \varphi & & \downarrow \varphi \\ H_{n+j+1}(B; \mathbb{Z}/k\mathbb{Z}) & \xrightarrow{\tau} & H_{n+j}(F; \mathbb{Z}/k\mathbb{Z}) \end{array}$$

commutes.

Now the strong form of the five lemma applies to show that if the mod k Hurewicz theorem is true for the fibre and base of the fibration sequence

$$K(F_\ell(\pi_m)/F_{\ell+1}(\pi_m), m) \to E_{m,\ell} \to E_{m,\ell-1}$$

then it is true for the total space. This completes the inductive step in the proof.

Hence the mod k Hurewicz theorem is true for all the Postnikov stages $E_{m,\ell}$. Since $X = \lim_{m \to \infty} E_{m,\ell}$ is an inverse limit which is finite in each degree, it follows that the mod k Hurewicz theorem is true for all X.

Exercise

(1) Suppose k and ℓ are positive integers. Suppose either that X is simply connected or that X is a connected H-space. Show that $\varphi : \pi_j(X; \mathbb{Z}/k\mathbb{Z}) \to H_j(X; \mathbb{Z}/k\mathbb{Z})$ is an isomorphism for all $1 \leq j < \ell$ and an epimorphism for $j = \ell$ if and only if the same is true for $\varphi : \pi_j(X; \mathbb{Z}/k^r\mathbb{Z}) \to H_j(X; \mathbb{Z}/k^r\mathbb{Z})$ where r is a fixed positive integer. (Hint: Use induction on r, the universal coefficient sequences and the general five lemma.)

1.9 The mod k Hurewicz isomorphism theorem for pairs

If (X, A) is a pair of spaces with $\pi_2(X, A)$ abelian, then the mod k homotopy group $\pi_2(X, A; \mathbb{Z}/k\mathbb{Z})$ is defined to be $\pi_2(X, A) \otimes \mathbb{Z}/k\mathbb{Z}$. The classical Hurewicz map

1.9 The mod k Hurewicz isomorphism theorem for pairs

induces a mod k Hurewicz map $\varphi : \pi_2(X, A; \mathbb{Z}/k\mathbb{Z}) \to H_2(X, A; \mathbb{Z}/k\mathbb{Z})$. For example, if A is simply connected, then $\pi_2(X, A)$ is abelian and this definition is valid. With these conventions, we assert:

Mod k Hurewicz theorem for pairs 1.9.1. *Let (X, A) be a pair of simply connected spaces and let $n \geq 2$. If $\pi_i(X, A; \mathbb{Z}/k\mathbb{Z}) = 0$ for $2 \geq i < n$, then $\varphi : \pi_i(X, A; \mathbb{Z}/k\mathbb{Z}) \to H_i(X, A; \mathbb{Z}/k\mathbb{Z})$ is a bijection for $2 \geq i \geq n$ and, if $n > 2$ it is an epimorphism for $i = n + 1$.*

This has the following corollary.

Corollary 1.9.2. *Let $f : X \to Y$ be a map between simply connected spaces. Then $f_* : \pi_i(X; \mathbb{Z}/k\mathbb{Z}) \to \pi_*(Y; \mathbb{Z}/k\mathbb{Z})$ is a bijection for all $i \geq 2$ if and only if $f_* : H_i(X; \mathbb{Z}/k\mathbb{Z}) \to H_*(Y; \mathbb{Z}/k\mathbb{Z})$ is a bijection for all $i \geq 2$.*

Proof: Use the mapping cylinder to convert the map f into an inclusion $X \to Y$. Then the mod k Hurewicz theorem for the pair (Y, X) asserts that the vanishing of all the relative homotopy groups $\pi_i(Y, X; \mathbb{Z}/k\mathbb{Z})$ is equivalent to the vanishing of all the relative homology groups $H_i(Y, X; \mathbb{Z}/k\mathbb{Z})$. □

Remark. The example of the inclusion of a circle into a fake circle shows that simple connectivity is necessary in the above results.

Proof of the mod k Hurewicz theorem for pairs: The mod k Hurewicz isomorphism theorem for a pair of simply connected spaces (X, A) is deduced from the mod k Hurewicz isomorphism theorem for a space by a method introduced by Serre.

Let $PX \to X$ be the path space fibration and let E be the subspace of PX consisting of all paths which terminate in A. Then $(PX, E) \to (X, A)$ is a relative fibration with fibre ΩX. In particular, the fibration sequence $\Omega X \to E \to A$ shows that E is a nilpotent space with abelian fundamental group.

We have isomorphisms $\pi_i(PX, E; \mathbb{Z}/k\mathbb{Z}) \to \pi_i(X, A; \mathbb{Z}/k\mathbb{Z})$, $\pi_i(PX, E; \mathbb{Z}/k\mathbb{Z}) \to \pi_{i-1}(E; \mathbb{Z}/k\mathbb{Z})$, and $H_i(PX, E; \mathbb{Z}/k\mathbb{Z}) \to H_{i-1}(E; \mathbb{Z}/k\mathbb{Z})$ for all $i \geq 2$.

Thus, $\pi_{i-1}(E; \mathbb{Z}/k\mathbb{Z}) = 0$ for all $2 \leq i < n$ and the mod k Hurewicz theorem applies to E. In particular, $H_{i-1}(E; \mathbb{Z}/k\mathbb{Z}) \cong H_i(PX, E; \mathbb{Z}/k\mathbb{Z}) = 0$ for all $2 \leq i < n$.

The Serre spectral sequence of the relative fibration sequence $\Omega X \to (PX, E) \to (X, A)$ shows that $H_i(PX, E; \mathbb{Z}/k\mathbb{Z}) \to H_i(X, A; \mathbb{Z}/k\mathbb{Z})$ is an isomorphism for all $i \leq n$ and an epimorphism for $i = n + 1$. Hence, the following commutative diagram completes the proof:

$$\begin{array}{ccccc}
\pi_{i-1}(E; \mathbb{Z}/k\mathbb{Z}) & \leftarrow & \pi_i(PX, E; \mathbb{Z}/k\mathbb{Z}) & \to & \pi_i(X, A; \mathbb{Z}/k\mathbb{Z}) \\
\downarrow & & \downarrow & & \downarrow \\
H_{i-1}(E; \mathbb{Z}/k\mathbb{Z}) & \leftarrow & H_i(PX, E; \mathbb{Z}/k\mathbb{Z}) & \to & H_i(X, A; \mathbb{Z}/k\mathbb{Z}).
\end{array}$$ □

1.10 The third homotopy group with odd coefficients is abelian

As an application of the Hurewicz theorem, we prove:

Proposition 1.10.1. *If k is odd, then $\pi_3(X; \mathbb{Z}/k\mathbb{Z})$ is an abelian group.*

Proof: We consider the isomorphic group $\pi_2(\Omega X; \mathbb{Z}/k\mathbb{Z})$.

Define the commutator $[\ ,\ ] : \Omega X \times \Omega X \to \Omega X$ by $[\omega, \gamma] = \omega \gamma \omega^{-1} \gamma^{-1}$. Note that $[\ ,\ ]$ is null homotopic on the bouquet $\Omega X \vee \Omega X$ and hence factors into

$$\Omega X \times \Omega X \to \Omega X \wedge \Omega X \xrightarrow{[\ ,\ ]} \Omega X.$$

Let $f : P^2(\mathbb{Z}/k\mathbb{Z}) \to \Omega X$ and $g : P^2(\mathbb{Z}/k\mathbb{Z}) \to \Omega X$ be two maps. Define the commutator $[f, g]$ by the composition

$$P^2(\mathbb{Z}/k\mathbb{Z}) \xrightarrow{\Delta} P^2(\mathbb{Z}/k\mathbb{Z}) \times P^2(\mathbb{Z}/k\mathbb{Z}) \xrightarrow{f \times g} \Omega X \times \Omega X \xrightarrow{[\ ,\ ]} \Omega X,$$

where Δ is the diagonal. If $\overline{\Delta}$ is the reduced diagonal, this is the same as the composition

$$P^2(\mathbb{Z}/k\mathbb{Z}) \xrightarrow{\overline{\Delta}} P^2(\mathbb{Z}/k\mathbb{Z}) \wedge P^2(\mathbb{Z}/k\mathbb{Z}) \xrightarrow{f \wedge g} \Omega X \wedge \Omega X \xrightarrow{[\ ,\ ]} \Omega X.$$

If e_1 and e_2 are generators of the reduced homology $\overline{H}_*(P^2(\mathbb{Z}/k\mathbb{Z}); \mathbb{Z}/k\mathbb{Z})$ of respective dimensions 1 and 2, then a result from Steenrod and Epstein's book asserts that

$$\overline{\Delta}_*(e_2) = \frac{k(k+1)}{2} e_1 \otimes e_1$$

and this equals 0 when k is odd. If k is odd, then the mod k Hurewicz image $\phi(\overline{\Delta}) = 0$.

The Hurewicz theorem implies that $\overline{\Delta}$ is null homotopic.

Hence $[f, g]$ is null homotopic and $\pi_2(\Omega X; \mathbb{Z}/k\mathbb{Z})$ is abelian if k is odd. $\square$

2 A general theory of localization

In this chapter we consider the general theory of localization which is due independently to Dror Farjoun [36] and to A.K. Bousfield [15, 16]. The theory is founded on the homotopy theoretic consequences of inverting a specific map μ of spaces. Those spaces for which the mapping space dual of μ is an equivalence are called local. In turn, the local spaces define a set of maps called local equivalences. The localization of a space X is defined to be a universal local space which is locally equivalent to X.

Localizations always exist. It is a nice fact that the localizations of simply connected spaces are also simply connected. This makes it possible to restrict the theory to simply connected spaces which is what we do in this chapter.

For simply connected spaces, the Dror Farjoun–Bousfield theory specializes to the classical example of localization of spaces at a subset of primes S. The complementary set of primes is inverted. We begin by inverting the maps $M \to *$ for all Moore spaces M with one nonzero first homology group isomorphic to $\mathbb{Z}/q\mathbb{Z}$ where q is a prime not in S. In this case, the equivalences are maps which induce an isomorphism of homology localized at S.

Localization of spaces first occurs in the works of Daniel Quillen [110], of Dennis Sullivan [128, 129] , and of A.K. Bousfield and D.M. Kan [17]. An early construction of this localization is due to D.W. Anderson [6] and is a special case of the construction of the Dror Farjoun localization. First applications occur in the theory of H-spaces and are due to Sullivan, to Peter Hilton and Joseph Roitberg [53], and to Alexander Zabrodsky [142].

There are two themes in localization theory. One is to study a space in more depth by inverting some or all primes. For example, Serre's result [118] that $\Omega S^{2n+2} \simeq S^{2n+1} \times \Omega S^{4n+3}$ is valid once 2 is inverted but not before unless $n = 0, 1, 3$. [2] The extreme example of this theme is rationalization, inverting all primes. After this is done, the simply connected homotopy category becomes equivalent to the category of rational simply connected differential coalgebras [110].

Another theme is to construct a space by piecing together complementary localizations. In essence, Hilton and Roitberg [55] constructed new H-spaces in this way. We will discuss these examples but, in this book, we will mostly restrict ourselves to the first theme.

For any prime p, another specialization of the Dror Farjoun-Bousfield theory is to p-completion where the equivalences are maps which induce isomorphisms of mod p homology. The process of p-completion begins by inverting the map $M \to *$ where M is a Moore space with nonvanishing first homology group isomorphic to $\mathbb{Z}[1/p]$.

The unstable Adams spectral sequence invented by Bousfield and Kan converges to the homotopy groups of p-complete spaces [17]. Completions have been vital to the theory of finite H-spaces with classifying spaces, the so-called p-compact groups studied by Dwyer and Wilkerson [38]. Before the development of the general theory of p-compact groups, Sullivan showed that certain completions of spheres have classifying spaces [128].

It is not always true that the localization of an n-connected space is also n-connected. This fact is true for the classical theory of localization of spaces at a set of primes and also true for the theory of p-completion of spaces. But it is certainly not true for the localization theory [103] based on inverting the map $K(\mathbb{Z}/p\mathbb{Z}, 1) \to *$. In this localization, all Eilenberg–MacLane spaces $K(G, n)$ with G a p-primary torsion abelian group are made locally equivalent to a point. Up to p-completion, simply connected spaces with π_2 torsion are locally equivalent to all their n-connected covers.

In addition, Miller's theorem [84] asserts that simply connected finite complexes are local in this theory with $K(\mathbb{Z}/p\mathbb{Z}, 1) \to *$ inverted. A lemma due to Zabrodsky shows that all $K(\pi, n) \to *$ are inverted with π a p-primary torsion abelian group. All these Eilenberg–MacLane spaces are equivalent to a point in this localization. In fact, if X is a simply connected finite complex with $\pi_2(X)$ torsion and $X\langle n \rangle$ is an n-connected cover of X, then the p-completion of this kind of localization of $X\langle n \rangle$ is just the p-completion of X. Up to p-completion, no information about such an X is lost by taking any connected cover.

One consequence of this fact is a simple proof of Serre's theorem [117] that a noncontractible simply connected finite complex always has infinitely many nonzero homotopy groups.

Serre conjectured that: if p is a prime and if X is a simply connected finite complex with nontrivial mod p reduced homology, that is, if $\overline{H}_*(X; \mathbb{Z}/p\mathbb{Z}) \neq 0$, then the p-torsion subgroups of the homotopy groups $\pi_n(X)$ are nonzero for infinitely many n. The proof of this conjecture by McGibbon and the author [83] is yet another application of Miller's theorem.

Modern forms of localization accommodate many of the classical results in homotopy theory but not all. The finite generation of the homotopy groups of a simply connected finite type complex, a result proved by Serre [118] with his C-theory, does not seem to yield to modern theories of localization. We include a brief presentation of C-theory in this chapter.

2.1 Dror Farjoun–Bousfield localization

Emmanuel Dror Farjoun has defined a notion of localization with respect to any continuous map $\mu : M \to N$. Independently, A.K. Bousfield has treated the special case of a constant map $\mu : M \to *$. We specialize to this case and refer to it as localization with respect to $M \to *$ or sometimes as M-nullification. The idea is that localization inverts the map $M \to *$, in other words, M is nullified.

We will work in the category of connected pointed spaces and pointed maps. But it still makes sense to consider $\mathrm{map}(A, B) =$ the space of all maps from A to B. And $\mathrm{map}_*(A, B) =$ the subspace of all pointed maps from A to B.

Definition 2.1.1. If M is a fixed connected pointed space, then a connected pointed space X is called M-null or local with respect to $M \to *$ if either of the following equivalent conditions hold:

(1) the map which evaluates a function at the basepoint, $\mathrm{map}(M, X) \to X$, is a weak equivalence.

(2) the space of pointed maps $\mathrm{map}_*(M, X)$ is weakly contractible.

The equivalence of the above two conditions is a consequence of the fibration sequence $\mathrm{map}_*(M, X) \to \mathrm{map}(M, X) \to X$.

Thus, X is M-null if and only if, for all $n \geq 0$, $\pi_n(\mathrm{map}_*(M, X)) = [\Sigma^n M, X]_* = *$, in other words, all pointed maps $\Sigma^n M \to X$ must be homotopic to the constant. In this sense, M looks like a point with respect to X.

It is convenient that the basic definitions of localization come in two equivalent versions, pointed and unpointed.

Definition 2.1.2. A pointed map $f : A \to B$ is called a local equivalence with respect to $M \to *$ if, for all spaces X which are local with respect to $M \to *$, either of the following equivalent conditions hold:

(1) the map of mapping spaces $f^* : \mathrm{map}(B, X) \to \mathrm{map}(A, X)$ is a weak equivalence.

(2) the map of pointed mapping spaces $f^* : \mathrm{map}_*(B, X) \to \mathrm{map}_*(A, X)$ is a weak equivalence.

The second condition means that for all integers $n \geq 0$ and maps $g : \Sigma^n A \to X$, there is a map h, unique up to homotopy, which makes the diagram below homotopy commutative.

$$\begin{array}{ccc} \Sigma^n A & \xrightarrow{\Sigma^n f} & \Sigma^n B \\ & {}_g \searrow & \downarrow h \\ & & X \end{array}$$

In particular, $M \to *$ is a local equivalence.

Definition 2.1.3. If A is a pointed space, a localization of A with respect to $M \to *$ is a pointed map $\iota : A \to \overline{A}$ such that:

(1) ι is an local equivalence with respect to $M \to *$ and

(2) $\overline{A}$ is local with respect to $M \to *$.

If the localization $\overline{A}$ exists, then it is unique up to homotopy and a functor on the homotopy category. This follows from the homotopy uniqueness of f_* in the homotopy commutative diagram:

$$\begin{array}{ccc} A & \xrightarrow{f} & B \\ \downarrow \iota & & \downarrow \iota \\ \overline{A} & \xrightarrow{f_*} & \overline{B}. \end{array}$$

We will see that localization with respect to $\mu : M \to *$ exists and can be constructed in a strictly functorial manner. We shall denote localization of A by $A \xrightarrow{\iota} L_\mu(A)$ or by $A \xrightarrow{\iota} L_M(A)$. We shall denote localization of a map $f : A \to B$ by $L_M(f) = f_* : L_M(A) \to L_M(B)$. Note that localization $A \xrightarrow{\iota} L_M(A)$ is the unique object up to homotopy which satisfies the following:

Universal mapping property 2.1.4.

(a) $L_M(A)$ is local with respect to $M \to *$.

(b) for all X which are local with respect to $M \to *$ and all maps $g : A \to X$, there is up to homotopy a unique map $h : L_M(A) \to X$ which makes the following homotopy commutative:

$$\begin{array}{ccc} A & \xrightarrow{\iota} & L_M(A) \\ \downarrow g & \swarrow h & \\ X & & \end{array}$$

Observe that L_M is an idempotent functor in the sense that

$$L_M(L_M(A)) \simeq L_M(A).$$

The next proposition says that localization commutes with finite products.

2.1 Dror Farjoun–Bousfield localization

Proposition 2.1.5.

(a) *If X and Y are local with respect to $M \to *$, then $X \times Y$ is local with respect to $M \to *$.*

(b) *If $A \to B$ is a local equivalence with respect to $M \to *$ and C is any space, then $A \times C \to B \times C$ is a local equivalence with respect to $M \to *$.*

(c) *If A and B are any spaces, then the natural map $L_M(A \times B) \to L_M(A) \times L_M(B)$ is a homotopy equivalence.*

Proof:

(a) If X and Y are local, then there are equivalences
$$\mathrm{map}(M, X \times Y) \cong \mathrm{map}(M, X) \times \mathrm{map}(M, Y) \simeq X \times Y.$$
Hence, $X \times Y$ is local.

(b) Let $A \to B$ be a local equivalence, C be any space, and X be any local space. The space $\mathrm{map}(C, X)$ is local since there are equivalences
$$\mathrm{map}(M, \mathrm{map}(C, X)) \cong \mathrm{map}(M \times C, X) \cong \mathrm{map}(C, \mathrm{map}(M, X))$$
$$\simeq \mathrm{map}(C, X).$$
It follows that the map $A \times C \to B \times C$ is a local equivalence since there are equivalences
$$\mathrm{map}(A \times C, X) \cong \mathrm{map}(A, \mathrm{map}(C, X)) \simeq \mathrm{map}(B, \mathrm{map}(C, X))$$
$$\cong \mathrm{map}(B \times C, X).$$

(c) Finally, we note that we can factor
$$A \times B \to L_M(A) \times B \to L_M(A) \times L_M(B)$$
into local equivalences and that $L_M(A) \times L_M(B)$ is local. The universal mapping property shows that $L_M(A \times B) \simeq L_M(A) \times L_M(B)$. □

We give a strictly functorial construction of localization with respect to $M \to *$:

Proposition 2.1.6. *For all pointed spaces A, there exists a localization $A \xrightarrow{\iota} L_M(A)$ with respect to $M \to *$. If A is simply connected, then so is the localization $L_M(A)$.*

Proof: In order to avoid even the mention of some very large sets, we shall make the simplifying assumption that M is a countable connected CW complex. In fact, all cases known to the author to be of interest satisfy this hypothesis.

Therefore, M can be expressed as a countable increasing union
$$M = \bigcup_n M_n$$
of finite complexes M_n, $n \geq 0$, with $M_n \subset M_{n+1}$.

For any pointed space B, we define $\overline{B}$ to be the mapping cone of the bouquet

$$\bigvee g : \bigvee \Sigma^{n_g}(M) \to B$$

of all the pointed maps $g : \Sigma^{n_g}(M) \to B$ to B from any suspension of M with $n_g \geq 0$. Since M is connected, it follows that $\overline{B}$ is simply connected if B is simply connected.

For later reference, we note that, if X is local with respect to $M \to *$, the long exact sequence associated to a cofibration sequence shows that there are bijections of homotopy classes of pointed maps $[\Sigma^k(\overline{B}), X]_* \to [\Sigma^k(B), X]_*$ for all $k \geq 0$.

Let Ω be the well ordered set of all ordinals which are less than the first uncountable ordinal; in other words, Ω is the first uncountable ordinal. For ordinals $\alpha \varepsilon \Omega$, we define $L_\alpha(A)$ by transfinite recursion as follows:

(1) $L_0(A) = A$ for the first ordinal $0 \varepsilon \Omega$.

(2) $L_{\alpha+1}(A) = \overline{L_\alpha(A)}$ for ordinals which are successors.

(3) $L_\alpha(A) = \bigcup_{\beta < \alpha} L_\beta(A)$ for limit ordinals.

Finally, we define $L_M(A) = L_\Omega(A) = \bigcup_{\beta < \Omega} L_\beta(A)$. We claim that $L_M(A)$ is a localization of A. Clearly, $L_M(A)$ is simply connected if A is simply connected.

First, we show that $L_M(A)$ is local. Suppose we have a pointed map $g : \Sigma^k(M) \to L_M(A)$. Since each $\Sigma^k(M_n)$ is a finite complex, its image is contained in some $L_{\alpha_n}(A)$ for an ordinal $\alpha_n \varepsilon \Omega$. Thus the image of $\Sigma^k(M)$ is contained in the countable limit $L_\gamma(A)$ with $\gamma = \sup \alpha_n \varepsilon \Omega$. Thus, g is null homotopic in the mapping cone $L_{\gamma+1}(A)$. Since $L_{\gamma+1}(A) \subset L_M(A)$, g is null homotopic in $L_M(A)$. Hence, $L_M(A)$ is local.

Second, we show that $A \to L_M(A)$ is a local equivalence. For all local X and $k \geq 0$, we need bijections $[\Sigma^k(L_M(A)), X]_* \to [\Sigma^k(A), X]_*$. But transfinite induction shows that there are bijections $[\Sigma^k(L_\alpha(A)), X]_* \to [\Sigma^k(A), X]_*$, even for the case $\alpha = \Omega$.

This completes the proof of the existence of localization. $\square$

The above proposition suggests that it is possible to restrict localization to the category of simply connected spaces. (On the other hand, there is no reason to believe that a general theory of localization can be restricted to, for example, nilpotent spaces.) To fully justify this restriction to simply connected spaces, we need the following lemma.

Lemma 2.1.7.

(1) Let $\tilde{X} \to X$ be any covering space of X. If X is local then $\tilde{X}$ is local.

(2) *A map $A \to B$ of simply connected spaces is a local equivalence if* $\mathrm{map}_*(B, W) \to \mathrm{map}_*(A, W)$ *is a weak equivalence for all simply connected local W.*

Proof:

(1) Unique path lifting for covering spaces asserts that $\mathrm{map}_*(M, \tilde{X})$ embeds in $\mathrm{map}_*(M, X)$ as the subspace consisting of the components of maps which lift to the covering. Hence, if $\mathrm{map}_*(M, X)$ is weakly contractible, so is $\mathrm{map}_*(M, \tilde{X})$.

(2) Let A be simply connected, W be local, and $\tilde{W}$ be the universal cover of W. All maps of A to W lift to $\tilde{W}$ and hence $\mathrm{map}_*(A, W) = \mathrm{map}_*(A, \tilde{W})$. Since $\tilde{W}$ is local, the proof of the lemma follows. □

Hence, for a fixed connected space M, the notions of local, local equivalence, and localization remain the same if we remain in the category of simply connected pointed spaces.

Finally, we record an important lemma due to Zabrodsky [143]. (See Miller's paper. [84])

The Zabrodsky Lemma 2.1.8. *If $p : E \to B$ is a fibre bundle with connected fibre F and the evaluation map $\mathrm{map}(F, X) \to X$ is a weak equivalence, then $p^* : \mathrm{map}(B, X) \to \mathrm{map}(E, X)$ is a weak equivalence.*

Proof: Given two towers of fibrations and a map from one to the other which is a weak homotopy equivalence at each level, the resulting map of inverse limits is a weak homotopy equivalence [17, 24, 37]. In other words, homotopy inverse limits are weakly homotopy invariant.

Hence, it follows from Zorn's lemma that there exists a maximal subcomplex $C \subseteq B$ for which the lemma is valid for the fibre bundle $p : p^{-1}(C) \to C$. If $C \neq B$, we can enlarge C by attaching a cell to get $C \cup e^n \subseteq B$. If we replace C by a thickening with the same homotopy type, we can assume that the boundary S^{n-1} of e^n embeds in C. Thus, $C \cup e^n$ is a union of C and e^n with intersection S^{n-1}. And $p^{-1}(C \cup e^n)$ is a union of $p^{-1}(C)$ and $p^{-1}(e^n) \cong e^n \times F$ with intersection $p^{-1}(S^{n-1}) \cong S^{n-1} \times F$. The adjoint equivalence $\mathrm{map}(A \times F, X) \cong \mathrm{map}(A, \mathrm{map}(F, X))$ shows that the lemma is true for the bundles over C, over e^n, and over S^{n-1}. Since the functor $\mathrm{map}(\ , X)$ converts pushout diagram of cofibrations into pullback diagrams of fibrations, it follows that the lemma is true for the fibre bundle over $C \cup e^n$ and that C is not maximal. Hence, C must be all of B. □

Remarks 2.1.9. In the proof of the Zabrodsky Lemma, we need that a map is a weak equivalence of pullbacks of fibrations if it is a weak equivalence of the total spaces and the base. Since these pullbacks are a type of homotopy inverse limit, this follows from the general form of the weak homotopy invariance of homotopy inverse limits. Alternatively, this is a consequence of the five-lemma and of the exactness of the Mayer–Vietoris homotopy sequences associated to these pullbacks.

Suppose that

$$\begin{array}{ccc} E & \xrightarrow{u} & X \\ \downarrow v & & \downarrow f \\ Y & \xrightarrow{g} & B \end{array}$$

is a homotopy pullback, that is, it is a pullback square with f and g being fibrations. The homotopy Mayer–Vietoris sequence [39] is, in dimensions $i \geq 1$, the long exact sequence

$$\cdots \to \pi_{i+1}B \xrightarrow{\partial} \pi_i E \xrightarrow{(u_*, v_*)} \pi_i X \oplus \pi_i Y \xrightarrow{f_* - g_*} \pi_i B \xrightarrow{\partial} \pi_{i-1} E \to \cdots.$$

In these dimensions, it is a long exact sequence of groups.

I am grateful to Emmanuel Dror-Farjoun for explaining the Mayer–Vietoris sequence to me in dimensions 0 and 1 and the sense in which it is exact.

In these low dimensions, the homotopy Mayer–Vietoris sequence is

$$\begin{array}{ccccc} \pi_1 X & & & & \pi_0 X \\ & \searrow f_* & & u_* \nearrow & & \searrow f_* \\ & & \pi_1 B \xrightarrow{\partial} \pi_0 E & & & \pi_0 B \\ & \nearrow g_* & & v_* \searrow & & \nearrow g_* \\ \pi_1 Y & & & & \pi_0 Y \end{array}$$

and it is exact in the following way:

(1) At $\pi_0 X$ and at $\pi_0 Y$, the exactness is: Given $\alpha \in \pi_0 X$ and $\beta \in \pi_0 Y$,

$$f_*\alpha = g_*\beta \quad \text{if and only if} \quad \exists \gamma \in \pi_0 E \quad \text{such that} \quad u_*\gamma = \alpha, v_*\gamma = \beta.$$

(2) At $\pi_0 E$, the exactness is: Given $\alpha \in \pi_0 E$, let $u_*\alpha = x$ and $v_*\alpha = y$, then

$$\exists \gamma \in \pi_1 B \quad \text{such that} \quad \partial \gamma = \alpha.$$

(Here, the basepoint of the loops in B is the common image of the basepoints x in X and y in Y.)

(3) At $\pi_1 B$, the exactness is: Given $\alpha, \beta \in \pi_1 B$,

$$\partial \alpha = \partial \beta \quad \text{if and only if} \quad \exists \gamma \in \pi_1 X, \delta \in \pi_1 Y \quad \text{such that}$$

$$\gamma * \alpha * \delta = (f_*\gamma)\alpha(g_*\delta) = \beta.$$

The following corollary of the Zabrodsky Lemma is an important source of local equivalences.

Corollary 2.1.10. *If $E \to B$ is a fibre bundle with connected fibre F, then $E \to B$ is a local equivalence with respect to $F \to *$.*

Exercises

(1) (a) Show that the loop space $\Omega(X)$ is local if X is.

   (b) Show that, if X is an H-space, then so is any localization $L_M(X)$.

(2) (a) Show that, for every pointed connected space X and $n \geq 1$, there exists a map $\iota : X \to Y$ with the property that $\iota_* : \pi_k(X) \to \pi_k(Y)$ is an isomorphism if $k \leq n-1$ and $\pi_k(Y) = 0$ if $k \geq n$.

   (b) If $M = S^n$, show that a space A is local with respect to $M \to *$ if and only if $\pi_k(A) = 0$ for all $k \geq n$.

   (c) Show that, for X and Y as in part (a), $L_M(X) = Y$.

   (d) Give an example of a fibration sequence $F \to E \to B$ such that $L_M(F) \to L_M(E) \to L_M(B)$ is not even a fibration sequence up to homotopy.

(3) Suppose $F \to E \to B$ is a fibration sequence and M is a pointed space.

   (a) Show that $\mathrm{map}(M, F) \to \mathrm{map}(M, E) \to \mathrm{map}(M, B)$ is a fibration sequence.

   (b) Show that $\mathrm{map}_*(M, F) \to \mathrm{map}_*(M, E) \to \mathrm{map}_*(M, B)$ is a fibration sequence.

   (c) Show that, if E and B are both local with respect to $M \to *$, then so is F.

   (d) Show that, if F and B are both local with respect to $M \to *$, then so is E.

   (e) Give an example to show that F and E can be local without B being local.

(4) Let $A \to B \to C$ be a cofibration sequence and X a pointed space.

   (a) Show that $\mathrm{map}(C, X) \to \mathrm{map}(B, X) \to \mathrm{map}(A, X)$ is a fibration sequence.

   (b) Show that $\mathrm{map}_*(C, X) \to \mathrm{map}_*(B, X) \to \mathrm{map}_*(A, X)$ is a fibration sequence.

(c) Suppose Y is the homotopy direct limit of a sequence X_n of spaces each of which is locally equivalent to a point with respect to $M \to *$. Show that Y is locally equivalent to a point.

(5) Suppose a space X is local with respect to $M \to *$ and with respect to $N \to *$. Show that X is local with respect to $M \times N \to *$.

(6) Let X be a space. Suppose that Γ is an ordinal and that for each ordinal $\alpha \leq \Gamma$, a space X_α is defined satisfying:

(a) $X_0 = X$

(b) $X_\alpha \subseteq X_{\alpha+1}$ is a cofibration whenever $\alpha + 1 \leq \Gamma$.

(c) $X_\beta = \bigcup_{\alpha < \beta} X_\alpha$ whenever β is a limit ordinal $\leq \Gamma$.

   (A) Show that the maps $X_\alpha \to X_\beta$ are cofibrations for all $\alpha < \beta \leq \Gamma$, that is, show that the homotopy extension property is satisfied.

   (B) If Y is a space, show that the maps of pointed mapping spaces
   $$\mathrm{map}_*(X_\beta, Y) \to \mathrm{map}_*(X_\alpha, Y)$$
   are fibrations for all $\alpha < \beta \leq \Gamma$, that is, show that the homotopy lifting property is satisfied.

   (C) If all the maps
   $$\mathrm{map}_*(X_{\alpha+1}, Y) \to \mathrm{map}_*(X_\alpha, Y)$$
   are weak equivalences, show that they are surjections and that the maps
   $$\mathrm{map}_*(X_\beta, Y) \to \mathrm{map}_*(X_\alpha, Y)$$
   are surjections for all $\alpha < \beta \leq \Gamma$.

   (D) Let $E \to B$ be a fibration and a weak equivalence and let $A \to W$ be a relative CW complex, that is, W is constructed from A by attaching cells. Given a commutative diagram of maps
   $$\begin{array}{ccc} A & \to & E \\ \downarrow & & \downarrow \\ W & \to & B \end{array}$$
   it is one of Quillen's axioms [108, 60] for a model category that there exists a map $W \to E$ which makes the diagram commute. Use this and (C) to show the following:

   If Y is a space such that the map of pointed homotopy classes $[X_{\alpha+1}, Y]_* \to [X_\alpha, Y]_*$ is a bijection whenever $\alpha + 1 \leq \Gamma$, then show that $[X_\Gamma, Y]_* \to [X, Y]_*$ is a bijection.

2.1 Dror Farjoun–Bousfield localization 45

(7) If $\mathcal{C}$ is an infinite cardinal, then $\mathcal{C} \cdot \mathcal{C} = \mathcal{C}$ [73]. Let Γ be the first ordinal with cardinal greater than $\mathcal{C}$, and let B be any set of ordinals such that the cardinality of B is less than $\mathcal{C}$ and such that all ordinals in B are less than Γ. Show that the supremum

$$\sup B = \bigcup_{\beta \in B} \beta$$

is less than Γ.

(8) Use problem 7 to remove the hypotheses that M be a countable CW complex in the proof of the existence of localization 2.1.6.

(9) Check the exactness in the Mayer–Vietoris sequence 2.1.9 at $\pi_0 X$ and at $\pi_0 Y$.

(10) Define the function $\partial : \Omega B \to E$ as follows: Let γ be a loop in B and write $\gamma = \gamma_1 * \gamma_2^{-1}$ by cutting the loop in the middle. Lift γ_1 to a path $\tilde{\gamma}_1$ in Y which starts at the basepoint and ends at $y \in Y$. Similarly, lift γ_2 to a path $\tilde{\gamma}_2$ in X which starts at the basepoint and ends at $x \in X$. Set $\partial \gamma = (x, y) \in E$ and check the exactness in the Mayer–Vietoris sequence 2.1.9 at $\pi_0 E$.

(11) Check the exactness in the Mayer–Vietoris sequence 2.1.9 at $\pi_1 B$.

(12) Given a pullback diagram as in 2.1.9, write

$$\begin{array}{ccccc} F & \xrightarrow{r} & E & \xrightarrow{v} & Y \\ \downarrow = & & \downarrow u & & \downarrow g \\ F & \xrightarrow{s} & X & \xrightarrow{f} & B \end{array}$$

with F the fibre of both f and v. Consider the two long exact homotopy sequences of these fibration sequences and define the Mayer–Vietoris connecting homomorphism $\partial : \pi_{i+1} B \to \pi_i E$ as the composition

$$\partial = r \cdot \partial : \pi_{i+1} B \xrightarrow{\partial} \pi_i F \xrightarrow{r} \pi_i E.$$

Use the two long exact homotopy sequences to show that the Mayer–Vietoris sequence

$$\cdots \to \pi_{i+1} B \xrightarrow{\partial} \pi_i E \xrightarrow{(u_*, v_*)} \pi_i X \oplus \pi_i Y \xrightarrow{f_* - g_*} \pi_i B$$

is exact for $i \geq 1$. (The homology version of this derivation of the Mayer–Vietoris sequence from long exact sequences is due to Michael Barratt.)

(13) Fill in the details of the proof of the Zabrodsky Lemma 2.1.8.

2.2 Localization of abelian groups

We describe the classical localization of abelian groups wherein a subset of the set of primes is inverted. Three special cases are particularly important, rationalization $A_{(0)}$ where all primes are inverted, localization $A_{(p)}$ at a prime p where all primes except p are inverted, and localization $A[\frac{1}{p}]$ away from p where p is the only prime which is inverted.

Let $\wp$ be the set of positive primes in the integers $\mathbb{Z}$ and let

$$\wp = S \bigcup T$$

be a decomposition into disjoint subsets.

Definition 2.2.1. An abelian group A is called S-local if every element of A is uniquely divisible by all elements of T, in other words, multiplication by q,

$$q : A \to A$$

is an isomorphism for all $q \varepsilon T$.

Let $\overline{T}$ be the multiplicative monoid generated by T. The fundamental example of an S-local abelian group is the subring of the rationals

$$\mathbb{Z}[T^{-1}] = \mathbb{Z}_{(S)} = \left\{ \frac{a}{q} \mid a \varepsilon \mathbb{Z}, \quad q \varepsilon \overline{T} \right\}.$$

An abelian group is S-local if and only if it is a $\mathbb{Z}_{(S)}$ module.

Definition 2.2.2. A map of abelian groups $f : A \to B$ is an S-local equivalence if $f^* : \hom(B, C) \to \hom(A, C)$ is a bijection for all S-local abelian groups C.

Definition 2.2.3. A map of abelian groups $\iota : A \to \overline{A}$ is called an S-localization of A if:

(1) $\overline{A}$ is S-local and

(2) $f : A \to \overline{A}$ is an S-local equivalence.

For any abelian group A, an S-localization exists and is given by

$$A \to \overline{A} = A[T^{-1}] = A_{(S)} = \left\{ \frac{a}{q} \mid a \varepsilon \mathbb{Z}, \quad q \varepsilon \overline{T} \right\}.$$

In this definition, $\frac{a}{q} = \frac{a_1}{q_1}$ if and only if there is an element $q_2 \varepsilon \overline{T}$ such that $q_2(aq_1 - a_1 q) = 0$.

Exercises

(1) Show that the definition of localization by a universal mapping property characterizes it uniquely up to isomorphism.

(2) Show that $A \to B$ is an S-local equivalence if and only if $A_{(S)} \to B_{(S)}$ is an isomorphism.

(3) For all abelian groups A, there is an isomorphism
$$A \otimes \mathbb{Z}_{(S)} \to A_{(S)}.$$

(4) $A_{(S)} = 0$ if and only if, for all elements $a \varepsilon A$, there is a $q \varepsilon \overline{T}$ such that $qa = 0$.

(5) (a) $\mathbb{Z}_{(S)}/q\mathbb{Z}_{(S)} \cong \mathbb{Z}/q\mathbb{Z}$ if $q \varepsilon \overline{S}$ = the multiplicative monoid generated by S.

(b) $\mathbb{Z}_{(S)}/q\mathbb{Z}_{(S)} = 0$ if $q \varepsilon \overline{T}$.

(6) If $0 \to A \to B \to C \to 0$ is a short exact sequence, then the sequence of localizations $0 \to A_{(S)} \to B_{(S)} \to C_{(S)} \to 0$ is also exact.

(7) Show that $H_*(X)_{(S)} \cong H_*(X; \mathbb{Z}_{(S)})$.

(8) (a) Show that $\mathbb{Z}_{(S)} \otimes_\mathbb{Z} \mathbb{Z}_{(S)} \cong \mathbb{Z}_{(S)}$.

(b) For all abelian groups M and N, show that $\operatorname{Tor}_i^\mathbb{Z}(M, N)_{(S)} \cong \operatorname{Tor}_i^{\mathbb{Z}_{(S)}}(M_{(S)}, N_{(S)})$ for $i = 0$ and $i = 1$.

(c) If $F \to E \to B$ is an orientable fibration sequence and $E_{p,q}^r$ is the Serre spectral sequence for integral homology, show that $(E_{p,q}^r)_{(S)}$ is the Serre spectral sequence for $\mathbb{Z}_{(S)}$ homology.

(9) Show that S-localization commutes with direct limits, that is, $(\lim_\to H_\alpha)_{(S)} \cong \lim_\to (H_{\alpha(S)})$.

(10) (a) Show that $A \to A_{(S)}$ is surjective if A is a torsion abelian group.

(b) Show that the kernel and cokernel of $A \to A_{(S)}$ are T-primary torsion groups where T is the set of inverted primes.

2.3 Classical localization of spaces: inverting primes

We develop the classical theory of localization for simply connected spaces. This includes the special cases of rationalization $X \to X_{(0)} = X \otimes Q$, localization at a prime $X \to X_{(p)} = X \otimes \mathbb{Z}_{(p)}$ and localization away from a prime $X \to X[1/p] = X \otimes \mathbb{Z}[1/p]$.

Let $\wp = S \bigcup T$ be a decomposition of the set of positive integral primes into disjoint subsets.

Definition 2.3.1. A simply connected pointed space X is S-local if the homotopy groups $\pi_k(X)$ are S-local for all $k \geq 1$.

For simply connected spaces X the universal coefficient exact sequence
$$0 \to \pi_*(X) \otimes \mathbb{Z}/q\mathbb{Z} \to \pi_*(X; \mathbb{Z}/q\mathbb{Z}) \to \operatorname{Tor}(\pi_{*-1}(X), \mathbb{Z}/q\mathbb{Z}) \to 0$$

48 A general theory of localization

shows that X is S-local if and only if the homotopy groups $\pi_k(X; \mathbb{Z}/q\mathbb{Z}) = 0$ for all $k \geq 2$ and for all $q \varepsilon T$. Thus

Proposition 2.3.2. *A simply connected pointed space X is S-local if and only if X is local with respect to $M \to *$ with*

$$M = \bigvee_{q \varepsilon T} P^2(\mathbb{Z}/q\mathbb{Z}),$$

$$P^2(\mathbb{Z}/q\mathbb{Z}) = S^1 \cup_q e^2.$$

Definition 2.3.3. A map $f : A \to B$ of simply connected spaces is an S-equivalence if the map of S-local homology $f_* : H_*(A; \mathbb{Z}_{(S)}) \to H_*(B; \mathbb{Z}_{(S)})$ is an isomorphism.

Proposition 2.3.4. *A pointed map $f : A \to B$ of simply connected spaces is an S-equivalence if and only if f is a local equivalence with respect to $M \to *$, that is, if and only if $f^* : \mathrm{map}_*(B, X) \to \mathrm{map}_*(A, X)$ is a weak equivalence for all S-local X. (Note: We know that it is sufficient to check it only for simply connected S-local X.)*

Proof: By using the mapping cylinder, we may suppose that $f : A \to B$ is an inclusion.

Suppose $f_* : H_*(A; \mathbb{Z}_{(S)}) \to H_*(B; \mathbb{Z}_{(S)})$ is an isomorphism. Thus, $H_*(B, A; \mathbb{Z}_{(S)}) = 0$. The universal coefficient exact sequence for cohomology

$$0 \to \mathrm{Ext}(H_{*-1}(B, A; \mathbb{Z}_{(S)}), D) \to H^*(B, A; D)$$
$$\to \mathrm{Hom}(H_*(B, A; \mathbb{Z}_{(S)}), D) \to 0$$

shows that $H^*(B, A; D) = 0$ for all $\mathbb{Z}_{(S)}$ modules D. Hence, if X is simply connected S-local, this gives the vanishing of the obstruction groups $H^{*+1}(\Sigma^n B, \Sigma^n A; \pi_*(X))$ and $H^*(\Sigma^n B, \Sigma^n A; \pi_*(X))$ to the existence and homotopy uniqueness of extending a map $\Sigma^n A \to X$ to a map $\Sigma^n B \to X$. Thus, $f : A \to B$ is a local equivalence with respect to $M \to *$.

Now suppose that $f : A \to B$ is a local equivalence with respect to $M \to *$. We note that any $\mathbb{Z}_{(S)}$ module D may be realized as the nonzero homotopy group of $X = K(D, n)$ and that this X is S-local. Since any nonzero cohomology class can be realized as an obstruction, the cohomology groups $H^*(B, A; D)$ must vanish for all such D. Hence, the homology groups $H_*(B, A; \mathbb{Z}_{(S)}) = 0$ and f is an S-equivalence. This completes the proof of the proposition. □

The general theory of Dror Farjoun presented in this chapter shows that S-localization exists and is unique up to homotopy equivalence, that is, for all simply connected X, there is a map $\iota : X \to X_{(S)}$ such that:

(1) $X_{(S)}$ is S-local.

(2) $\iota : X \to X_{(S)}$ is an S-equivalence.

2.3 Classical localization of spaces: inverting primes 49

(3) for all maps $f : X \to Y$ with Y an S-local space, there is up to homotopy a unique extension of f to a map $\bar{f} : X_{(S)} \to Y$.

Alternate notations for S-localization are:

$$X_{(S)} = X \otimes \mathbb{Z}_{(S)} = X \otimes \mathbb{Z}[T^{-1}] = L_M(X)$$

with T a complementary set of primes to S and $M = \bigvee_{q \varepsilon T} P^2(\mathbb{Z}/q\mathbb{Z})$.

It is useful to know that S-local spaces may also be defined in terms of homology groups.

Lemma 2.3.5. *A simply connected space X is S-local if and only if the reduced homology groups $\overline{H}_*(X)$ are S-local.*

Proof: This follows from the universal coefficient theorems for homotopy and homology and the mod q Hurewicz theorems: By definition, X is S-local if and only if all $\pi_*(X)$ are S-local. This is equivalent to the vanishing $\pi_*(X; \mathbb{Z}/q\mathbb{Z}) = 0$ for all $q \varepsilon T$ which in turn is equivalent to the vanishing $\overline{H}_*(X; \mathbb{Z}/q\mathbb{Z}) = 0$ for all $q \varepsilon T$. This is equivalent to $\overline{H}_*(X)$ being S-local. □

Since S-localization of abelian groups is characterized by being a local equivalence with a local target, we have:

Corollary 2.3.6. *A map of simply connected spaces $X \to Y$ is S-localization if and only if the map of reduced homology $\overline{H}_*(X) \to \overline{H}_*(Y)$ is S-localization.*

We shall prove a homotopy version.

Proposition 2.3.7. *A map of simply connected spaces $X \to Y$ is S-localization if and only if the map of homotopy groups $\pi_*(X) \to \pi_*(Y)$ is S-localization.*

We start with the statement of the following $K(G, 1)$ localization lemma.

Lemma 2.3.8. *$G \to H$ is an S-localization of abelian groups if and only if $K(G, 1) \to K(H, 1)$ is an S-localization of homology.*

Proof: First of all, S-localization of homology implies S-localization of fundamental groups, hence, that $G \to H$ is an S-localization.

Now let $H = G_{(S)}$ be S-localization. We consider the case of cyclic groups.

(1) $G = \mathbb{Z}$ and $H = \mathbb{Z}_{(S)}$: Write $H = \lim_\to H_\alpha$ as a direct limit of finitely generated torsion free subgroups. Then

$$\overline{H}_*(K(H, 1)) = \lim_\to \overline{H}_*(K(H_\alpha, 1)) = \lim_\to \overline{E}[H_\alpha]$$
$$= \overline{E}[\lim_\to H_\alpha] = \overline{E}[H] = \overline{E}[G_{(S)}] = \overline{E}[G]_{(S)} = \overline{H}_*(K(G, 1))_{(S)}$$

where $\overline{E}$ denotes the exterior algebra without unit.

(2) $G = \mathbb{Z}/q\mathbb{Z}$, $q\varepsilon \overline{T}$, $H = 0$: Since the integral homology of $K(\mathbb{Z}/q\mathbb{Z}, 1)$ is $\mathbb{Z}/q\mathbb{Z}$ in odd dimensions, $\overline{H}_*(K(G,1))_{(S)} = 0 = \overline{H}_*(K(H,1))$.

(3) $G = \mathbb{Z}/q\mathbb{Z}$, $q\varepsilon \overline{S}$, $H = \mathbb{Z}/q\mathbb{Z}$: For the same reason, $\overline{H}_*(K(G,1))_{(S)} = \overline{H}_*(K(H,1))$.

We recall the following theorem which we do not prove [144].

Zeeman comparison theorem 2.3.9. *Suppose we have a map of orientable fibration sequences*

$$\begin{array}{ccc} F & \xrightarrow{f} & F_1 \\ \downarrow & & \downarrow \\ E & \xrightarrow{g} & E_1 \\ \downarrow & & \downarrow \\ B & \xrightarrow{h} & B_1. \end{array}$$

If any two of f, g, h are homology isomorphisms, then so is the third.

We also note that the characterization of S-local by vanishing mod q homotopy groups gives:

Lemma 2.3.10. *Let $F \to E \to B$ be a fibration sequence of simply connected spaces. If two of F, E, B are S-local, then so is the third.*

Now it follows that the $K(G, 1)$ localization lemma is true for all finitely generated abelian groups. Just use the fact that any finitely generated abelian group is a direct sum of cyclic groups and the fact that we have a fibration sequence $K(G, 1) \to K(G \oplus H, 1) \to K(H, 1)$.

Finally, expressing G as a direct limit of finitely generated groups and using the fact that direct limits commute with localization shows that the $K(G, 1)$ localization lemma is true for all abelian groups. □

We continue with the statement of the $K(G, n)$ localization lemma.

Lemma 2.3.11. *For all $n \geq 1$, $G \to H$ is an S-localization of abelian groups if and only if $K(G, n) \to K(H, n)$ is an S-localization of homology.*

Proof: As before, the Hurewicz theorem implies that S-localization of homology implies that $G \to H$ is S-localization.

The other implication follows by applying the Zeeman comparison theorem and induction to the map of pathspace fibrations

$$\begin{array}{ccc} K(G,n) & \to & K(H,n) \\ \downarrow & & \downarrow \\ PK(G,n+1) & \to & PK(H,n+1). \\ \downarrow & & \downarrow \\ K(G,n+1) & \to & K(H,n+1) \end{array}$$

Since all the $K(H, n+1)$ are S-local, we need only show that $K(G, n+1) \to K(H, n+1)$ is an S-equivalence of homology. □

In fact, the same argument shows that:

Local comparison lemma 2.3.12. *Suppose we have a map of orientable fibration sequences*

$$\begin{array}{ccc} F & \xrightarrow{f} & F_1 \\ \downarrow & & \downarrow \\ E & \xrightarrow{g} & E_1 \\ \downarrow & & \downarrow \\ B & \xrightarrow{h} & B_1. \end{array}$$

If any two of f, g, h are S-localizations, then so is the third.

Let $X \to Y$ be a map of simply connected spaces which is S-localization. Since this induces localization of homology, it follows that it induces localization on the bottom nonvanishing homotopy groups of $\pi_n(X) \to \pi_n(Y)$. The local comparison lemma shows that the map $X\langle n \rangle \to Y\langle n \rangle$ of n-connected covers is S-localization. This provides the inductive step to show that $\pi_k(X) \to \pi_k(Y)$ is S-localization for all $k \geq 2$.

Now suppose that $\pi_k(X) \to \pi_k(Y)$ is S-localization for all $k \geq 2$. We need to show that $X \to Y$ is S-localization. But since Y is S-local, all we need to show is that $H_*(X) \to H_*(Y)$ is an S-local equivalence.

We have already done this if X has only one nonvanishing homotopy group. Induction on Postnikov systems and the local comparison lemma proves it when X has only finitely many homotopy groups.

In general, $X \to Y$ is the inverse limit of the maps of the stages of the Postnikov system $X_n \to Y_n$. In each fixed degree, the map of homology $H_*(X) \to H_*(Y)$ is isomorphic to $H_*(X_n) \to H_*(Y_n)$ for n large enough. Therefore, $H_*(X) \to H_*(Y)$ is an S-local equivalence.

This completes the proof that S-localization of simply connected spaces is equivalent to S-localization of homotopy groups and also to S-localization of reduced homology groups. □

The following is an immediate corollary.

Corollary 2.3.13. *If $X \to Y$ is a map of simply connected spaces, the following are equivalent:*

(1) $H_*(X) \to H_*(Y)$ *is an S-local equivalence.*

(2) $\pi_*(X) \to \pi_*(Y)$ *is an S-local equivalence.*

(3) $X_{(S)} \to Y_{(S)}$ *is a homotopy equivalence.*

The classical Hurewicz theorem applied to $X_{(S)}$ gives:

Local Hurewicz theorem 2.3.14. *If X is simply connected and $\pi_k(X)_{(S)} = 0$ for all $1 \leq k \leq n-1$, then $H_k(X)_{(S)} = 0$ for all $k \leq n-1$ and the localized Hurewicz map $\pi_n(X)_{(S)} \to H_n(X)_{(S)}$ is an isomorphism.*

Remarks. Although we have developed an S-localization theory restricted to simply connected spaces, it is possible to extend this theory to connected loop spaces with no effort. Let $X = \Omega Y$ where Y is simply connected. Then $Y_{(S)}$ is defined and we set $X_{(S)} = \Omega(Y_{(S)})$. Then $X \to X_{(S)}$ induces localization of homotopy groups and, by 2.3.12, localization of homology groups. In other words, $X \to X_{(S)}$ possesses the characterizing properties of localization, that is, it is a local homology equivalence and the homotopy groups of the range are local. In particular, $X_{(S)}$ is local in the sense that: Any map $A \to B$ which is a local homology equivalence between possibly nonsimply connected spaces induces a weak equivalence $\mathrm{map}_*(A, X_{(S)}) \leftarrow \mathrm{map}_*(B, X_{(S)})$.

An additional property is valid for this extension of localization. Consider the standard fibration sequence $Y\langle 2\rangle \to Y \to K(\pi, 2)$ which defines the two-connected cover $Y\langle 2\rangle$. The loops on it is the orientable fibration sequence $X\langle 1\rangle \to X \to K(\pi, 1)$ which defines the universal cover $X\langle 1\rangle$. Thus, if $f : X \to X'$ is a map of connected loop spaces which is possibly not a loop map, it is still the case that the following statements are equivalent:

(a) f induces an isomorphism of localized homology groups.

(b) f induces an isomorphism of localized fundamental groups and on universal covers an isomorphism of localized homology groups.

(c) f induces an isomorphism of localized fundamental groups and on universal covers an isomorphism of localized homotopy groups.

(d) f induces an isomorphism of localized homotopy groups.

In other words, for maps between connected loop spaces, a local homology isomorphism is equivalent to a local homotopy isomorphism.

Exercises

(1) If $F \to E \to B$ is a fibration sequence of simply connected spaces, then $F_{(S)} \to E_{(S)} \to B_{(S)}$ is a fibration sequence up to homotopy, in other words, the homotopy theoretic fibre of $E_{(S)} \to B_{(S)}$ is $F_{(S)}$.

(2) If $X \to Y \to Z$ is a cofibration sequence of simply connected spaces, then $X_{(S)} \to Y_{(S)} \to Z_{(S)}$ is a cofibration sequence up to homotopy.

(3) (a) Show that S-localization preserves homotopy pullback diagrams of simply connected spaces.

(b) Show that S-localization preserves homotopy pushout diagrams of simply connected spaces.

2.4 Limits and derived functors

In order to discuss the concept of completion we need to recall some facts concerning inverse limits. Let R be a fixed commutative ring.

A sequential inverse system of R modules $\{A_n, p_n\}$ is a collection of R modules A_n and homomorphisms $p_n : A_{n+1} \to A_n$ for $n \geq 1$. Morphisms of inverse systems $A_n \to B_n$ are defined in the obvious way by a commutative diagram of homomorphisms

$$\begin{array}{ccccccccc} A_1 & \leftarrow & A_2 & \leftarrow & A_3 & \leftarrow & A_4 & \leftarrow & \cdots \\ \downarrow & & \downarrow & & \downarrow & & \downarrow & & \\ B_1 & \leftarrow & B_2 & \leftarrow & B_3 & \leftarrow & B_4 & \leftarrow & \cdots \end{array}.$$

Consider the cochain complex

$$0 \to \prod_n A_n \xrightarrow{\Phi} \prod_n A_n \to 0$$

where $\Phi(a_n) = (a_n - p_n(a_{n+1}))$. The cohomology of this complex defines the inverse limit functor and its derived functor.

Definition 2.4.1. The inverse limit functor is

$$\varprojlim A_n = \ker(\Phi)$$

and the first derived functor of inverse limit is

$$\varprojlim{}^1 A_n = \operatorname{coker}(\Phi).$$

Thus $(a_n) \varepsilon \varprojlim A_n$ if and only if $a_n = p_n(a_{n+1})$ for all $n \geq 1$.

Given a short exact sequence of inverse systems

$$0 \to A_n \to B_n \to C_n \to 0$$

we get a short exact sequence of cochain complexes and hence a long exact sequence of cohomology groups

$$0 \to \varprojlim A_n \to \varprojlim B_n \to \varprojlim C_n \xrightarrow{\delta} \varprojlim{}^1 A_n \to \varprojlim{}^1 B_n \to \varprojlim{}^1 C_n \to 0.$$

We shall say that an inverse system A_n is eventually zero if for all n there is some k so that the composition $p_n \circ p_{n+1} \circ \cdots \circ p_{n+k-1} \circ p_{n+k} = p^k$:

$$A_n \leftarrow A_{n+1} \leftarrow A_{n+2} \leftarrow \cdots \leftarrow A_{n+k+1}$$

is zero.

Lemma 2.4.2. *If an inverse system of R modules A_n is eventually zero then*
$$\varprojlim A_n = \varprojlim{}^1 A_n = 0.$$

Proof: The inverse limit vanishes since $(a_n) \varepsilon \varprojlim A_n$ implies that
$$a_n = p_n(a_{n+1}) = p_n \circ p_{n+1}(a_{n+2})$$
$$= \cdots = p_n \circ p_{n+1} \circ \cdots \circ p_{n+k-1} \circ p_{n+k}(a_{n+k+1}) = 0$$
for all n.

To show that the derived functor is zero we need to show that the map $\Phi : \prod_n A_n \to \prod_n A_n$ is surjective. Let $(b_n) \varepsilon \prod_n A_n$ be any element. Since the system is eventually zero, the following infinite sums terminate and make sense:
$$a_1 = b_1 + pb_2 + p^2 b_3 + p^3 b_4 + \cdots$$
$$a_2 = b_2 + pb_3 + p^2 b_4 + p^3 b_5 + \cdots$$
$$a_3 = b_3 + pb_4 + p^2 b_5 + p^3 b_6 + \cdots$$
$$\cdots$$

Then $\Phi(a_n) = (b_n)$ and Φ is surjective. □

For an inverse system A_n and $k \geq 0$, let $A_{n,k} = \mathrm{im}(A_{n+k} \to \cdots \to A_n)$ be the k-th image inverse system and set
$$A_{n,\infty} = \bigcap_{k \geq 0} A_{n,k}$$
= the infinite image inverse system.

Definition 2.4.3. *The inverse system A_n satisfies the Mittag–Leffler condition if the image inverse system converges in the sense that, for each n, there is a k such that $A_{n,k} = A_{n,\infty}$.*

Since finite groups satisfy the descending chain condition, every inverse system of finite abelian groups satisfies the Mittag–Leffler condition.

An inverse system A_n is called epimorphic if every map $A_{n+1} \to A_n$ is an epimorphism. Clearly, an epimorphic inverse system satisfies the Mittag–Leffler condition. A simple exercise shows that to be true.

Lemma 2.4.4. *If A_n is an epimorphic inverse system, then*
$$\varprojlim{}^1 A_n = 0.$$

Proposition 2.4.5. *If A_n is an inverse system which satisfies the Mittag–Leffler condition then*
$$\varprojlim{}^1 A_n = 0.$$

Proof: Consider the short exact sequence of inverse systems
$$0 \to A_{n,\infty} \to A_n \to \frac{A_n}{A_{n,\infty}} \to 0.$$
The left-hand system is epimorphic and the Mittag–Leffler condition implies that the right-hand system is eventually zero. Since the derived functor vanishes for the systems on the ends, it vanishes for the middle system. $\square$

Exercise

(1) Define a sequential direct system of R modules to be a collection of R modules A_n and homomorphisms $\iota_n : A_n \to A_{n+1}$ for all $n \geq 1$. Consider the complex
$$0 \to \bigoplus_n A_n \xrightarrow{\Psi} \bigoplus_n A_n \to 0$$
with $\Psi a_n = a_n - \iota_n(a_n)$. Define the direct limit to be
$$\varinjlim A_n = \mathrm{coker}(\Psi).$$

(a) Show that Ψ is a monomorphism.

(b) Show that if $0 \to A_n \to B_n \to C_n \to 0$ is a short exact sequence of direct systems, then
$$0 \to \varinjlim A_n \to \varinjlim B_n \to \varinjlim C_n \to 0$$
is a short exact sequence of R modules. In other words, the direct limit functor is exact.

2.5 Hom and Ext

Let A and B be R modules and consider the functor of two variables $\hom(A, B)$. First we recall that an R module P is projective if and only if the covariant functor $\hom(P, \ )$ is exact. An R module Q is injective if and only if the contravariant functor $\hom(\ , Q)$ is exact. This leads to the fact that the derived functors of $\hom(\ , \ )$ may be defined in two different ways:

(1) If $P_* \to A \to 0$ is a projective resolution of A, then $\mathrm{Ext}^q(A, B)$ is the q–the cohomology group of the cochain complex $\hom(P_*, B)$. If $A_1 \to A_2 \to A_3 \to 0$ is exact, then
$$\hom(A_1, B) \leftarrow \hom(A_2, B) \leftarrow \hom(A_3, B) \leftarrow 0$$
is exact. This shows that $\mathrm{Ext}^0(A, B) = \hom(A, B)$.

Of course, if the ground ring R is any principal ideal domain, for example $R = \mathbb{Z}$, then projective resolutions can be chosen to have length ≤ 1 and thus $\mathrm{Ext}^q(A, B) = 0$ if $q \geq 2$. In this case, we write $\mathrm{Ext}^1(A, B) = \mathrm{Ext}(A, B)$.

(2) Alternatively, if $0 \to B \to Q_*$ is an injective resolution of B, then $\operatorname{Ext}^q(A,B)$ is the q–the cohomology group of the cochain complex $\hom(A, Q_*)$. The equivalence of this definition with the preceding definition is shown in exercise 1 below. If $0 \to B_1 \to B_2 \to B_3$ is exact, then

$$0 \to \hom(A, B_1) \to \hom(A, B_2) \to \hom(A, B_3)$$

is exact. This also shows that $\operatorname{Ext}^0(A,B) = \hom(A,B)$.

Proposition 2.5.1. *If A_n is a direct system of R modules, then*

$$\hom(\varinjlim A_n, B) = \varprojlim \hom(A_n, B)$$

and for all $q \geq 1$ there are short exact sequences

$$0 \to \varprojlim{}^1 \operatorname{Ext}^{q-1}(A_n, B) \to \operatorname{Ext}^q(\varinjlim A_n, B) \to \varprojlim \operatorname{Ext}^q(A_n, B) \to 0$$

Proof: Apply the long exact exact sequence in Exercise 3 below to the short exact sequence

$$0 \to \oplus A_n \xrightarrow{\Psi} \oplus A_n \to \varinjlim A_n \to 0$$

and use Exercise 1. $\square$

We conclude this section with a result of Cartan–Eilenberg. Suppose A, B, C are R modules, $P_* \to A \to 0$ is a projective resolution, and $0 \to C \to Q_*$ is an injective resolution. Consider the double complex

$$\hom(P_* \otimes B, Q_*) = \hom(P_*, \hom(B, Q_*)).$$

There are two spectral sequences converging to the cohomology of the associated total complex.

(1) If we filter the associated total complex by the injective resolution degree, we get a first quadrant spectral sequence with

$$E_1^{p,q} = \hom(\operatorname{Tor}_p(A,B), Q_q)$$
$$E_2^{p,q} = \operatorname{Ext}^q(\operatorname{Tor}_p(A,B), C)$$

and differentials $d_r : E_r^{p,q} \to E_r^{p+1-r, q+r}$.

(2) If we filter the associated total complex by the projective resolution degree, we get a first quadrant spectral sequence with

$$\overline{E}_1^{p,q} = \hom(P_p, \operatorname{Ext}^q(B,C))$$
$$\overline{E}_2^{p,q} = \operatorname{Ext}^p(A, \operatorname{Ext}^q(B,C))$$

and differentials $\overline{d}_r : \overline{E}_r^{p,q} \to \overline{E}_r^{p+r, q+1-r}$.

If the ground ring R is a principal ideal domain, then derived functors vanish beyond degree 1 and thus $E_2^{p,q} = E_\infty^{p,q}$, $\overline{E}_2^{p,q} = \overline{E}_\infty^{p,q}$. Hence,

Corollary 2.5.2. *For modules over a principal ideal domain R,*

$$\text{Ext}^q(\text{Tor}_p(A,B),C) = 0 \text{ for all } p,q \geq 0$$

if and only if

$$\text{Ext}^p(A, \text{Ext}^q(B,C)) = 0 \text{ for all } p,q \geq 0.$$

Exercises

(1) Show that

$$\text{Ext}^q(\oplus_\alpha A_\alpha, B) = \prod_\alpha \text{Ext}^q(A_\alpha, B)$$

and

$$\text{Ext}^q(A, \prod_\alpha A_\alpha) = \prod_\alpha \text{Ext}^q(A, B_\alpha).$$

(2) Let $P_* \to A \to 0$ be a projective resolution and let $0 \to B \to Q_*$ be an injective resolution. Consider the double complex $\hom(P_*, Q_*)$.

(a) Show that there is a spectral sequence with

$$E_1^{p,q} = \begin{cases} \hom(A, Q_q) & \text{if } p = 0, \\ 0 & \text{if } p \neq 0. \end{cases}$$

(b) Show that there is a spectral sequence with

$$\overline{E}_1^{p,q} = \begin{cases} \hom(P_p, B) & \text{if } q = 0, \\ 0 & \text{if } q \neq 0. \end{cases}$$

(c) Show that $\text{Ext}^q(A, B)$ can be defined by using either projective resolutions of A or by using injective resolutions of B.

(3) (a) If $0 \to A_1 \to A_2 \to A_3 \to 0$ is a short exact sequence, show that there is a short exact sequence of projective resolutions

$$\begin{array}{ccccccccc}
0 & \to & P_* & \to & P_* \oplus Q_* & \to & Q_* & \to & 0 \\
& & \downarrow & & \downarrow & & \downarrow & & \\
0 & \to & A_1 & \to & A_2 & \to & A_3 & \to & 0 \\
& & \downarrow & & \downarrow & & \downarrow & & \\
& & 0 & & 0 & & 0 & &
\end{array}$$

(Hint: The easiest way to do this is to construct $P_*, P_* \oplus Q_*, Q_*$ in that order.)

(b) Show that, if $0 \to A_1 \to A_2 \to A_3 \to 0$ is a short exact sequence, then there is a long exact sequence

$$0 \to \hom(A_3, B) \to \hom(A_2, B) \to \hom(A_1, B) \to$$
$$\operatorname{Ext}^1(A_3, B) \to \operatorname{Ext}^1(A_2, B) \to \operatorname{Ext}^1(A_1, B) \to$$
$$\operatorname{Ext}^2(A_3, B) \to \operatorname{Ext}^2(A_2, B) \to \operatorname{Ext}^2(A_1, B) \to \cdots.$$

(4) (a) If $0 \to B_1 \to B_2 \to B_3 \to 0$ is a short exact sequence, show that there is a short exact sequence of injective resolutions

$$\begin{array}{ccccccccc}
 & & 0 & & 0 & & 0 & & \\
 & & \downarrow & & \downarrow & & \downarrow & & \\
0 & \to & B_1 & \to & B_2 & \to & B_3 & \to & 0 \\
 & & \downarrow & & \downarrow & & \downarrow & & \\
0 & \to & P_* & \to & P_* \oplus Q_* & \to & Q_* & \to & 0.
\end{array}$$

(b) Show that, if $0 \to B_1 \to B_2 \to B_3 \to 0$ is a short exact sequence, then there is a long exact sequence

$$0 \to \hom(A, B_1) \to \hom(A, B_2) \to \hom(A, B_3) \to$$
$$\operatorname{Ext}^1(A, B_1) \to \operatorname{Ext}^1(A, B_2) \operatorname{Ext}^1(A, B_3) \to$$
$$\operatorname{Ext}^2(A, B_1) \to \operatorname{Ext}^2(A, B_2) \to \operatorname{Ext}^2(A, \to \cdots.$$

2.6 p-completion of abelian groups

Dennis Sullivan introduced completions into homotopy theory. The notion of p-completion occurs in the seminal work of Bousfield and Kan. The exposition here is influenced by the thesis of Stephen Shiffman.

Let p be a prime. Recall that

$$\mathbb{Z}\left[\frac{1}{p}\right] = \varinjlim \mathbb{Z}$$

with respect to the direct system

$$\mathbb{Z} \xrightarrow{p} \mathbb{Z} \xrightarrow{p} \mathbb{Z} \xrightarrow{p} \mathbb{Z} \xrightarrow{p} \cdots$$

and

$$\mathbb{Z}(p^\infty) = \mathbb{Z}\left[\frac{1}{p}\right]/\mathbb{Z} = \varinjlim \mathbb{Z}/p^r\mathbb{Z}.$$

2.6 p-completion of abelian groups

Definition 2.6.1. An abelian group A is p-complete if
$$\hom\left(\mathbb{Z}\left[\frac{1}{p}\right], A\right) = \operatorname{Ext}\left(\mathbb{Z}\left[\frac{1}{p}\right], A\right) = 0.$$

Since $\mathbb{Z}\left[\frac{1}{p}\right] = \varinjlim \mathbb{Z}$, it follows that
$$\hom\left(\mathbb{Z}\left[\frac{1}{p}\right], A\right) \cong \varprojlim A$$
$$\operatorname{Ext}\left(\mathbb{Z}\left[\frac{1}{p}\right], A\right) \cong \varprojlim{}^1 A$$

with respect to the inverse system
$$A \xleftarrow{p} A \xleftarrow{p} A \xleftarrow{p} A \xleftarrow{p} \cdots.$$

Thus, A is p-complete if and only if this limit and this derived functor vanish.

If B is any $\mathbb{Z}[\frac{1}{p}]$ module, then the long exact sequence associated to a free $\mathbb{Z}[\frac{1}{p}]$ resolution
$$0 \to F_1 \to F_0 \to B \to 0$$
gives

Lemma 2.6.2. *An abelian group A is p-complete if and only if*
$$\hom(B, A) = \operatorname{Ext}(B, A) = 0$$
for all $\mathbb{Z}[\frac{1}{p}]$ modules B.

If we recall that $\mathbb{Z}/p\mathbb{Z}$ is a p-complete module we see that the following result is related.

Lemma 2.6.3. *An abelian group A is a $\mathbb{Z}[\frac{1}{p}]$ module if* $\hom(A, \mathbb{Z}/p\mathbb{Z}) = \operatorname{Ext}(A, \mathbb{Z}/p\mathbb{Z}) = 0$.

Proof: If $\hom(A, \mathbb{Z}/p\mathbb{Z}) = 0$, then $A \to A/pA$ must be zero, hence $A/pA = 0$ and A is p-divisible. Thus, there is a short exact sequence $0 \to {}_pA \to A \xrightarrow{p} A \to 0$ and a long exact sequence
$$0 \to \hom(A, \mathbb{Z}/p\mathbb{Z}) \to \hom(A, \mathbb{Z}/p\mathbb{Z}) \to \hom({}_pA, \mathbb{Z}/p\mathbb{Z}) \to$$
$$\operatorname{Ext}(A, \mathbb{Z}/p\mathbb{Z}) \to \operatorname{Ext}(A, \mathbb{Z}/p\mathbb{Z}) \to \operatorname{Ext}({}_pA, \mathbb{Z}/p\mathbb{Z}) \to 0.$$

Thus $\hom({}_pA, \mathbb{Z}/p\mathbb{Z}) = 0$ and this implies ${}_pA = 0$, in other words, A is uniquely p-divisible, a $\mathbb{Z}[\frac{1}{p}]$ module. $\square$

If
$$\mathbb{Z}(p^\infty) = \mathbb{Z}\left[\frac{1}{p}\right]\Big/\mathbb{Z} = \varinjlim \mathbb{Z}/p^r\mathbb{Z},$$
then

Proposition 2.6.4. *For any abelian group A,*
$$\hom(\mathbb{Z}(p^\infty), A) \text{ and } \mathrm{Ext}(\mathbb{Z}(p^\infty), A)$$
are p-complete.

Proof: Since $\mathbb{Z}[\frac{1}{p}] \otimes \mathbb{Z}(p^\infty) = \mathrm{Tor}(\mathbb{Z}[\frac{1}{p}], \mathbb{Z}(p^\infty)) = 0$, it follows from the result of Cartan–Eilenberg in the previous section that
$$\hom\left(\mathbb{Z}\left[\frac{1}{p}\right], \hom(\mathbb{Z}(p^\infty), A)\right) = \mathrm{Ext}\left(\mathbb{Z}\left[\frac{1}{p}\right], \hom(\mathbb{Z}(p^\infty), A)\right) = 0$$
and
$$\hom\left(\mathbb{Z}\left[\frac{1}{p}\right], \mathrm{Ext}(\mathbb{Z}(p^\infty), A)\right) = \mathrm{Ext}\left(\mathbb{Z}\left[\frac{1}{p}\right], \mathrm{Ext}(\mathbb{Z}(p^\infty), A)\right) = 0.$$
$\square$

Consider the short exact sequence $0 \to \mathbb{Z} \to \mathbb{Z}[\frac{1}{p}] \to \mathbb{Z}(p^\infty) \to 0$ and the resulting long exact sequence
$$0 \to \hom(\mathbb{Z}(p^\infty), A) \to \hom\left(\mathbb{Z}\left[\frac{1}{p}\right], A\right) \to \hom(\mathbb{Z}, A)(= A)$$
$$\xrightarrow{\delta} \mathrm{Ext}(\mathbb{Z}(p^\infty), A) \to \mathrm{Ext}\left(\mathbb{Z}\left[\frac{1}{p}\right], A\right) \to \mathrm{Ext}(\mathbb{Z}, A)(= 0) \to 0.$$
We define:

Definition 2.6.5. *The p-completion of A is the map*
$$\delta : A \to \mathrm{Ext}(\mathbb{Z}(p^\infty), A).$$

We shall also write $\delta : A \to \hat{A}_p$ for p-completion.

Proposition 2.6.6. *Up to isomorphism the map $\delta : A \to \hat{A}_p$ is the unique map with the properties:*

(1) $\hat{A}_p = \mathrm{Ext}(\mathbb{Z}(p^\infty), A)$ *is p-complete.*

(2) *For all homomorphisms $f : A \to B$ with B p-complete, there is a unique homomorphism $\hat{f} : \hat{A}_p \to B$ such that $\hat{f} \circ \delta = f$.*

Proof: Clearly, the two properties characterize the p-completion $\hat{A}_p = \mathrm{Ext}(\mathbb{Z}(p^\infty), A)$ up to isomorphism. So it suffices to verify the property (2).

But consider the exact sequence
$$\hom\left(\mathbb{Z}\left[\frac{1}{p}\right], A\right) \to A \to \mathrm{Ext}(\mathbb{Z}(p^\infty), A) \to \mathrm{Ext}\left(\mathbb{Z}\left[\frac{1}{p}\right], A\right) \to 0.$$
This factors into two exact sequences as follows:
$$\hom\left(\mathbb{Z}\left[\frac{1}{p}\right], A\right) \to A \to C \to 0$$
$$0 \to C \to \mathrm{Ext}(\mathbb{Z}(p^\infty), A) \to \mathrm{Ext}\left(\mathbb{Z}\left[\frac{1}{p}\right], A\right) \to 0.$$
Since $\mathrm{Ext}(\mathbb{Z}[\frac{1}{p}], A)$ is a $\mathbb{Z}[\frac{1}{p}]$-module and B is p-complete,
$$\hom\left(\mathrm{Ext}\left(\mathbb{Z}\left[\frac{1}{p}\right], A\right), B\right) = \mathrm{Ext}\left(\left(\mathrm{Ext}\left(\mathbb{Z}\left[\frac{1}{p}\right], A\right), B\right)\right) = 0$$
and therefore $\hom(\mathrm{Ext}(\mathbb{Z}(p^\infty), A), B) \to \hom(C, A)$ is an isomorphism.
Since $\hom(\mathbb{Z}[\frac{1}{p}], A)$ is a $\mathbb{Z}[\frac{1}{p}]$-module and B is p-complete,
$$\hom\left(\hom\left(\mathbb{Z}\left[\frac{1}{p}\right], A\right), B\right) = 0$$
and therefore $\hom(C, B) \to \hom(A, B)$ is an isomorphism. Thus, property (2) is verified. $\square$

The defining long exact sequence for the p-completion shows that p-completion map is often a monomorphism.

Lemma 2.6.7. *If* $\hom(\mathbb{Z}[\frac{1}{p}], A) = 0$, *that is, if A has no nontrivial elements which are infinitely divisible by p, then there is a short exact sequence* $0 \to A \xrightarrow{\delta} \mathrm{Ext}(\mathbb{Z}(p^\infty), A) \to \mathrm{Ext}(\mathbb{Z}[\frac{1}{p}], A) \to 0$.

Since
$$\mathbb{Z}(p^\infty) = \varinjlim \mathbb{Z}/p^r\mathbb{Z},$$
we have a short exact sequence
$$0 \to \varprojlim{}^1 \hom(\mathbb{Z}/p^r\mathbb{Z}, A) \to \mathrm{Ext}(\mathbb{Z}(p^\infty), A) \to \varprojlim \mathrm{Ext}(\mathbb{Z}/p^r\mathbb{Z}, A) \to 0,$$
that is:

Lemma 2.6.8. *For any abelian group A, there is a short exact sequence*
$$0 \to \varprojlim{}^1 {}_{p^r}A \to \hat{A}_p \to \varprojlim A/p^r A \to 0$$
where ${}_{p^r}A = \{a \varepsilon A \mid p^r(a) = 0\}$ *is part of the inverse system*
$${}_pA \xleftarrow{p} {}_{p^2}A \xleftarrow{p} {}_{p^3}A \xleftarrow{p} {}_{p^4}A \xleftarrow{p} \cdots .$$

and $A/p^r A$ is part of the epimorphic inverse system

$$A/pA \leftarrow A/p^2 A \leftarrow A/p^3 A \leftarrow A/p^4 A \leftarrow \cdots .$$

Corollary 2.6.9. $\hat{A}_p = \mathrm{Ext}(\mathbb{Z}(p^\infty), A) = 0$ if and only if A is a p-divisible abelian group.

Now it is easy to compute p-completion for cyclic groups and thus for all finitely generated abelian groups. We get

$$\hat{\mathbb{Z}}_p = \varprojlim \mathbb{Z}/p^r \mathbb{Z}, \qquad \hat{G}_p = G, \qquad \hat{H}_p = 0$$

if G is a finite abelian p group and H is a finite abelian group of torsion relatively prime to p.

In order to get further information about p-completion, we recall without proof a theorem in Kaplansky's book on infinite abelian groups [72].

Proposition 2.6.10. If A is any abelian group, there is a direct sum decomposition $A = D \oplus R$ where D is a divisible abelian group and R has no divisible subgroups.

Suppose A is any p-complete group. Then $A_p = D \oplus R$ where D is p-complete and divisible and R is p-complete and has no divisible subgroups. Since D is p-complete and divisible, $D = 0$. Thus $A = R$. Since R has no divisible subgroups,

$$\bigcap_{r \geq 0} p^r R = 0$$

and

$$R \to \varprojlim R/p^r R$$

is an isomorphism. Now it follows from 2.6.8 that

$$\varprojlim{}^1_{p^r} R = \bigcap_{r \geq 0} p^r R = 0.$$

Hence

Proposition 2.6.11. If A is any p-complete group, then A has no divisible subgroups,

$$A \to \varprojlim A/p^n A$$

is an isomorphism and

$$\varprojlim{}^1_{p^r} A = \bigcap_{r \geq 0} p^r A = 0.$$

Exercises

(1) If G is a p-complete abelian group, then show that G is localized at p, that is, for all integers q prime to p, the map $q : G \to G$ is an isomorphism.

(2) If G is an abelian group and p is a prime, then the p-completion of localization at (p) is p-completion, that is,
$$(G_{(p)})\hat{}_p = \hat{G}_p.$$

2.7 p-completion of simply connected spaces

Let p be a prime.

Definition 2.7.1. A simply connected space X is p-complete if its homotopy groups $\pi_k(X)$ are p-complete for all k.

Let $M = M(\mathbb{Z}[\frac{1}{p}], 1) = $ a Moore space with exactly one nonzero reduced integral homology group isomorphic to $\mathbb{Z}[\frac{1}{p}]$ in dimension 1, that is, $M(\mathbb{Z}[\frac{1}{p}], 1)$ is the homotopy direct limit of the degree p maps on a circle,
$$S^1 \xrightarrow{p} S^1 \xrightarrow{p} S^1 \xrightarrow{p} S^1 \xrightarrow{p} \cdots.$$

Thus, there is a cofibration sequence
$$\bigvee_{n \geq 1} S^1 \xrightarrow{\Psi} \bigvee_{n \geq 1} S^1 \to M\left(\mathbb{Z}\left[\frac{1}{p}\right], 1\right) \to \bigvee_{n \geq 1} S^2 \xrightarrow{\Sigma(\Psi)} \bigvee_{n \geq 1} S^2.$$

From this we get immediately that for all $k \geq 0$ there are short exact sequences
$$0 \to \underleftarrow{\lim}^1 \pi_{k+2}(X) \to \left[\Sigma^k M\left(\mathbb{Z}\left[\frac{1}{p}\right], 1\right), X\right]_* \to \underleftarrow{\lim} \pi_{k+1}(X) \to 0.$$

Hence

Lemma 2.7.2. *A simply connected space X is p-complete if and only if X is local with respect to the map $M(\mathbb{Z}[\frac{1}{p}], 1) \to *$.*

Recall that a map $f : X \to Y$ is called a mod p equivalence if it induces an isomorphism in mod p homology.

Proposition 2.7.3. *A map $f : X \to Y$ between simply connected spaces is a mod p equivalence if and only if f is a local equivalence with respect to $M(\mathbb{Z}[\frac{1}{p}], 1) \to *$.*

Proof: We may suppose that f is an inclusion. Assume f is a mod p equivalence. This is the case if and only if $H_*(Y, X; \mathbb{Z}/p\mathbb{Z}) = 0$ which is so if and only if $H_*(Y, X)$ is a $\mathbb{Z}[\frac{1}{p}]$ module. But this is equivalent to $0 = \hom(H_*(Y, X), D) = \text{Ext}(H_*(Y, X), D)$ for all p-complete abelian groups D. Thus, if W is any simply

connected p-complete space, the obstruction groups $H^*(\Sigma^k Y, \Sigma^k X; \pi_*(W))$ all vanish and
$$f^* : \mathrm{map}_*(Y, W) \to \mathrm{map}_*(X, W)$$
is a weak equivalence. Hence, f is a local equivalence with respect to $M(\mathbb{Z}[\frac{1}{p}], 1) \to *$.

To show the equivalence the other way, it suffices to note that $\mathrm{map}_*(Y, W) \to \mathrm{map}_*(X, W)$ can be a weak equivalence only if all obstruction groups $H^*(Y, X; \pi_*(W))$ vanish. But for any p-complete abelian group D, the space $W = K(D, n)$ is also p-complete. It follows that $H_*(Y, X)$ is a $\mathbb{Z}[\frac{1}{p}]$ module and f is a mod p equivalence. $\square$

We already know that, for simply connected spaces, mod p homology isomorphisms are equivalent to mod p homotopy isomorphisms.

Once again, for any simply connected X, the general theory of Dror Farjoun presented in this chapter implies the existence of a simply connected p-completion
$$\iota : X \to \hat{X}_p = L_{M(\mathbb{Z}[\frac{1}{p}], 1)}(X)$$
characterized uniquely up to homotopy by:

(1) $\hat{X}_p$ is p-complete, and

(2) $\iota : X \to \hat{X}_p$ is a mod p equivalence.

Just as with localization, we have that p-completion preserves fibrations.

Proposition 2.7.4. *If $F \to E \to B$ is a fibration sequence of simply connected spaces, then $\hat{F}_p \to \hat{E}_p \to \hat{B}_p$ is a fibration sequence up to homotopy.*

Proof: Let G be the homotopy theoretic fibre of $\hat{E}_p \to \hat{B}_p$. Then G is p-complete. Apply the Zeeman comparison theorem to the map of fibration sequences

$$\begin{array}{ccccc} F & \to & E & \to & B \\ \downarrow & & \downarrow & & \downarrow \\ G & \to & \hat{E}_p & \to & \hat{B}_p. \end{array}$$

Thus $F \to G$ is a mod p equivalence and $G \simeq \hat{F}_p$. $\square$

We now consider p-completion of Eilenberg–MacLane spaces.

Proposition 2.7.5. *If G is a finitely generated abelian group, then $K(G, n) \to K(\hat{G}_p, n)$ is p-completion for all $n \geq 2$.*

Proof: Since the target is complete, it is sufficient to check that the map is a mod p equivalence. Since the spaces are simply connected, a mod p homology isomorphism is equivalent to a mod p homotopy isomorphism, but the only nonzero mod p homotopy groups are in dimensions n and $n+1$. Since G is finitely generated,

$$\pi_n(K(G,n); \mathbb{Z}/p\mathbb{Z}) \cong G/pG \cong \hat{G}_p/p\hat{G}_p \cong \pi_n(K(G,n); \mathbb{Z}/p\mathbb{Z}),$$
$$\pi_{n+1}(K(G,n); \mathbb{Z}/p\mathbb{Z}) \cong_p G \cong_p \hat{G}_p \cong \pi_{n+1}(K(\hat{G}_p,n); \mathbb{Z}/p\mathbb{Z}).$$

$\square$

The above proposition is false if G is not finitely generated, for example, it is false when $G = \mathbb{Z}(p^\infty)$. The short exact sequence $0 \to \mathbb{Z} \to \mathbb{Z}[\frac{1}{p}] \to \mathbb{Z}(p^\infty) \to 0$ yields the fibration sequence

$$K\left(\mathbb{Z}\left[\frac{1}{p}\right], 1\right) \to K(\mathbb{Z}(p^\infty), 1) \to K(\mathbb{Z}, 2).$$

Since $\mathbb{Z}[\frac{1}{p}]$ has trivial mod p homology, the map $K(\mathbb{Z}(p^\infty), 1) \to K(\mathbb{Z}, 2)$ is a mod p equivalence and thus the composition

$$K(\mathbb{Z}(p^\infty), 1) \to K(\mathbb{Z}, 2) \to K(\hat{\mathbb{Z}}_p, 2)$$

is the p-completion. We want to understand the p-completion of Eilenberg–MacLane spaces.

Proposition 2.7.6.

(a) *If F is an abelian group with no nontrivial elements which are infinitely divisible by p, then*

$$K(F,n) \to K(\hat{F}_p, n)$$

is the p-completion for all $n \geq 2$.

(b) *In general, the p-completion of $K(G,n)$ for $n \geq 2$ is a space Y with two possibly nonzero homotopy groups $\pi_n(Y) = \hat{G}_p$ and $\pi_{n+1}(Y) = \hom(\mathbb{Z}(p^\infty), G)$.*

Proof: The short exact sequence $0 \to F \to \hat{F}_p \to \operatorname{Ext}(\mathbb{Z}[\frac{1}{p}], F) \to 0$ yields a fibration sequence

$$K(F,n) \to K(\hat{F}_p, n) \to K\left(\operatorname{Ext}\left(\mathbb{Z}\left[\frac{1}{p}\right], F\right), n\right).$$

The mod p homotopy and hence the mod p homology of the base is trivial. We know that $K(F,n) \to K(\hat{F}_p, n)$ is a mod p equivalence with target having p-complete homotopy groups. Hence, this is the p-completion.

For any abelian group G let $0 \to F_1 \to F_0 \to G \to 0$ be a free resolution. Let Y be the homotopy theoretic fibre of $K(\hat{F}_{1p}, n+1) \to K(\hat{F}_{0p}, n+1)$ and consider the map of fibration sequences

$$\begin{array}{ccc} K(G,n) & \to & Y \\ \downarrow & & \downarrow \\ K(F_1, n+1) & \to & K(\hat{F}_{1p}, n+1) \\ \downarrow & & \downarrow \\ K(F_0, n+1) & \to & K(\hat{F}_{0p}, n+1). \end{array}$$

Since $K(\hat{F}_{1p}, n+1)$ and $K(\hat{F}_{0p}, n+1)$ are p-complete spaces, so is Y and the Zeeman comparison theorem shows that Y is the p-completion of $K(G, 1)$.

We have a long exact sequence associated to hom and Ext:

$$0 \to \hom(\mathbb{Z}(p^\infty), G) \to \mathrm{Ext}(\mathbb{Z}(p^\infty), F_1) \to \mathrm{Ext}(\mathbb{Z}(p^\infty), F_0)$$
$$\to \mathrm{Ext}(\mathbb{Z}(p^\infty), G) \to 0.$$

Hence the homotopy groups of Y are as indicated. $\square$

Having determined the homotopy groups of the p-completion of an Eilenberg–MacLane space, it is now easy to determine the homotopy groups of the p-completion of any simply connected space.

Proposition 2.7.7. *If X is a simply connected space, then we have short exact sequences*

$$0 \to \mathrm{Ext}(\mathbb{Z}(p^\infty), \pi_n(X)) \to \pi_n(\hat{X}_p) \to \hom(\mathbb{Z}(p^\infty), \pi_{n-1}(X)) \to 0.$$

Proof: We are given the result for Eilenberg–MacLane spaces.

For spaces X with finitely many nonzero homotopy groups, the highest dimensional one being $\pi_n(X)$, the result follows by induction using the fibration

$$K(\pi_n(X), n) \to X \to Y,$$

where Y is a space with one less nonvanishing homotopy group than X.

For an arbitrary simply connected space, write X as an inverse limit of its Postnikov stages, $X = \lim_\leftarrow X_\ell$, and note that $\hat{X}_p = \lim_\leftarrow \hat{X}_{\ell p}$. (The fact that p-completion commutes with this inverse limit process is a consequence of the fact that, in this case, the limit is finite in each degree and thus homology commutes with the inverse limit.) The result follows for general simply connected X. $\square$

In the case where the homotopy groups are finitely generated, the preceding proposition takes the following simple form.

Corollary 2.7.8. *If X is a simply connected space with finitely generated homotopy groups in every degree, then there are isomorphisms*

$$\pi_n(\hat{X}_p) \cong \hat{\pi}_n(X)_p \cong \lim_{\leftarrow r} \pi_n(X)/p^r \pi_n(X).$$

2.7 p-completion of simply connected spaces

Remarks. If $L_M(X)$ is the localization of X with respect to $M \to *$, it may not be the case that $L_M(X) = \lim_\leftarrow L_M(X_\ell)$ where X_ℓ is the Postnikov system of X. It is the case for the localizations which are p-completion and localization at a set of primes but it is not the case for the localization functor L_M in Section 2.10. Here $M = B\mathbb{Z}/p\mathbb{Z}$ and the localization has the following properties: $L_M(X) = *$ whenever X has only finitely many nonzero homotopy groups, all of which are p-primary abelian. Also, any simply connected finite complex is local in this localization. Hence, if $X = P^m(\mathbb{Z}/p\mathbb{Z}) = $ the mod p Moore space with $m \geq 3$ and X_ℓ is its Postnikov system, $X = L_M(X) \neq \lim_\leftarrow L_M(X_\ell) = \lim_\leftarrow(*) = *$.

Remarks. Just as with localization at a set of primes, it is also possible to extend p-completion to connected loop spaces $X = \Omega Y$ via the definition $\hat{X}_p = \Omega(\hat{Y}_p)$.

Exercises

(1) Suppose X is a simply connected space with finitely generated homotopy groups. Show that the homotopy groups of the p-completion $\hat{X}_p$ are the p-completion of the homotopy groups of X, that is, for all $k \geq 2$,
$$\pi_k(\hat{X}_p) \cong \hat{\pi}_k(X)_p.$$

(2) Show that p-completion does not commute with taking n-connected covers, that is, give a counterexample to $\hat{X}\langle n\rangle_p \simeq (\hat{X}_p)\langle n\rangle$.

(3) Let X be a simply connected space and let S be a set of primes.

   (a) Show that X and the localization $X_{(S)}$ have the same p-completion if $p \in S$.

   (b) Show that the p-completion of the localization $X_{(S)}$ is contractible if $p \notin S$.

(4) Show that, if a space X is n-connected with $n \geq 1$, then so is the p-completion $\hat{X}_p$.

(5) (a) Show that a p-local equivalence is always a mod p equivalence.

   (b) Show that a p-complete space is always a p-local space.

(6) Let $X \to Y$ be a map of simply connected spaces.

   (a) If the map of localizations at p, $X_{(p)} \to Y_{(p)}$, is a homotopy equivalence, show that the map of p-completions, $\hat{X}_p \to \hat{Y}_p$, is a homotopy equivalence.

   (b) Give an example to show that the converse to (a) is false.

(7) Show that the p-completion of a pullback diagram of fibrations of simply connected spaces is a pullback diagram of fibrations.

(8) If q and p are distinct primes, show that $\pi_*(\hat{X}_p; \mathbb{Z}/q\mathbb{Z}) = 0$ for all simply connected X.

2.8 Completion implies the mod k Hurewicz isomorphism

In this section we shall show that the existence of p-completion implies the truth of the mod k Hurewicz isomorphism theorem, at least for simply connected spaces. Since we have already proven this theorem for all nilpotent spaces in the previous chapter, the point of this exercise is to show that it can be done independently using the existence of p-completion. Accordingly, we do not want to assume any consequences of the mod k Hurewicz isomorphism theorem in this section. We shall show:

Proposition 2.8.1. *Let* $n \geq 2$. *If* X *is any simply connected space and* $\pi_i(X; \mathbb{Z}/k\mathbb{Z}) = 0$ *for all* $2 \leq i < n$, *then*

(a) $H_i(X; \mathbb{Z}/k\mathbb{Z}) = 0$ *for all* $2 \leq i < n$ *and*

(b) *the Hurewicz map* $\varphi : \pi_n(X : \mathbb{Z}/k\mathbb{Z}) \to H_n(X; \mathbb{Z}/k\mathbb{Z})$ *is an isomorphism.*

First of all we note that, if $k = p_1^{\alpha_1} p_2^{\alpha_2} \ldots p_\ell^{\alpha_\ell}$ is a factorization into powers of distinct primes, then

$$\mathbb{Z}/k\mathbb{Z} \cong \mathbb{Z}/\mathbb{Z}p_1^{\alpha_1}\mathbb{Z} \oplus \mathbb{Z}/p_2^{\alpha_2}\mathbb{Z} \oplus \cdots \oplus \mathbb{Z}/p_\ell^{\alpha_\ell}\mathbb{Z}$$
$$\pi_i(X; \mathbb{Z}/k\mathbb{Z}) \cong \pi_i(X; \mathbb{Z}/\mathbb{Z}p_1^{\alpha_1}\mathbb{Z}) \oplus \pi_i(X; \mathbb{Z}/p_2^{\alpha_2}\mathbb{Z}) \oplus \cdots \oplus \pi_i(X; \mathbb{Z}/p_\ell^{\alpha_\ell}\mathbb{Z})$$
$$H_i(X; \mathbb{Z}/k\mathbb{Z}) \cong H_i(X; \mathbb{Z}/\mathbb{Z}p_1^{\alpha_1}\mathbb{Z}) \oplus H_i(X; \mathbb{Z}/p_2^{\alpha_2}\mathbb{Z}) \oplus \ldots \oplus H_i(X; \mathbb{Z}/p_\ell^{\alpha_\ell}\mathbb{Z})$$

and thus it is sufficient to prove the Hurewicz theorem when $k = p^\alpha$ is a power of a prime.

Now we show:

Lemma 2.8.2. *For all simply connected* X, *the p-completion map* $\iota : X \to \hat{X}_p$ *induces an isomorphism of all mod* p^α *homotopy groups.*

Remark. By definition, the p-completion map induces an isomorphism of all mod p homology groups and thus an isomorphism of all mod p^α homology groups. The Hurewicz isomorphism theorem for pairs implies that it induces an isomorphism of all mod p^α homotopy groups. But, in this section, we cannot use this fact.

Proof: The proof of the lemma is done in successive steps:

(1) $X = K(F, n)$ with F free abelian, $n \geq 2$.
(2) $X = K(G, n)$ with G abelian, $n \geq 2$.

2.8 Completion implies the mod k Hurewicz isomorphism

(3) X has finitely many nonzero homotopy groups.

(4) X is an arbitrary simply connected space.

In case (1), $\hat{X}_p = K(\hat{F}_p, n)$ and the only nonzero mod p^α homotopy groups are in dimension n, isomorphic to $F \otimes \mathbb{Z}/p^\alpha \mathbb{Z}$ and $\hat{F}_p \otimes \mathbb{Z}/p^\alpha \mathbb{Z}$, respectively. The tensor-Tor exact sequence associated to the short exact sequence

$$0 \to F \to \hat{F}_p \to \text{Ext}\left(\mathbb{Z}\left[\frac{1}{p}\right], F\right) \to 0$$

shows that these two are isomorphic.

In case (2), we take a free resolution $0 \to F_1 \to F_0 \to G \to 0$ and apply the five-lemma to the map of fibration sequences

$$\begin{array}{ccccc}
K(G, n) & \to & K(F_1, n+1) & \to & K(F_0, n+1) \\
\downarrow \iota & & \downarrow \iota & & \downarrow \iota \\
\widehat{K(G, n)}_p & \to & K(\hat{F}_{1p}, n+1) & \to & K(\hat{F}_{0p}, n+1).
\end{array}$$

In case (3), we apply the five-lemma and induction to the p-completion maps related to the fibration sequence $Y \to X \to K(G, n)$, where Y has one less nonzero homotopy group than X.

Finally, in case (4), we note that X is the inverse limit of its Postnikov stages X_n and $\hat{X}_p$ is the inverse limit of $(\hat{X_n})_p$, both limits being finite in each degree. The lemma follows. $\square$

We now finish the proof of the mod k Hurewicz isomorphism theorem. Suppose $\pi_i(X; \mathbb{Z}/p^\alpha \mathbb{Z}) = 0$ for all $2 \leq i < n$. By induction we can assume that $H_i(X; \mathbb{Z}/p^\alpha \mathbb{Z}) = 0$ for all $2 \leq i < n$. We need to show that the Hurewicz map $\varphi: \pi_n(X; \mathbb{Z}/p^\alpha \mathbb{Z}) \to H_n(X; \mathbb{Z}/p^\alpha \mathbb{Z})$ is an isomorphism.

But $\pi_i(X) \otimes \mathbb{Z}/p^\alpha \mathbb{Z} = 0$ for $i < n$ and $\text{Tor}(\pi_i(X), \mathbb{Z}/p^\alpha \mathbb{Z}) = 0$ for $i < n - 1$. Thus $\pi_i(X)$ is p-divisible if $i < n$ and has no nontrivial p torsion if $i < n - 1$.

With $\pi = \pi_i(X)$, p divisibility and the short exact sequence

$$0 \to \varprojlim{}^1(_{p^\ell}\pi) \to \text{Ext}(\mathbb{Z}(p^\infty), \pi) \to \varprojlim(\pi/p^\ell \pi) \to 0$$

shows that $\text{Ext}(\mathbb{Z}(p^\infty), \pi) = \hat{\pi}_p = \hat{\pi}_i(X)_p = 0$ for all $i < n$. (The inverse system

$$_p\pi \xleftarrow{p}{}_{p^2}\pi \xleftarrow{p}{}_{p^3}\pi \xleftarrow{p} \cdots$$

is epimorphic and thus has a vanishing first derived functor.)

Now no nontrivial p torsion and the short exact sequence

$$0 \to \text{Ext}(\mathbb{Z}(p^\infty), \pi_i(X)) \to \pi_i(\hat{X}_p) \to \hom(\mathbb{Z}(p^\infty), \pi_{i-1}(X)) \to 0$$

shows that $\pi_i(\hat{X}_p) = 0$ for all $i < n$. We record this fact.

Lemma 2.8.3. *If X is a simply connected space with $\pi_i(X; \mathbb{Z}/p^\alpha \mathbb{Z}) = 0$ for all $2 \leq i < n$ then $\pi_i(\hat{X}_p) = 0$ for all $2 \leq i < n$.*

The classical Hurewicz theorem asserts that $H_i(\hat{X}_p) = 0$ for all $i < n$ and $\varphi : \pi_n(\hat{X}_p) \cong H_n(\hat{X}_p)$. Thus,

$$\pi_n(X; \mathbb{Z}/p^\alpha \mathbb{Z}) \cong \pi_n(\hat{X}_p; \mathbb{Z}/p^\alpha \mathbb{Z}) \cong \pi_n(\hat{X}_p) \otimes \mathbb{Z}/p^\alpha \mathbb{Z}$$

and these are isomorphic to

$$H_n(\hat{X}_p) \otimes \mathbb{Z}/p^\alpha \mathbb{Z} \cong H_n(\hat{X}_p; \mathbb{Z}/p^\alpha \mathbb{Z}) \cong H_n(X; \mathbb{Z}/p^\alpha \mathbb{Z}).$$

This completes the inductive step. □

2.9 Fracture lemmas

Dennis Sullivan [128, 129] called the next result the topological Hasse–Minkowski principle.

Proposition 2.9.1. *For any simply connected space X, there is up to homotopy a pullback diagram of fibrations*

$$\begin{array}{ccc} X & \to & \prod_p \hat{X}_p \\ \downarrow & & \downarrow \\ X \otimes Q & \to & (\prod_p \hat{X}_p) \otimes Q, \end{array}$$

the product being taken over all primes p.

Since p-completions and localizations are functors, the diagram is automatically strictly commutative. That the diagram is a homotopy pullback is recognized by the fact that the homotopy theoretic fibres of $X \to \prod_p \hat{X}_p$ and $X \otimes Q \to (\prod_p \hat{X}_p) \otimes Q$ are identical. Equivalently, the homotopy theoretic fibres of $X \to X \otimes Q$ and $(\prod_p \hat{X}_p) \to (\prod_p \hat{X}_p) \otimes Q$ are identical.

Proof: We adopt the standard strategy. We will prove this in several steps:

(1) where $X = K(F, n)$ with F free abelian.

(2) where $X = K(G, n)$ with G abelian.

(3) where X has only finitely many nonvanishing homotopy groups.

(4) where X is any simply connected space.

Step (1): $X = K(F, n)$ with F free abelian.

2.9 Fracture lemmas

Consider the short exact sequence

$$0 \to \mathbb{Z} \to Q \to Q/\mathbb{Z} \to 0.$$

The long exact sequence associated to hom and Ext collapses to a short exact sequence

$$0 \to F \to \operatorname{Ext}(Q/\mathbb{Z}, F) \to \operatorname{Ext}(Q, F) \to 0.$$

We observe two things, $\operatorname{Ext}(Q, F)$ is a rational vector space and Q is torsion free. Hence, $\operatorname{Ext}(Q, F) \otimes Q \cong \operatorname{Ext}(Q, F)$ and we get a short exact sequence

$$0 \to F \otimes Q \to \operatorname{Ext}(Q/\mathbb{Z}, F) \otimes Q \to \operatorname{Ext}(Q, F) \otimes Q \to 0.$$

In fact, we have a commutative diagram in which the rows and columns are short exact sequences:

$$\begin{array}{ccccccccc}
& & 0 & & 0 & & 0 & & \\
& & \downarrow & & \downarrow & & \downarrow & & \\
0 & \to & F & \to & \operatorname{Ext}(Q/\mathbb{Z}, F) & \to & \operatorname{Ext}(Q, F) & \to & 0 \\
& & \downarrow & & \downarrow & & \downarrow & & \\
0 & \to & F \otimes Q & \to & \operatorname{Ext}(Q/\mathbb{Z}, F) \otimes Q & \to & \operatorname{Ext}(Q, F) \otimes Q & \to & 0 \\
& & \downarrow & & \downarrow & & \downarrow & & \\
0 & \to & F \otimes Q/\mathbb{Z} & \to & \operatorname{Ext}(Q/\mathbb{Z}, F) \otimes Q/\mathbb{Z} & \to & 0 & \to & 0 \\
& & \downarrow & & \downarrow & & \downarrow & & \\
& & 0 & & 0 & & 0 & &
\end{array}$$

The first two columns are short exact since F and $\operatorname{Ext}(Q/\mathbb{Z}, F)$ are torsion free. The last column is short exact since $\operatorname{Ext}(Q, F)$ is a rational vector space. We already know that the first two rows are short exact and so it follows that the bottom (quotient) row is short exact.

It follows that there is a commutative diagram in which rows and columns are fibration sequences:

$$\begin{array}{ccccc}
* & \to & K(F \otimes Q/\mathbb{Z}, n+1) & \to & K(\operatorname{Ext}(Q/\mathbb{Z}, F) \otimes Q/\mathbb{Z}, n+1) \\
\downarrow & & \downarrow & & \downarrow \\
K(\operatorname{Ext}(Q, F), n+1) & \to & K(F, n) & \to & K(\operatorname{Ext}(Q/\mathbb{Z}, F), n) \\
\downarrow & & \downarrow & & \downarrow \\
K(\operatorname{Ext}(Q, F) \otimes Q, n+1) & \to & K(F \otimes Q, n) & \to & K(\operatorname{Ext}(Q/\mathbb{Z}, F) \otimes Q, n)
\end{array}$$

in other words the lower right-hand square is a pullback diagram of fibrations.

Now we need only observe that

$$Q/\mathbb{Z} = \oplus_p \mathbb{Z}(p^\infty),$$
$$\mathrm{Ext}(Q/\mathbb{Z}, F) = \prod_p \mathrm{Ext}(\mathbb{Z}(p^\infty), F),$$
$$K(\mathrm{Ext}(Q/\mathbb{Z}, F), n) = \prod_p K(\mathrm{Ext}(\mathbb{Z}(p^\infty), F), n) = \prod_p \hat{K}(F, n)_p$$

and

$$K(G \otimes Q, n) = K(G, n) \otimes Q$$

to show that we have demonstrated the required pullback square in this case.

Step (2): $X = K(G, n)$ with G abelian.

We need the following lemma which will be proved in Chapter 3.

Lemma 2.9.2. *Let A and B be pullback squares of fibrations and let $\phi : A \to B$ be a map of squares. If C is the square formed by taking the homotopy theoretic fibres of ϕ, then C is also a pullback square of fibrations.*

If $0 \to F_1 \to F_0 \to G \to 0$ is a free resolution of G, then the two pullback squares for $K(F_1, n+1)$ and $K(F_0, n+1)$ have a map for which the fibres form the pullback square for $K(G, n)$.

Step (3): X has only finitely many nonzero homotopy groups with the top nonzero one being $\pi_n(X)$.

We apply induction and the above lemma to the fibration $K(\pi_n(X), n) \to X \to Y$ where Y has one less nonzero homotopy group than X.

Step (4): For any simply connected X, the required diagram is the inverse limit of the pullback diagrams of the Postnikov stages and thus is itself a pullback diagram. □

There is a pullback diagram for S-localizations [128, 129, 35]:

Proposition 2.9.3. *Let S_1 and S_2 be two sets of primes. For any simply connected space X, there is up to homotopy a pullback diagram of fibrations*

$$\begin{array}{ccc} X_{(S_1 \cup S_2)} & \to & X_{(S_1)} \\ \downarrow & & \downarrow \\ X_{(S_2)} & \to & X_{(S_1 \cap S_2)}. \end{array}$$

Proof: Let T_1 and T_2 be complementary sets of primes to S_1 and S_2, respectively. The theory of partial fraction decompositions asserts that

$$\mathbb{Z}_{(S_1 \cap S_2)} = \mathbb{Z}[(T_1 \cup T_2)^{-1}] = \mathbb{Z}[T_1^{-1}] + \mathbb{Z}[T_2^{-1}] = \mathbb{Z}_{(S_1)} + \mathbb{Z}_{(S_2)},$$

$$\mathbb{Z}_{(S_1 \cup S_2)} = \mathbb{Z}[(T_1 \cap T_2)^{-1}] = \mathbb{Z}[T_1^{-1}] \cap \mathbb{Z}[T_2^{-1}] = \mathbb{Z}_{(S_1)} \cap \mathbb{Z}_{(S_2)}.$$

The Noether isomorphisms give a diagram with short exact rows and columns:

$$\begin{array}{ccccccccc}
& & 0 & & 0 & & 0 & & \\
& & \downarrow & & \downarrow & & \downarrow & & \\
0 & \to & \mathbb{Z}_{(S_1 \cup S_2)} & \to & \mathbb{Z}_{(S_1)} & \to & \mathbb{Z}_{(S_1)}/\mathbb{Z}_{(S_1 \cup S_2)} & \to & 0 \\
& & \downarrow & & \downarrow & & \downarrow & & \\
0 & \to & \mathbb{Z}_{(S_2)} & \to & \mathbb{Z}_{(S_1 \cap S_2)} & \to & \mathbb{Z}_{(S_1 \cap S_2)}/\mathbb{Z}_{(S_2)} & \to & 0 \\
& & \downarrow & & \downarrow & & \downarrow & & \\
0 & \to & \mathbb{Z}_{(S_2)}/\mathbb{Z}(S_1 \cup S_2) & \to & \mathbb{Z}_{(S_1 \cap S_2)}/\mathbb{Z}_{(S_1)} & \to & 0 & \to & 0 \\
& & \downarrow & & \downarrow & & \downarrow & & \\
& & 0 & & 0 & & 0 & &
\end{array}$$

For any abelian group G, since any localization of $\mathbb{Z}$ is torsion free, the rows and columns remain short exact if we tensor this diagram with G. But this just replaces $\mathbb{Z}$ by G in the diagram and leads in the same way as before to the validity of the pullback diagram for $X = K(G, n)$. Without any essential change from before the proof continues to establish the validity of the pullback diagram for finite Postnikov systems and then for all simply connected spaces. □

Corollary 2.9.4. *If S_1 and S_2 are complementary sets of primes and X is simply connected, there is up to homotopy a pullback diagram of fibrations*

$$\begin{array}{ccc}
X & \to & X_{(S_1)} \\
\downarrow & & \downarrow \\
X_{(S_2)} & \to & X \otimes Q.
\end{array}$$

Exercises

(1) Let S be a set of primes. For any simply connected space X, show that there is up to homotopy a pullback diagram of fibrations

$$\begin{array}{ccc}
X_{(S)} & \to & \prod_{p \in S} \hat{X}_p \\
\downarrow & & \downarrow \\
X \otimes Q & \to & (\prod_{p \in S} \hat{X}_p) \otimes Q.
\end{array}$$

(2) Suppose that $M \otimes Q \simeq *$ and X is simply connected. Show that X is local with respect to $M \to *$ if and only if, for all primes p, $\hat{X}_p$ is local with respect to $M \to *$.

(3) If S_1 and S_2 are complementary sets of primes and if X is simply connected with $X \otimes Q \simeq *$, then show that
$$X \simeq X_{(S_1)} \times X_{(S_2)}.$$
(4) Let $f : X \to Y$ be a map of spaces.

  (a) If X and Y are simply connected, show that the Hasse–Minkowski principle implies that f is a homotopy equivalence if the induced maps in rational homology and in mod p homology are isomorphisms for all primes p.

  (b) For arbitrary spaces X and Y, show that f induces an isomorphism in integral homology if it induces isomorphisms in rational homology and in mod p homology for all primes p.

(5) (a) For any abelian group A, show that $A = 0$ is and only if the localizations $A_{(p)} = 0$ for all primes p.

  (b) For any homomorphism $f : A \to B$ of abelian groups, show that f is an isomorphism if and only if the localizations $f_{(p)} : A_{(p)} \to B_{(p)}$ are isomorphisms for all primes p.

(6) Let $f : X \to Y$ be a map of spaces.

  (a) Show that f induces an isomorphism in integral homology if and only if f induces isomorphisms in homology localized at all primes.

  (b) Suppose X and Y are both simply connected or both H-spaces. Show that f is a (weak) homotopy equivalence if and only if f induces isomorphisms in homology localized at all primes.

2.10 Killing Eilenberg–MacLane spaces: Miller's theorem

We now consider localization $L_{B\mathbb{Z}/p\mathbb{Z}}$ with respect to the map $f : B\mathbb{Z}/p\mathbb{Z} \to *$. We begin with some consequences of the Zabrodsky Lemma.

Lemma 2.10.1. *If G is any abelian p-primary torsion group and $n \geq 1$, then $K(G, n) \to *$ is a local equivalence with respect to f.*

Proof: If G is a finite p group, then G has a nontrivial center and hence a central subgroup isomorphic H to $\mathbb{Z}/p\mathbb{Z}$. The Zabrodsky Lemma applied to the bundle sequence
$$BH \to BG \to B(G/H)$$
leads to an inductive proof that $BG \to *$ is a local equivalence for all finite p groups G.

2.10 Killing Eilenberg–MacLane spaces: Miller's theorem

Since any abelian p-primary torsion group G is a union of its finite subgroups H, we have a pushout diagram

$$\begin{array}{ccc} \bigcup_{H_1 \subseteq H_2} H_1 & \xrightarrow{S} & \bigcup_{H_1 \subseteq H_2} H_2 \\ \downarrow T & & \downarrow \\ \bigcup_{H_1 \subseteq H_2} H_1 & \rightarrow & G \end{array}$$

where $S : H_1 \to H_2$ is inclusion and $T : H_1 \to H_1$ is the identity, the unions being taken over all pairs of finite subgroups $H_1 \subseteq H_2 \subseteq G$.

After applying the classifying space functor B to all the groups, we get a corresponding pushout diagram of classifying spaces where the maps are cofibrations. Now for any local space X we apply the functor $\mathrm{map}(\ , X)$ and get a pullback diagram where the maps are fibrations

$$\begin{array}{ccc} \mathrm{map}(\bigcup_{H_1 \subseteq H_2} BH_1, X) & \xleftarrow{S^*} & \mathrm{map}(\bigcup_{H_1 \subseteq H_2} BH_2, X) \\ \uparrow T^* & & \uparrow \\ \mathrm{map}(\bigcup_{H_1 \subseteq H_2} BH_1, X) & \leftarrow & \mathrm{map}(BG, X). \end{array}$$

This shows that

$$\mathrm{map}(BG, X) = \varprojlim X = X,$$

and $BG \to *$ is a local equivalence.

Finally, applying the Zabrodsky Lemma to the principal bundle sequence

$$K(G, n) \to PK(G, n+1) \to K(G, n+1)$$

completes the inductive step to show that $K(G, n) \to *$ is a local equivalence for all abelian p-primary torsion groups G and all $n \geq 1$. $\square$

Recall that Moore–Postnikov factorizations are factorizations of maps between simply connected spaces $X \to Y$ into an inverse limit of a sequence of principal bundles

$$X_n \to X_{n-1} \to X_{n-2} \to \cdots \to X_2 \to X_1 \to X_0 = Y$$

with

$$X = \varprojlim X_n$$

via maps $X \to X_n$.

If F is the homotopy theoretic fibre of $X \to Y$, then each $X_n \to X_{n-1}$ is a principal bundle with fibre group $K(\pi_n(F), n)$. If F_n is the homotopy theoretic fibre of $X_n \to Y$ and G_n is the homotopy theoretic fibre of $X \to X_n$, then

$$\pi_q(F_n) = \begin{cases} \pi_q(F) & \text{if } q \leq n \\ 0 & \text{if } q > n \end{cases}$$

and
$$\pi_q(G_n) = \begin{cases} \pi_q(F) & \text{if } q > n \\ 0 & \text{if } q \leq n. \end{cases}$$

In the diagram below, rows and columns are fibration sequences up to homotopy

$$\begin{array}{ccccc} G_n & \to & G_n & \to & * \\ \downarrow & & \downarrow & & \downarrow \\ F & \to & X & \to & Y \\ \downarrow & & \downarrow & & \downarrow \\ F_n & \to & X_n & \to & Y. \end{array}$$

Applying the Zabrodsky Lemma to a Moore–Postnikov factorization yields

Proposition 2.10.2. *Suppose $X \to Y$ is a map of simply connected spaces where the homotopy theoretic fibre F has all $\pi_*(F)$ abelian p-primary torsion and only finitely many $\pi_*(F)$ nonzero. Then $X \to Y$ is a local equivalence with respect to the map $B\mathbb{Z}/p\mathbb{Z} \to *$.*

Lemma 2.10.3. *Suppose $X \to Y$ is a mod p equivalence of simply connected spaces. Then $\mathrm{map}_*(B\mathbb{Z}/p\mathbb{Z}, X) \to \mathrm{map}_*(B\mathbb{Z}/p\mathbb{Z}, Y)$ is a weak equivalence.*

Proof: Let F be the homotopy theoretic fibre of $X \to Y$. Since $X \to Y$ is a mod p equivalence, F has trivial mod p homotopy and therefore $\pi_*(F)$ is a $\mathbb{Z}[\frac{1}{p}]$ module.

Let $X \to X_1 \to Y$ be the first stage of a Moore–Postnikov factorization with $X_1 \to Y$ being a principal bundle with fibre a $K(\pi_1(F), 1)$ and $X \to X_1$ being a map of simply connected spaces with a simply connected fibre G_1.

Note that $\pi_q \mathrm{map}_*(B\mathbb{Z}/p\mathbb{Z}, K(\pi_1(F), 1)) = H^1(\Sigma^q(B\mathbb{Z}/p\mathbb{Z}); \pi_1(F)) = 0$. This follows from $\overline{H}_*(B\mathbb{Z}/p\mathbb{Z}) = 0$ or $= \mathbb{Z}/p\mathbb{Z}$ and

$$\hom(\mathbb{Z}/p\mathbb{Z}, \pi_1(F)) = \mathrm{Ext}(\mathbb{Z}/p\mathbb{Z}, \pi_1(F)) = 0$$

since $\pi_1(F))$ is a $\mathbb{Z}[\frac{1}{p}]$ module. Thus $\mathrm{map}_*(B\mathbb{Z}/p\mathbb{Z}, K(\pi_1(F), 1))$ is weakly contractible and $\mathrm{map}_*(B\mathbb{Z}/p\mathbb{Z}, X_1) \to \mathrm{map}_*(B\mathbb{Z}/p\mathbb{Z}, Y)$ is a weak equivalence.

On the other hand, G_1 is local away from p and $B\mathbb{Z}/p\mathbb{Z} \to *$ is a local equivalence away from p. Hence, $\mathrm{map}_*(B\mathbb{Z}/p\mathbb{Z}, G_1)$ is weakly contractible and $\mathrm{map}_*(B\mathbb{Z}/p\mathbb{Z}, X) \to \mathrm{map}_*(B\mathbb{Z}/p\mathbb{Z}, X_1)$ is a weak equivalence. □

Hence, if a simply connected space is local with respect to $B\mathbb{Z}/p\mathbb{Z} \to *$, then the same is true for any simply connected space which is mod p equivalent to it.

We come to the main result of this section.

Proposition 2.10.4. *Let X be a simply connected space which is local with respect to the map $f : B\mathbb{Z}/p\mathbb{Z} \to *$ and assume that $\pi_2(X)$ is a torsion group. Up*

2.10 Killing Eilenberg–MacLane spaces: Miller's theorem

to homotopy, there is a pullback diagram

$$\begin{array}{ccc} L_{B\mathbb{Z}/p\mathbb{Z}}(X\langle n\rangle) & \to & (X\langle n\rangle)[\tfrac{1}{p}] \\ \downarrow & & \downarrow \\ X & \to & X[\tfrac{1}{p}]. \end{array}$$

Proof: In general, let Y_τ denote the fibre of the localization away from p, that is, $Y_\tau \to Y \to Y[\tfrac{1}{p}]$ is a fibration sequence. The homotopy groups $\pi_*(Y_\tau)$ are always p-primary torsion.

Consider the following diagram in which the row and columns and the southeast arrows are fibration sequences and E is the pullback of X and $X\langle n\rangle[\tfrac{1}{p}]$ over $X[\tfrac{1}{p}]$.

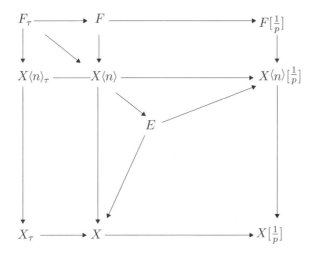

The hypothesis on $\pi_2(X) \cong \pi_1(F)$ guarantees that F_τ is connected. Thus $X\langle n\rangle \to E$ is a map of simply connected spaces with fibre F_τ having finitely many nonzero homotopy groups all of which are p-primary torsion. It follows that $X\langle n\rangle \to E$ is a local equivalence with respect to the map $f : B\mathbb{Z}/p\mathbb{Z} \to *$.

Since $F[\tfrac{1}{p}] \to E \to X$ is a fibration, $E \to X$ is a mod p equivalence and the fact that X is local with respect to $B\mathbb{Z}/p\mathbb{Z} \to *$ shows that E is also. Hence, $E \simeq L_{B\mathbb{Z}/p\mathbb{Z}}(X\langle n\rangle)$. □

The form of the proposition above was suggested by Jesper Møller. The original form of the proposition [103] was the corollary we are about to state. Let $\hat{L}_{B\mathbb{Z}/p\mathbb{Z}}(X)$ denote the p-completion of $L_{B\mathbb{Z}/p\mathbb{Z}}(X)$. Then:

Corollary 2.10.5. *Let X be a simply connected space which is local with respect to the map $f : B\mathbb{Z}/p\mathbb{Z} \to *$ and assume that $\pi_2(X)$ is a torsion group. Then we have $\hat{L}_{B\mathbb{Z}/p\mathbb{Z}}(X\langle n\rangle) \simeq \hat{X}_p$.*

These results derive their interest from the following deep result of Haynes Miller [84] which we will not prove here.

Miller's theorem 2.10.6. *If X is a finite complex (or any loop space thereof), then $\mathrm{map}_*(B\mathbb{Z}/p\mathbb{Z}, X)$ is weakly contractible, that is, X is local with respect to $B\mathbb{Z}/p\mathbb{Z} \to *$.*

Joseph Roitberg has suggested that a simply connected space with π_2 torsion be called $1\frac{1}{2}$-connected. Thus, up to p-completion, a $1\frac{1}{2}$-connected finite complex can be recovered from any n-connected cover. The $1\frac{1}{2}$-connected hypothesis is necessary since $S^2\langle 2\rangle = S^3$. For example, a corollary is [117]:

Serre's theorem 2.10.7. *If X is a simply connected finite complex with nontrivial reduced mod p homology, then $\pi_n(X) \otimes \mathbb{Z}/p\mathbb{Z} \neq 0$ for infinitely many n.*

Proof: First of all, recall that the homotopy groups of a simply connected finite complex X are finitely generated. If $\pi_2(X)$ has a nontrivial free summand, we can form a pullback bundle sequence

$$S^1 \times S^1 \times \cdots \times S^1 \to Y \to X$$

via a map

$$X \to CP^\infty \times CP^\infty \times \cdots \times CP^\infty$$

which is a isomorphism on the torsion free summand of π_2. We can replace X by Y and assume that X is a simply connected finite complex with $\pi_2(X)$ torsion.

Finally we can replace X by its p-completion. If $\pi_q(X) \otimes \mathbb{Z}/p\mathbb{Z} = 0$ for all sufficient large q, then $X\langle n\rangle \simeq *$ for a large n and thus

$$* \simeq \hat{L}_{B\mathbb{Z}/p\mathbb{Z}}(X\langle n\rangle) \simeq \hat{X}_p \simeq X,$$

a contradiction. $\square$

In fact there is a simpler argument which uses Miller's theorem to prove Serre's theorem but does not use the existence of localization. This argument will now be given. It is due to Charles McGibbon and the author and also to Alexander Zabrodsky independently. It also serves as an introduction to the method used by McGibbon and the author [83] to demonstrate the validity of Serre's conjecture [117, 83] that a nontrivial simply connected complex has infinitely many homotopy groups with nonvanishing torsion.

We will drop the language of localization since it is really irrelevant for this argument. It is instructive to translate facts known in the language of localization into simpler terms. Suppose X is a simply connected finite complex. Then Miller's theorem says that $\mathrm{map}_*(B\mathbb{Z}/p\mathbb{Z}, X)$ is weakly contractible. Hence, $\mathrm{map}_*(B\mathbb{Z}/p\mathbb{Z}, \hat{X}_p)$ is also weakly contractible. We have shown above that this implies that $\mathrm{map}_*(K(G, n), \hat{X}_p)$ is weakly contractible for any abelian p-group

2.10 Killing Eilenberg–MacLane spaces: Miller's theorem

G and all $n \geq 1$. The short exact sequence

$$0 \to \mathbb{Z} \to \mathbb{Z}\left[\frac{1}{p}\right] \to \mathbb{Z}(p^\infty) \to 0$$

shows that we have mod p equivalences

$$K(\mathbb{Z}(p^\infty), n) \to K(\mathbb{Z}, n+1)$$

for all $n \geq 1$. Thus $\mathrm{map}_*(K(\mathbb{Z}, n+1), \hat{X}_p)$ is weakly contractible for all $n \geq 1$. Therefore, $\mathrm{map}_*(K(G, n), \hat{X}_p)$ is weakly contractible for all $n \geq 2$ and all finitely generated abelian groups G.

Suppose now that X is a simply connected finite complex for which there is some n such that $\pi_n(X) \otimes \mathbb{Z}/p\mathbb{Z} \neq 0$ but $\pi_k(X) \otimes \mathbb{Z}/p\mathbb{Z} = 0$ for all $k > n$. Since $\pi_*(X)$ is finitely generated in every degree, we have the isomorphism

$$\pi_n(\hat{X}_p) \cong \hat{\pi}_n(X)_p \cong \varprojlim_r \pi_n(X)/p^r \pi_n(X).$$

Let $Y = (\Omega^{n-2}\hat{X}_p)_o$ be the component of the basepoint in the $n-2$-fold loop space of the p-completion. Then the universal cover of Y is $\tilde{Y} = K(\hat{\pi}_n(X)_p, 2)$. The natural map $K(\pi_n(X)_p, 2) \to \tilde{Y} \to Y \to \Omega^{n-2}\hat{X}_p$ is non-trivial and thus so is the adjoint $\Sigma^{n-2} K(\pi_n(X)_p, 2) \to \hat{X}_p$. This means that $\pi_{n-2}(\mathrm{map}_*(K(\pi_n(X), 2), X)) \neq 0$, so this is a contradiction. □

Serre's conjecture 2.10.8. *If X is a simply connected finite complex with non-trivial reduced mod p homology, then the p-torsion subgroup of $\pi_n(X)$ is nonzero for infinitely many n.*

Proof: We begin by recalling a well known fact concerning the rationalization $X \otimes \mathbb{Q}$ of a simply connected H-space X. □

Lemma 2.10.9. *If X is a simply connected H-space, then the rationalization $X \otimes \mathbb{Q}$ has the homotopy type of a product of Eilenberg–MacLane spaces, that is, there is a homotopy equivalence*

$$X \otimes \mathbb{Q} \simeq \prod_{n \geq 2} K(\pi_n, n).$$

This result is due to Cartan and Serre who in fact proved it in the following more convenient form: if X is a simply connected H-space, then all the rational k-invariants are zero, for all i, $k_i \otimes \mathbb{Q} = 0$. (See [90].)

Recall the theory of Postnikov systems and k-invariants. If X is a simply connected space, then X can be written up to homotopy as an inverse limit of fibrations, that is,

$$X = \varprojlim X_n$$

where:

(1) for all $n \geq 2$, there are maps $X \xrightarrow{q_n} X_n$ and fibrations $X_{n+1} \xrightarrow{p_{n+1}} X_n$ such that $q_n = p_{n+1} \cdot q_{n+1}$. These maps induce the equivalence $X = \lim_{\leftarrow} X_n$.

(2) $X_2 = K(\pi_2, 2)$ and, for all $n \geq 2$, there are fibration sequences up to homotopy

$$K(\pi_{n+1}, n+1) \to X_{n+1} \xrightarrow{p_{n+1}} X_n \xrightarrow{k_{n+1}} K(\pi_{n+1}, n+2).$$

(3)
$$\pi_i X_n = \begin{cases} 0, & i > n \\ \pi_i X, & i \leq n. \end{cases}$$

Hence, $k_i \in H^{i+1}(X_{i-1}, \pi_i)$ and we have that $0 = k_i$ implies that $X_i \simeq X_{i-1} \times K(\pi_i, i)$.

In fact, the lifting problem of the existence of f:

$$\begin{array}{ccc} X & \to & X_2 \\ \exists f \nwarrow & & \uparrow f_2 \\ & Y & \end{array}$$

is equivalent to a succession of lifting problems, namely, the existence of the f_i:

$$\begin{array}{ccc} X_i & \to & X_{i-1} \\ \exists f_i \nwarrow & & \uparrow f_{i-1} \\ & Y & \end{array}$$

and this in turn is equivalent to $k_i \cdot f_{i-1} = 0 \in H^{i+1}(Y, \pi_i)$.

It is convenient to state the next lemma in the form in which X and Y are localized at a prime p.

Lemma 2.10.11. *If X and Y are localized at p, then the lift f exists if $H_*(Y; \mathbb{Z}_{(p)})$ is free over $\mathbb{Z}_{(p)}$, $\pi_*(X)$ is torsion free, and X is an H-space.*

The lemma follows from the fact that the obstructions are $k_i \cdot f_{i-1} \in H^{i+1}(Y, \pi_i)$. We see that they are zero by noting that:

(a) $H_*(Y, \mathbb{Z}_{(p)})$ free implies that $H^{i+1}(Y, \pi_i) = \hom(H_{i+1}Y, \pi_i)$. Of course, $H^{i+1}(Y, \pi_i \otimes Q) = \hom(H_{i+1}Y, \pi_i \otimes Q)$.

(b) π_i torsion free implies that $\pi_i \to \pi_i \otimes Q$ and $\hom(H_{i+1}Y, \pi_i) \to \hom(H_{i+1}Y, \pi_i \otimes Q)$ are monomorphisms.

Via the monomorphisms in (a) and (b), Lemma 2.10.9 implies that $k_i \cdot f_{i-1}$ maps to $0 = (k_i \otimes Q) \cdot f_{i-1}$. Hence, the obstructions are 0 and the lift exists.

We can now conclude the proof of Serre's conjecture.

2.10 Killing Eilenberg–MacLane spaces: Miller's theorem

Let X be the localization at p of a simply connected finite complex with $\overline{H}_*(X; \mathbb{Z}/p\mathbb{Z}) \neq 0$ and suppose that the p-torsion of π_*X vanishes for $*$ sufficiently large. We shall derive a contradiction.

We already know that there are larger and larger homotopy groups for which π_*X does not vanish. Hence, there exists an $n \geq 3$ such that $\pi_n X$ has a split $\mathbb{Z}_{(p)}$ summand and $\pi_i X$ has no p-torsion for $i \geq n$.

Let $W = $ the universal cover of the basepoint component of the iterated loop space $\Omega^{n-2}X$. Then W is simply connected, $\pi_2 W$ contains a split $\mathbb{Z}_{(p)}$ summand, and $[B\mathbb{Z}/p\mathbb{Z}, W]_* = $ the subset of $[B\mathbb{Z}/p\mathbb{Z}, \Omega^{n-2}X]_*$ of homotopy classes which induce $\mathbb{Z}/p\mathbb{Z} \to 0$ on π_1. Recall that $[B\mathbb{Z}/p\mathbb{Z}, \Omega^{n-2}X]_* = [\Sigma^{n-2}B\mathbb{Z}/p\mathbb{Z}, X]_* = 0$. Thus $[B\mathbb{Z}/p\mathbb{Z}, W]_* = 0$.

But the splitting $\mathbb{Z}_{(p)} \to \pi_2 W \to \mathbb{Z}_{(p)}$ and the nontrivial two-dimensional cohomology class $B\mathbb{Z}/p\mathbb{Z} \to K(\mathbb{Z}_{(p)}, 2)$ show that this is a contradiction via

$$\begin{array}{ccccc} W & \to & W_2 = K(\pi_2, 2) & \to & K(\mathbb{Z}_{(p)}, 2) \\ \exists \searrow & & \uparrow & \nearrow = & \\ & & K(\mathbb{Z}_{(p)}, 2) & & \\ & & \uparrow & & \\ & & B\mathbb{Z}/p\mathbb{Z} & & \end{array}$$

Exercises

(1) Prove that, no matter how large k is, the p^k-th power map

$$p^k : \Omega^{2n-2}(S^{2n+1}\langle 2n+1 \rangle_{(p)}) \to \Omega^{2n-2}(S^{2n+1}\langle 2n+1 \rangle_{(p)})$$

on this many loopings of the localized connected cover of an odd dimensional sphere is not null homotopic.

(2) Let g be the map $M = \vee_p B\mathbb{Z}/p\mathbb{Z} \to *$ with the bouquet M being taken over all primes p. Let X be a simply connected space which is local with respect to the map $g: M \to *$ and assume that $\pi_2(X)$ is a torsion group. Up to homotopy, there is a pullback diagram

$$\begin{array}{ccc} L_M(X\langle n \rangle) & \to & (X\langle n \rangle) \otimes \mathbb{Q} \\ \downarrow & & \downarrow \\ X & \to & X \otimes \mathbb{Q}. \end{array}$$

(3) Suppose $X \to Y$ is a mod p equivalence of simply connected spaces. Show that X is local with respect to $B\mathbb{Z}/p\mathbb{Z} \to *$ if and only if Y is.

(4) Let G be a locally finite abelian p-group, that is, G is an abelian group which is a union of its finite p-subgroups. Assume X is a simply connected space which is local with respect to $B\mathbb{Z}/p\mathbb{Z} \to *$.

82   A general theory of localization

   (a) Show that X is local with respect to $BG \to *$.
   (b) Show that X is local with respect to $K(G,n) \to *$ for all $n \geq 1$.
   (c) Show that, if X is also p-complete, then X is local with respect to $K(\mathbb{Z},n) \to *$ and $K(\hat{\mathbb{Z}}_p, 2) \to *$ for all $n \geq 2$.

(5) This problem is an alternate approach to a localization theorem and was suggested by Jesper Grodal.

   Consider localization L of simply connected spaces X with respect to the map $B\mathbb{Z}/p\mathbb{Z} \vee M(\mathbb{Z}[\frac{1}{p}], 1) \to *$.

   (a) Show that X is local if and only if X is p-complete and $\mathrm{map}_*(B\mathbb{Z}/p\mathbb{Z}, X)$ is weakly contractible, that is, X is Miller local.
   (b) Show that any mod p homology isomorphism $f : A \to B$ is a local equivalence.
   (c) Show that, if X is Miller local, then $\mathrm{map}_*(BG, X)$ is weakly contractible for all locally finite p-groups G.
   (d) Show that, if X is local and G is torsion free abelian, then $\mathrm{map}_*(K(G,2), X)$ is weakly contractible. (Hint: Use the fact that
   $$0 \to G \to \mathbb{Z}\left[\frac{1}{p}\right] \otimes G \to \mathbb{Z}(p^\infty) \otimes G \to 0$$
   is exact and that $B(\mathbb{Z}[\frac{1}{p}] \otimes G)$ has trivial mod p homology.)
   (e) Show that, if X is p-complete and H is an abelian torsion group with all torsion of order relatively prime to p, then $\mathrm{map}_*(BH, X)$ is weakly contractible.
   (f) Suppose that $\mathrm{map}_*(G, X)$ is weakly contractible. Show that $\mathrm{map}_*(BG, X)$ is weakly contractible.
   (g) Suppose that there is an exact sequence $1 \to H \to T \to G \to 1$ and that $\mathrm{map}_*(BG, X)$ and $\mathrm{map}_*(BH, X)$ are both weakly contractible. Show that both $\mathrm{map}_*(BT, X)$ and $\mathrm{map}_*(B(G \times H), X)$ are weakly contractible.
   (h) Show that, if X is local and G is abelian, then $\mathrm{map}_*(K(G,2), X)$ is weakly contractible.
   (i) Show that, if X is local and G is abelian, then $\mathrm{map}_*(K(G,n), X)$ is weakly contractible for all $n \geq 2$.
   (j) Show that, if X is local and G is torsion abelian, then $\mathrm{map}_*(K(G,n), X)$ is weakly contractible for all $n \geq 1$.

(k) Prove the localization theorem: Let L be localization with respect to inverting the map $B\mathbb{Z}p\mathbb{Z} \vee M(\mathbb{Z}[\frac{1}{p}], 1) \to *$. If Y is a simply connected space, $H_k(Y, \mathbb{Z}/p\mathbb{Z}) = 0$ for all sufficiently large k, and $\pi_2 Y$ is torsion, then $LY\langle n\rangle = \hat{Y}_p$ for all n. Hint: Show that the composition $Y\langle n\rangle \to Y \to \hat{Y}_p$ is localization.

(6) The original proof of Corollary 2.10.5 is based on localization in the category of simply connected p-complete spaces. In this category, we shall say that X is p-complete local with respect to the map $B\mathbb{Z}/p\mathbb{Z} \to *$ if X is simply connected, p-complete, and the evaluation at the basepoint $\mathrm{map}(B\mathbb{Z}/p\mathbb{Z}, X) \to X$ is a weak equivalence. A map $A \to B$ of simply connected p-complete spaces is called a p-complete local equivalence if $\mathrm{map}(B, X) \to \mathrm{map}(A, X)$ is a weak equivalence for all p-complete local X. A p-complete localization of X is a p-complete local equivalence $\iota : X \to LX$ such that LX is p-complete local.

(a) Show that LX exists, satisfies the appropriate universal mapping property, and is unique up to homotopy.

(b) If X is the p-completion of a finite type simply connected complex Y with $\pi_2(Y)$ torsion, show that the n-connected covering $X\langle n\rangle \to X$ is a p-complete local equivalence for all n.

(c) If the above Y is also a finite complex, show that the p-complete localization $L(X\langle n\rangle)$ is homotopy equivalent to X for all n.

2.11 Zabrodsky mixing: the Hilton–Roitberg examples

We begin by defining three families of Lie groups, the orthogonal groups $O(n)$, the unitary groups $U(n)$, and the symplectic groups $Sp(n)$.

Let F be any one of three fields, the real numbers R, the complex numbers C, or the quaternions H and let d be the dimension of F as a real vector space, that is, $d = 1, 2$ or 4. Regard F^n as an inner product space via

$$\langle (x_1, \ldots, x_n), (y_1, \ldots, y_n) \rangle = x_1 \overline{y_1} + \cdots + x_n \overline{y_n} = \sum_{i=1}^{n} x_i \overline{y_i}.$$

Now let $O(F^n) = $ the group of all F-linear transformations T on F_n which preserve the inner product, that is, $\langle T(x), T(y)\rangle = \langle x, y\rangle$ for all $x, y \varepsilon F^n$.

Then $O(R^n) = O(n)$, $O(C^n) = U(n)$, and $O(H^n) = Sp(n)$. Notice that there is an embedding $Sp(n) \subseteq U(2n)$.

If $e_1 = (1, 0, \ldots, 0), \ldots, e_n = (0, 0, \ldots, 1)$ is the standard basis of F^n over F, then the map $p_n : O(F^n) \to S^{nd-1}$ with $p_n(T) = T(e_n)$ defines a fibration

sequence
$$O(F^{n-1}) \to O(F^n) \xrightarrow{p_n} S^{nd-1}.$$

Note that $O(F^1) = S^{d-1}$.

Observe that, if $d = 1$ or $d = 2$, then the determinant defines a homomorphism to the group of units $det : O(F^n) \to F^*$ and we let $SO(F^n) =$ the kernel of this map. For $n \geq 2$ we have a fibration sequence

$$SO(F^{n-1}) \to SO(F^n) \to S^{nd-1}.$$

Note the following:

$$O(n) = SO(n) \times \{\pm 1\}$$

for all $n \geq 0$, and

$$O(1) = \{\pm 1\}, SO(2) = S^1.$$

It is easy to see that any element T of $SO(3)$ has 1 as an eigenvector and that the 1-eigenspace has dimension 1 or 3. Let v be an element of $S^2 \subseteq R^3$ and let $0 \leq \theta \leq \pi$ represent an angle of positive rotation in the plane perpendicular to v. This defines an element $T \varepsilon SO(3)$ and, if we regard (v, θ) as an element of the ball $D^3(\pi) \subseteq R^3$ of radius π, then the assignment $(v, \theta) \mapsto T$ defines a homeomorphism $RP^3 \to SO(3)$.

Note that

$$SU(n) \to U(n) \to S^1$$

has a section and

$$U(1) = S^1, SU(2) = S^3.$$

Note that

$$Sp(1) = S^3.$$

We denote by O, SO, U, SU, Sp the stable groups obtained as the limit when n goes to ∞. Note the embedding $Sp \subseteq U$.

Recall without proof [87]

2.11 Zabrodsky mixing: the Hilton–Roitberg examples

Bott periodicity 2.11.1. There are homotopy equivalences

$$U \to \Omega^2(U)$$
$$U/Sp \to \Omega^2(O)$$
$$Sp \to \Omega^2(U/Sp)$$
$$O \to \Omega^4(Sp)$$
$$Sp \to \Omega^4(O)$$

Thus the homotopy groups satisfy the relations

$$\pi_k(U) \cong \pi_{k+2}(U)$$
$$\pi_k(U/Sp) \cong \pi_{k+2}(O)$$
$$\pi_k(Sp) \cong \pi_{k+2}(U/Sp)$$
$$\pi_k(O) \cong \pi_{k+4}(Sp)$$
$$\pi_k(Sp) \cong \pi_{k+4}(O),$$

the last two generating a periodicity mod 8.

We have the following tables

k modulo 2	$\pi_k(U)$
0	0
1	$\mathbb{Z}$

k modulo 8	$\pi_k(O)$	$\pi_k(Sp)$
0	$\mathbb{Z}/2\mathbb{Z}$	0
1	$\mathbb{Z}/2\mathbb{Z}$	0
2	0	0
3	$\mathbb{Z}$	$\mathbb{Z}$
4	0	$\mathbb{Z}/2\mathbb{Z}$
5	0	$\mathbb{Z}/2\mathbb{Z}$
6	0	0
7	$\mathbb{Z}$	$\mathbb{Z}$

The first table follows from the above fibrations and the fact that $U(1) = S^1$. Except for $\pi_3(O) = \mathbb{Z}$ and thus $\pi_7(Sp) = \mathbb{Z}$, the second table follows from the above fibrations, the fact that $Sp(1) = S^3$, and the fact that the universal cover of RP^3 is S^3. To complete the second table, observe that $\pi_3(O) \cong \pi_1(U/Sp) \cong \mathbb{Z}$.

Consider the Lie group $Sp(2)$ and the S^3 bundle

$$S^3 \to Sp(2) \to S^7.$$

It is classified by a map $\omega : S^7 \to BS^3$, that is, $Sp(2)$ is the pullback of the universal S^3 bundle ES^3 via the pullback diagram

$$\begin{array}{ccc} Sp(2) & \to & ES^3 \\ \downarrow & & \downarrow \\ S^7 & \stackrel{\omega}{\to} & BS^3. \end{array}$$

In a subsequent section we shall prove that

$$\pi_7(BS^3) = \pi_6(S^3) = \mathbb{Z}/12\mathbb{Z}.$$

For now we note that the fact that $\pi_6(Sp(2)) = \pi_6(Sp) = 0$ implies that the classifying map ω is a generator of $\pi_7(BS^3)$.

If $\alpha : S^7 \to BS^3$ is any map, we shall denote by E_α the S^3 bundle over S^7 which is the pullback via α. These examples were first studied by Hilton and Roitberg [55, 53]. We shall apply the technique called Zabrodsky [142] mixing to prove their result:

Hilton–Roitberg 2.11.2. *If $\alpha = k\omega$ with $k = 0, \pm 1, \pm 3, \pm 4, \pm 5$, then E_α is an H-space.*

Remark. Zabrodsky has shown that these are the only maps α for which E_α is an H-space.

The technique called Zabrodsky mixing is essentially the following result.

Proposition 2.11.3. *Suppose S_1 and S_2 are complementary sets of primes and consider the homotopy pullback diagram*

$$\begin{array}{ccc} X & \stackrel{\alpha_1}{\to} & X_{S_1} \\ \downarrow \alpha_2 & & \downarrow \beta_1 \\ X_{S_2} & \stackrel{\beta_2}{\to} & X \otimes Q. \end{array}$$

Then X is an H-space if and only if all of $X_{S_1}, X_{S_2}, X \otimes Q$ are H-spaces and both of β_1, β_2 are H-maps.

Proof: The forward implication is true since localizations are functors which preserve products.

As for the reverse implication, we may assume that, in the diagram, the maps β_1, β_2 are fibrations. If

$$\mu_1 : X_{S_1} \times X_{S_1} \to X_{S_1}, \mu_2 : X_{S_2} \times X_{S_2} \to X_{S_2}, \mu : (X \otimes Q) \times (X \otimes Q) \to (X \otimes Q)$$

are the multiplications, then we can alter the multiplications μ_1, μ_2 so that β_1, β_2 become strict H-maps. Let X be the strict pullback via the maps β_1, β_2. Then the three multiplications μ_1, μ_2, μ define a multiplication $X \times X \to X$. $\square$

2.11 Zabrodsky mixing: the Hilton–Roitberg examples

Recall the following fundamental result [90].

Hopf-Borel 2.11.4. *If Y is an H-space with finite type homology, then $H^*(Y;Q)$ is an exterior algebra on odd degree generators tensored with a polynomial algebra on even degree generators.*

For Y as above, it follows that $Y \otimes Q$ has the homotopy type of a product of Eilenberg–MacLane spaces.

This leads to the following improvement in Proposition 2.11.3.

If $X \otimes Q$ has finite type, then, in the reverse implication of Proposition 2.11.2, the condition that β_1, β_2 be H-maps is automatic when $X \otimes Q$ is an H-space for which either of the the following two equivalent conditions hold:

(a) Any homotopy self equivalence of $X \otimes Q$ is an H-map.

(b) Any automorphism of the rational cohomology algebra $H_*(X;Q)$ is a map of Hopf algebras.

Clearly this is satisfied for the spaces $E_\alpha \otimes Q \simeq (S^3 \otimes Q) \times (S^7 \otimes Q)$.

The diagram

$$\begin{array}{ccccc} S^3 & \xrightarrow{=} & S^3 & \xrightarrow{=} & S^3 \\ \downarrow & & \downarrow & & \downarrow \\ E_{k\alpha} & \to & E_\alpha & \to & ES^3 \\ \downarrow & & \downarrow & & \downarrow \\ S^7 & \xrightarrow{k} & S^7 & \xrightarrow{\alpha} & BS^3 \end{array}$$

shows that: if the degree k map is a local self equivalence of S^7, then $E_{k\alpha} \to E_\alpha$ is a local equivalence.

Hence, if E_α is a local H-space, then so is $E_{k\alpha}$. In particular, if E_α is an H-space, then so is $E_{-\alpha}$.

The Hilton–Roitberg examples arise from mixing the H-space structures of $Sp(2)$ and $S^3 \times S^7$ at localizations of complementary sets of primes.

For example, since $E_\omega = Sp(2)$ is certainly an H-space and the degree 5 map is a local equivalence away from 5, we have that $E_{5\omega}$ is an H-space localized away from 5. Localized at 5, ω and 5ω are trivial. Hence, $(E_{5\omega})_{(5)} - S^3_{(5)} \times S^7_{(5)}$ is an H-space. Therefore, $E_{5\omega}$ is an H-space.

Localized away from 2, 4 is a unit and hence $E_{4\omega}$ is an H-space. Localized at 2, $4\omega = 0$ and $E_{4\omega}$ is equivalent to the product $S^3 \times S^7$ and hence is an H-space. Therefore, $E_{4\omega}$ is an H-space.

Similarly, $E_{3\omega}$ is an H-space since 3 is a unit localized away from 3 and $3\omega = 0$ localized at 3.

This completes the proof of the existence of the Hilton–Roitberg examples.

Exercises

(1) Show that the unit tangent sphere bundle sequence
$$S^3 = SO(4)/SO(3) \to SO(5)/SO(3) \to S^4$$
has no section by considering the degree of the antipodal map on S^4.

(2) Show that the connecting homomorphism for the fibration in Exercise 1,
$$\partial : \pi_4(S^4) \to \pi_3(S^3),$$
is nonzero. Conclude that $\pi_3(SO(5)/SO(3))$ and $\pi_4(SO(5)/SO(3))$ are finite.

(3) Even though there is a fibration sequence $S^3 \to SU(3) \to S^5$, the Zabrodsky mixing technique yields no new H-spaces when applied to $SU(3)$. Explain.

2.12 Loop structures on p-completions of spheres

One of the first applications of p-completions was in showing that certain p-completions of spheres were loop spaces. This was a precursor to the work of Dwyer–Wilkerson [38] which has resulted in the successful extension of much of the theory of Lie groups to the homotopy theoretic setting of p-compact groups. But in the beginning, we had the following result due to Sullivan [128, 129]:

The Sullivan examples 2.12.1. Let p be an odd prime and let d be a positive divisor of $p - 1$. Then there exists a space B_d such that the p-completion $\hat{S}_p^{2d-1}$ has the homotopy type of the loop space $\Omega(B_d)$. Furthermore, these are the only spheres for which this is true.

Proof: If p is an odd prime, then the group of units of the p-adic integers $\hat{\mathbb{Z}}_p$ is isomorphic to the cyclic group $\mathbb{Z}/(p-1)\mathbb{Z}$. If π is the subgroup of order d, then π acts as a group of homeomorphisms on the Eilenberg–MacLane space $K(\hat{\mathbb{Z}}_p, 2)$. In general, this is a larger group of automorphisms than the group of units $\{\pm 1\}$ which acts on $K(\mathbb{Z}, 2)$.

We use a standard construction to replace $K(\hat{\mathbb{Z}}_p, 2)$ by a space of the same homotopy type on which π acts freely and properly discontinuously. The group π acts freely and properly discontinuously on the contractible universal bundle $E\pi$ and hence freely and properly discontinuously on the product $Y = E\pi \times K(\hat{\mathbb{Z}}_p, 2)$ via $g(e, x) = (eg^{-1}, gx)$ for $g \varepsilon \pi, (e, x) \varepsilon Y$.

Form the orbit space Y/π under the action of π on Y. Since Y is simply connected, the map $Y \to Y/\pi = E\pi \times_\pi K(\hat{\mathbb{Z}}_p, 2)$ is a covering space with fundamental

2.12 Loop structures on p-completions of spheres

group $\pi_1(Y/\pi) = \pi$. The orbit space $Y/\pi = E\pi \times_\pi K(\hat{\mathbb{Z}}_p, 2)$ is called the Borel construction.

We claim that the mod p cohomology of the orbit space can be identified with the fixed points in the mod p cohomology of $K(\hat{\mathbb{Z}}_p, 2)$ under the action of π.

The projection $Y \to E\pi$ induces a fibration sequence

$$K(\hat{\mathbb{Z}}_p, 2) \to E\pi \times_\pi K(\hat{\mathbb{Z}}_p, 2) \to B\pi.$$

We need to describe the mod p cohomology Serre spectral sequence of this fibration.

The mod p cohomology $H^s(K(\hat{\mathbb{Z}}_p, 2); \mathbb{Z}/p\mathbb{Z}) = P^s[u]$ where $P[u] = P^*[u]$ denotes a polynomial algebra generated by an element u of degree 2. The action of π on $P[u]$ is given as follows. If λ is a generator of π, then $\lambda : K(\hat{\mathbb{Z}}_p, 2) \to K(\hat{\mathbb{Z}}_p, 2)$ induces the algebra isomorphism $\lambda^* : P[u] \to P[u]$ determined by $\lambda^*(u) = \overline{\lambda} u$ where $\overline{\lambda}$ is the image of λ in $\mathbb{Z}/p\mathbb{Z}$. Thus, $\lambda^*(u^k) = \overline{\lambda}^k u^k$.

The mod p cohomology Serre spectral sequence has

$$E_2^{r,s} = H^r(B\pi; H^s(K(\hat{\mathbb{Z}}_p, 2))).$$

The local coefficients identify this with

$$E_2^{r,s} = \mathrm{Ext}^r_{\mathbb{Z}/p\mathbb{Z}[\pi]}(\mathbb{Z}/p\mathbb{Z}, H^s(K(\hat{\mathbb{Z}}_p, 2))) = H^*(\pi; P[u]).$$

It converges to the mod p cohomology

$$H^{r+s}(E\pi \times_\pi K(\hat{\mathbb{Z}}_p, 2); \mathbb{Z}/p\mathbb{Z}) = H^{r+s}(Y/\pi).$$

If $\Delta = 1 - \lambda$ and $N = 1 + \lambda + \lambda^2 + \cdots + \lambda^{d-1}$ in the group algebra $\mathbb{Z}/p\mathbb{Z}[\pi]]$, then, since d is relatively prime to p, it is easily verified that there is a free $\mathbb{Z}/p\mathbb{Z}[\pi]$ resolution

$$0 \leftarrow \mathbb{Z}/p\mathbb{Z} \xleftarrow{\epsilon} \mathbb{Z}/p\mathbb{Z}[\pi] \xleftarrow{\Delta} \mathbb{Z}/p\mathbb{Z}[\pi] \xleftarrow{N} \mathbb{Z}/p\mathbb{Z}[\pi] \xleftarrow{\Delta} \mathbb{Z}/p\mathbb{Z}[\pi] \xleftarrow{N} \cdots.$$

It follows that $E_2^{*,*}$ is the cohomology of the complex obtained by applying the functor $\mathrm{Hom}_{\mathbb{Z}/p\mathbb{Z}[\pi]}(\ , P[u])$, that is,

$$0 \to P[u] \xrightarrow{\Delta^*} P[u] \xrightarrow{N^*} P[u] \xrightarrow{\Delta^*} P[u] \xrightarrow{N^*} \cdots.$$

In other words,

$$E_2^{r,*} = \begin{cases} \mathrm{kernel} \ \Delta^* : P^*[u] \to P^*[u] & \text{if } r = 0 \\ 0 & \text{if } r > 0. \end{cases}$$

Hence the spectral sequence collapses at E_2 and

$$H^*(E\pi \times_\pi K(\hat{\mathbb{Z}}_p, 2); \mathbb{Z}/p\mathbb{Z}) = H^*(Y/\pi; \mathbb{Z}/p\mathbb{Z}) = \text{kernel}(\Delta^*)$$
$$= P[u]^\pi = P[u^d] =$$

a polynomial algebra generated by an element u^d of degree $2d$.

The space B_d that we seek is the p-completion of the result of killing the fundamental group in the orbit space Y/π. More precisely, let $M(\pi, 1)$ be a Moore space with one nonzero integral homology group isomorphic to π in dimension 1. The space $B_d = \hat{X}_p$ is the p-completion of the mapping cone X of a map $M(\pi, 1) \to Y/\pi$ which is an isomorphism on fundamental groups. We see that X is simply connected and has the same mod p cohomology ring as Y/π. The mod p cohomology and mod p homology of B_d both vanish up to and including dimension $2d - 1$. Since B_d is p-complete, it is in fact connected up to dimension $2d - 1$. Excluding the trivial case when $d = 1$, we see that the loop space $\Omega(B_d)$ is simply connected.

The Serre spectral sequence of the path space fibration shows that

$$H^*(\Omega(B_d); \mathbb{Z}/p\mathbb{Z}) = E[v] =$$

an exterior algebra where v is a generator of degree $2d - 1$. Hence, the mod p homology $H_*(\Omega(B_d); \mathbb{Z}/p\mathbb{Z}) = E[v^*]$ is an exterior algebra on the dual generator v^*. Since the Hurewicz map gives an isomorphism $\pi_{2d-1}(\Omega(B_d); \mathbb{Z}/p\mathbb{Z}) \to H_{2d-1}(\Omega(B_d); \mathbb{Z}/p\mathbb{Z})$ and since $\mathbb{Z}/p\mathbb{Z} \cong \pi_{2d-1}(\Omega(B_d); \mathbb{Z}/p\mathbb{Z}) \cong \pi_{2d-1}(\Omega(B_d)) \otimes \mathbb{Z}/p\mathbb{Z}$ we can pick a map $S^{2d-1} \to \Omega(B_d)$ which induces an equivalence in mod p homology. Since $\Omega(B_d)$ is p-complete, the map $\hat{S}_p^{2d-1} \to \Omega(B_d)$ is a homotopy equivalence.

Let B_d be any classifying space for the p-completion $\hat{S}_p^{2d-1}$, that is, $\Omega(B_d) \simeq \hat{S}_p^{2d-1}$. The mod p cohomology Serre spectral sequence shows that the mod p cohomology $H^*(B_d)$ must be a polynomial algebra on a generator x of degree $2d$. Thus, this space has torsion free cohomology, in particular, the degree one Bockstein differential β is zero. Now we recall without proof a strong form of the nonexistence of mod p Hopf invariant one. It is a consequence of the Adem relations and of the result of Liulevicius [76] and Shimada–Yamanoshita [121] which decomposes Steenrod operations via secondary cohomology operations:

Liulevicius–Shimada–Yamanoshita vanishing theorem 2.12.2. *Suppose p is an odd prime. If X is a space such that the degree one Bockstein β and the first Steenrod operation P^1 of degree $2p - 2$ both vanish in the mod p cohomology of X, then all Steenrod operations vanish.*

In particular, since $P^d(x) = x^p \neq 0$, it must be the case that $P^1(x) \neq 0$ which is only possible if the degree $2d$ of x divides $2p - 2$. □

2.13 Serre's C-theory and finite generation

In this section we recall Serre's theory of classes of abelian groups. We do this in order to prove the basic theorem that, for simply connected spaces, homology groups being finitely generated in each dimension is equivalent to homotopy groups being finitely generated in each dimension. The basic definition is motivated by its compatibility with the Serre spectral sequence for the homology of an (orientable) fibration.

Definition 2.13.1. A class of abelian groups C is called a Serre class if: C contains 0, C is closed under isomorphism, subgroups, quotients, and extensions. Thus

(a) if A is in C, then any group B isomorphic to A is in C.

(b) if A is in C and $B \subseteq A$ is a subgroup, then B is in C.

(c) if A is in C and $A \to B$ is an epimorphism, then B is in C.

(d) if $0 \to A \to B \to C \to 0$ is a short exact sequence of abelian groups and A and C are in **C**, then B is in **C**.

If A is an abelian group in a Serre class C, then any subquotient B/C of A is in C. Furthermore, if A is an abelian group with a finite filtration

$$0 = F_0 \subseteq F_1 \subseteq \cdots \subseteq F_n = A$$

then A is in C if and only if all filtration quotients F_k/F_{k-1} are in C.

Definition 2.13.2. A Serre class C is called a Serre ring if A and B in C implies that $A \otimes B$ and $\text{Tor}(A, B)$ are in C. It is called a Serre ideal if this is true if only one of A and B are required to be in C.

Definition 2.13.3. A Serre class C is called acyclic if A in C implies that all the reduced integral homology groups $\overline{H}_*(K(A, 1))$ are in C.

For example, the following are acyclic Serre classes. The class of all finitely generated abelian groups is an acyclic Serre ring. The class of all torsion abelian groups is an acyclic Serre ideal. The class of all p-primary torsion abelian groups is also an acyclic Serre ideal. The main result on Serre classes is the following proposition.

Proposition 2.13.4. *Let* C *be a Serre ring. Suppose* $F \to E \to B$ *is an orientable fibration sequence. Consider the reduced integral homology groups*

$$\overline{H}_*(F), \overline{H}_*(E), \overline{H}_*(B).$$

If two out of three of them are in C then so is the third.

Proof: First of all recall the universal coefficient exact sequence
$$0 \to H_*(B) \otimes H_*(F) \to H_*(B; H_*(F)) \to \text{Tor}(H_*(B), H_*(F)) \to 0$$
and the fact that in the Serre spectral sequence $E_{p,q}^2 = H_p(B; H_q(F))$.

Suppose $\overline{H_*}(B)$ and $\overline{H_*}(F)$ are in C. Then $E_{p,q}^2$ is in C for all $(p,q) \neq (0,0)$. Hence, $E_{p,q}^r$ is in C for all $(p,q) \neq (0,0)$ and for all $2 \leq r \leq \infty$. Since $E_{p,q}^\infty$ are the filtration quotients of $\overline{H_{p+q}}(E)$, it follows that $\overline{H_*}(E)$ is in C.

Now suppose that $\overline{H_*}(E)$ and $\overline{H_*}(B)$ are in C. We may assume that $\overline{H_*}(F)$ is in C for $* \leq q-1$. We know that $E_{0,q}^\infty$ is in C. The edge $E_{0,q}^r$ is hit by differentials coming from groups in C. Hence, all $E_{0,q}^r$ must be in C. In particular, $E_{0,q}^2 = H_q(F)$ is in C. By induction, $\overline{H_*}(F)$ is in C.

Suppose that $\overline{H_*}(E)$ and $\overline{H_*}(F)$ are in C. Now $E_{p,0}^\infty$ is in C and the edge $E_{p,0}^r$ is the source of differentials going to groups in C. A similar induction shows that $E_{p,0}^2 = H_p(B)$ is in C for all $p > 0$. □

The proposition gives two immediate corollaries.

Corollary 2.13.5. *Let C be a Serre ring. If X is a simply connected space, then $\overline{H_*}(X)$ is in C if and only if $\overline{H_*}(\Omega X)$ is in C.*

Corollary 2.13.6. *Let C be an acyclic Serre ring. Then an abelian group π is in C if and only $\overline{H_*}(K(\pi, n))$ is in C for all (or any) $n \geq 1$.*

Finally, we have the result below which applies to at least three Serre classes, namely, finitely generated abelian groups, torsion abelian groups, and p-primary torsion abelian groups:

Proposition 2.13.7. *Let C be an acyclic Serre ring and suppose X is a simply connected space. Then the homotopy groups $\pi_*(X)$ are in C if and only if the reduced integral homology groups $\overline{H_*}(X)$ are in C.*

Proof: To see that homotopy in C implies reduced homology in C consider the Postnikov system X_n with
$$X = \varprojlim X_n$$
and use induction on the fibration sequences
$$X_n \to X_{n-1} \to K(\pi_n(X), n+1).$$

To see that reduced integral homology in C implies homotopy in C consider the connected covers $X\langle n \rangle$ with
$$X = X\langle 1 \rangle$$

and use induction on the fibration sequences
$$X\langle n+1\rangle \to X\langle n\rangle \to K(\pi_n(X), n).$$

□

Exercises

(1) Suppose C is a Serre ideal and $F \to E \to B$ is an orientable fibration sequence. Suppose $\overline{H}_*(F)$ is in C. Show that $H_*(E) \to H_*(B)$ is a Serre isomorphism, that is, the kernel and cokernel of this map are both in C.

(2) (a) Let A be a finitely generated module over a principal ideal domain. Suppose that $0 = \mathrm{Hom}(A, \mathbb{Z}) = \mathrm{Ext}(A, \mathbb{Z})$. Show that $A = 0$.

   (b) Suppose $f : X \to Y$ is a map between spaces both with homology finitely generated in each degree over a principal ideal domain. Show that f induces a homology equivalence if and only if f induces a cohomology equivalence.

3 Fibre extensions of squares and the Peterson–Stein formula

In this chapter we discuss fibre extensions of squares, a notion which is dual to the cofibration squares which appear in Lemma 1.5.1 and which are used in the construction of the Bockstein long exact sequences for homotopy groups with coefficients. A fibre extension is formed by starting with a commutative square of maps, replacing the maps by fibrations, and extending the square to a larger square where all the rows and columns are fibration sequences.

Fibre extensions of squares are used in the study of homotopy pullbacks, for example, to prove the fracture lemmas for localization. The main result here is that the homotopy theoretic fibre of a map of homotopy pullbacks is itself a homotopy pullback. For this purpose we need to study the higher dimensional fibre extensions of cubes.

The treatment we give of fibre extensions of squares has the advantage that it is efficient and self contained. It has the disadvantage that it does not embed it as part of a larger theory of homotopy inverse limits, as it could be if we were willing to develop that theory here. In this more general context, the result concerning fibre extensions of squares would be a trivial consequence of the commutativity of homotopy inverse limits over a product category.

Fibre extensions of squares have appeared in the work of Cohen–Moore–Neisendorfer and also in the work of Goodwillie on analytic functors. In the first case, the defining cofibration sequence of a Moore space

$$S^{2n} \xrightarrow{p^r} S^{2n} \to P^{2n+1} \to S^{2n+1} \xrightarrow{p^r} S^{2n+1}$$

leads to the square of maps

$$\begin{array}{ccc} P^{2n+1}(p^r) & \to & S^{2n+1} \\ \downarrow & & \downarrow p^r \\ * & \to & S^{2n+1} \end{array}$$

whose fibre extension is the main object of study in the proof of the odd primary exponent theorem for the homotopy groups of the odd dimensional spheres.

In the context of this chapter it is natural to discuss homotopy theoretic fibres and the Peterson–Stein formula [105, 50], the latter being referred to by H. Cartan as the compatibility with the connecting homomorphism. In the next chapter, we shall use the Peterson–Stein formula to prove that $\pi_6(S^3) = \mathbb{Z}/12\mathbb{Z}$.

Finally, we recall some results of J. H. C. Whitehead [135] and J. W. Milnor[89] which assert that the category of spaces with the homotopy type of CW complexes is closed under homotopy pushouts and homotopy pullbacks. In particular, loop spaces of CW complexes have the homotopy type of CW complexes.

3.1 Homotopy theoretic fibres

Recall that a cofibration is an inclusion $A \subseteq X$ which satisfies the homotopy extension property: for all spaces Y we can complete by a continuous map F all diagrams

$$\begin{array}{ccc} X \times I & \xrightarrow{F} & Y \\ \uparrow i & \nearrow & \\ X \times 0 \cup A \times I & & \end{array}.$$

Given the cofibrations, fibrations are defined to be continuous maps $p : E \to B$ with the homotopy lifting property: for all cofibrations $A \subseteq X$ we can complete by a continuous map G all diagrams

$$\begin{array}{ccc} A \times I \cup X \times 0 & \longrightarrow & E \\ \downarrow & \nearrow G & \downarrow p \\ X \times I & \longrightarrow & B. \end{array}$$

Given a continuous map $f : X \to Y$, there are several natural ways to replace f by a fibration. For example, let $E_f = \{(x, \omega)\varepsilon X \times Y^I | f(x) = \omega(1)\}$. Then $p : E_f \to Y$ with $p(x, \omega) = \omega(1)$ defines a fibration. Furthermore, if ω_c is the constant path, $\omega_c(t) = c$ for all $t\varepsilon I$, then $\iota : X \to E_f$ defined by $\iota(x) = (x, \omega_{fx})$ is a homotopy equivalence and we have a commutative diagram

$$\begin{array}{ccc} X & \xrightarrow{\iota} & E_f \\ & \searrow f & \downarrow p \\ & & Y \end{array}$$

Replacing $f : X \to Y$ by $p : E_f \to Y$ is what we shall mean by replacing a map by a fibration. Note that we have factored f into a composite $p \circ \iota : X \to E_f \to Y$ with p a fibration and ι both a cofibration and homotopy equivalence. The existence of such a factorization is one of Quillen's axioms for a model category.

Definition 3.1.1. The homotopy theoretic fibre of f is the fibre F_f of the map p, that is, $F_f = \{(x,\omega)\varepsilon X \times Y^I | f(x) = \omega(1), \omega(0) = y_0\}$ where y_0 is the basepoint of Y.

Exercises

(1) Suppose a continuous map $f: X \to Y$ is factored into $p \circ \iota: X \to E \to Y$ where p is a fibration and ι is both a cofibration and homotopy equivalence. If F is the fibre of p show that there is a homotopy equivalence $F_f \to F$. In other words, the homotopy theoretic fibre of a map is well defined up to homotopy equivalence.

(2) Let G be a not necessarily associative H-space and let $k: G \to G$, with $k(x) = x^k$, be any k-th power map.

   (a) If $G\{k\} = F_k$ is the homotopy theoretic fibre of k, show that there are homotopy equivalences
   $$\Omega(G\{k\}) \simeq (\Omega G)\{k\} \simeq \mathrm{map}_*(P^2(k), G).$$

   (b) Show that $\pi_{*+1}(G\{k\}) \cong \pi_{*+2}(G; \mathbb{Z}/k\mathbb{Z})$.

3.2 Fibre extensions of squares

Suppose we start with a homotopy commutative square

$$\begin{array}{ccc} X & \xrightarrow{g_2} & Y \\ \downarrow g_1 & & \downarrow f_2 \\ Z & \xrightarrow{f_1} & D. \end{array}$$

If we replace f_1 and f_2 by fibrations p_1 and p_2, we can alter g_1 by a homotopy to make the diagram strictly commutative:

$$\begin{array}{ccc} X & \xrightarrow{g_2} & B \\ \downarrow g_1 & & \downarrow p_2 \\ C & \xrightarrow{p_1} & D. \end{array}$$

Having done this, let

$$\begin{array}{ccc} E & \xrightarrow{h_2} & B \\ \downarrow h_1 & & \downarrow p_2 \\ C & \xrightarrow{p_1} & D. \end{array}$$

be the pullback diagram of B and C over D, with $E = \{(b,c)\,\varepsilon\,B \times C | p_2(b) = p_1(c)\}$. Then we have a map $g: X \to E$ which we can replace by a fibration

$p : A \to E$. We get a strictly commutative diagram

$$\begin{array}{ccc} A & \xrightarrow{g_2} & B \\ \downarrow g_1 & & \downarrow p_2 \\ C & \xrightarrow{p_1} & D \end{array}$$

with $g_1 = h_1 \circ p : A \to E \to C$, and $g_2 = h_2 \circ p : A \to E \to B$. Now all seven maps $p, h_1, h_2, g_1, g_2, p_1, p_2$ are fibrations.

Definition 3.2.1. A strictly commutative square in which all of the above seven maps are fibrations (including the map from the upper right corner to the pullback) is called a totally fibred square.

We have just shown the following proposition.

Proposition 3.2.2. *Every homotopy commutative square is homotopy equivalent to a totally fibred square. Even better, if the square is strictly commutative to start with, then there is a map of this square to a totally fibred square which is a homotopy equivalence on each vertex.*

Suppose we have a totally fibred square

$$\begin{array}{ccc} A & \to & B \\ \downarrow & & \downarrow \\ C & \to & D \end{array}$$

with pullback E of B and C over D. The fibre extension of this square is the larger commutative diagram

$$\begin{array}{ccccc} H & \to & F_2 & \to & F_1 \\ \downarrow & & \downarrow & & \downarrow \\ G_2 & \to & A & \to & B \\ \downarrow & & \downarrow & & \downarrow \\ G_1 & \to & C & \to & D \end{array}$$

where

$$\begin{array}{ccccc} F_1 & \to & B & \to & D, \\ F_2 & \to & A & \to & C, \\ G_1 & \to & C & \to & D, \\ G_2 & \to & A & \to & B \end{array}$$

are all fibration sequences and

$$H \to A \to E$$

is also a fibration sequence. The main result about totally fibred squares is

Proposition 3.2.3. $H \to F_2 \to F_1$ *and* $H \to G_2 \to G_1$ *are both fibration sequences.*

Proof: We need to check that $F_2 \to F_1$ is a fibration with fibre H. The case $G_2 \to G_1$ is identical.

First we check the homotopy lifting property for a cofibration $X \subseteq Y$. Suppose we are given a homotopy $h_t : Y \to F_1$ and compatible lifts to a homotopy $\overline{k_t} : X \to F_2$ and a lift of one end of h_t to a map $\overline{h_0} : Y \to F_2$. Since h_t defines a homotopy in B, together with the constant basepoint homotopy in C, we get a homotopy $H_t : Y \to E$, where E is the pullback. Note that $\overline{h_0}$ and $\overline{k_t}$ define maps $\overline{H_0}$ and $\overline{K_t}$ which are partial lifts of H_t up to A. Since $A \to E$ is a fibration, we can extend these partial lifts to a lift of $H_t : Y \to E$ up to $\overline{H_t} : Y \to A$. Clearly, H_t defines a homotopy $\overline{h_t} : Y \to F_2$ which lifts h_t and extends $\overline{h_0}$ and $\overline{k_t}$.

Thus $F_2 \to F_1$ sastisfies the homotopy lifting property and is a fibration.

Finally, the mapping properties of the pullback show that the fibre of $F_2 \to F_1$ is H. □

If the totally fibred square is itself a pullback, then the maps $F_2 \to F_1$ and $G_2 \to G_1$ are homeomorphisms or, equivalently, H is a point. This leads to the following definition.

Definition 3.2.4. The totally fibred square is a homotopy pullback if any of the following four equivalent conditions are satisfied:

(1) $F_2 \to F_1$ is a homotopy equivalence.

(2) H is contractible.

(3) $G_2 \to G_1$ is a homotopy equivalence.

(4) The map $A \to E$ from the upper right corner to the pullback is a homotopy equivalence.

More generally, a strictly commutative square is called a homotopy pullback if there is a map Θ from it to a totally fibred homotopy pullback such that Θ is a homotopy equivalence on each vertex. It is easy to see that:

Lemma 3.2.5. *The pullback of a fibration gives a homotopy pullback square.*

The main theorem about homotopy pullbacks will be proved in the last section of this chapter and is as follows:

Proposition 3.2.6. *If Θ is a map between strictly commutative homotopy pullback squares, then the homotopy theoretic fibres of Θ on the vertices form a homotopy pullback square. In other words, the homotopy theoretic fibre of a map between homotopy pullbacks is itself a homotopy pullback.*

Exercises

(1) Prove Lemma 3.2.5.

(2) Given maps $f : X \to Y$ and $g : Y \to \mathbb{Z}$, show that there is a fibration sequence of homotopy theoretic fibres

$$F_f \to F_{gf} \to F_g.$$

3.3 The Peterson–Stein formula

In this section we prove a geometric version of a formula due to Peterson and Stein [105]. This geometric version has been advocated by Harper [50], who stressed the adjoint relationship based on special cases known to Toda and independently to Cartan. The proof becomes very easy if we prepare by exhibiting alternate identifications of homotopy theoretic fibres. The demonstrations of these alternatives use Proposition 3.2.6 above.

First, a homotopy theoretic fibre may also be identified as a pullback of a path fibration. Given a map $f : X \to Y$, let $F_f \to E_f \to Y$ be the fibration sequence which defines the homotopy theoretic fibre F_f. If $p : PY \to Y$ with $p(\omega) = \omega(1)$ is the path space fibration, then we define W_f as the pullback

$$\begin{array}{ccc} W_f & \to & PY \\ \downarrow & & \downarrow \\ X & \to & Y. \end{array}$$

Hence we have a map Θ of pullback squares

$$\begin{array}{ccccccc} F_f & \to & * & & W_f & \to & PY \\ \downarrow & & \downarrow & \xleftarrow{\Theta} & \downarrow & & \downarrow \\ E_f & \to & Y & & X & \to & Y \end{array}$$

and the homotopy theoretic fibre of Θ is a homotopy pullback square with three contractible corners. Hence, the fourth corner, the homotopy theoretic fibre of $W_f \to F_f$, is contractible. It follows that $W_f \to F_f$ is a homotopy equivalence.

Now suppose that $F \xrightarrow{\iota} E \xrightarrow{p} B$ is a fibration sequence. We are going to show that there is a homotopy equivalence $W_\iota \to \Omega B$. In other words, the homotopy theoretic fibre of $\iota : F \to E$ is homotopy equivalent to the loop space ΩB. The map p defines a map Ψ of pullback squares

$$\begin{array}{ccccccc} \Omega B & \to & PB & & W_\iota & \to & PE \\ \downarrow & & \downarrow & \xleftarrow{\Psi} & \downarrow & & \downarrow \\ * & \to & B & & F & \xrightarrow{\iota} & E. \end{array}$$

The homotopy theoretic fibre of Ψ is the homotopy pullback square

$$\begin{array}{ccc} G & \to & PF \\ \downarrow & & \downarrow \\ F & \xrightarrow{=} & F \end{array}$$

with G being the homotopy theoretic fibre of $\Psi : W_\iota \to \Omega B$. Since G is contractible, $\Psi : W_\iota \to \Omega B$ is a homotopy equivalence.

Now suppose that $A \subseteq X$ is a cofibration and $F \xrightarrow{\iota} E \xrightarrow{p} B$ is a fibration sequence. The Peterson–Stein formula relates the cofibration sequence of $A \subseteq X$ to the fibration sequence of $E \to B$.

Peterson–Stein formula 3.3.1. *Given compatible maps $H : A \to W_\iota$ and $G : X \to F$, let $h = \Psi \circ H : A \to W_\iota \to \Omega B$. Then we have a commutative diagram*

$$\begin{array}{ccccccc} A & \to & X & \to & X \cup CA & \to & X \cup CA \cup CX \\ \downarrow H & & \downarrow G & & \downarrow K & & \downarrow L \\ W_\iota & \to & F & \xrightarrow{\iota} & E & \xrightarrow{p} & B \\ \downarrow \Psi & & & & & & \\ \Omega B & & & & & & \end{array}$$

Furthermore, $L = *$ on CX and L factors as $X \cup CA \cup CX \xrightarrow{\simeq} \Sigma A \xrightarrow{\overline{h}} B$ where $\overline{h}$ is the adjoint of h.

Proof: The maps H and G constitute a set of compatible maps $H_1 : A \to F$, $H_2 : A \to PE, G : X \to F$. The adjoint of H_2 defines an extension of $\iota \circ G$ to $K : X \cup CA \to E$. Clearly $p \circ K$ is trivial on X and it extends to $L : X \cup CA \cup CX$ by making it trivial on CX. Finally, since h is defined by projecting H_2 from W_i to ΩB and since $\overline{h}$ is defined by projecting the adjoint of H_2 to B, it is clear that $\overline{h}$ is the adjoint of h. □

Exercise

(1) Let $F \xrightarrow{\iota} E \xrightarrow{p} B$ be a fibration and let W_f be the above pullback construction of the homotopy theoretic fibre of $f : X \to Y$. Show that we have a commutative diagram

$$\begin{array}{ccccccccc} W_\partial & \to & W_\iota & \xrightarrow{\partial} & F & \xrightarrow{\iota} & E & \xrightarrow{p} & B \\ \downarrow & & \downarrow & & & & & & \\ \Omega E & \xrightarrow{\Omega p} & \Omega B & & & & & & \end{array}$$

in which the vertical maps are homotopy equivalences.

3.4 Totally fibred cubes

In this section we generalize the concept of totally fibred squares to higher dimensions. Our reason for doing this is to use the higher dimensions, especially three, to prove things about squares.

Let $C(n, k)$ be the category with objects $\{(a_1, a_2, \ldots, a_n) | 0 \leq a_i \leq k, a_i \varepsilon \mathbb{Z}\}$ and exactly one morphism denoted $(a_1, a_2, \ldots, a_k) \geq (b_1, b_2, \ldots, b_k)$ whenever $a_i \geq b_i$ for all $1 \leq i \leq n$.

Definition 3.4.1. An n-dimensional cube of sidelength k is a covariant functor F from $C(n, k)$ to the category of pointed topological spaces. The spaces $F(a_1, a_2, \ldots, a_n)$ are called the vertices of the cube. Cubes of sidelength 1 are simply called cubes.

For simplicity of notation, we will often shuffle the coordinates in a cube so that they are in descending order. For example, we write $F(1, \ldots, 1, 0, \ldots, 0)$ to represent $F(1, 0, 1, 0, \ldots, 1, 0)$. Properly understood there is no loss of generality.

Given a string $(1, \ldots, 1)$, we shall use $(*, \ldots, *)$ to represent all points with at least one $0 = * < 1$. Then we have maps $F(1, \ldots, 1, 1, \ldots, 1, 0, \ldots 0) \to F(*, \ldots, *, 1, \ldots, 1, 0, \ldots, 0)$ and thus a map to the inverse limit

$$F(1, \ldots, 1, 1, \ldots, 1, 0, \ldots 0) \to \varprojlim F(*, \ldots, *, 1, \ldots, 1, 0, \ldots, 0).$$

Definition 3.4.2. An n-dimensional cube of sidelength 1 is a totally fibred cube if all maps

$$F(1, \ldots, 1, 1, \ldots, 1, 0, \ldots 0) \to \varprojlim F(*, \ldots, *, 1, \ldots, 1, 0, \ldots, 0)$$

are fibrations. Of course, this is understood to be true for all shufflings of 0's and 1's.

This generalizes the concept of totally fibred square and we have the following useful lemma:

Lemma 3.4.3. *If all maps $F(1, \ldots, 1, 0, \ldots 0) \to \varprojlim F(*, \ldots, *, 0, \ldots, 0)$ are fibrations for all shufflings of 0's and 1's, then the cube is totally fibred.*

Proof: Consider the map $F(1, \ldots, 1, 1, \ldots, 1, 0, \ldots 0) \to \varprojlim F(*, \ldots, *, 1, \ldots, 1, 0, \ldots 0)$. Let b the length of the second string of 1's. We can assume that we have a fibration whenever b is decreased. We need to show that $F(1, \ldots, 1, 1, \ldots, 1, 0, \ldots 0) \to \varprojlim F(*, \ldots, *, 1, \ldots, 1, 0, \ldots 0)$ is a fibration. As usual we start with a homotopy in the base and a partial lift to the total space. Clearly this gives a homotopy in $\varprojlim F(*, \ldots, *, 0, \ldots, 1, 0, \ldots 0)$ and a partial lift to $F(1, \ldots, 1, 0, \ldots, 1, 0, \ldots 0)$. Since b has decreased by

1, we can lift the homotopy to $F(1,\ldots,1,0,\ldots,1,0,\ldots,0)$. Now we get a homotopy in $\lim_{\leftarrow} F(*,\ldots,*,*,\ldots,1,0,\ldots,0)$. Since b is again decreased by 1 here, we can lift this homotopy to $F(1,\ldots,1,1,\ldots,1,0,\ldots,0)$. Thus, $F(1,\ldots,1,1,\ldots,1,0,\ldots 0) \to \lim_{\leftarrow} F(*,\ldots,*,1,\ldots,1,0,\ldots 0)$ is a fibration. $\square$

Proposition 3.4.4. *If F is any cube there is a totally fibred cube G and a natural transformation $\Theta : F \to G$ such that Θ is a homotopy equivalence on all vertices.*

Proof: We replace a map $f : X \to Y$ by a fibration via

$$\begin{array}{ccc} X & \xrightarrow{\simeq} & E_f \\ {}_{f}\searrow & & \downarrow p \\ & Y. & \end{array}$$

Note that X maps to its replacement E_f and that Y is unchanged.

First, replace all maps $F(1,0,\ldots,0) \to F(0,0,\ldots,0)$ by fibrations. Having replaced all maps $F(1,\ldots,1,0,\ldots,0) \to \lim_{\leftarrow} F(*,\ldots,*,0,\ldots 0)$ by fibrations for all shufflings of 0's and 1's, then replace all maps $F(1,\ldots,1,1,\ldots,0) \to \lim_{\leftarrow} F(*,\ldots,*,*,\ldots 0)$ by fibrations for all shufflings.

Lemma 3.4.3 then shows that the final cube G is totally fibred. $\square$

Remark. If F and G are n-dimensional cubes and $\Theta : F \to G$ is a natural transformation, then this clearly defines a new cube H of dimension $n + 1$. If G is a totally fibred cube, then a slight variation of the above proof shows that the identity natural transformation of G extends to a natural transformation $\Theta : H \to K$ where K is totally fibred and Θ is a homotopy equivalence on the vertices.

Definition 3.4.5. If F is an n-dimensional cube of sidelength 1, the fibre extension of F is the extension of F to an n-dimensional cube of sidelength 2, also denoted F, and defined by

$$F(2,\ldots,2,a_1,\ldots,a_k) = \mathrm{fibre}(F(1,\ldots,1,a_1,\ldots,a_k)$$
$$\to \lim_{\leftarrow} F(*,\ldots,*,a_1,\ldots,a_k))$$

where $0 \leq a_i \leq 1$. Of course this definition represents all shufflings of $(2,\ldots,2,a_1,\ldots,a_k)$.

A section of a cube is defined by choosing a subset of coordinates to be constants, that is, up to shuffling, restrict to $F(a_1,\ldots,a_r,c_1,\ldots,c_s)$ where $(c_1,\ldots,c_s)$ are constants.

Proposition 3.4.6. *A section of a fibre extension is a fibre extension of a section.*

Proof: Suppose we have a section of a fibre extension

$$F(a_1,\ldots,a_r,2,\ldots,2,1,\ldots,1,0,\ldots,0)$$

3.4 Totally fibred cubes

with the a_i variable. We must show that each

$$F(1, \ldots, 1, 0, \ldots, 0, 2, \ldots, 2, 1, \ldots, 1, 0, \ldots, 0)$$
$$\to \varprojlim F(*, \ldots, *, 0, \ldots, 0, 2, \ldots, 2, 1, \ldots, 1, 0, \ldots, 0),$$

where $*, \ldots *$ indicates that at least one coordinate $*$ is < 1, is a fibration with fibre

$$F(2, \ldots, 2, 0, \ldots, 0, 2, \ldots, 2, 1, \ldots, 1, 0, \ldots, 0).$$

Suppose we have a homotopy H in

$$\varprojlim F(*, \ldots, *, 0, \ldots, 0, 2, \ldots, 2, 1, \ldots, 1, 0, \ldots, 0)$$

with a partial lift K to

$$F(1, \ldots, 1, 0, \ldots, 0, 2, \ldots, 2, 1, \ldots, 1, 0, \ldots, 0).$$

So H is a homotopy in

$$\varprojlim F(*, \ldots, *, 0, \ldots, 0, 1, \ldots, 1, 1, \ldots, 1, 0, \ldots, 0)$$

which projects to the basepoint in

$$\varprojlim \varprojlim F(*, \ldots, *, 0, \ldots, 0, *, \ldots, *, 1, \ldots, 1, 0, \ldots, 0),$$

where each group contains a coordinate $* < 1$, and K is a partial lift to

$$F(1, \ldots, 1, 0, \ldots, 0, 1, \ldots, 1, 1, \ldots, 1, 0, \ldots, 0)$$

which projects to the basepoint in

$$\varprojlim F(1, \ldots, 1, 0, \ldots, 0, *, \ldots, *, 1, \ldots, 1, 0, \ldots, 0).$$

First we lift H to the basepoint homotopy in

$$\varprojlim F(1, \ldots, 1, 0, \ldots, 0, *, \ldots, *, 1, \ldots, 1, 0, \ldots, 0)$$

and get a homotopy in

$$\varprojlim F(*, \ldots, *, 0, \ldots, 0, *, \ldots, *, 1, \ldots, 1, 0, \ldots, 0)$$

where only one coordinate $*$ is required to be < 1. Since

$$F(1, \ldots, 1, 0, \ldots, 0, 1, \ldots, 1, 1, \ldots, 1, 0, \ldots, 0)$$
$$\to \varprojlim F(*, \ldots, *, 0, \ldots, 0, *, \ldots, *, 1, \ldots, 1, 0, \ldots, 0)$$

is a fibration, we can lift this homotopy to an extension of K in

$$F(1, \ldots, 1, 0, \ldots, 0, 2, \ldots, 2, 1, \ldots, 1, 0, \ldots, 0)$$
$$\subseteq F(1, \ldots, 1, 0, \ldots, 0, 1, \ldots, 1, 1, \ldots, 1, 0, \ldots, 0).$$

Thus,
$$F(1,\ldots,1,0,\ldots,0,2,\ldots,2,1,\ldots,1,0,\ldots,0)$$
$$\to \varprojlim F(*,\ldots,*,0,\ldots,0,2,\ldots,2,1,\ldots,1,0,\ldots,0)$$

is a fibration. It is clear that its fibre is
$$F(2,\ldots,2,0,\ldots,0,2,\ldots,2,1,\ldots,1,0,\ldots,0). \qquad \square$$

In a fibre extension of a square all rows and columns are fibration sequences. This fact is generalized by the following immediate consequence of Proposition 3.4.6:

Corollary 3.4.7. *In a fibre extension, all one-dimensional sections are fibration sequences.*

Definition 3.4.8. Let F be a totally fibred n-dimensional cube of side length 1. It is a homotopy pullback if its fibre extension has $F(2,\ldots,2)$ contractible.

More generally, an arbitrary cube F is called a homotopy pullback if there is a natural transformation $\Theta : F \to G$ where G is a totally fibred homotopy pullback. The following is an immediate consequence of Corollary 3.4.7:

Proposition 3.4.9. *Suppose $\Theta : F \to G$ is a natural transformation of homotopy pullbacks. If H is the cube formed from the homotopy theoretic fibres of Θ on the vertices, then H is a homotopy pullback.*

Exercises

(1) Consider the commutative array Δ of maps

$$\begin{array}{ccccc} A & \to & B & \leftarrow & * \\ \downarrow & & \downarrow & & \downarrow \\ C & \to & D & \leftarrow & * \\ \uparrow & & \uparrow & & \uparrow \\ * & \to & * & \leftarrow & *. \end{array}$$

Let V represent taking the homotopy pullback of the vertical maps and let H represent taking the homotopy pullback of the horizontal maps. Show that $V(H(\Delta)) = H(V(\Delta))$.

(2) Extend Exercise 1 to three dimensions.

3.5 Spaces of the homotopy type of a CW complex

The following theorem of J. H. C. Whitehead [135] is one of the reasons why CW complexes are useful in homotopy theory.

3.5 Spaces of the homotopy type of a CW complex

Theorem 3.5.1. *Let $f : X \to Y$ be a map between connected CW complexes. If f induces an isomorphism on all homotopy groups, then f is a homotopy equivalence.*

Hence, it is important to know simple theorems which imply that the category of CW complexes is closed under various operations. For example, if X and Y are CW complexes, then so is the bouquet $X \vee Y$ and, if one takes the product in the category of compactly generated spaces, then so is the product $X \times Y$.

Other facts concerning closure follow from the following theorem which is also due to J. H. C. Whitehead:

Theorem 3.5.2. *If $f : X \to Y$ is a map between CW complexes, then f is homotopic to a skeletal map $g : X \to Y$, that is, $g(X^{(n)}) \subseteq Y^{(n)}$ for all n.*

For example, suppose we have two maps $f : A \to X$ and $h : A \to Y$. Consider the mapping torus $T_{f,h} =$

$$X \cup (A \times I) \cup Y$$

with identifications $(a, 0) = f(a)$ and $(a, 1) = h(a)$. Then the homotopy pushout of $X \xleftarrow{f} A \xrightarrow{g} Y$ is $T_{f,h}$. We have:

Proposition 3.5.3. *If $f, g : A \to X$ and $h, k : A \to Y$ are homotopic maps then there is a homotopy equivalence $T_{f,h} \simeq T_{g,k}$. [136, 87]*

Thus, Theorem 3.5.2 and Proposition 3.5.3 imply that homotopy pushouts of maps of CW complexes have the homotopy type of CW complexes. In particular, this applies to the mapping cone of a map between CW complexes.

Recall that an an n-ad is an ordered n-tuple of spaces $\overline{X} = (X; A_1, \ldots, A_{n-1})$ where all the A_i are closed subspaces of X. If X is a CW complex and all the A_i are subcomplexes, then it is called a CW n-ad. Maps of n-ads $f : \overline{X} \to \overline{Y}$ are required to preserve subspaces and the set of all such f forms a topological space $\mathrm{map}(\overline{X}, \overline{Y})$. There is the following theorem of J. W. Milnor [89]:

Theorem 3.5.4. *If $\overline{C}$ is a compact n-ad and $\overline{X}$ is a CW n-ad, then the mapping space $\mathrm{map}(\overline{C}, \overline{X})$ has the homotopy type of a CW complex.*

For example, we get the following:

(a) If X is a CW complex with basepoint x_0, then the space of loops $\Omega(X, x_0) = \mathrm{map}((I; 0, 1), (X, x_0, x_0))$ has the homotopy type of a CW complex.

(b) If $(X; A, B)$ is a CW 3-ad, then the homotopy pullback of $A \to X \leftarrow B$ is the mapping space $\mathrm{map}((I; 0, 1), (X; A, B))$ and has the homotopy type of a CW complex. In particular, if B is a point, it follows that the homotopy theoretic fibre of the inclusion $A \subseteq X$ has the homotopy type of a CW complex.

Exercises

(1) Prove Proposition 3.5.3.

(2) Check that $\mathrm{map}((I; 0, 1), (X; A, B))$ is the homotopy pullback of the inclusions.

4 Hilton–Hopf invariants and the EHP sequence

In this chapter we introduce the Bott–Samelson theorem and explore the consequences of this result on the homology of loop suspensions. Among the consequences are the James construction [66] and the Hilton–Milnor theorem. The James construction and the Hilton–Milnor theorem both lead to Hopf invariants and to the EHP sequence.

The Hilton–Hopf invariants are particularly well suited to study the distributive properties of compositions and this is why we use them in our proof of the 2-primary exponent theorem of James [67].

We prove the following theorem of James: 4^n annihilates the 2-primary component of all the homotopy groups of S^{2n+1}. The exponent result which we prove was first formulated in a geometric form involving loop spaces by John Moore in a graduate course at Princeton. A necessary lemma on the vanishing of twice the second Hilton–Hopf invariant was supplied by Michael Barratt.

The result we prove is not the best possible 2-primary exponent. A considerable improvement was made by Paul Selick but there is no reason to believe that his result is the best possible. The Barratt–Mahowald conjecture for the best possible 2-primary exponent is: The 2-primary component of all the homotopy groups of S^{2n+1} is annihilated by $2^{n+\epsilon}$ where

$$\epsilon = \begin{cases} 0 & \text{if} \quad n=0 \quad \text{or} \quad n=3 \mod 4 \\ 1 & \text{if} \quad n=1 \quad \text{or} \quad n=2 \mod 4. \end{cases}$$

It remains unproved.

We do a few EHP sequence computations related to the three-dimensional sphere in order to show that it does possess at least one homotopy class of order 4. In order to do this, we look at the 3-connected cover of the three-dimensional sphere and compute its homology and a little bit of its homotopy including the first nontrivial 2-primary element η_3 and the first nontrivial odd primary element α_1.

We prove a lemma which shows one way in which η produces elements of order 4. This is related to the fact that homotopy groups with coefficients mod 2 have exponent 4.

4.1 The Bott–Samelson theorem

The fact that the suspension of a space is the union of two contractible cones implies that fibrations over a suspension have a simple structure. Since the fibration is trivial over each of the cones, it can be described as the union of two trivial bundles with an identification on the intersection.

In this section we use this to determine the homology of the loop space $\Omega\Sigma X$ when X is connected and the homology $H_*(X)$ is free over the coefficient ring.

Define the suspension map $\Sigma = \Sigma_X : X \to \Omega\Sigma X$ by $\Sigma(x)(t) = \langle x, t \rangle$ for all $x \varepsilon X$ and $0 \le t \le 1$. This is just the adjoint of the identity map $1_{\Sigma X} : \Sigma X \to \Sigma X$. The other adjoint is the evaluation map $e = e_X : \Sigma\Omega X \to X$ with $e(\langle t, \omega \rangle) = \omega(t)$. The space $\Omega\Sigma X$ has an important universal multiplicative propery:

Definition 4.1.1. If $f : X \to \Omega Y$ is a continuous map, then there is a unique loop map $\overline{f} : \Omega\Sigma X \to \Omega Y$ such that $\overline{f} \circ \Sigma_X = f$. The map is $\overline{f} = \Omega(e_Y) \circ \Omega\Sigma f : \Omega\Sigma X \to \Omega\Sigma\Omega Y \to \Omega Y$ and is called the multipicative extension of f.

Note that, if $X \xrightarrow{f} Y \xrightarrow{g} \Omega\mathbb{Z}$ are maps, then $\overline{g \circ f} = \overline{g} \circ \Omega\Sigma(f) : \Omega\Sigma(X) \to \Omega\Sigma(Y) \to \Omega\mathbb{Z}$.

The Bott–Samelson theorem below says that Definition 4.1.1 is consistent with an analogous universal property of the Pontrjagin ring $H_*(\Omega\Sigma X)$, at least with field coefficients.

If $\mu : \Omega\Sigma X \times \Omega\Sigma X \to \Omega\Sigma X$ is the multiplication of loops, $\mu(\omega, \gamma) = \omega * \gamma$,

$$(\omega * \gamma)(t) = \begin{cases} \omega(2t) & \text{if } 0 \le t \le \frac{1}{2} \\ \gamma(2t - 1) & \text{if } \frac{1}{2} \le t \le 1, \end{cases}$$

define the clutching function $\nu : \Omega\Sigma X \times X \to \Omega\Sigma X \times X$ by $\nu(\omega, x) = (\omega * \Sigma(x), x)$.

Write $\Sigma X = C_- X \cup C_+ X$ with

$$C_- X = \left\{ \langle x, t \rangle | x \varepsilon X, 0 \le t \le \frac{1}{2} \right\},$$

$$C_+ X = \left\{ \langle x, t \rangle | x \varepsilon X, \frac{1}{2} \le t \le 1 \right\},$$

4.1 The Bott–Samelson theorem

and
$$C_-X \cap C_+X = X \times \frac{1}{2} = X.$$

Form the quotient space $E = (\Omega\Sigma X \times C_-X) \cup (\Omega\Sigma X \times C_+X)$ with identification of the boundary of the first product to the boundary of the second product given by the clutching function $\nu : \Omega\Sigma X \times X \to \Omega\Sigma X \times X$. In other words, E is defined by a pushout diagram

$$\begin{array}{ccc} \Omega\Sigma X \times X & \xrightarrow{\nu} & \Omega\Sigma X \times C_+X \\ \downarrow 1 \times \iota & & \downarrow \\ \Omega\Sigma X \times C_-X & \to & E. \end{array}$$

We note that there is a map $\tau : E \to \Sigma X$ well defined by $\tau(\omega, \alpha) = \alpha$. We claim that this is model for the path space fibration $\pi : P\Sigma X \to X$ with $\pi(\omega) = \omega(1)$. More precisely, all we need is

Proposition 4.1.2. *There is a homotopy equivalence $P\Sigma X \to E$, that is, E is contractible.*

Proof: Write $P\Sigma X = E_- \cup E_+$ where $E_- = \pi^{-1}(C_-X)$, $E_+ = \pi^{-1}(C_+X)$, and thus $E_- \cap E_+ = \pi^{-1}(C_-X \cap C_+X)$.

There is a pushout diagram

$$\begin{array}{ccc} E_- \cap E_+ & \to & E_+ \\ \downarrow & & \downarrow \\ E_- & \to & P\Sigma X. \end{array}$$

If $\alpha = \langle x, t \rangle \varepsilon C_-X$, let $\gamma_-(\alpha)$ be the path in C_-X which goes linearly from the basepoint $\langle x, 0 \rangle$ to $\langle x, t \rangle$. Similarly, if $\alpha = \langle x, t \rangle \varepsilon C_+X$, let $\gamma_+(\alpha)$ be the path in C_+X which goes linearly from the basepoint $\langle x, 1 \rangle$ to $\langle x, t \rangle$. Note that, if $\alpha \varepsilon (C_-X \cap C_+X) = X$, then $\gamma_-(x) * \gamma_+^{-1}(x) \simeq \Sigma(x)$.

We have equivalences

$$\Omega\Sigma X \times C_-X \xrightarrow{\psi_-} E_-, \quad \psi_-(\omega, \alpha) = \omega * (\gamma_-(\alpha))$$

$$E_- \xrightarrow{\phi_-} \Omega\Sigma X \times C_-X, \quad \phi_-(\omega) = (\omega * (\gamma_-(\omega(1))^{-1}), \omega(1))$$

$$\Omega\Sigma X \times C_+X \xrightarrow{\psi_+} E_+, \quad \psi_+(\omega, \alpha) = \omega * (\gamma_+(\alpha))$$

$$E_+ \xrightarrow{\phi_+} \Omega\Sigma X \times C_+X, \quad \phi_+(\omega) = (\omega * (\gamma_+(\omega(1))^{-1}), \omega(1))$$

with fibre homotopies

$$\phi_- \circ \psi_- \simeq 1, \ \psi_- \circ \phi_- \simeq 1, \ \phi_+ \circ \psi_+ \simeq 1, \ \psi_+ \circ \phi_+ \simeq 1.$$

We have a homotopy commutative diagram

$$\begin{array}{ccccc} E_- & \leftarrow & E_- \cap E_+ & \to & E_+ \\ \downarrow \phi_- & & \downarrow \phi_- & & \downarrow \phi_+ \\ \Omega\Sigma X \times C_-X & \xleftarrow{1 \times \iota} & \Omega\Sigma X \times X & \xrightarrow{\nu} & \Omega\Sigma X \times C_+X. \end{array}$$

Since the maps in the top row are cofibrations we can alter ϕ_+ by a homotopy and make the diagram strictly commutative. Hence we get a map of pushouts $P\Sigma X \to E$. The Seifert–van Kampen theorem and the Mayer–Vietoris sequence show that this map is an isomorphism on fundamental groups and homology. It follows that E is contractible. $\square$

Let R be a commutative ring and let V be an R-module. The tensor algebra

$$T(V) = R \oplus V \oplus (V \otimes V) \oplus (V \otimes V \otimes V) \oplus \cdots =$$

the free associative R-algebra generated by the module V. The tensor algebra is characterized by the universal mapping property:

Given an associative algebra A and a linear map $f : V \to A$, there is a unique extension of f to an algebra homomorphism $\overline{f} : T(V) \to A$.

Bott–Samelson theorem 4.1.3[13]. *Let the coefficient ring R be a principal ideal domain and let X be a connected topological space such that $H_*(X)$ is a free R-module, then the map $\Sigma_* : \overline{H}_*(X) \to H_*(\Omega\Sigma X)$ induces an algebra isomorphism $T(\overline{H}_*(X)) \to H_*(\Omega\Sigma X)$.*

Proof: A simple induction proves the following.

Lemma 4.1.4. *Let A be a connected associative algebra and let V be a connected module which is free. Suppose there is a linear map $\iota : V \to \overline{A}$. Then the extension $\overline{\iota} : T(V) \to A$ is an isomorphism if and only if the composition $A \otimes V \xrightarrow{1 \otimes \iota} A \otimes \overline{A} \xrightarrow{\mu} \overline{A}$ is an isomorphism (where μ is the multiplication).*

Since $E \simeq P\Sigma X$ is contractible, the Mayer–Vietoris sequence of the pushout diagram

$$\begin{array}{ccc} \Omega\Sigma X \times X & \xrightarrow{\nu} & \Omega\Sigma X \times C_+X \\ \downarrow \iota \times 1 & & \downarrow \\ \Omega\Sigma X \times C_-X & \to & E \end{array}$$

is short exact as follows:

$$0 \to H_*(\Omega\Sigma X) \otimes H_*(X)$$
$$\xrightarrow{\Delta} H_*(\Omega\Sigma X) \otimes H_*(C_-X) \oplus H_*(\Omega\Sigma X) \otimes H_*(C_+X) \to R \to 0$$

and if $\epsilon : H_*(X) \to R$ is the augmentation, then

$$\Delta(\beta \otimes \alpha) = (\beta \otimes \epsilon(\alpha), \beta\alpha \otimes 1).$$

It is immediate that the map

$$H_*(\Omega\Sigma X) \otimes \overline{H}_*(X) \to \overline{H}_*(\Omega\Sigma X)$$

is an isomorphism and Lemma 4.1.4 finishes the proof. □

Corollary 4.1.5. *Let X be a connected space with $H_*(X)$ a free R-module. There is a Hopf algebra isomorphism $T(\overline{H}_*(X)) \to H_*(\Omega\Sigma X)$ with the comultiplication of the tensor algebra determined on generators by the comultiplication of $H_*(X)$.*

The following two cases will be useful.

Let μ_n be a generator of $H_n(S^n)$. With any principal ideal domain R as coefficient ring, there is an isomorphism of primitively generated Hopf algebras $T(\mu_n) \to H_*(\Omega S^{n+1})$.

On the other hand, let p be a prime, $1 \leq s \leq r$, and let $R = \mathbb{Z}/p^s\mathbb{Z}$ be the coefficient ring. For $n \geq 2$, note that $\overline{H}_*(P^n(p^r))$ is a free R-module with generators ν_n and μ_{n-1} of degrees n and $n-1$. Thus there is an isomorphism of algebras

$$T(\nu_n, \mu_{n-1}) \to H_*(\Omega P^{n+1}(p^r)).$$

If p is odd or $n > 2$, both generators are primitive. If $n = 2, p = 2, r = s = 1$, then the comultiplication is given by

$$\Delta_*(\nu_2) = \nu_2 \otimes 1 + 1 \otimes \nu_2 + \mu_1 \otimes \mu_1.$$

Exercises

(1) Prove Lemma 4.1.4.

(2) Prove that the maps $\Sigma_X : X \to \Omega\Sigma X$ and $e_X : \Sigma\Omega X \to X$ satisfy:

  (a) Σ_X and e_X are natural transformations.

  (b) $\Omega(e_X) \circ \Sigma_{\Omega X} = 1_{\Omega X}$ and $e_{\Sigma x} \circ \Sigma(\Sigma_X) = 1_{\Sigma X}$.

  (c) Verify the universal property of the multiplicative extension in definition 4.1.1.

4.2 The James construction

Let X be a space with a nondegenerate basepoint x_0. The James construction [66] provides a homotopy equivalent model for the loops on a suspension. It is the free associative topological monoid generated by the points of X with the single relation that the basepoint x_0 is the unit. More precisely, the James construction

$J(X)$ is the direct limit of subspaces
$$J_1(X) \subseteq J_2(X) \subseteq J_3(X) \subseteq \cdots \subseteq J_n(X) \subseteq J_{n+1} \subseteq \cdots,$$
where $J_n(X)$ consists of the words of length $\leq n$ and the topology is defined by the quotient map $\pi_n : X \times \cdots \times X \to J_n(X)$.

Thus $J_1(X) = X$, and, if $W_n(X) \subseteq X \times \cdots \times X$ is the subset of the n-fold product with at least one coordinate equal to the basepoint, there are pushout diagrams

$$\begin{array}{ccc} W_n(X) & \xrightarrow{\pi_n} & J_{n-1}(X) \\ \downarrow\subseteq & & \downarrow\subseteq \\ X \times \cdots \times X & \xrightarrow{\pi} & J_n(X) \end{array}$$

and, if X has a nondegenerate basepoint, there are cofibration sequences
$$J_{n-1}(X) \to J_n(X) \to X \wedge \cdots \wedge X.$$

Universal property 4.2.1. Let M is a strictly associative topological monoid with the unit as the basepoint and let $f : X \to M$ be a continuous basepoint preserving map. Then f has a unique extension to a continuous homomorphism $\overline{f} : J(X) \to M$. Thus, $\overline{f}(x_1 x_2 \ldots x_n) = f(x_1)f(x_2)\ldots f(x_n)$.

For example, for any pointed space Y, the inclusion is a homotopy equivalence $\Omega Y \to \Omega_\mu Y$ from the space of loops in Y parametrized by the unit interval to the Moore space of loops of variable length in Y. The Moore loops $\Omega_\mu Y$ form a strictly associative monoid and hence the suspension map $\Sigma : X \to \Omega_\mu \Sigma X$ has a unique multiplicative extension $\overline{\Sigma} : J(X) \to \Omega_\mu \Sigma X$.

Proposition 4.2.2. *If X is a space with a nondegenerate basepoint, then the map $\overline{f} : J(X) \to \Omega_\mu \Sigma X$ is a weak equivalence.*

Remark. If X is a CW complex, then $J(X)$ is a CW complex and $\Omega_\mu \Sigma X$ has the homotopy type of a CW complex. Hence the above map is a homotopy equivalence.

Proof: Since both sides are H-spaces, $\overline{f}$ is a weak homotopy equivalence if it is an integral homology equivalence. Hence it is sufficient to check that $\overline{f}$ is both a rational homology equivalence and a mod p homology equivalence for all primes p.

By the Bott–Samelson theorem, it is sufficient to check that $X \to J(X)$ induces an algebra isomorphism $T(\overline{H}_*(X)) \to H_*(J(X))$.

This is proved inductively. We claim that there is an isomorphism from tensors of length n,
$$T_n(\overline{H}_*(X)) \to H_*(J_n(X))$$
to the homology of the n-th filtration. For this, we need the following lemma.

Lemma 4.2.3. *For connected X and Y, there is a weak homotopy equivalence*
$$\Sigma(X \times Y) \to \Sigma X \vee \Sigma Y \vee \Sigma(X \wedge Y).$$

Proof: The three maps
$$\Sigma(X \times Y) \to \Sigma X, \quad \Sigma(X \times Y) \to \Sigma Y, \quad \Sigma(X \times Y) \to \Sigma(X \wedge Y)$$
can be added using the co-H structure to get a homology equivalence
$$\Sigma(X \times Y) \to \Sigma X \vee \Sigma Y \vee \Sigma(X \wedge Y).$$
Since the spaces involved are simply connected, this is a weak equivalence. $\square$

It follows that the composition $X \times \cdots \times X \to J_n(X) \to X \wedge \cdots \wedge X$ is surjective in homology. Thus the cofibration sequences $J_{n-1}(X) \to J_n(X) \to X \wedge \cdots \wedge X$ show that there is a homology isomorphism $T_n(\overline{H}_*(X)) \to H_*(J_n(X))$. $\square$

An immediate corollary of the above is [45].

The Freudenthal suspension theorem 4.2.4. *If X is an $n-1$ connected space with $n-1 \geq 1$, then the pair $(J(X), X)$ is $2n-1$ connected. Thus, $\Sigma_* : \pi_k(X) \to \pi_{k+1}(\Sigma X)$ is an epimorphism if $k \leq 2n-1$ and an isomorphism if $k \leq 2n-2$.*

4.3 The Hilton–Milnor theorem

An H-space G is called grouplike if it is homotopy associative and has a homotopy inverse. For pointed maps $f : X \to G$ and $g : Y \to G$ into a grouplike space, we can define a Samelson product $[f, g] : X \wedge Y \to G$ as follows:

Consider the commutator map
$$X \times Y \xrightarrow{f \times g} G \times G \xrightarrow{[\ ,\ ]} G, (x, y) \to f(x)g(y)f(x)^{-1}g(y)^{-1}.$$

It is homotopically trivial on $G \vee G$ and hence we have a factorization up to homotopy
$$X \times Y \to X \wedge Y \xrightarrow{[f,g]} G.$$

The cofibration sequence $X \vee Y \to X \times Y \to X \wedge Y \to \Sigma X \vee \Sigma Y \to \cdots$ leads to the short exact sequence
$$0 \leftarrow [X, G]_* \times [Y, G]_* \leftarrow [X \times Y, G]_* \leftarrow [X \wedge Y, G]_* \leftarrow 0$$
and it follows that the Samelson product $[f, g]$ is unique up to homotopy.

In Chapter 6, we will make a more detailed study of Samelson products. In particular, if $f : S^n \to G$ and $g : S^m \to G$, then the Samelson product $[f,g] : S^{n+m} \to G$ gives the homotopy groups $\pi_*(G)$ the structure of a graded Lie algebra. That is for $x, y, z \in \pi_*(G)$ of respective degrees a, b, c we have bilinearity, antisymmetry, and the Jacobi identity:

$$[x+y, z] = [x, z] + [y, z], \quad [x, y] = -(-1)^{ab}[y, x],$$
$$[x, [y, z]] = [[x, y], z] + (-1)^{ab}[y, [x, z]].$$

The next lemma relates the Samelson product to the commutator in the Pontjagin ring $H_*(G)$. It is an easy exercise.

Lemma 4.3.1. *If $\alpha \varepsilon \overline{H}_*(X)$ and $\beta \varepsilon \overline{H}_*(Y)$ are primitive homology classes, then*

$$[f, g]_*(\alpha \otimes \beta) = [f_*(\alpha), g_*(\beta)] = f_*(\alpha)g_*(\beta) - (-1)^{|\alpha||\beta|}g_*(\beta)f_*(\alpha).$$

Now we state the Hilton–Milnor theorem in its compact form. If X and Y are spaces, let $\iota_X : X \to X \vee Y$ and $\iota_Y : Y \to X \vee Y$ be the two inclusions. There should be no confusion if we identify $\iota_X = \Sigma \circ \iota_X : X \to X \vee Y \xrightarrow{\Sigma} \Omega\Sigma(X \vee Y)$ and $\iota_Y = \Sigma \circ \iota_Y : Y \to X \vee Y \xrightarrow{\Sigma} \Omega\Sigma(X \vee Y)$.

Write $\mathrm{ad}(\alpha)(\beta) = [\alpha, \beta]$. We can form iterated Samelson products

$$\mathrm{ad}(\iota_X)^i(\iota_Y) : X^{\wedge i} \wedge Y \to \Omega\Sigma(X \vee Y)$$

and add them up to get a map of the infinite bouquet

$$\bigvee_{i \geq 0} \mathrm{ad}(\iota_X)^i(\iota_Y) : \bigvee_{i \geq 0} X^{\wedge i} \wedge Y \to \Omega\Sigma(X \vee Y).$$

Now form the multiplicative extensions

$$\overline{\iota_X} = \Omega\Sigma(\iota_X) : \Omega\Sigma X \to \Omega\Sigma(X \vee Y)$$

and

$$\overline{\bigvee_{i \geq 0} \mathrm{ad}(\iota_X)^i(\iota_Y)} : \Omega\Sigma \left(\bigvee_{i \geq 0} X^{\wedge i} \wedge Y \right) \to \Omega\Sigma(X \vee Y).$$

Finally use the multiplication of $\Omega\Sigma(X \vee Y)$ to multiply these maps and get:

The Hilton–Milnor theorem 4.3.2 [54, 89]. *If X and Y are connected, then there is a weak equivalence*

$$\Omega\Sigma X \times \Omega\Sigma \left(\bigvee_{i \geq 0} X^{\wedge i} \wedge Y \right) \to \Omega\Sigma(X \vee Y).$$

Of course, if X and Y are both CW complexes then the above map is actually a homotopy equivalence.

4.3 The Hilton–Milnor theorem

We will prove the Hilton–Milnor theorem in a subsequent chapter when we have developed more algebra. For now we will just say that the Hilton–Milnor theorem is the topological reflection of the algebraic isomophism of right modules

$$T(V \oplus W) \cong T(V) \otimes T\left(\bigoplus_{i \geq 0} V^{\otimes i} \otimes W\right),$$

where $T(V \oplus W)$ is the homology with field coefficients of $\Omega\Sigma(X \vee Y)$. This decomposition is also valid if the homology of $\Omega\Sigma(X \vee Y)$ is torsion free. In this decomposition $V^{\otimes i} \otimes W$ is embedded in $T(V \oplus W)$ via the commutator isomorphism $V^{\otimes i} \otimes W \xrightarrow{\cong} \mathrm{ad}(V)^i(W)$.

The Hilton–Milnor theorem can be iterated by expanding the second factor and continuing this process countably many times. In order to describe the final result, we need the notion of a Hall basis.

Given a finite-ordered list of elements $L = L_1$ with first element x_1, we define a Hall basis as follows:

(a) Let $B_1 = \{x_1\}$ and let $L_2 = \{\mathrm{ad}(x_1)^i(x) | i \geq 0, x \varepsilon L_1, x \neq x_1\}$.

(b) Order L_2 so that elements of smaller length come before elements of greater length and let x_2 be the first element of L_2.

(c) Let $B_2 = B_1 \cup \{x_2\}$ and let $L_3 = \{\mathrm{ad}(x_2)^i(x) | i \geq 0, x \varepsilon L_2, x \neq x_2\}$.

(d) Repeat b) and c) countably often and let

$$B = \bigcup_{n=1}^{\infty} B_n.$$

B is called a Hall basis generated by L. It is an ordered basis for the (ungraded) free Lie algebra generated by the set L.

Using the notation of Theorem 4.3.2, let B be a Hall basis generated by the set $\{\iota_X, \iota_Y\}$. For example, $B =$

$$\{\iota_X, \quad \iota_Y, \quad [\iota_X, \iota_Y], \quad [\iota_X, [\iota_X, \iota_Y]], \quad [\iota_Y, [\iota_X, \iota_Y]],$$
$$[\iota_X, [\iota_X, [\iota_X, \iota_Y]]], \quad [\iota_Y, [\iota_X, [\iota_X, \iota_Y]]], \quad [\iota_Y, [\iota_Y, [\iota_X, \iota_Y]]], \quad \ldots \}.$$

If $\omega = \omega(\iota_X, \iota_Y)$ is an element of B, we shall write $\omega(X, Y)$ for the domain of this Samelson product. For example, if $\omega(\iota_X, \iota_Y) = [\iota_X, [\iota_X, \iota_Y]]$, then $\omega(X, Y) = X \wedge (X \wedge Y)$. Now the Hilton–Milnor theorem may be restated as:

The expanded Hilton–Milnor theorem 4.3.3. *There is a weak equivalence*

$$\Theta : \prod_{\omega \varepsilon B} \Omega\Sigma(\omega(X, Y)) \to \Omega\Sigma(X \vee Y)$$

which is the multiplicative extension of $\omega(\iota_X, \iota_Y)$ on the ω factor and which is defined by multiplying these maps in the order determined by B.

Remark. The topology on the infinite product above is not the product topology. It is the direct limit topology coming from the topology on the finite products. This is called the weak product. In order to make the map Θ well defined we need to reparametrize a product of maps $f_1 * f_2 * f_3 * \ldots$ so that the first factor runs on the interval $[0, \frac{1}{2}]$, the second on the interval $[\frac{1}{2}, \frac{3}{4}]$, the third on the interval $[\frac{3}{4}, \frac{7}{8}]$, etc.

Thus $\Omega\Sigma(X \vee Y)$ is weakly equivalent to the weak infinite product

$$\Omega\Sigma(X) \times \Omega\Sigma(Y) \times \Omega\Sigma(X \wedge Y) \times \Omega\Sigma(X \wedge X \wedge Y)$$
$$\times \Omega\Sigma(Y \wedge X \wedge Y) \times \cdots$$

If $H_*\Omega\Sigma(X \vee Y) = T(V \oplus W)$ is torsion free, this corresponds to the homology decomposition

$$T(V \oplus W) \cong T(V) \otimes T(W) \otimes T(V \otimes W) \otimes T(V \otimes V \otimes W)$$
$$\otimes T(W \otimes V \otimes W) \otimes \cdots$$

In this decomposition, $V \otimes W$ embeds in $T(V \otimes W)$ via the commutator isomorphism

$$V \otimes W \xrightarrow{\cong} [V, W]$$

and similarly for

$$V \otimes V \otimes W \xrightarrow{\cong} [V, [V, W]],$$
$$W \otimes V \otimes W \xrightarrow{\cong} [W, [V, W]], \quad \cdots$$

Let A be a connected CW co-H-space. Thus we can add the two inclusions into summands, $\iota_1 : A \to A \vee A$ and $\iota_2 : A \to A \vee A$, and we can use the co-H structure to add them to get $\iota_1 + \iota_2 : A \to A \vee A$. Of course, this is just a compression of the diagonal $\delta : A \to A \times A$.

For each element ω of a Hall basis B generated by $\{\iota_1, \iota_2\}$, the Hilton–Hopf invariants h_ω are defined to be the composition

$$h_\omega = p_\omega \circ \Theta^{-1} \circ \Omega\Sigma(\iota_1 + \iota_2) : \Omega\Sigma A \to \Omega\Sigma(A \vee A) \to \Omega\Sigma(\omega(A, A))$$

where

$$p_\omega : \prod_{\gamma \varepsilon B} \Omega\Sigma(\gamma(A, A)) \to \Omega\Sigma(\omega(A, A))$$

4.3 The Hilton–Milnor theorem 117

is the projection and

$$\Theta^{-1} : \Omega\Sigma(A \vee A) \to \prod_{\gamma \varepsilon B} \Omega\Sigma(\gamma(A, A))$$

is the homotopy inverse to the homotopy equivalence in Theorem 4.3.3. For example, with respect to the Hall basis $\{\iota_1, \iota_2, [\iota_1, \iota_2], [\iota_1, [\iota_1, \iota_2]], \ldots\}$, we have countably many Hilton–Hopf invariants, among them

$$h_{[\iota_1, \iota_2]} : \Omega\Sigma A \to \Omega\Sigma(A \wedge A),$$

$$h_{[\iota_1, [\iota_1, \iota_2]]} : \Omega\Sigma A \to \Omega\Sigma(A \wedge A \wedge A).$$

Note that the Hilton–Hopf invariants depend on the Hall basis and its ordering. The following is an immediate consequence of the Hilton–Milnor decomposition:

Proposition 4.3.4. *Let $\alpha : X \to \Omega\Sigma A$ be any map. Then we have the formula*

$$\Omega\Sigma(\iota_1 + \iota_2) \circ \alpha \simeq \Sigma_{\omega \varepsilon B} \overline{\omega(\iota_1, \iota_2)} \circ h_\omega \circ \alpha$$

$$= \overline{\iota_1} \circ \alpha + \overline{\iota_2} \circ \alpha + \overline{[\iota_1, \iota_2]} \circ h_{[\iota_1, \iota_2]} \circ \alpha$$

$$+ \overline{[\iota_1, [\iota_1, \iota_2]]} \circ h_{[\iota_1, [\iota_1, \iota_2]]} \circ \alpha$$

$$+ \overline{[\iota_2, [\iota_1, \iota_2]]} \circ h_{[\iota_2, [\iota_1, \iota_2]]} \circ \alpha + \cdots$$

in which we denote the noncommutative multiplication in $\Omega\Sigma(A \vee A)$ by $+$.

Remark. We note that the multiplication would become commutative if we added another loop, that is, if we consider the formula in $\Omega^2 \Sigma A$. Equivalently, we could require the domain X to be a suspension.

Naturality allows us to extend this Proposition 4.3.4 as follows: Let $\beta : A \to B$ and $\gamma : A \to B$ be any maps. Use the co-H space structure to add them to get $\beta + \gamma : A \to A \vee A \to B$. Then we have

Corollary 4.3.5. *In $[X, \Omega\Sigma B]_*$, we have*

$$\Omega\Sigma(\beta + \gamma) \circ \alpha \simeq \overline{\beta} \circ \alpha + \overline{\gamma} \circ \alpha + \overline{[\beta, \gamma]} \circ h_{[\iota_1, \iota_2]} \circ \alpha$$

$$+ \overline{[\beta, [\beta, \gamma]]} \circ h_{[\iota_1, [\iota_1, \iota_2]]} \circ \alpha$$

$$+ \overline{[\gamma, [\beta, \gamma]]} \circ h_{[\iota_2, [\iota_1, \iota_2]]} \circ \alpha + \cdots.$$

The Hilton–Hopf invariants are natural in the sense that

Lemma 4.3.6. *If $\beta : A \to B$ is a map, then*

$$h_\omega \circ \Omega\Sigma(\beta) \circ \alpha = \Omega\Sigma(\wedge \beta) \circ h_\omega \circ \alpha$$

where $\wedge\beta : \omega(A, A) = A \wedge \cdots \wedge A \to B \wedge \cdots \wedge B \to \omega(B, B)$ is the natural map.

Exercises

(1) Show that in a graded Lie algebra as defined in this section, the following identities hold:

   (a) if x is an even degree element, then
$$2[x,x] = 0, \qquad [x,[x,x]] = 0.$$
   (b) if x is an odd degree element, then
$$3[x,[x,x]] = 0.$$

(2) If $\beta : S^n \to S^n$ and $\gamma : S^n \to S^n$ are self maps of spheres and $h = h_2 = h_{[\iota_1, \iota_2]} : \Omega\Sigma S^n \to \Omega\Sigma S^{2n}$ is the so-called second Hilton–Hopf invariant, then show that
$$\Omega\Sigma(\beta + \gamma) \simeq \Omega\Sigma(\beta) + \Omega\Sigma(\gamma) + \overline{[\beta,\gamma]} \circ h$$
where, if n is odd, we localize at 2.

(3) Let G be a grouplike space and suppose $f : X \to G$ and $g \to G$ are pointed maps.

   (a) if G is homotopy commutative show that the Samelson product $[f,g]$ is null homotopic.
   (b) if $G = \Omega\mathbb{Z}$ and $\Omega(\Sigma_\mathbb{Z}) : \Omega\mathbb{Z} \to \Omega^2\Sigma\mathbb{Z}$ is the loop of the suspension map, show that $\Omega(\Sigma_\mathbb{Z}) \circ [f,g]$ is null homotopic. (If the Whitehead product $\Sigma(X \wedge Y) \to \mathbb{Z}$ is the adjoint of the Samelson product, this says that the suspension of the Whitehead product is 0.)

4.4 The James fibrations and the EHP sequence

Let A be a co-H space. By definition the second Hilton–Hopf invariant is
$$h = h_2 = h_{[\iota_1, \iota_2]} : \Omega\Sigma A \to \Omega\Sigma(A \wedge A).$$
In this section we specialize to the case $A = S^n$ and consider the sequences
$$S^n \xrightarrow{\Sigma} \Omega S^{n+1} \xrightarrow{h} \Omega S^{2n+1}.$$

We shall consider two cases. When n is an odd integer, we shall show that the above is a fibration sequence and that it splits into a product when we localize away from 2. When n is an even integer, we shall show that the above is a fibration sequence if we localize at 2.

Let $H_*(\Omega S^{n+1}) = T[u_n] = $ the tensor algebra generated by an element u_n of degree n. Write
$$(\iota_1 + \iota_2)_*(u_n) = z + w \ \varepsilon H_*(\Omega(S^{n+1} \vee S^{n+1})) = T[z,w].$$

4.4 The James fibrations and the EHP sequence

We compute
$$(z+w)^2 = z^2 + w^2 + zw + wz$$
$$= \begin{cases} z^2 + w^2 + [z,w] & \text{if } n \text{ is odd} \\ z^2 + w^2 + 2zw - [z,w] & \text{if } n \text{ is even.} \end{cases}$$

Thus for both parities of n we have
$$h(u_n^2) = (p_{[\iota_1,\iota_2]})_*((z+w)^2) = \pm u_{2n}\varepsilon H_*(\Omega S^{2n+1}) = T[u_{2n}].$$

In a subsequent section we shall prove that

Proposition 4.4.1. *Suppose n is odd. If $h : \Omega S^{n+1} \to \Omega S^{2n+1}$ is any map such that $h_*(u_n^2) = \pm u_{2n}$, then, with any coefficient ring, h^* makes $H^*(\Omega S^{n+1})$ into a free module over $H^*(\Omega S^{2n+1})$ with basis $1, u_n$.*

We will also delay the proof of:

Proposition 4.4.2. *Suppose that $F \xrightarrow{\iota} E \xrightarrow{p} B$ is a fibration sequence such that p^* makes the cohomology ring $H^*(E)$ into a free $H^*(B)$ module with a basis $\{b_\alpha\}$. Then $\{b_\alpha\}$ restricts to a basis for $H^*(F)$. Thus*
$$H^*(E) \cong H^*(B) \otimes H^*(F)$$
as an $H^*(B)$ module.

The above propositions have the following corollary:

Corollary 4.4.3. *If n is odd, there is up to homotopy a fibration sequence*
$$S^n \xrightarrow{\Sigma} \Omega S^{n+1} \xrightarrow{h} \Omega S^{2n+1}.$$

Proof: It is clear that the composition $S^n \xrightarrow{\Sigma} \Omega S^{n+1} \xrightarrow{h} \Omega S^{2n+1}$ is null homotopic. Thus, if F is the homotopy theoretic fibre of h, we have a map $S^n \to F$. Since F has the cohomology and homology of S^n, it is easy to see that this map is a homology equivalence, hence a homotopy equivalence. $\square$

Still supposing that n is odd, consider the Samelson product $[\iota, \iota] : S^{2n} \to \Omega S^{n+1}$ where $\iota = \Sigma : S^n \to \Omega S^{n+1}$ is a generator. Its Hurewicz image is $[u_n, u_n] = 2u_n^2$. Thus the multiplicative extension $\overline{[\iota, \iota]} : \Omega\Sigma(S^{2n}) \to \Omega S^{n+1}$ induces in homology an isomorphism of $T[u_{2n}]$ onto the subalgebra of $T[u_n]$ generated by $2u_n^2$.

If we localize away from 2, for example, with coefficients $\mathbb{Z}[\tfrac{1}{2}]$ or $\mathbb{Z}_{(p)}$ for p an odd prime, then 2 is a unit and we have an isomorphism of $T[u_{2n}]$ onto the subalgebra generated by u_n^2. Thus, composition with $h : \Omega S^{n+1} \to \Omega S^{2n+1}$ is a homology bijection $\Omega\Sigma(S^{2n}) \to \Omega S^{n+1} \to \Omega S^{2n+1}$. Hence we get Serre's result which says that, at odd primes, the homotopy groups of even dimensional spheres can be expressed in terms of those of odd dimensional spheres:

Proposition 4.4.4. *If $n = 2m + 1$ is odd and we localize away from 2, then the fibration sequence*

$$S^{2m+1} \xrightarrow{\Sigma} \Omega S^{2m+2} \xrightarrow{h} \Omega S^{4m+3}$$

has a section, that is, there is a homotopy equivalence of spaces localized away from 2:

$$\Omega S^{2m+2} \simeq S^{2m+1} \times \Omega S^{4m+3}.$$

Since a retract of an H-space is an H-space, we have the following corollary:

Corollary 4.4.5. *Localized away from 2, any odd dimensional sphere S^{2n+1} is an H-space.*

Now we switch to the other parity. We will need to localize at 2.

Proposition 4.4.6. *Suppose n is even and use coefficients localized at 2. If $h : \Omega S^{n+1} \to \Omega S^{2n+1}$ is any map such that $h_*(u_n^2) = \pm u_{2n}$, then h^* makes $H^*(\Omega S^{n+1})$ into a free module over $H^*(\Omega S^{2n+1})$ with basis $1, u_n$.*

As before this gives

Corollary 4.4.7. *If n is even, there is up to homotopy a fibration sequence*

$$S^n \xrightarrow{\Sigma} \Omega S^{n+1} \xrightarrow{h} \Omega S^{2n+1}$$

localized at 2.

Thus, if we localize at 2 we have fibration sequences in both parities. The resulting long exact homotopy sequences of a fibration are called the EHP sequences:

$$\cdots \xrightarrow{\partial = P} \pi_{k+1} S^n \xrightarrow{\Sigma_* = E} \pi_{k+1} \Omega S^{n+1} \xrightarrow{h_* = H} \pi_{k+1} \Omega S^{2n+1} \xrightarrow{\partial = P}$$
$$\pi_k S^n \xrightarrow{\Sigma_* = E} \pi_k \Omega S^{n+1} \xrightarrow{h = H} \pi_k \Omega S^{2n+1} \xrightarrow{\partial = P} \cdots.$$

We have the following test for desuspension.

Corollary 4.4.8. *Suppose that spaces are localized at 2 or that n is odd. Let $\alpha : X \to \Omega S^{n+1}$ be any map. Then α desuspends up to homotopy (that is factors through $X \to S^n$) if and only if the composition $h \circ \alpha : X \to \Omega S^{n+1} \to \Omega S^{2n+1}$ is null homotopic.*

If $\Sigma : S^n \to \Omega S^{n+1}$ is the suspension, we can apply the above to decide whether the Samelson product $[\Sigma, \Sigma] : S^{2n} \to \Omega S^{n+1}$ desuspends. Recall that h is the composition

$$\Omega\Sigma(S^n) \xrightarrow{\Omega\Sigma\nu} \Omega\Sigma(S^n \vee S^n) \xrightarrow{p = \Theta^{-1} \circ p_{[\iota_1, \iota_2]}} \Omega\Sigma S^{2n}$$

where we write $\Omega\Sigma(\nu)_*(\Sigma) = \Sigma_1 + \Sigma_2$. Hence,

$$h \circ [\Sigma, \Sigma] = p([\Sigma_1 + \Sigma_2, \Sigma_1 + \Sigma_2])$$
$$= p([\Sigma_1, \Sigma_1] + [\Sigma_2, \Sigma_2] + [\Sigma_1, \Sigma_2] + [\Sigma_2, \Sigma_1])$$
$$= \begin{cases} p([\Sigma_1, \Sigma_1] + [\Sigma_2, \Sigma_2] + 2[\Sigma_1, \Sigma_2]) = 2\Sigma : S^{2n} \to \Omega\Sigma S^{2n} & \text{if } n \text{ is odd} \\ p([\Sigma_1, \Sigma_1] + [\Sigma_2, \Sigma_2]) = 0 : S^{2n} \to \Omega\Sigma S^{2n} & \text{if } n \text{ is even.} \end{cases}$$

Hence

Corollary 4.4.9. *The Samelson product* $[\Sigma, \Sigma] : S^{2n} \to \Omega\Sigma S^n$ *does not desuspend if n is odd. It does desuspend if n is even and we localize at 2.*

Exercises

(1) Show that it is not necessary to localize at 2 in Corollary 4.4.8.

4.5 James's 2-primary exponent theorem

In this section we prove the 2-primary exponent theorem of James [67]: 4^n annihilates the 2-primary components of $\pi_k(S^{2n+1})$ for all $k > 2n + 1$.

First of all, we need to make a clear distinction between the addition of maps using co-H spaces structures and the multiplication of maps using H-space structures. Suppose $f : X \to Y$ is a map. If X is a co-H space and k is an integer, then we can add the identity map of X to itself k times (using the inverse if k is negative) to get a map $[k] : X \to X$ with the property that $f \circ [k] = kf = k$ times f. The map $[k]$ is called k times the identity. For example, $[k] : S^n \to S^n$ is just the degree k map.

On the other hand, if Y is an H-space, we can multiply the identity of Y with itself k times to get a map $k : Y \to Y$ such that $k \circ f = kf = k$ times f. The map k is called the k–th power map. In the rare case when S^n is an H-space, $k : S^n \to S^n$ is the degree k map. If n is odd and we localize away from 2, then S^n is an H-space and the k-th power map will be the degree k map.

The theorem of James depends heavily on the following result. The proof of this is due to Michael Barratt who did not publish it.

Proposition 4.5.1. *If n is even then*

$$2 \circ \Omega(h) : \Omega\Sigma S^n \to \Omega\Sigma S^{2n}$$

is null homotopic, that is, 2 times the loops on the 2-nd Hilton–Hopf invariant is null.

Proof: Begin by considering the Hilton–Hopf invariant expansion of zero:
$$0 = \Omega\Sigma([1] + [-1]) = \Omega\Sigma([1]) + \Omega\Sigma([-1]) + \overline{[[1],[-1]]} \circ h.$$

Apply the Hilton–Hopf invariant to each piece:
$$h \circ \Omega\Sigma([-1]) = \Omega\Sigma([-1] \wedge [-1]) \circ h$$
$$= \Omega\Sigma([1]) \circ h$$
$$= h.$$

Recall that $[[1],[-1]]$ desuspends, that is, $[[1],[-1]] = \Sigma \circ \beta : S^{2n} \to S^n \to \Omega\Sigma S^n$ where $\beta : S^{2n} \to S^n$.

$$h \circ \overline{[[1],[-1]]} = h \circ \Omega\Sigma(\beta)$$
$$= \Omega\Sigma(\beta \wedge \beta) \circ h$$
$$= \Omega\Sigma(\beta \wedge [1_{2n}]) \circ ([1_n] \wedge \beta)) \circ h$$
$$= \Omega\Sigma(\pm\Sigma^{2n}\beta) \circ (\Sigma^n\beta)) \circ h$$
$$= 0$$

since $\Sigma^2(\beta)$ is adjoint to the null composition

$$\begin{array}{ccccc} S^{2n} & \xrightarrow{[[1],[-1]]} & \Omega\Sigma S^n & \xrightarrow{\Omega(\Sigma_\Sigma S^n)} & \Omega^2\Sigma^2 S^n \\ \uparrow= & & \uparrow \Sigma_{S^n} & \nearrow \Sigma^2_{S^n} & \\ S^{2n} & \xrightarrow{\beta} & S^n & & \end{array}$$

If we now loop the above expansion of zero and apply the H-map $\Omega(h)$ we get
$$2 \circ \Omega(h) = 0.$$

□

The James exponent theorem now follows from the following geometric desuspension theorem formulated by John Moore.

Proposition 4.5.2. *Localized at 2, there is a factorization of the 4-th power map* 4:

$$\Omega^3 S^{2n+1} \xrightarrow{\gamma} \Omega S^{2n-1} \xrightarrow{\Sigma^2} \Omega^3 S^{2n+1}.$$

Proof: We can summarize the ideas in this proof as follows: The power map 2 on $\Omega^3 S^{2n+1}$ has Hilton–Hopf invariant $\Omega^2(h) \circ 2 = 0$. Therefore, 2 desuspends to β on $\Omega^2 S^{2n}$. Since the second Hilton–Hopf invariant Ωh is quadratic, we have $\Omega h(\Omega\Sigma[-1]) \circ \beta = \Omega h \circ \beta$. Therefore $\beta - \Omega\Sigma[-1]) \circ \beta$ desuspends to γ on ΩS^{2n-1}. The double suspension $\Sigma^2(\gamma) = 4$. In more detail the argument is as follows here.

4.5 James's 2-primary exponent theorem

We consider the following diagram which incorporates loopings of James fibrations

$$\Omega S^{2n-1} \xrightarrow{\Omega(\Sigma)} \Omega^2 \Sigma S^{2n-1} \xrightarrow{\Omega^2(\Sigma)} \Omega^3 \Sigma S^{2n}$$
$$\downarrow \Omega(h) \qquad\qquad \downarrow \Omega^2(h)$$
$$\Omega^2 \Sigma S^{4n-2} \qquad\qquad \Omega^3 \Sigma S^{4n}$$

We note that Proposition 4.5.1 and $\Omega^2(h) \circ 2 = 2 \circ \Omega^2(h)$ imply that we can factor the squaring map 2 as

$$\Omega^3 \Sigma S^{2n} \xrightarrow{\beta} \Omega^2 \Sigma S^{2n-1} \xrightarrow{\Omega^2(\Sigma)} \Omega^3 \Sigma S^{2n}.$$

We know that $h \circ \Omega\Sigma([-1]) = \Omega\Sigma([-1] \wedge [-1]) \circ h = h$. Hence $\Omega(h) \circ \beta = \Omega(h) \circ \Omega^2\Sigma([-1]) \circ \beta$ and there exists a map $\gamma : \Omega^3\Sigma S^{2n} \to \Omega S^{2n-1}$ such that $\Omega(\Sigma) \circ \gamma = \beta - \Omega^2\Sigma([-1]) \circ \beta$.

Notice the following commutative diagram which relates the degree -1 map $[-1]$ and the multiplicative inverse -1.

$$S^{2n-1} \xrightarrow{\Sigma} \Omega\Sigma S^{2n}$$
$$\downarrow [-1] \qquad \downarrow -1$$
$$S^{2n-1} \xrightarrow{\Sigma} \Omega\Sigma S^{2n}.$$

Hence we get

$$\Omega^2(\Sigma) \circ \Omega(\Sigma) \circ \gamma = \Omega^2(\Sigma) \circ (\beta - \Omega^2\Sigma([-1]) \circ \beta)$$
$$= 2 - \Omega^2(\Sigma \circ [-1]) \circ \beta$$
$$= 2 - \Omega^2((-1) \circ \Sigma) \circ \beta$$
$$= 2 - \Omega^2((-1)) \circ \Omega^2(\Sigma)\beta$$
$$= 2 - (-1) \circ 2$$
$$= 4.$$

$\square$

The above proposition has an immediate corollary.

Corollary 4.5.3. *Localized at 2, we have*

(a) *The 4^n-th power map factors as*

$$\Omega^{2n+1} S^{2n+1} \to \Omega S^1 \xrightarrow{\Sigma^{2n}} \Omega^{2n+1} S^{2n+1}.$$

(b) *If $S^{2n+1}\langle 2n+1 \rangle$ is the $2n+1$ connected cover of S^{2n+1}, then the 4^n-th power is null homotopic on the $2n+1$ fold loop space*

$$\Omega^{2n+1}(S^{2n+1}\langle 2n+1 \rangle).$$

(c) *If $k > 2n+1$, then $4^n \pi_k(S^{2n+1}) = 0$.*

Exercises

(1) On the space ΩS^{2n+1} the following formulas are valid without localization:

   (a) $\Omega[2] = 2 + \overline{[[1],[1]]} \circ h$

   (b) $\Omega[3] = 3 + \overline{[[1],[1]]} \circ h$

(2) On the space $\Omega^2 S^{2n+1}$ the following formulas are valid localized at 2:

   (a) $\Omega^2[4] = 4$

   (b) For all integers $\Omega^2([k+4] = \Omega^2[k] + 4$

   (c) If r is congruent to 0 or 1 mod 4, then $\Omega^2[r] = r$. If r is congruent to 2 or 3 mod 4, then $\Omega^2[r] = r + \Omega(\overline{[[1],[1]]}) \circ \Omega(h)$.

4.6 The 3-connected cover of S^3 and its loop space

The first step in understanding the homotopy groups of S^3 is to compute the homology of its 3-connected cover $S^3\langle 3\rangle$. It is a simple consequence of the cohomology Serre spectral sequence of the fibration sequence

$$K(\mathbb{Z},2) \to S^3\langle 3\rangle \to S^3.$$

Proposition 4.6.1. *The reduced integral homology* $\overline{H}_k(S^3\langle 3\rangle; \mathbb{Z}) =$
$$\begin{cases} \mathbb{Z}/n\mathbb{Z} & \text{if } k = 2n \\ 0 & \text{if } k = 2n+1. \end{cases}$$

Proof: In the integral cohomology Serre spectral sequence we have

$$E_2^{*,*} = H^*(S^3) \otimes H^*(K(\mathbb{Z},2)) = E[v] \otimes P[u]$$

with the bidegree of $v = (3,0)$ and the bidegree of $u = (0,2)$. The only nonzero differential is d_3 and we have $d_3(u) = v$. Hence, the derivation property shows that $d_3(u^n) = nu^{n-1}v$. We get that the reduced integral cohomology $\overline{H}^k(S^3\langle 3\rangle; \mathbb{Z}) =$
$$\begin{cases} \mathbb{Z}/n\mathbb{Z} & \text{if } k = 2n+1 \\ 0 & \text{if } k = 2n. \end{cases}$$

Since the homology is finitely generated in each degree, the universal coefficient theorem implies the stated result for the homology. $\square$

Localization at a prime and the relative Hurewicz theorem yields

4.6 The 3-connected cover of S^3 and its loop space

Corollary 4.6.2. *Localized at a prime p, there are maps $P^{2p+1}(p) \to S^3\langle 3\rangle$ which induce isomorphisms of homotopy groups in dimensions $\leq 4p - 2$ and an epimorphism in dimension $4p - 1$.*

In particular, localized at a prime p, the first nonzero homotopy group of $S^3\langle 3\rangle$ is $\pi_{2p}(S^3\langle 3\rangle) = \pi_{2p}(S^3) = \mathbb{Z}/p\mathbb{Z}$.

A generator of $\pi_4(S^3) = \mathbb{Z}/2\mathbb{Z}$ is given by an element which we will see to be $\Sigma(\eta)$, the suspension of the Hopf map $\eta : S^3 \to S^2$.

If p is an odd prime, a generator of the localized group $\pi_{2p}(S^3\langle 3\rangle) = \pi_{2p}(S^3) = \mathbb{Z}/p\mathbb{Z}$ is called α_1.

The same information is conveyed by the mod p cohomology computation:

Proposition 4.6.3. *With coefficients $\mathbb{Z}/p\mathbb{Z}$, p a prime, the cohomology $H^*(S^3\langle 3\rangle; \mathbb{Z}/p\mathbb{Z}) = P[u_{2p}] \otimes E[v_{2p+1}] = $ a polynomial algebra generated by u_{2p} of degree $2p$ tensor an exterior algebra generated by v_{2p+1} of degree $2p + 1$.*

Proof: In the mod p cohomology Serre spectral sequence of the fibration sequence we get
$$E_2^{*,*} = H^*(S^3) \otimes H^*(K(\mathbb{Z}, 2) = E[v] \otimes P[u]$$
with the only nonzero differential $d_3(u^n) = nu^{n-1}v$ as before. Thus
$$E_4^{*,*} = E[u^{p-1}v] \otimes P[u^p].$$
Clearly, $E_4^{*,*} = E_\infty^{*,*}$ and the result follows. $\square$

On the other hand, we can look at the fibration of loop spaces
$$S^1 \to \Omega(S^3\langle 3\rangle) \to \Omega S^3.$$
In this case the homology Serre spectral sequence is a spectral sequence of algebras with
$$E_{*,*}^2 = H_*(S^1) \otimes H_*(\Omega S^3) = E[u_1] \otimes P[v_2]$$
and $d^2(v_2) = u_1$. Hence:

Proposition 4.6.4.

(a) *The reduced integral homology*
$$\overline{H_k}(\Omega(S^3\langle 3\rangle); \mathbb{Z}) = \begin{cases} \mathbb{Z}/n\mathbb{Z} & \text{if } k = 2n - 1 \\ 0 & \text{if } k = 2n. \end{cases}$$

(b) *If p is a prime, the mod p homology*
$$H_*(\Omega(S^3\langle 3\rangle); \mathbb{Z}/p\mathbb{Z}) = P[v_{2p}] \otimes E[u_{2p-1}]$$

with v_{2p} a polynomial generator of degree $2p$ and u_{2p-1} an exterior generator of degree $2p - 1$.

(c) Localized at a prime p, there are maps $P^{2p}(p) \to \Omega(S^3\langle 3\rangle)$ which induce isomorphisms of homotopy groups in dimensions $\leq 4p - 3$ and an epimorphism in dimension $4p - 2$.

Exercises

(1) For the complex projective plane there is a cell structure $CP^2 = S^2 \cup_\eta e^4$ where $\eta : S^3 \to S^2$. We have mod 2 cohomology ring $H^*(CP^2) = \{1, u, u^2\}$ with mod 2 Steenrod operation $Sq^2(u) = u^2$. Show that no suspension $\Sigma^k(\eta) : S^{k+3} \to S^{k+2}$ is homotopically trivial. In particular, $\Sigma(\eta)$ is a generator of $\pi_4(S^3)$.

(2) (a) Show that the iterated suspension $\Sigma^k : \pi_4(S^3) \to \pi_{k+4}(S^{k+3})$ is an isomorphism.

(b) The element $\Sigma^k(\eta)$ is said to be detected in mod 2 cohomology by Sq^2, that is, in the adjunction complex

$$S^{k+3} \cup_\beta e^{k+5},$$

the homotopy class β is homotopic to $\Sigma^k(\eta)$ if and only if Sq^2 is nontrivial in the mod 2 cohomology.

4.7 The first odd primary homotopy class

Let p be an odd prime. In this section we show that the element $\alpha_1 \varepsilon \pi_{2p}(S^3)$ is detected by the Steenrod operation P^1 in the mod p cohomology of the adjunction space $S^3 \cup_{\alpha_1} e^{2p+1}$.

We begin with an algebraic lemma. Let V be a vector space over $\mathbb{Z}/p\mathbb{Z}$ and let $T : V \to V$ be a linear operator. Suppose we are given relatively prime mod p polynomials $p_1, p_2, \ldots, p_k$ such that $(p_1 p_2 \ldots p_k)(T) = 0$ on V. Let $q_i = p_1 p_2 \ldots \hat{p_i} \ldots p_k =$ the product with the i-th factor omitted. Then we have

Lemma 4.7.1.

$$V = \bigoplus_{i=1}^{k} \ker p_i(T)$$

and there is an isomorphism

$$\ker p_i(T) \subseteq V \to \lim_{n \to \infty} (q_i(T)^n : V \to V).$$

For example let X be a co-H space and let $\tau : X \to X$ be a self map. Suppose we have integral polynomials $p_1, p_2, \ldots, p_k$ which are relatively prime mod p. Let

4.7 The first odd primary homotopy class   127

$f_1 = p_1(\tau), f_2 = p_2(\tau), \ldots, f_k = p_k(\tau) : X \to X$ be such that $(f_1)_* \circ (f_2)_* \circ \ldots (f_k)_* = 0$ in reduced mod p homology. As before, let $q_i = p_1 p_2 \ldots \hat{p}_i \ldots p_k =$ the product with the i-th factor omitted. For $1 \leq i \leq k$, let

$$X_i = \lim_{n \to \infty} (q_i(\tau)^n : X \to X)$$

be the mapping telescope. Then we can use the co-H stucture on X to add maps and we get:

Proposition 4.7.2. *The is a mod p homology isomorphism*

$$X \to \bigoplus_{i=1}^{k} X_i \ .$$

The example we want here is $X = \Sigma K(\mathbb{Z}, 2) = \Sigma CP^\infty$. Let λ be an integer which is a generator mod p of the group of units $(\mathbb{Z}/p\mathbb{Z})^*$ and let $\Lambda : CP^\infty \to CP^\infty$ be a map which induces multiplication by λ in dimension 2 integral cohomology and homology. Let $\tau = \Sigma(\Lambda) : X \to X$ be the suspension of this map. Now let $f_i = \tau - \Lambda^i$ and note that

$$(f_1)_* \circ (f_2)_* \circ \cdots (f_{p-1})_* = \tau_*^{p-1} - 1 = 0 \mod p.$$

Hence we get

Proposition 4.7.3. *There is a equivalence of spaces completed at p:*

$$\Sigma(CP^\infty) \to \bigvee_{i=1}^{p-1} X_i$$

where, if $H^*(CP^\infty; \mathbb{Z}/p\mathbb{Z}) = P[u]$ with u of degree 2, then $\overline{H^*}(X_i; \mathbb{Z}/p\mathbb{Z})$ is generated by the suspensions of

$$\{u^i, u^{i+p-1}, u^{i+2p-2}, u^{i+3p-3}, \cdots \}.$$

A closer look at the effect of the map

$$q_1(\tau) = (\tau - \Lambda^2) \circ (\tau - \Lambda^3) \circ \cdots \circ (\tau - \Lambda^{p-1})$$

in homology shows that this map is a p local equivalence in dimensions congruent to 3 mod $2p - 2$ and is zero in dimensions k with $3 < k < 2p + 1$. (It induces multiplication by a unit times a power of p in other higher dimensions. This creates copies of the rationals in the homology in these dimensions of the mapping telescope.)

Thus, localized at p, we have that the $2p + 2$ skeleton

$$X_1^{[2p+2]} = S^3 \cup e^{2p+1}$$

and, since the Steenrod operation $P^1(u) = u^p$, we have that $P^1 : H^3(X_1; \mathbb{Z}/p\mathbb{Z}) \to H^{2p+1}(X_1; \mathbb{Z}/p\mathbb{Z})$ is an isomorphism.

If we localize at p, the group $\pi_{2p}(S^3) = \mathbb{Z}/p\mathbb{Z}$ is cyclic and thus the nontrivial attaching map in $S^3 \cup e^{2p+1} \subseteq X_1$ must be a generator which we shall call α_1.

Of course the fact that α_1 is detected by the Steenrod operation P^1 shows that no suspension of α_1 is trivial.

Exercises

(1) Prove Lemma 4.7.1.

(2) Show that, in $BS^3 = HP^\infty$ localized at 3, the eight-dimensional cell is attached to the four-dimensional cell by $\pm \alpha_1$.

4.8 Elements of order 4

For an integer k and a co-H-space X recall that $[k] : X \to X$ denotes k times the identity.

The following proposition is fundamental for this section.

Proposition 4.8.1. *For all $n \geq 2$, the map $[2] : P^{n+1}(2) \to P^{n+1}(2)$ is not null homotopic.*

Proof: The cofibration sequence $S^1 \xrightarrow{[2]} S^1 \to P^2(2)$ yields the cofibration sequence $S^1 \wedge P^n(2) \xrightarrow{[2] \wedge 1} S^1 \wedge P^n(2) \to P^2(2) \wedge P^n(2)$.

Suppose $[2] : \Sigma P^n \to \Sigma P^n(2)$ is null homotopic. Then the cofibration sequence is split, that is, $P^2(2) \wedge P^n(2) \simeq P^{n+1}(2) \vee P^{n+2}(2)$.

But this is inconsistent with the Cartan formula for mod 2 Steenrod operations. Let u generate $H^1(P^2(2))$ and let $\overline{u}$ generate $H^{n-1}(P^n(2))$. Then

$$Sq^2(u \otimes \overline{u}) = Sq^1(u) \otimes Sq^1(\overline{u}) \neq 0 \varepsilon H^4(P^2(2) \wedge P^n(2)).$$

Thus the cofibration sequence cannot be split. $\square$

Proposition 4.8.2. *If $n \geq 2$, then the map $[2]$ factors as*

$$P^{n+1}(2) \xrightarrow{q} S^{n+1} \xrightarrow{\beta} S^n \xrightarrow{\iota} P^{n+1}(2)$$

where q is the map which pinches the bottom cell to a point and ι is the inclusion of the bottom cell. If $n > 2$, then $\beta = \Sigma^{k-2}(\eta) = \eta_n =$ the iterated suspension of the Hopf map.

4.8 Elements of order 4

Proof: There is a commutative diagram

$$\begin{array}{ccccc} S^n & \stackrel{\iota}{\to} & P^{n+1}(2) & \stackrel{q}{\to} & S^{n+1} \\ \downarrow [2] & & \downarrow [2] & & \downarrow [2] \\ S^n & \stackrel{\iota}{\to} & P^{n+1}(2) & \stackrel{q}{\to} & S^{n+1} \end{array}$$

where the horizontal rows are cofibration sequences. Since $\iota \circ [2] \simeq *$, there is a map γ which factors $[2]$ as

$$P^{n+1}(2) \stackrel{q}{\to} S^{n+1} \stackrel{\gamma}{\to} P^{n+1}(2).$$

Since the composition $S^{n+1} \stackrel{\gamma}{\to} P^{n+1}(2) \stackrel{q}{\to} S^{n+1}$ is degree 0 and therefore null homotopic, there is a factorization of γ as

$$S^{n+1} \stackrel{\delta}{\to} F \to P^{n+1}(2),$$

where F is the homotopy theoretic fibre of the pinch map q.

A computation with the integral homology Serre spectral sequence of the fibration sequence $F \to P^{n+1}(2) \to S^{n+1}$ shows that $\overline{H_k}(F) =$

$$\begin{cases} \mathbb{Z} & \text{if } k = mn \\ 0 & \text{otherwise.} \end{cases}$$

For more details see the exercises at the end of this section.

The $2n - 1$ skeleton of F is just S^n and hence δ factors as

$$S^{n+1} \stackrel{\beta}{\to} S^n \to F.$$

Finally, $[2]$ factors as

$$P^{n+1}(2) \stackrel{q}{\to} S^{n+1} \stackrel{\beta}{\to} S^n \stackrel{\iota}{\to} P^{n+1}(2).$$

Since we know that $[2]$ is not null homotopic, we must have that β is not null homotopic. If $n > 2$ then $\pi_{n+1}(S^n) \cong \mathbb{Z}/2\mathbb{Z}$ and $\beta \simeq \eta_n$. $\square$

An immediate consequence is

Corollary 4.8.3. *If $n > 2$, the map $[4] : P^{n+1}(2) \to P^{n+1}(2)$ is null homotopic and hence the mod 2 homotopy groups satisfy $4\pi_{n+1}(X; \mathbb{Z}/2\mathbb{Z}) = 0$ for all spaces X.*

Let G be an H-space and let $G\{2\}$ be the homotopy theoretic fibre of the squaring map $2 : G \to G$. Thus there is a fibration sequence

$$\Omega G \xrightarrow{2 = \Omega 2} \Omega G \stackrel{\partial}{\to} G\{2\} \stackrel{\iota}{\to} G \stackrel{2}{\to} G.$$

We write $\eta_n = \Sigma^{n-2}\eta : S^{n+1} \to S^n$ for the iterated suspension of the Hopf map.

Proposition 4.8.4. *Suppose that $n > 3$ and there exists $\beta \varepsilon \pi_n G$ such that $2\beta = 0$ and $\beta \circ \eta_n$ is not divisible by 2. Then there exists an element of order 4 in $\pi_n G\{2\}$.*

Proof: We shall construct a commutative diagram in which the top row is a cofibration sequence and the bottom row is the above fibration sequence

$$\begin{array}{ccccccc}
P^n(2) & \to & P^{n+1}(2) & \to & P^{n+1}(4) & \to & P^{n+1}(2) \\
\downarrow \bar{\gamma} & & \downarrow \delta & & \downarrow \epsilon & & \downarrow \gamma \\
\Omega G & \xrightarrow{2=\Omega 2} & \Omega G & \xrightarrow{\partial} & G\{2\} & \xrightarrow{\iota} & G \xrightarrow{2} G.
\end{array}$$

By the Peterson–Stein formula, we can choose $\bar{\gamma}$ and γ to be adjoint maps. Since $2\beta = 0$, there is an extension of β to $\gamma : P^{n+1}(2) \to G$.

There are commutative diagrams

$$\begin{array}{ccc}
P^n(2) & \xrightarrow{q} & S^n \\
\downarrow = & & \downarrow \eta_{n-1} \\
P^n(2) & & S^{n-1} \\
\downarrow = & & \downarrow \iota \\
P^n(2) & \xrightarrow{[2]} & P^n(2) \\
\downarrow \bar{\gamma} & & \downarrow \bar{\gamma} \\
\Omega G & \xrightarrow{2} & \Omega G.
\end{array}$$

Since $2\eta_{n-1} = 0$ we can extend the right-hand vertical composition to a map $\delta : P^{n+1}(2) \to \Omega G$.

Recall the diagram below in which all rows and columns are cofibration sequences

$$\begin{array}{ccccccc}
* & \to & P^n(2) & \xrightarrow{=} & P^n(2) \\
\downarrow & & \downarrow q & & \downarrow \\
S^n & \xrightarrow{[2]} & S^n & \xrightarrow{\iota} & P^{n+1}(2) \\
\downarrow = & & \downarrow [2] & & \downarrow \\
S^n & \xrightarrow{[4]} & S^n & \xrightarrow{\iota} & P^{n+1}(2).
\end{array}$$

Hence we have constructed the left-hand square in the first diagram of this proof. But, by the Peterson–Stein formula, the remainder of the diagram follows.

If we apply π_n to this diagram we get horizontal exact sequences

$$\begin{array}{ccccc}
 & & \mathbb{Z}/2\mathbb{Z} & \to & \mathbb{Z}/4\mathbb{Z} & \to & \mathbb{Z}/2\mathbb{Z} \\
 & & \downarrow \delta_* & & \downarrow \epsilon_* & & \downarrow \gamma_* \\
\pi_n \Omega G & \xrightarrow{2} & \pi_n \Omega G & \xrightarrow{\partial_*} & \pi_n G\{2\} & \xrightarrow{\iota_*} & \pi_n G.
\end{array}$$

Since $\delta_*(1)$ = the adjoint of $\beta \circ \eta_n$ is not divisible by 2, it follows that $2\epsilon_*(1) \neq 0$ and $\epsilon_*(1)$ has order 4. $\square$

We now formulate a convenient stable version of the preceding proposition.

Recall the definition of the stable homotopy group
$$\pi_n^S(X) = \lim_{k \to \infty} \pi_n(\Omega^k \Sigma^k X) = \pi_n(\Omega^\infty \Sigma^\infty X)$$
where $\Omega^\infty \Sigma^\infty X$ is the direct limit
$$X \to \Omega \Sigma X \Omega^2 \Sigma^2 X \to \Omega^3 \Sigma^3 X \to \cdots$$
Note that $\Omega^\infty \Sigma^\infty X$ is an H-space so that Proposition 4.8.4 can be applied to it. If X is a co-H-space and $X[k]$ is the cofibre of $[k] : X \to X$, then the Peterson–Stein formula gives a commutative diagram

$$\begin{array}{ccccccc} X & \xrightarrow{[k]} & X & \to & X[k] & \to & \Sigma X & \xrightarrow{[k]} & \Sigma X \\ \downarrow & & \downarrow & & \downarrow & & \downarrow & & \downarrow \\ \Omega^\infty \Sigma^\infty X & \xrightarrow{k} & \Omega^\infty \Sigma^\infty X & \to & (\Omega^\infty \Sigma^{\infty+1} X)\{k\} & \to & \Omega^\infty \Sigma^{\infty+1} X & \xrightarrow{k} & \Omega^\infty \Sigma^{\infty+1} X. \end{array}$$

If X is $k - 1$ connected, then $X \to \Omega^\infty \Sigma^\infty X$ induces a homotopy isomorphism in dimensions $< 2k - 1$. Hence we have

Corollary 4.8.5. *Let X be a $k - 1$ connected co-H-space. Suppose that $3 < n < 2k - 1$ and there exists $\beta \in \pi_{n-1} X$ such that $2\beta = 0$ and $\beta \circ \eta_n$ is not divisible by 2. Then there exists an element of order 4 in $\pi_n(X[2])$.*

Exercises

(1) Let k be an integer and let $F^m\{k\}$ be the homotopy theoretic fibre of the pinch map $P^m(k) \xrightarrow{q} S^m$.

  (a) Show that the Peterson–Stein formula gives a commutative diagram

  $$\begin{array}{ccccccccc} P^{m-1} & \xrightarrow{q} & S^{m-1} & \xrightarrow{[k]} & S^{m-1} & \to & P^m(k) & \xrightarrow{q} & S^m \\ \downarrow \Sigma & & \downarrow \Sigma & & \downarrow \iota & & \downarrow = & & \downarrow = \\ \Omega P^m(k) & \xrightarrow{\Omega(q)} & \Omega S^m & \to & F^m\{k\} & \to & P^m(k) & \xrightarrow{q} & S^m \end{array}$$

  in which the top row is a cofibration sequence and the bottom row is a fibration sequence.

  (b) Use the above to show that $\pi_{m-1}(F^m\{k\})$ is infinite cyclic with generator represented by $\iota : S^{m-1} \to F^m\{k\}$.

(2) Use the integral homology Serre spectral sequence of the fibration sequence $F^m\{k\} \to P^m(k) \xrightarrow{q} S^m$ to show that $\overline{H_k}(F^m\{k\}) =$
$$\begin{cases} \mathbb{Z} & \text{if } k = r(m-1), r > 0 \\ 0 & \text{otherwise.} \end{cases}$$

4.9 Computations with the EHP sequence

In this section we use the EHP sequence to compute the following little table of homotopy groups of the spheres S^3 and S^2 localized at 2. Of particular interest is the fact that $\pi_6(S^3) = \mathbb{Z}/4\mathbb{Z}$ localized at 2, showing that in this one case the result of James on the 2-primary exponent is the best possible. Until we say otherwise, everything in this section is localized at 2.

	$\pi_k(S^3)$	generator	$\pi_k(S^2)$	generator
$k=3$	$\mathbb{Z}_{(2)}$	ι_3	$\mathbb{Z}_{(2)}$	η_2
$k=4$	$\mathbb{Z}/2\mathbb{Z}$	η_3	$\mathbb{Z}/2\mathbb{Z}$	$\eta_2 \circ \eta_3$
$k=5$	$\mathbb{Z}/2\mathbb{Z}$	$\eta_3 \circ \eta_4$	$\mathbb{Z}/2\mathbb{Z}$	$\eta_2 \circ \eta_3 \circ \eta_4$
$k=6$	$\mathbb{Z}/4\mathbb{Z}$	ν	$\mathbb{Z}/4\mathbb{Z}$	$\eta_2 \circ \nu$

In this table, $\eta_k = \Sigma^{k-2}\eta : S^{k+1} \to S^k$ denotes the $(k-2)$-nd suspension of the Hopf map $\eta = \eta_2 : S^3 \to S^2$.

Note that the Hopf fibration $S^1 \to S^3 \xrightarrow{\eta} S^2$ makes the half of this table referring to S^2 automatic since $\eta_* : \pi_k S^3 \to \pi_k S^2$ is an isomorphism for $k \geq 3$.

What follows is the verification of this table.

Localized at 2, we have the fibration sequence
$$S^2 \xrightarrow{\Sigma} \Omega S^3 \xrightarrow{h} \Omega S^5$$
and its long exact sequence
$$\pi_5 S^2 \xrightarrow{\Sigma} \pi_6 S^3 \xrightarrow{h} \pi_6 S^5 \xrightarrow{P}$$
$$\pi_4 S^2 \xrightarrow{\Sigma} \pi_5 S^3 \xrightarrow{h} \pi_5 S^5 \xrightarrow{P}$$
$$\pi_3 S^2 \xrightarrow{\Sigma} \pi_4 S^3 \xrightarrow{h} \pi_4 S^5 = 0.$$

We start by knowing that $\pi_3 S^2$ is $\mathbb{Z}_{(2)}$ generated by the Hopf map η. From looking at the 3-connected cover $S^3\langle 3 \rangle$, we know that $\pi_4 S^3$ is $\mathbb{Z}/2\mathbb{Z}$ and the above exact sequence shows that it is generated by the suspension $\Sigma \eta = \eta_3$.

Of course, it follows that $\pi_4 S^2 = \mathbb{Z}/2\mathbb{Z}$ generated by $\eta_2 \circ \eta_3$.

The map $P : \pi_5 S^5 = \mathbb{Z}_{(2)} \to \pi_3 S^2 = \mathbb{Z}_{(2)}$ is multiplication by 2, a monomorphism. Hence $\Sigma : \pi_4 S^2 \to \pi_5 S^3$ is an epimorphism and $\Sigma(\eta_2 \circ \eta_3) = \eta_3 \circ \eta_4$ generates the image. The next lemma shows that $\eta_3 \circ \eta_4 \neq 0$ and hence $\pi_5 S^3 = \mathbb{Z}/2\mathbb{Z}$ generated by $\eta_3 \circ \eta_4$.

Lemma 4.9.1. *For all $k \geq 2$, the composition*
$$\eta_k \circ \eta_{k+1} : S^{k+2} \to S^{k+1} \to S^k$$
is nontrivial.

4.9 Computations with the EHP sequence   133

Proof: Suppose it is trivial. Then we have a homotopy commutative diagram

$$\begin{array}{ccc} S^{k+2} & \to & * \\ \downarrow \eta_{k+1} & & \downarrow \\ S^{k+1} & \xrightarrow{\eta_k} & S^k. \end{array}$$

We expand this diagram to a homotopy commutative diagram in which the rows and columns are cofibration sequences

$$\begin{array}{ccccc} S^{k+2} & \to & * & \to & S^{k+3} \\ \downarrow \eta_{k+1} & & \downarrow & & \downarrow \\ S^{k+1} & \xrightarrow{\eta_k} & S^k & \to & S^k \cup_{\eta_k} e^{k+2} \\ \downarrow & & \downarrow = & & \downarrow \\ S^{k+1} \cup_{\eta_{k+1}} e^{k+3} & \to & S^k & \to & S^k \cup e^{k+2} \cup e^{k+4}. \end{array}$$

In the mod 2 cohomology of both of the above 2-cell complexes, we have a nontrivial Sq^2. It follows that in the mod 2 cohomology of the above 3-cell complex, we have a nontrivial $Sq^2 Sq^2$. But the Adem relation $Sq^2 Sq^2 = Sq^3 Sq^1$ is a contradiction. Hence, the composition $\eta_k \circ \eta_{k+1} \neq 0$. □

Once again we automatically have that $\pi_5 S^2 = \mathbb{Z}/2\mathbb{Z}$ generated by $\eta_2 \circ \eta_3 \circ \eta_4$.

We need to recall the Freudenthal suspension theorem: the suspension $\Sigma : S^3 \to \Omega S^4$ induces isomorphisms of homotopy groups in dimensions ≤ 4 and an epimorphism in dimension 5. An even larger range is true for the suspensions between higher dimensional spheres. But we know that $\pi_5 S^3 = \mathbb{Z}/2\mathbb{Z}$ generated by $\eta_3 \circ \eta_4$ and Lemma 4.9.1 asserts that all suspensions of this are nontrivial. Hence we certainly know that $\pi_{k+1} S^k = \mathbb{Z}/2\mathbb{Z}$ generated by η_k and $\pi_{k+2} S^k = \mathbb{Z}/2\mathbb{Z}$ generated by $\eta_k \circ \eta_{k+1}$ for all $k \geq 3$.

Recall that Corollary 4.6.2 says that we have a map $P^5(2) \to S^3\langle 3 \rangle$ which induces an isomorphism of homotopy groups in dimensions ≤ 6 and an epimorphism in dimension 7. Since $P^5(2)$ is the cofibre of $[2] : S^4 \to S^4$, Corollary 4.7.5 applies. In particular, the homotopy fibration sequence of the pinch map, $F \to P^5(2) \xrightarrow{q} S^5$, has F being homologically equivalent to S^4 in dimensions less than 8 and leads to a short exact sequence

$$0 \to (\pi_6 S^4 = \mathbb{Z}/2\mathbb{Z}) \to \pi_6 P^5(2) \to (\pi_6 S^5 = \mathbb{Z}/2\mathbb{Z}) \to 0.$$

This exact sequence depends on the facts: $2\eta_4 = 0$ and $\eta_4 \circ \eta_5$ is not divisible by 2. The corollary implies that there exists an element of order 4 in $\pi_6(P^5(2)) = \mathbb{Z}/4\mathbb{Z}$.

The corresponding element of order 4 in $\pi_6 S^3 = \mathbb{Z}/4\mathbb{Z}$ is called ν and we also have that $\pi_6 S^2 = \mathbb{Z}/4\mathbb{Z}$ generated by $\eta_2 \circ \nu$. The completes the confirmation of the table localized at 2.

We remark that, if we do not localize, we have that
$$\pi_6 S^3 = \mathbb{Z}/4\mathbb{Z} \oplus \mathbb{Z}/3\mathbb{Z} = \mathbb{Z}/12\mathbb{Z}$$
with generators ν and α_1.

Exercises

(1) If p is an odd prime, show that the composition
$$\alpha_1 \circ (\Sigma^{2p-3}\alpha_1) : S^{4p-3} \to S^{2p} \to S^3$$
is not null homotopic. (Hint: Use the Adem relation
$$P^1 \circ P^1 = 2P^2$$
and the fact that P^2 must vanish on a three-dimensional class.)

5 James–Hopf invariants and Toda–Hopf invariants

We begin this chapter by introducing the important concept of a divided power algebra. Divided power algebras arise in the description of the cohomology algebras of several important spaces, among them loops on spheres. These algebras have interesting properties when localized at primes. In particular, these properties give rise to Hopf invariants of two types. The first type is due to James [66] and Hilton [54]. The second type is due to Toda [130].

The James construction allows two definitions of Hopf invariants. The so-called combinatorial definition is probably the simplest and most attractive. Another one, the so-called decomposition definition, is based on James' natural splitting of the suspension of the James construction.

As we have seen in the previous chapter, the Hilton–Milnor theorem leads to the Hilton–Hopf invariants. Certain Lie identities are required to control the homological properties of the Hilton–Hopf invariants.

These three constructions of Hopf invariants are for the most part interchangeable for all practical purposes (although the Hilton–Hopf invariants are more tractable concerning the issues which arise in the 2-primary exponent theorem for the homotopy groups of spheres). The most important properties which they share are the homological properties necessary to construct the James fibrations and the p-adic naturality which is used in the proof of exponent theorems.

Localized at odd primes, the James fibrations involve filtrations of the James constructions. The homology of the loops on these filtrations was first studied by Toda. This leads to the Toda–Hopf invariants and to the Toda fibrations. The Toda fibrations also possess a p-adic naturality which comes into the proof of Toda's odd primary exponent theorem. If p is an odd prime, then p^{2n} annihilates the p-primary component of the homotopy groups $\pi_k(S^{2n+1})$ for all $k > 2n + 1$. We prove this in the geometric form introduced by John Moore. In this form it is a theorem about power maps on loop spaces.

As we shall see later in this book, this result is not the best possible exponent. First, Paul Selick showed that, for p an odd prime, the p-primary component of all the homotopy groups of S^3 are annihilated by p. Almost immediately after Selick's result, Cohen, Moore, and Neisendorfer used completely different methods to prove that p^n annihilates the odd primary components of S^{2n+1}, at least when $p > 3$. This proof will be presented later in this book. The technical difficulties that were overcome by Neisendorfer to extend this result to the prime 3 will not be in this book.

Finally, we note that Brayton Gray showed that, for p an odd prime, there are elements of order p^n in the homotopy groups of S^{2n+1}. Serre's odd primary splitting,

$$\Omega S^{2n+2} \simeq S^{2n+1} \times \Omega S^{4n+3},$$

shows that the case of even dimensional spheres reduces to the case of odd dimensional spheres. Thus, the best possible odd primary exponent for the spheres is known.

5.1 Divided power algebras

The cohomology of the loops on an odd dimensional sphere is given by a divided power algebra. The algebraic structure of these algebras is fundamental to the existence of fibrations related to Hopf invariants.

Definition 5.1.1. A divided power algebra $\Gamma[x]$ generated by an element x of even degree $2n$ is the algebra with basis

$$\{1 = \gamma_0(x), x = \gamma_1(x), \gamma_2(x), \ldots, \gamma_k(x), \ldots\}$$

with the degree of $\gamma_k(x)$ equal to $2nk$ and multplication given by $\gamma_i(x)\gamma_j(x) = (i,j)\gamma_{i+j}(x)$. The divided power algebra has the structure of a Hopf algebra with the diagonal given by

$$\Delta(\gamma_k(x)) = \Sigma_{i=0}^{k} \gamma_i(x) \otimes \gamma_{k-i}(x).$$

Remark. The symbol (i,j) is the symmetric representation of the binomial coefficient

$$(i,j) = \binom{i+j}{i} = \frac{(i+j)!}{(i!)(j!)}.$$

It is immediate that $\gamma_1(x)\gamma_k(x) = (k+1)\gamma_{k+1}(x)$ and thus

$$x^k = \gamma_1(x)^k = k!\gamma_k(x).$$

If $k!$ is a unit in the ground ring, then $\gamma_k(x) = \frac{x}{k!}$. Hence, if the ground ring is the rationals, $\Gamma[x]$ is isomorphic to the polynomial algebra $P[x]$, even as a Hopf algebra. But over a field of characteristic p, it is very different, for example, $\gamma_1(x)^p = 0$.

5.1 Divided power algebras

Lemma 5.1.2.

(a) *If $P[y] = T[y]$ is the primitively generated polynomial algebra generated by an element y of even degree $2n$, then the dual Hopf algebra is*

$$P[y]^* = \Gamma[x]$$

where x has degree $2n$ and

$$\{\gamma_0(x), \gamma_1(x), \gamma_2(x) \ldots, \gamma_k(x), \ldots\}$$

is the dual basis to

$$\{1, y, y^2, \ldots, y^k, \ldots\},$$

that is, $\langle \gamma_i(x), y^j \rangle = \delta_{ij}$.

(b) *if $T[x]$ is the primitively generated tensor algebra generated by an element y of odd degree $2n + 1$, then the dual Hopf algebra is*

$$T[y]^* = E[x] \otimes \Gamma[w]$$

as an algebra where x is a primitive element of degree $2n + 1$, w has degree $4n + 2$, and the comultiplication satisfies

$$\Delta(w) = w \otimes 1 + x \otimes x + 1 \otimes w.$$

Proof:

(a) Let $\mu : P[y] \otimes P[y] \to P[y]$ be the multiplication and let $\nu : P[y] \to P[y] \otimes P[y]$ be the comultiplication of the polynomial algebra. Then μ^* and ν^* are the comultiplication and multiplication, respectively, of the dual Hopf algebra. We let

$$\{\gamma_0(x), \gamma_1(x), \gamma_2(x) \ldots, \gamma_k(x), \ldots\}$$

represent the dual basis to the basis of powers of y and we shall determine the formulas for the multiplication and comultiplication.

We have

$$\langle \mu^*(\gamma_k(x)), y^i \otimes y^j \rangle = \langle \gamma_k(x), \mu(y^i \otimes y^j) \rangle$$
$$= \langle \gamma_k(x), y^{i+j} \rangle = \delta_{k, i+j}$$

which implies that

$$\mu^*(\gamma_k(x)) = \Sigma_{i+j=k} \gamma_i(x) \otimes \gamma_j(x).$$

We have

$$\langle \nu * (\gamma_i(x) \otimes \gamma_j(x)), y^k \rangle = \langle \gamma_i(x) \otimes \gamma_j(x), \nu(y^k) \rangle$$
$$= \langle \gamma_i(x) \otimes \gamma_j(x), (\nu(y))^k) \rangle = \langle \gamma_i(x) \otimes \gamma_j(x), (y \otimes 1 + 1 \otimes y)^k \rangle$$
$$= \langle \gamma_i(x) \otimes \gamma_j(x), \Sigma_{a+b=k}(a,b) y^a \otimes y^b \rangle = (i,j) \delta_{ia} \delta_{jb}$$

which implies that
$$\nu * (\gamma_i(x) \otimes \gamma_j(x)) = (i,j)\gamma_{i+j}(x).$$
It follows that the dual Hopf algebra of the primitively generated polynomial algebra $P[y]$ is exactly the divided power algebra $\Gamma[x]$.

(b) As for the second part, we note that
$$T[y] = E[y] \otimes P[y^2]$$
as a coalgebra. This is the case since the composition
$$E[y] \otimes P[y^2] \to T[y] \otimes T[y] \xrightarrow{\text{mult}} T[y]$$
is a map of coalgebras and an isomorphism.
Hence, the dual is
$$(T[y])^* = (E[y])^* \otimes (P[y^2])^* = E[x] \otimes \Gamma[w]$$
as an algebra. Here, $\langle x, y \rangle = 1 = \langle w, y^2 \rangle$.
The element x is clearly primitive and $\mu(y \otimes y) = y^2$ implies that
$$\Delta(w) = \mu^*(w) = w \otimes 1 + x \otimes x + 1 \otimes w.$$

$\square$

Since the homology Hopf algebra of the loops on a sphere is given by a tensor algebra generated by a single element, we get:

Corollary 5.1.3.

(a) *The cohomology Hopf algebra $H^*(\Omega S^{2n+1})$ is isomorphic to $\Gamma[x]$ with degree x equal to $2n$.*

(b) *The cohomology Hopf algebra $H^*(\Omega S^{2n+2})$ is isomorphic to $E[x] \otimes \Gamma[w]$ where x is a primitive element of degree $2n+1$ and w is an element of degree $4n+2$ with comultiplication $\Delta(w) = w \otimes 1 + x \otimes x + 1 \otimes w$.*

Recall the second Hopf invariant $h: \Omega S^{2n+2} \to \Omega S^{4n+3}$. The above corollary implies that the induced cohomology map h^* makes $H^*(\Omega S^{2n+2})$ into a free module over $H^*(\Omega S^{4n+3})$ with $\{1, x\}$ as a basis. Thus it implies the existence of the fibration sequence up to homotopy as in Proposition 4.4.1, Proposition 4.4.2, and Corollary 4.4.3:
$$S^{2n+1} \to \Omega S^{2n+2} \xrightarrow{h} \Omega S^{4n+3}.$$

In order to show the existence of the fibration sequence localized at 2,
$$S^{2n} \to \Omega S^{2n+1} \xrightarrow{h} \Omega S^{4n+1},$$
we need a bit more algebra. First, induction on n shows

Lemma 5.1.4. *Over a subring of the rationals, we have*
$$(\gamma_k(x))^n = \frac{(kn)!}{(k!)^n} \gamma_{kn}(x).$$

A special case of the above is $x^n = n! \gamma_n(x)$.

There is a well known lemma on binomial coefficients mod a prime:

Let p be a prime and let
$$b = \Sigma_{i=0}^m b_i p^i, \qquad a = \Sigma_{i=0}^m a_i p^i$$
be p-adic expansions with $0 \leq a_i, b_i \leq p-1$. Then

Lemma 5.1.5. *Mod p,*
$$\binom{b}{a} = \prod_{i=0}^m \binom{b_i}{a_i}$$

Proof: The convention is that
$$\binom{b}{a} = 0$$
if $a > b$.

In the polynomial ring $\mathbb{Z}/p\mathbb{Z}[x]$
$$(1+x)^b = \Sigma_{a=0}^b \binom{b}{a} x^a.$$

But
$$(1+x)^b = \prod_i (1+x^{p^i})^{b_i}$$
$$= \prod_i \left(\Sigma_{j=0}^{b_i} \binom{b_i}{j} x^{jp^i} \right)$$

and the result follows from the uniqueness of the p-adic expansion of a. □

Elementary number theory includes the result that the highest power p^α of a prime p that divides a factorial $b!$ is given by
$$\alpha = \nu_p(b!) = \left[\frac{b}{p}\right] + \left[\frac{b}{p^2}\right] + \cdots + \left[\frac{b}{p^k}\right] + \cdots.$$

This implies that the following improvement of a special case of Lemma 5.1.5:

Lemma 5.1.6.
$$\binom{p^k b}{p^k a} = u \binom{b}{a}$$

where u is a unit in $\mathbb{Z}_{(p)}$, the integers localized at p.

We apply the above lemmas to the following subalgebras of the divided power algebra.

Definition 5.1.7. If $\Gamma[x]$ is the divided power algebra generated by an even degree element x and r is a positive integer, let $\Gamma_r[x]$ denote the subalgebra concentrated in degrees which are multiples of r and spanned by the set $\{1, \gamma_r(x), \gamma_{2r}(x), \ldots, \gamma_{kr}(x), \ldots\}$.

The lemmas imply that

Proposition 5.1.7. *For all $r \geq 0$ and with ground ring $\mathbb{Z}_{(p)}$*

(a) *The divided power algebra*
$$\Gamma[\gamma_{p^r}(x)] = \Gamma_{p^r}[x].$$

(b) $\Gamma[x]$ *is a free $\Gamma_{p^r}[x]$ module with a basis*
$$\{1, \gamma_1(x), \gamma_2(x), \ldots, \gamma_{p^r - 1}(x)\}.$$

Proof: A basis for $\Gamma[\gamma_{p^r}(x)]$ is
$$\gamma_n(\gamma_{p^r}(x)) = \frac{1}{n!}\{\gamma_{p^r}(x)\}^n = \frac{(np^r)!}{(p^r!)^n(n!)}\gamma_{np^r}(x)$$

and
$$\nu_p\left(\frac{(np^r)!}{(p^r!)^n(n!)}\right) = 0$$

shows that the coefficient is a unit in $\mathbb{Z}_{(p)}$. This proves part (a).

Part (b) follows from the fact that, if $0 \leq i < p^r$, then
$$\gamma_{kp^r}(x)\gamma_i(x) = \binom{kp^r + i}{i}\gamma_{kp^r + i}(x) = u\gamma_{kp^r + i}(x)$$

where u is a unit in $\mathbb{Z}_{(p)}$. □

Setting $p = 2$ and $r = 1$ gives Corollary 4.4.6, that there is a fibration sequence localized at 2,
$$S^{2n} \to \Omega S^{2n+1} \xrightarrow{h} \Omega S^{4n+1}.$$

We leave the next lemma as an exercise

Lemma 5.1.8. *With ground ring $\mathbb{Z}_{(p)}$:*

(a) $\{\gamma_{p^k}(x)\}^p = pu\gamma_{p^{k+1}}(x)$ *where u is a unit.*

(b) $i < p$ *implies that* $\{\gamma_{p^k}(x)\}^i = v\gamma_{ip^k}(x)$ *where v is a unit.*

(c) If $n = i_0 + i_1 p + i_2 p^2 + \cdots + i_k p^k$ is the p-adic expansion with $0 \leq i_j < p$ for all $0 \leq j \leq k$, then

$$\gamma_n(x) = u\gamma_{i_0}(x)\gamma_{i_1 p}(x)\ldots\gamma_{i_k p^k}(x)$$
$$= v\gamma_1(x)^{i_0}\gamma_p(x)^{i_1}\ldots\gamma_{p^k}(x)^{i_k}.$$

where u and v are units.

If we reduce mod p, the above lemma says the following:

Proposition 5.1.9. *With $\mathbb{Z}/p\mathbb{Z}$ as ground ring, there is an isomorphism of algebras with a tensor product of truncated polynomial algebras*

$$\Gamma[x] = \bigotimes_{k=0}^{\infty} P[\gamma_{p^k}(x)]/\{\gamma_{p^k}(x)^p = 0\}$$
$$= P[x]/\{x^p = 0\} \otimes P[\gamma_p(x)]/\{\gamma_p(x)^p = 0\} \otimes \cdots \otimes P[\gamma_{p^k}(x)]/\{\gamma_{p^k}(x)^p = 0\} \otimes \cdots.$$

Exercises

(1) Prove Lemma 5.1.8.

5.2 James–Hopf invariants

The James construction leads to two definitions of Hopf invariants. They are both natural constructions and both share with the Hilton–Hopf invariants the fact that they lead to EHP fibration sequences. The Hilton–Hopf invariants have the one advantage that they are more suited to the distributivity questions that arise in the proof of the 2-primary exponent theorem for the homotopy groups of spheres. On the other hand, the Hopf invariants that arise from the James construction have an attractive simplicity. In order to be complete, we shall in this and in the subsequent section present three different definitions of Hopf invariants.

In this and subsequent sections of this chapter we shall assume that all spaces have the homotopy type of CW complexes.

Let k be a positive integer. First we give James' combinatorial definition of the k-th Hopf invariant $h_k : J(X) \to J(X^{\wedge k})$ where $X^{\wedge k} = X \wedge X \wedge \cdots \wedge X$ is the k-fold smash product. We shall denote a point in $X^{\wedge k}$ by $\overline{x_1 \wedge x_2 \wedge \cdots \wedge x_k}$.

Recall that $J(X) = \{x_1 x_2 \ldots x_n | x_i \, \varepsilon X\}$ is the free monoid generated by X with the single relation that the basepoint $*$ of X is the unit. It has an increasing filtration given by $J_n(X) =$ the subspace of words of length $\leq n$.

We define
$$h_k(x_1 x_2 \ldots x_n) = * \quad \text{if } n < k$$
$$= \overline{x_1 \wedge x_2 \wedge \cdots \wedge x_k} \quad \text{if } n = k$$
$$= \prod \overline{x_{i_1} \wedge x_{i_2} \wedge \cdots \wedge x_{i_k}} \quad \text{if } n > k$$

where the product is taken over all length k subsequences of $x_1 x_2 \ldots x_n$ and the product of the subsequences is taken in the lexicographic ordering.

For example,
$$h_2(x_1 x_2 x_3) = (\overline{x_1 x_2})(\overline{x_1 x_3})(\overline{x_2 x_3})$$
$$h_2(x_1 x_2 x_3 x_4) = (\overline{x_1 x_2})(\overline{x_1 x_3})(\overline{x_1 x_4})(\overline{x_2 x_3})(\overline{x_2 x_4})(\overline{x_3 x_4})$$

It is easy to check that the so-called k-th combinatorial James–Hopf invariant $h_k : J(X) \to J(X^{\wedge k})$ is a well defined continuous map. It is an extension of the canonical map
$$h_k : J_k(X) \to X^{\wedge k}.$$

Furthermore, it is a natural transformation of functors: if $f : X \to Y$ is a continuous pointed map, then

$$\begin{array}{ccc} J(X) & \xrightarrow{h_k} & J(X^{\wedge k}) \\ \downarrow J(f) & & \downarrow J(f^{\wedge n}) \\ J(Y) & \xrightarrow{h_k} & J(Y^{\wedge k}) \end{array}$$

commutes.

Since

$$\begin{array}{ccc} J(X) & \xrightarrow{h_k} & J(X^{\wedge k}) \\ \uparrow \iota & & \uparrow \iota \\ J_k(X) & \xrightarrow{h_k} & X^{\wedge k} \end{array}$$

commutes, we get

Proposition 5.2.1. *If $H_*(X)$ is torsion free, the k-th combinatorial James–Hopf invariant satisfies*
$$h_{k*}(\alpha_1 \otimes \alpha_2 \otimes \cdots \otimes \alpha_k) = \overline{\alpha_1 \otimes \alpha_2 \otimes \cdots \otimes \alpha_k}$$
where $\alpha_1 \otimes \alpha_2 \otimes \cdots \otimes \alpha_k \, \varepsilon H_*(J(X))$ and $\overline{\alpha_1 \otimes \alpha_2 \otimes \cdots \otimes \alpha_k} \, \varepsilon H_*(X^k))$.

We always have that the composition
$$J_{k-1}(X) \xrightarrow{\iota} J(X) \xrightarrow{h_k} J(X^{\wedge k})$$

5.2 James–Hopf invariants   143

is the trivial map to the basepoint. It is natural to ask when it is a fibration sequence up to homotopy.

Consider the case $k = 2$ and $X = S^n$. Although we used the second Hilton–Hopf invariant in the previous chapter, the very same arguments show that the second combinatorial James–Hopf invariant satisfies:

$$S^n = J_1(S^n) \xrightarrow{\iota} J(S^n) \xrightarrow{h_2} J(S^{4n})$$

is a fibration sequence up to homotopy if n is odd and a fibration sequence up to homotopy localized at 2 if n is even.

Consider the case $k = p^r$ where $p = $ a prime and $X = S^{2n}$. If we localize at p, then Proposition 5.1.7 shows that the p^r-th combinatorial James–Hopf invariant makes $H^*(J(S^{2n}))$ into a free $H^*(J(S^{2np^r}))$ module, that is, there is an isomorphism of $H^*(J(S^{2np^r}))$ modules:

$$H^*(J(S^{2n})) \cong H^*(J(S^{2np^r})) \otimes H^*(J_{p^r-1}(S^{2n}))$$

Proposition 4.4.2 shows that $J_{p^r-1}(S^{2n})$ has the homotopy type of the homotopy theoretic fibre of h_{p^r}. Hence we get the James fibration sequences at the prime p:

Proposition 5.2.2. *Localized at a prime p the combinatorial James–Hopf invariant gives fibration sequences up to homotopy*

$$J_{p^r-1}(S^{2n}) \xrightarrow{\iota} J(S^{2n}) \xrightarrow{h_{p^r}} J(S^{2np^r}).$$

Another construction of Hopf invariants starts with the splitting of the suspension of a product:

Recall that, for connected X and Y, there is a homotopy equivalence

$$\Sigma(X \times Y) \to \Sigma X \vee \Sigma Y \vee \Sigma(X \wedge Y).$$

Hence the map $\Sigma(X \times Y) \to \Sigma(X \wedge Y)$ has a section (right inverse) $\Sigma(X \wedge Y) \to \Sigma(X \times Y)$. If we iterate this in the following way

$$\Sigma(X_1 \times X_2 \times \cdots X_{k-1} \times X_k) \to \Sigma(X_1 \times X_2 \times \cdots X_{k-1}) \wedge X_k$$
$$\to \Sigma(X_1 \times X_2 \times \cdots X_{k-2}) \wedge X_{k-1} \wedge X_k$$
$$\to \cdots \to \Sigma X_1 \wedge X_2 \wedge \cdots X_{k-1} \wedge X_k$$

then we get a map

$$\Sigma(X_1 \times X_2 \times \cdots X_{k-1} \times X_k) \to \Sigma(X_1 \wedge X_2 \wedge \cdots X_{k-1} \wedge X_k)$$

with a section.

Now consider the natural surjections
$$X \times X \times \cdots \times X \to J_k(X) \to X \wedge X \wedge \cdots \wedge X.$$
Since there is a section the cofibration sequence
$$\Sigma J_{k-1}(X) \to \Sigma J(X) \to \Sigma(X \wedge X \wedge \cdots \wedge X)$$
is split, and we get

Lemma 5.2.3. *For connected X, there is a homotopy equivalence*
$$\Sigma J_{k-1}(X) \vee \Sigma(X^{\wedge k}) \to \Sigma J_k(X).$$

Iteration of this decomposition yields

The James splitting 5.2.4. *For connected X and all positive integers n, there is a homotopy equivalence*
$$\bigvee_{k=1}^{n} \Sigma(X^{\wedge k}) \to \Sigma J_n(X).$$

This result is also valid for $n = \infty$.

But the above lemma also implies that any map $\Sigma J_k(X) \to Y$ extends to a map $\Sigma J(X) \to Y$. In particular, the map $\Sigma J_k(X) \to \Sigma X^{\wedge k}$ extends to a map $H: \Sigma J(X) \to \Sigma X^{\wedge k}$. Note that H restricts to the trivial map on $\Sigma J_{k-1}(X)$.

We define the k-th decomposition James–Hopf invariant $\overline{h_k}$ to be the composite
$$J(X) \xrightarrow{\Sigma} \Omega\Sigma J(X) \xrightarrow{\Omega H} \Omega\Sigma(X^{\wedge k}).$$

We have a lemma which is similar to Proposition 5.2.1:

Lemma 5.2.5. *If $H_*(X)$ is torsion free, then the decomposition James–Hopf invariant satisfies*
$$\overline{h_k}_*(\alpha_1 \otimes \alpha_2 \otimes \cdots \otimes \alpha_k) = \overline{\alpha_1 \otimes \alpha_2 \otimes \cdots \otimes \alpha_k}$$
where $\alpha_1 \otimes \alpha_2 \otimes \cdots \otimes \alpha_k \, \varepsilon H_*(J(X))$ and $\overline{\alpha_1 \otimes \alpha_2 \otimes \cdots \otimes \alpha_k} \, \varepsilon H_*(X^k))$.

Proof: It follows from the commutative diagram

$$\begin{array}{ccccc}
J(X) & \xrightarrow{\Sigma} & \Omega\Sigma J(X) & \xrightarrow{\Omega H} & \Omega\Sigma X^{\wedge k} \\
\uparrow & & \uparrow & & \uparrow \\
J_k(X) & \xrightarrow{\Sigma} & \Omega\Sigma J_k(X) & \to & \Omega\Sigma X^{\wedge k} \\
\uparrow = & & \uparrow \Sigma & & \uparrow \Sigma \\
J_k(X) & \leftarrow & J_k(X) & \to & X^{\wedge k}.
\end{array}$$

$\square$

Just as with the p^r-th combinatorial James–Hopf invariant, this lemma implies

Proposition 5.2.6. *Localized at a prime p, there is a fibration sequence up to homotopy*

$$J_{p-1}(S^{2n}) \to J(S^{2n}) \xrightarrow{\overline{h_{p^r}}} \Omega\Sigma S^{2np^r}$$

where $\overline{h_{p^r}}$ is the p^r-th decomposition James–Hopf invariant.

Exercises

(1) Show that, up to homotopy equivalence localized at a prime p, the loops on the p^r-th combinatorial James–Hopf invariant Ωh_{p^r} is the same as the loops on the p^r-th decomposition James–Hopf invariant $\Omega \overline{h_{p^r}}$.

(2) (a) Show that the section $\Sigma(X \wedge Y) \to \Sigma(X \times Y)$ is a natural transformation in the homotopy category.

(b) Show that the section $\Sigma(X_1 \wedge X_2 \wedge \cdots \wedge X_k) \to \Sigma(X_1 \times X_2 \times \cdots \times X_k)$ is a natural transformation in the homotopy category.

(c) Show that the James splitting is natural in the homotopy category.

(d) Show that the k-th decomposition James–Hopf invariant $\overline{h_k} : J(X) \to \Omega\Sigma X^{\wedge k}$ is a natural transformation in the homotopy category.

(3) (a) Show that, localized at a prime p, the p-th James–Hopf invariant $h_p : \Omega S^3 \to \Omega S^{2p+1}$ restricts to a map

$$\overline{h} : \Omega(S^3\langle 3\rangle) \to \Omega S^{2p+1}$$

with the homotopy theoretic fibre of $\overline{h}$ equal to S^{2p-1}. (It does not matter whether h_p is the combinatorial James–Hopf invariant, the decomposition James–Hopf invariant, or the Hilton–Hopf invariant of the next section.)

(b) If $p = 2$ show that the inclusion $S^{2p-1} \to \Omega(S^3\langle 3\rangle)$ is the adjoint of η.

(c) If p is an odd prime, show that the inclusion $S^{2p-1} \to \Omega(S^3\langle 3\rangle)$ is the adjoint of α_1.

5.3 p-th Hilton–Hopf invariants

In this section we show that the Hilton–Hopf invariants can also be used to obtain James fibration sequences localized at a prime p.

If A is a co-H-space, the p-th Hilton–Hopf invariant is a map

$$h_p : \Omega\Sigma A \to \Omega\Sigma(\wedge^p A).$$

More precisely, let $\{\iota_1, \iota_2\}$ generate the Hall basis

$$\{\iota_1,\ \iota_2,\ [\iota_1, [\iota_1, \iota_2]],\ \ldots,\ \mathrm{ad}(\iota_1)^{p-1}(\iota_2),\ \ldots\}.$$

It does not much matter in which order the elements occur. The important thing is that the element $\mathrm{ad}(\iota_1)^{p-1}(\iota_2)$ occurs.

Definition 5.3.1. The p-th Hilton–Hopf invariant is $h_p = h_{\mathrm{ad}(\iota_1)^{p-1}(\iota_2)}$.

Thus, h_p is the composition

$$\Omega\Sigma A \xrightarrow{\Omega\Sigma(\iota_1 + \iota_2)} \Omega\Sigma(A \vee A)$$

$$\xrightarrow{\Theta^{-1}} \Omega\Sigma A \times \Omega\Sigma A \times \cdots \times \Omega\Sigma(\wedge^p A) \times \cdots$$

$$\xrightarrow{p_{\mathrm{ad}(\iota_1)^p(\iota_2)}} \Omega\Sigma(\wedge^p A),$$

where the last map is the projection on a factor of the product.

We need to investigate the homological properties of this map, at least in the case of $A = S^{2n}$ with mod p coefficients. To this end we present some identities which hold mod p. Our presentation follows the presentation by Jacobson in his book on Lie algebras.

Let F be a field of characteristic p and consider the polynomial algebra $F[x, y]$ generated by two commuting variables x, y. We begin with the familiar identity

$$(x - y)^p = x^p - y^p.$$

We can factor

$$x^p - y^p = (x - y)(\Sigma_{i=0}^{p-1} x^i y^{p-1-i})$$

and thus

$$(x - y)^{p-1} = \Sigma_{i=0}^{p-1} x^i y^{p-1-i}.$$

Now consider the tensor algebra $T = T[z, w]$ generated by two noncommuting variables z, w over the field F. We have two commuting operators z_R, z_L on T given by right multiplication by z, that is, $z_R(a) = az$, and by left multiplication by z, that is, $z_L(a) = za$ for all $a \,\varepsilon\, T$. We note that

$$(z_L - z_R)(a) = za - az = [z, a] = \mathrm{ad}(z)(a).$$

Thus,

$$(z_L - z_R)^p = (z_L)^p - (z_R)^p = (z^p)_L - (z^p)_R$$

and

$$(z_L - z_R)^{p-1} = \Sigma_{i=0}^{p-1} z_L^i z_R^{p-1-i}.$$

In other words,
$$\mathrm{ad}(z)^p(w) = \mathrm{ad}(z^p)(w)$$
and
$$\mathrm{ad}(z)^{p-1}(w) = \Sigma_{i=0}^{p-1} z^i w z^{p-1-i}.$$
The following identity is the key to the homological properties of p-th Hilton–Hopf invariants:

Lemma 5.3.2.
$$(z+w)^p = z^p + w^p + \Sigma_{i=0}^{p-1} P_i(z,w)$$
where $iP_i(z,w) =$ the coefficient of t^{i-1} in $\mathrm{ad}(z+tw)^{p-1}(w)$.

Proof: Introduce the commuting variable t and write the more general formula
$$(z+tw)^p = z^p + t^p w^p + \Sigma_{i=0}^{p-1} t^i P_i(z,w)$$
where $P_i(z,w)$ is a polynomial with weights $p-i$ and i in the noncommuting variables z and w. Differentiate this formula using the product rule
$$\frac{d}{dt}(fg) = \frac{df}{dt}g + f\frac{dg}{dt}$$
and thus
$$\frac{d}{dt}(f^p) = \Sigma_{i=0}^{p-1} f^i \frac{df}{dt} f^{p-1-i}.$$
We get
$$\mathrm{ad}(z+tw)^{p-1}(w) = \Sigma_{i=0}^{p-1}(z+tw)^i w (z+tw)^{p-1-i} = \Sigma_{i=0}^{p-1} t^{i-1} i P_i(z,w).$$
□

In particular, when $p=2$:
$$P_1(z,w) = [z,w]$$
and when $p=3$:
$$P_1(z,w) = [z,[z,w]], \quad 2P_2(z,w) = [w,[z,w]]$$
On the other hand, for all p it is clear that:
$$P_1(z,w) = \mathrm{ad}(w)^{p-1}(z)$$

Corollary 5.3.3. *The map induced by the p-th Hilton–Hopf invariant in mod p homology,*
$$h_{p*} : H_*(\Omega\Sigma S^{2n}) = T[u_{2n}] \to H_*(\Omega\Sigma S^{2np}) = T[u_{2np}]$$
satisfies $h_{p*}(x_{2n}^p) = x_{2np}$.

Proof: If $\iota_{1*}u_{2n} = z$, $\iota_{2*}u_{2n} = w$, we compute
$$h_{p*}(x_{2n}^p) = (p_{\mathrm{ad}(\iota_1)^{p-1}(\iota_2)})_*(\Theta^{-1})_*(z+w)^p$$
and note that $(\Theta^{-1})_*$ writes $(z+w)^p$ in terms of the Hall basis (and products thereof if necessary) while $(p_{\mathrm{ad}(\iota_1)^{p-1}(\iota_2)})_*$ has the effect of picking off the coefficient of the $\mathrm{ad}(z)^{p-1}(w)$ term and making it the coefficient of u_{2np}. Hence, the result folows from Lemma 5.3.2. $\square$

Now it is immediate that

Corollary 5.3.4. *Localized at a prime p, the p-th Hilton–Hopf invariant h_p : $\Omega\Sigma S^{2n} \to \Omega\Sigma S^{2np}$ makes the cohomology of the domain into a free module with basis $\{1, \gamma_1(x), \gamma_2(x), \ldots, \gamma_{p-1}(x)\}$ over the cohomology of the range. Thus there is a fibration sequence up to homotopy localized at p,*
$$F \to \Omega\Sigma S^{2n} \xrightarrow{h_p} \Omega\Sigma S^{2np}$$
where $F \simeq J_{p-1}(S^{2n})$ is the localization of the $p-1$ skeleton of $\Omega\Sigma S^{2n}$.

Exercises

(1) Show that, up to homotopy equivalence at a prime p, the loops on all of the three p-th Hopf invariants (the combinatorial, the decomposition, and the Hilton) are the same.

5.4 Loops on filtrations of the James construction

Let p be an odd prime. The James fibrations at the prime p encourage further study of the spaces $J_{p-1}(S^{2n})$. We shall show that, up to homotopy, there exist Toda fibration sequences localized at p of the form
$$S^{2n-1} \to \Omega J_{p-1}(S^{2n}) \xrightarrow{H} \Omega S^{2np-1}.$$

If we combine the Toda fibration sequences with the loop of the James fibration sequences we get a factorization of the double suspension
$$\begin{array}{ccccc} S^{2n-1} & \to & \Omega J_{p-1}(S^{2n}) & \to & \Omega^2 S^{2n+1} \\ & & \downarrow H & & \downarrow \Omega h_p \\ & & \Omega S^{2np-1} & & \Omega^2 S^{2np+1} \end{array}$$

which we can use to obtain results on the p-primary components of the homotopy groups of spheres. In particular, we shall eventually use these to prove Toda's p-primary exponent theorem.

We begin by computing the homology localized at p of the loop space.

5.4 Loops on filtrations of the James construction   149

Proposition 5.4.1. *Localized at a prime p, there is an isomorphism of Hopf algebras*
$$H^*(\Omega J_{p-1}(S^{2n})) \cong E[x_{2n-1}] \otimes \Gamma[y_{2np-2}]$$
where $E[x_{2n-1}]$ is the exterior algebra on a primitive generator x_{2n-1} of degree $2n-1$ and $\Gamma[y_{2np-2}]$ is the divided power algebra on a primitive generator y_{2np-2} of degree $2np-2$.

Proof: Consider the cohomology Serre spectral sequence of the path space fibration sequence
$$\Omega J_{p-1}(S^{2n}) \to P J_{p-1}(S^{2n}) \to J_{p-1}(S^{2n}).$$
Of course, it must abut to the cohomology of the acyclic path space and this has consequences for the cohomology ring $H^*(\Omega J_{p-1}(S^{2n}))$.

The cohomology ring $H^*(J_{p-1}(S^{2n}))$ is a truncated divided power algebra spanned by
$$\{1,\ \gamma_1(x_{2n}),\ \gamma_2(x_{2n}),\ \ldots,\gamma_{p-1}(x_{2n})\}.$$
The idea of the proof is to show that there is a unique acyclic algebra
$$E_2 = E[x_{2n-1}] \otimes \Gamma[y_{2np-2}] \otimes H^*(J_{p-1}(S^{2n}))$$
with differentials
$$d_{2n}(x_{2n-1}) = x_{2n} = \gamma_1(x_{2n}),\quad d_{2np-2n}(y_{2np-2}) = x_{2n-1}\gamma_{p-1}(x_{2n}).$$
Here are the details.

There exists
$$x_{2n-1}\ \varepsilon H^{2n-1}(\Omega J_{p-1}(S^{2n})),$$
which transgresses to $\gamma_1(x_{2n}) = x_{2n}$, that is,
$$d_{2n}(x_{2n-1}) = x_{2n}.$$
This is the first nontrivial element in the reduced cohomology of the fibre.

Thus,
$$d_{2n}(x_{2n-1}\gamma_i(x_{2n})) = (i+1)\gamma_{i+1}(x_{2n}).$$
In particular,
$$d_{2n}(x_{2n-1}\gamma_{p-1}(x_{2n})) = 0$$
and there exists $y_{2np-2}\ \varepsilon H^{2np-2}(\Omega J_{p-1}(S^{2n}))$ such that
$$d_{2np-2n}(y_{2np-2}) = x_{2n-1}\gamma_{p-1}(x_{2n}).$$
This is the second nontrivial element in the reduced cohomology of the fibre.

It follows that
$$d_{2n}(y_{2np-2}x_{2n-1}) = y_{2np-2}x_{2n}$$
and hence $y_{2np-2}x_{2n-1} \neq 0$. This is the third nontrivial element in the reduced cohomology of the fibre.

Furthermore,
$$d_{2np-2}(y_{2np-2}^2) = 2y_{2np-2}x_{2n-1}\gamma_{p-1}(x_{2n}).$$
There must exist a nonzero element
$$\frac{1}{2}y_{2np-2}^2 = \gamma_2(y_{2np-2})$$
such that
$$d_{2np-2n}(\gamma_2(y_{2np-2})) = y_{2np-2}x_{2n-1} = \gamma_1(y_{2np-2})x_{2n-1}.$$

As before,
$$d_{2n}(\gamma_2(y_{2np-2})x_{2n-1}) = \gamma_2(y_{2np-2})x_{2n}$$
and it follows that $\gamma_2(y_{2np-2})x_{2n-1} \neq 0$. It is the fourth nontrivial element in the reduced cohomology of the fibre.

Once again,
$$d_{2np-2}(\gamma_1(y_{2np-2})\gamma_2(y_{2np-2})) = 3\gamma_2(y_{2np-2})x_{2n-1}\gamma_{p-1}(x_{2n}).$$
There must exist a nonzero element
$$\frac{1}{3}\gamma_1(y_{2np-2})\gamma_2(y_{2np-2}) = \gamma_3(y_{2np-2})$$
such that
$$d_{2np-2n}(\gamma_3(y_{2np-2})) = \gamma_2(y_{2np-2})x_{2n-1}.$$

Continuing in this way, we see that the divided powers $\gamma_i(y_{2np-2})$ exist in $H^{(2np-2)i}(\Omega J_{p-1}(S^{2n}))$ and satisfy
$$d_{2np-2n}(\gamma_i(y_{2np-2})) = \gamma_{i-1}(y_{2np-2})x_{2n-1}.$$

The elements
$$\{1, \gamma_1(y_{2np-2}), \ldots, \gamma_i(y_{2np-2}), \ldots;$$
$$x_{2n-1}, \gamma_1(y_{2np-2})x_{2n-1}, \ldots, \gamma_i(y_{2np-2})x_{2n-1}, \ldots\}$$
are nonzero and form a basis for $H^*(\Omega J_{p-1}(S^{2n}))$.

Finally, we see that the elements x_{2n-1} and y_{2np-2} are primitive for obvious dimension reasons. □

Since the previous result is valid over an integral domain $\mathbb{Z}_{(p)}$ of characteristic zero, it follows that the comultiplication is completely determined in the Hopf algebra $H^*(\Omega J_{p-1}(S^{2n}))$ and we get

Proposition 5.4.2. *Localized at a prime p, there is an isomorphism of Hopf algebras*

$$H_*(\Omega J_{p-1}(S^{2n})) \cong E[u_{2n-1}] \otimes P[v_{2np-2}]$$

where $E[u_{2n-1}]$ is the exterior algebra on a primitive generator u_{2n-1} of degree $2n-1$ and $P[v_{2np-2}]$ is the polynomial algebra on a primitive generator v_{2np-2} of degree $2np-2$.

5.5 Toda–Hopf invariants

In this section we construct the Toda–Hopf invariants $H : \Omega J_{p-1}(S^{2n}) \to \Omega S^{2np-1}$. First of all we recall a basic result.

Proposition 5.5.1. *The homotopy theoretic fibre of the inclusion*

$$A \vee B \to A \times B$$

is the join of the loop spaces $(\Omega A) * (\Omega B)$.

Proof: The homotopy theoretic fibre F is the space of paths in $A \times B$ which begin in $A \vee B$ and end at the basepoint $(*,*)$. This is the union of the paths which begin in $A \times *$ and end at $(*,*)$ with the paths which begin in $* \times B$ and end at $(*,*)$. In other words, F is the union

$$(PA \times \Omega B) \cup (\Omega A \times PB)$$

with intersection

$$\Omega A \times \Omega B.$$

Thus F is the homotopy pushout

$$\begin{array}{ccc} \Omega A \times \Omega B & \to & \Omega A \\ \downarrow & & \downarrow \\ \Omega B & \to & F. \end{array}$$

Hence up to homotopy F is the join $(\Omega A) * (\Omega B)$. □

We also record:

Lemma 5.5.2. *There is a homotopy equivalence $A * B \xrightarrow{\simeq} \Sigma(A \wedge B)$.*

Proof: Represent $A * B$ as the strict pushout

$$\begin{array}{ccc} A \times B & \to & CA \times B \\ \downarrow & & \downarrow \\ A \times CB & \to & A * B. \end{array}$$

If we collapse the contractible subspace $CA \vee CB$ to a point, we get the strict pushout

$$\begin{array}{ccc} A \wedge B & \to & CA \wedge B \\ \downarrow & & \downarrow \\ A \wedge CB & \to & (A * B)/(CA \vee CB). \end{array}$$

But this has the homotopy type of the pushout

$$\begin{array}{ccc} A \wedge B & \to & C(A \wedge B) \\ \downarrow & & \downarrow \\ C(A \wedge B) & \to & \Sigma(A \wedge B). \end{array}$$

Thus $A * B \simeq (A * B)/(CA \vee CB) \simeq \Sigma(A \wedge B)$. □

If we multiply the maps $\Omega A \to \Omega(A \vee B)$ and $\Omega B \to \Omega(A \vee B)$, we get a section to the map

$$\Omega(A \vee B) \to \Omega(A \times B) = (\Omega A) \times (\Omega B).$$

Therefore, up to homotopy we have an equivalence.

Lemma 5.5.3.

$$\Omega(A \vee B) \xrightarrow{\simeq} (\Omega A) \times (\Omega B) \times \Omega\Sigma(\Omega A \wedge \Omega B).$$

We can now define the pre-Toda–Hopf invariant as follows:

Since $J_{p-1}(S^{2n}) = J_{p-2}(S^{2n}) \cup e^{2n(p-1)}$, we can pinch the top cell in half and get a map

$$T : J_{p-1}(S^{2n}) \to J_{p-1}(S^{2n}) \vee S^{2n(p-1)}.$$

The pre-Toda–Hopf invariant is the composition

$$\overline{H} = \pi \circ \Omega T : \Omega(J_{p-1}(S^{2n})) \to \Omega(J_{p-1}(S^{2n}) \vee S^{2n(p-1)})$$
$$\xrightarrow{\pi} \Omega\Sigma(\Omega J_{p-1}(S^{2n}) \wedge \Omega S^{2n(p-1)}),$$

where π is the projection in the product decomposition Lemma 5.5.3.

5.5 Toda–Hopf invariants 153

Since we have the evaluation natural transformations $\Sigma\Omega A \to A$, we get a composition of evaluation maps

$$R : \Sigma(\Omega J_{p-1}(S^{2n}) \wedge \Omega S^{2n(p-1)}) \to (\Omega J_{p-1}(S^{2n}) \wedge S^{2n(p-1)})$$
$$\to J_{p-1}(S^{2n}) \wedge S^{2n(p-1)-1}.$$

The James splitting

$$\Sigma J_{p-1}(S^{2n}) \xrightarrow{\simeq} \bigvee_{i=1}^{p-1} \Sigma S^{2ni}$$

allows us to project to the bottom sphere

$$J_{p-1}(S^{2n}) \wedge S^{2n(p-1)-1} \xrightarrow{q} S^{2np-1}.$$

Finally, we can state

Definition 5.5.4. *The Toda–Hopf invariant is the composition* $H = \Omega q \circ \Omega R \circ \overline{H}$:

$$\Omega J_{p-1}(S^{2n}) \xrightarrow{\Omega T} \Omega(J_{p-1}(S^{2n}) \vee S^{2n(p-1)})$$
$$\xrightarrow{\pi} \Omega\Sigma(\Omega J_{p-1}(S^{2n}) \wedge \Omega S^{2n(p-1)})$$
$$\xrightarrow{\Omega R} \Omega(J_{p-1}(S^{2n}) \wedge S^{2n(p-1)-1})$$
$$\xrightarrow{\Omega q} \Omega S^{2np-1}.$$

We claim that

Lemma 5.5.5. *Localized at p, the Toda–Hopf invariant makes the cohomology ring*

$$H^*(\Omega J_{p-1}(S^{2n})) = E[x_{2n-1}] \otimes \Gamma[y_{2np-2}]$$

into a free module with basis $\{1, x_{2n-1}\}$ *over the the cohomology ring*

$$H^*(\Omega S^{2np-1}) = \Gamma[\overline{y}_{2np-2}].$$

Proof: It is sufficient to show that $H^*(\overline{y}_{2np-2}) = y_{2np-2}$.

We begin by determining

$$\pi^* \circ (\Omega R)^* \circ (\Omega q)^*(\overline{y}_{2np-2}) = \overline{y}.$$

But this is the generator of the cohomology in dimension $2np - 2$ of the factor $\Omega\Sigma(\Omega J_{p-1}(S^{2n}) \wedge \Omega S^{2n(p-1)})$ in the product decomposition

$$\Omega(J_{p-1}(S^{2n}) \vee S^{2n(p-1)}) \simeq \Omega J_{p-1}(S^{2n}) \times \Omega S^{2n(p-1)}$$
$$\times \Omega\Sigma(\Omega J_{p-1}(S^{2n}) \wedge \Omega S^{2n(p-1)}).$$

The cohomology ring of $J_{p-1}(S^{2n} \vee S^{2n(p-1)})$ has a basis

$$\{1, x = x_{2n} = \gamma_1(x_{2n}), \ldots, \gamma_{p-1}(x_{2n}), \ e = e_{2n(p-1)}\}$$

with $x\gamma_{p-1}(x) = 0 = xe$. In the cohomology Serre spectral sequence of the path space fibration of the bouquet $J_{p-1}(S^{2n}) \vee S^{2n(p-1)}$ we have a transgression

$$d_{2n}(x_{2n-1}) = x_{2n}$$

and the class $\overline{y}$ is determined by

$$d_{2np-2n}(\overline{y}) = x_{2n-1}e.$$

Recall that, in the cohomology Serre spectral sequence of the path space fibration of $J_{p-1}(S^{2n})$ we have a transgression

$$d_{2n}(x_{2n-1}) = x_{2n}$$

and the class y_{2np-2} is determined by

$$d_{2np-2n}(y_{2np-2}) = x_{2n-1}\gamma_{p-1}(x_{2n}).$$

The map $T : J_{p-1}(S^{2n}) \to J_{p-1}(S^{2n}) \vee S^{2n(p-1)}$ induces a map of cohomology Serre spectral sequences of the path space fibrations and we have that

$$T^*(x_{2n}) = x_{2n}, \ T^*(\gamma_i(x_{2n})) = \gamma_i(x_{2n}), \ T^*(e) = \gamma_{p-1}(x_{2n})$$
$$(\Omega T)^*(x_{2n-1}) = x_{2n-1}$$

and thus

$$d_{2np-2n} H^*(\overline{y}_{2np-2}) = d_{2np-2n}(\Omega T)^*(\overline{y})$$
$$= T^* d_{2np-2n}(\overline{y})$$
$$= (\Omega T)^* x_{2n-1} \ T^*e$$
$$= x_{2n-1}\gamma_{p-1}(x_{2n}).$$

Hence, $H^*(\overline{y}_{2np-2}) = y_{2np-2}$. $\square$

As before the above proposition gives a corollary. These are the Toda fibrations:

Proposition 5.5.6. *Localized at a odd prime p, the Toda–Hopf invariant H gives fibration sequences up to homotopy*

$$S^{2n-1} \to \Omega J_{p-1}(S^{2n}) \xrightarrow{H} \Omega S^{2np-1}.$$

We note that it is immediate that the Toda–Hopf invariant satisfies the following naturality condition:

Lemma 5.5.7. *If $k : S^{2n} \to S^{2n}$ is the degree k map, then the following diagram commutes*

$$\begin{array}{ccccc} S^{2n} & \xrightarrow{\iota} & \Omega J_{p-1}(S^{2n}) & \xrightarrow{H} & \Omega S^{2np-1} \\ \downarrow k & & \downarrow J_{p-1}(k) & & \downarrow \Omega k^p \\ S^{2n} & \xrightarrow{\iota} & \Omega J_{p-1}(S^{2n}) & \xrightarrow{H} & \Omega S^{2np-1}. \end{array}$$

5.6 Toda's odd primary exponent theorem

In this section we prove Toda's odd primary exponent theorem: If p is an odd prime, then p^{2n} annihilates the p-primary component of $\pi_k(S^{2n+1})$ for all k.

We start with some preliminary observations. Localized at an odd prime, S^{2n+1} is an H-space. The simplest way to see this is that, localized away from 2, we have a homotopy equivalence

$$\Omega S^{2n+2} \simeq S^{2n+1} \times \Omega S^{4n+3}.$$

The result follows by noting that any retract of an H-space is also an H-space.

In addition, localized at an odd prime, we can identify the degree k map as a k-th power map $k : S^{2n+1} \to S^{2n+1}$. Since the two multiplications on ΩS^{2n+1} are homotopic, it follows that the following are homotopic: $\Omega(k) \simeq k : \Omega S^{2n+1} \to \Omega S^{2n+1}$. Furthermore, any iterated loop on a degree k is also a k-th power map. This is different from the case of localization at the prime 2 when it is important to distinguish between k-th powers and maps induced by the degree k map.

John Moore's geometric formulation of Toda's result is the following.

Proposition 5.6.1. *Localized at a odd prime p, there exists a factorization of the p^2-nd power map*

$$p^2 : \Omega^3 S^{2n+1} \xrightarrow{\pi} \Omega S^{2n-1} \xrightarrow{\Omega(\Sigma)} \Omega^3 S^{2n+1}.$$

We point out the following immediate corollaries.

Corollary 5.6.2. *Localized at a odd prime p, there exists a factorization of the p^{2n} power map*

$$p^{2n} : \Omega^{2n+1} S^{2n+1} \xrightarrow{\tau} \Omega S^1 \xrightarrow{\Omega(\Sigma^{2n})} \Omega^{2n+1} S^{2n+1}.$$

Corollary 5.6.3. *If $S^{2n+1}\langle 2n+1 \rangle$ is the $2n+1$ connected cover and we localize at an odd prime p, the p^{2n} power map is null homotopic on $\Omega^{2n+1}(S^{2n+1}\langle 2n+1 \rangle)$ and thus all the homotopy groups are annihilated by p^{2n}.*

Proof of 5.6.1: We can use any of three Hopf invariants but the p-th combinatorial James–Hopf invariant h_p is most transparently natural in the sense that the following commutes:

$$\begin{array}{ccc} \Omega^3 S^{2n+1} & \xrightarrow{\Omega^2 h_p} & \Omega^3 S^{2np+1} \\ \downarrow p & & \downarrow p^p \\ \Omega^3 S^{2n+1} & \xrightarrow{\Omega^2 h_p} & \Omega^3 S^{2np+1}. \end{array}$$

Let $\alpha : X \to \Omega^3 S^{2n+1}$ be any map. Then

$$(\Omega^2 h_p) \circ p \circ \alpha = p^p \circ (\Omega^2 h_p) \circ \alpha.$$

Since $\Omega^2 h_p$ is certainly an H-map, it commutes with power maps and inverses and therefore,

$$\Omega^2 h_p \circ (p^p - p) \circ \alpha = 0.$$

It follows from the James fibration sequence that there exists

$$(p^p - p) \circ \alpha = \Omega^2(\sigma) \circ \beta$$

where

$$\sigma : J_{p-1}(S^{2n}) \to \Omega S^{2n+1}$$

and

$$\beta : X \to \Omega^2 J_{p-1}(S^{2n}).$$

We also have the naturality of the loops of the Toda fibrations:

$$\begin{array}{ccc} \Omega^2 J_{p-1}(S^{2n}) & \xrightarrow{\Omega H} & \Omega^2 S^{2np-1} \\ \downarrow \Omega^2 J_{p-1}(p) & & \downarrow p^p \\ \Omega^2 J_{p-1}(S^{2n}) & \xrightarrow{\Omega H} & \Omega^2 S^{2np-1} \end{array}$$

Thus

$$p^p \circ \Omega H = \Omega H \circ \Omega^2 J_{p-1}(p)$$

and

$$\Omega H \circ p^p = \Omega H \circ \Omega^2 J_{p-1}(p).$$

It follows that there exists a factorization

$$(p^p - \Omega^2 J_{p-1}(p)) \circ \beta = \Omega(\tau) \circ \gamma$$

where

$$\tau : S^{2n-1} \to \Omega J_{p-1}(S^{2n})$$

and
$$\gamma: X \to \Omega S^{2n-1}.$$
Hence $\Omega^2(\sigma) \circ \Omega^2 J_{p-1}(p) = p \circ \Omega^2(\sigma)$ implies that
$$\begin{aligned}
\Omega^2(\sigma) \circ \Omega(\tau) \circ \gamma &= \Omega^2(\sigma) \circ (p^p - \Omega^2 J_{p-1}(p)) \circ \beta \\
&= (p^p - p) \circ \Omega^2(\sigma) \circ \beta \\
&= (p^p - p) \circ (p^p - p) \circ \alpha \\
&= p^2 \circ (p^{p-1} - 1)^2 \circ \alpha.
\end{aligned}$$
Finally, since $(p^{p-1} - 1): \Omega S^{2n+1} \to \Omega S^{2n+1}$ is a homotopy equivalence, there exists $\alpha: \Omega^3 S^{2n+1} \to \Omega^3 S^{2n+1}$ such that
$$(p^{p-1} - 1)^2 \circ \alpha = 1$$
and we conclude that
$$\Omega^2(\sigma) \circ \Omega(\tau) \circ \gamma = p^2$$
which was to be demonstrated. $\square$

6 Samelson products

The first step in studing self maps of cyclic Moore spaces, $P^m(k) = P^m(\mathbb{Z}/k\mathbb{Z})$, is to look at what is induced in mod k homology. If the map induces 0 in mod k homology then it will factor through the homotopy theoretic fibre $F^m\{k\}$ of the pinch map $q : P^m(k) \to S^m$. We begin this chapter by studying the homology of the fibre of the pinch map and then use it to determine the homotopy classes of self maps of cyclic Moore spaces.

We use the results on self maps to determine the precise exponents of the homotopy groups with cyclic coefficients. Furthermore, if the self map of a Moore space $k : P^m(k) \to P^m(k)$, that is, k times the identity, is null homotopic, it implies that the smash products $P^m(k) \wedge P^n(k)$ have the homotopy type of a bouquet, $P^{m+n}(k) \vee P^{m+n-1}(k)$.

These decompositions of smash products into bouquets are uniquely determined up to compositions with Whitehead products by the decomposition of the mod k homology of the smash into a direct sum. In order to see this, we use the fact that, if p is an odd prime, then the odd-dimensional Moore space $P^{2n+1}(p^r)$ is equivalent modulo Whitehead products through a range of dimensions with the fibre $S^{2n+1}\{p^r\}$ of the degree p^r map $p^r : S^{2n+1} \to S^{2n+1}$. Surprisingly, we also have the stronger result that the spaces $\Omega P^{2n+2}(p^r)$ and $S^{2n+1}\{p^r\}$ are equivalent through all dimensions up to compositions with multiplicative extensions of Samelson products.

In the case of maps out of spheres, George Whitehead was the first to realize that the Lie identities for Samelson products were consequences of Lie identities for commutators in groups. [133, 134] With slight modifications, his treatment yields Lie identities for more general (external) Samelson products. When these external Samelson products are combined with certain decompositions of smash products, the result is a theory of internal Samelson products which makes homotopy with coefficients into a graded Lie algebra, at least if the coefficients are $\mathbb{Z}/p^r\mathbb{Z}$ with p a prime greater than 3. [99]

The adjoint relation between Samelson products and Whitehead products was first established by Hans Samelson. [111] The homology suspension was studied by George Whitehead [134] and John Moore [95]. One application of this is to the proof that Whitehead products vanish in homology with any coefficients.

We give a brief treatment of group models for loop spaces. Hans Samelson first pointed out the connection between the H-space structure of the loop space and the multiplication in the group of an acyclic principle bundle. John Milnor [86] was the first to produce a group model for the loop space but, in order to achieve full functoriality, it is necessary to use the simplicial loop group introduced by Dan Kan. [69, 81]

These group models for loop spaces enable us to eliminate some of the homotopies in the definition of Samelson products. In fact, with these models, the external Samelson products are exactly the same as the commutators in the group of maps. This simplification makes possible a treatment of relative Samelson products for fibration sequences.

Suppose we have a fibration sequence of loop spaces. Then the Samelson product of homotopy classes with one in the total space and one in the fibre can be regarded as a relative Samelson product in the fibre. This is almost automatic. The group models for loop spaces allow one to replace the fibre by the kernel of the map from the total space to the base space. Then this relative Samelson product is the geometric version of the fact that a normal subgroup is closed under commutators with all elements of the larger group. It is only a little harder to show that it can be done in such a way that the crucial identities of bilinearity, anti-symmetry, the Jacobi identity, and the derivation formula for Bocksteins remain valid for these relative Samelson products. (Of course, in homotopy theory with mod 3 coefficients, the Jacobi identity is not always valid.)

We also give a treatment of relative Samelson products via universal models. These universal models reduce the theory of relative Samelson products to the usual theory of Samelson products in loop spaces. Universal models also make possible a theory of relative Samelson products when the base of a loop space fibration sequence is the loops on an H-space. In this case, the relative Samelson product lands in the homotopy of the fibre even if both of the classes are in the total space. Again, this is an almost automatic consequence of the fact that the base, being the loops on an H-space, is homotopy commutative. This is a loose analog of the algebraic fact that commutators always land in a normal subgroup if the quotient group is abelian. But, since this is only a loose analog, it is not automatic that it can be done in such a way that the crucial identities remain valid.

Relative Samelson products have a relation to the usual Samelson products that is similar to the relation of relative cup products to the usual cup products in cohomology. Recall that relative cup products are bilinear pairings $H^n(X, A) \otimes H^m(X, B) \to H^{n+m}(X, A \cup B)$ of the cohomology groups of pairs. These are related via naturality to the usual cup product pairings of spaces.

That is, on the cochain level, these relative products are compressions of the usual cup products into subspaces of the cochains. These relative pairings satisfy the usual properties of associativity and (graded) commutativity. For cup products, we have the advantage of definitions via cochains. Associativity is true on the cochain level and commutativity can be proved with the help of acyclic models to get natural cochain homotopies.

For relative Samelson products, especially in the case of those over the loops on an H-space, we have nothing like cochains or an acyclic model theorem. Instead, we have the Hilton–Milnor theorem which allows us to use universal models and the usual theory of nonrelative Samelson products to both define relative Samelson products and to prove that they satisfy the usual identities of bilinearity, antisymmetry, the Jacobi identity, and the derivation formula for Bocksteins. In the case of relative Samelson products over the loops on an H-space, there is a twisting of the anti-symmetry identity which disappears when the base is the loops on a homotopy commutative H-space. Localized at odd primes, this is not a problem since multiplications on H-spaces can be replaced by homotopy commutative multiplications.

I want to take this opportunity to thank Brayton Gray for resurrecting my faith in universal models for relative Samelson products.

6.1 The fibre of the pinch map and self maps of Moore spaces

Let k be an integer and let $P^m(k) = P^m(\mathbb{Z}/k\mathbb{Z}) = S^{m-1} \cup_k e^m$, $m \geq 3$, be a cyclic Moore space. Recall the pinch map $q : P^m(k) \to S^m$ and its homotopy theoretic fibre $F^m\{k\}$. The exercises in Section 4.8 assert the following.

Proposition 6.1.1. *The integral homology groups $H_*(F^m\{k\})$ are torsion free and*

$$\overline{H}_*(F^m\{k\}) = \begin{cases} \mathbb{Z} & \text{if } * = s(m-1), s > 0 \\ 0 & \text{otherwise.} \end{cases}$$

In order to investigate this further, we note that the fibration sequence

$$F^m\{k\} \to P^m(k) \to S^m$$

has a translate which is a principle bundle sequence

$$\Omega S^m \xrightarrow{\iota} F^m\{k\} \to P^m(k)$$

with a right action

$$\mu : \Omega S^m \times F^m\{k\} \to F^m\{k\}.$$

6.1 The fibre of the pinch map and self maps of Moore spaces

The next result determines the coalgebra structure on the homology of the fibre of the pinch map.

Proposition 6.1.2.

(a) *The map* $\iota_* : H_*(\Omega S^m) \to H_*(F^m\{k\})$ *is a monomorphism.*

(b) *In each nonzero positive degree $s(m-1)$, the image of ι_* has index k. In other words, if we let u_{m-1} be a generator of $H_{m-1}(\Omega S^m)$ and if we abuse notation by also writing $u_{m-1} = \iota_*(u_{m-1})$ for its image in $H_{m-1}(F^m\{k\})$, then $H_*(F^m\{k\})$ has a basis*

$$1, \frac{u_{m-1}}{k}, \frac{u_{m-1}^2}{k}, \frac{u_{m-1}^3}{k}, \ldots, \frac{u_{m-1}^k}{k}, \ldots$$

(c) *Thus, if $a_n = u_{m-1}^n/k, n \geq 1$, is the basis for the reduced homology $\overline{H}_*(F^m\{k\})$, we have the formula for the diagonal*

$$\Delta(a_n) = a_n \otimes 1 + 1 \otimes a_n + \sum_{i+j=n, i>0, j>0} k(i,j) a_i \otimes a_j.$$

Proof: Consider the integral homology Serre spectral sequence of the principal bundle sequence $\Omega S^m \xrightarrow{\iota} F^m\{k\} \to P^m(k)$.

We note that

$$E_{s,*}^2 \cong H_s(P^m(k); H_*(\Omega S^m))$$

$$\cong \begin{cases} \mathbb{Z} \otimes H_*(\Omega S^m) \cong T(u_{m-1}) & \text{if } s = 0 \\ \mathbb{Z}/k\mathbb{Z} \otimes H_*(\Omega S^m) \cong e_{m-1} T(u_{m-1}) & \text{if } s = m-1 \\ 0 & \text{otherwise.} \end{cases}$$

where $e_{m-1} \varepsilon H_{m-1}(P^m(k))$ is a generator.

Furthermore, the entire spectral sequence is a right module over the homology algebra $H_*(\Omega S^m) \cong T(u_{m-1})$.

Since this spectral sequence is confined to two vertical lines, it is easy to see that it collapses.

Since we already know that $H_*(F^m\{k\})$ is either 0 or infinite cyclic, we conclude that the image of ι_* has index k in every nonzero positive degree. The proposition follows. □

Let $K_m(k)$ be the kernel of the mod k Hurewicz map $\phi : \pi_m(P^m(k); \mathbb{Z}/k\mathbb{Z}) \to H_m(P^m(k); \mathbb{Z}/k\mathbb{Z})$.

Lemma 6.1.3. *If $m \geq 3$ and $f : P^m(k) \to P^m(k)$ is a map which represents a class in $K_m(k)$, then f induces the 0 map on all nontrivial mod k homology* groups and there is a factorization

$$f : P^m(k) \to S^m \xrightarrow{\bar{f}} S^{m-1} \to P^m(k)$$

where $P^m(k) \to S^m$ is the pinch map onto the top cell and $S^{m-1} \to P^m(k)$ is the inclusion map of the bottom cell.

Proof: We may assume f to be skeletal and hence that it induces a map of cofibration sequences:

$$\begin{array}{ccccccc} S^{m-1} & \to & P^m(k) & \to & S^m & \xrightarrow{k} & S^m \\ \downarrow g & & \downarrow f & & \downarrow h & & \downarrow \Sigma g \\ S^{m-1} & \to & P^m(p^r) & \to & S^m & \xrightarrow{k} & S^m \end{array}$$

Since f is in $K_m(k)$, h induces 0 on mod k homology. Hence h is a map of degree divisible by k. Since the degrees of Σg and of g are the same as the degree of h, it follows that f induces 0 in all mod p^r homology groups in nonzero degrees.

Consider the map of the bottom cell $S^{m-1} \to F^m\{k\}$ into the homotopy theoretic fibre of the pinch map $P^m(k) \to S^m$. Since $S^{m-1} \to F^m\{k\}$ induces an isomorphism of integral homology groups in dimensions less than $2m-2$, it induces an isomorphism of homotopy in dimensions less than $2m-3$ and an epimorphism in homotopy in dimensions less than $2m-2$.

If f is in K_m then f factors through $F^m\{k\}$ and since $m < 2m-2$, f factors through a map $\bar{f} : P^m(k) \to S^{m-1}$. Since the restriction of $\bar{f}$ to S^{m-1} has degree 0, we have a factorization

$$P^m(k) \to S^m \xrightarrow{\bar{f}} S^{m-1} \to P^(k).$$

$\square$

Since the homotopy classes of maps $S^m \to S^{m-1}$ are well understood, Lemma 6.1.3 gives a strong hold on the groups $K_m(k)$.

Let p be a prime. We now determine the set of homotopy classes $[P^m(p^r), P^m(p^r)]_* = \pi_m(P^m(p^r); \mathbb{Z}/p^r\mathbb{Z})$ for p primary Moore spaces. If we consider self maps $[k] : P^m(p^r) \to P^m(p^r)$ which are degree k on the bottom and top cells, we see that the mod p^r Hurewicz map provides an epimorphism

$$\phi : \pi_m(P^m(p^r); \mathbb{Z}/p^r\mathbb{Z}) \to H_m(P^m(p^r); \mathbb{Z}/p^r\mathbb{Z})$$

onto the group $H_m(P^m(p^r); \mathbb{Z}/p^r\mathbb{Z}) \cong \mathbb{Z}/p^r\mathbb{Z}$. Consider the exact sequence

$$0 \to K_m(p^r) \to \pi_m(P^m(p^r); \mathbb{Z}/p^r\mathbb{Z}) \xrightarrow{\phi} \mathbb{Z}/p^r\mathbb{Z} \to 0$$

which is a short exact sequence of groups if $m \geq 3$. The main results are:

6.1 The fibre of the pinch map and self maps of Moore spaces

Proposition 6.1.4.

(a) $K_3(p^r)$ is isomorphic to $\mathbb{Z}/p^r\mathbb{Z}$.

(b) If $m \geq 4$ and p is an odd prime, then $K_m(p^r) = 0$.

(c) If $m \geq 4$ and $p = 2$, then $K_m(2^r) = \mathbb{Z}/2\mathbb{Z}$.

Proposition 6.1.5.

(a) If $m \geq 3$ and p is an odd prime, then the sequence is split.

(b) If $m \geq 3$ and $r \geq 2$, then the sequence is split.

(c) If $m \geq 3$ and $p^r = 2$, then the sequence is not split.

Proofs: We begin with the case $m = 3$.

We have that $\pi_3(S^2) = \mathbb{Z}\eta$ = the infinite cyclic group generated by the Hopf map η. Recall that the Hopf invariant $H(f)$ of a map $F : S^3 \to S^2$ is defined as follows:

Let the nontrivial integral cohomology of the mapping cone $C_F = S^2 \cup_F e^4$ be generated by classes u_2 and u_4 in dimensions 2 and 4, respectively. In particular, $C_\eta = CP^2$ = the complex projective plane. We define the integral Hopf invariant $H(F)$ by

$$u_2 \cup u_2 = H(F)u_4,$$

with orientation chosen so that $H(\eta) = 1$.

The diagram with horizontal cofibration sequences

$$\begin{array}{ccccccc}
S^3 & \xrightarrow{\eta} & S^2 & \to & CP^2 & \to & S^4 \\
\uparrow k & & \uparrow = & & \uparrow & & \uparrow k \\
S^3 & \xrightarrow{k\eta = \eta \circ k} & S^2 & \to & C_{k\eta} & \to & S^4
\end{array}$$

shows that $H(k\eta) = H(\eta \circ k) = k$. In other words, maps $S^3 \to S^2$ are detected by this Hopf invariant in the integral cohomology of the mapping cone.

Now let $f : P^3(p^r) \to P^3(p^r)$ be in $K_3(p^r)$ and define the mod p^r Hopf invariant $\overline{H}(f)$ by

$$u_2 \cup u_2 = \overline{H}(f)u_4,$$

where u_2 and u_4 are generators of the mod p^r cohomology of the mapping cone $C_f = P^3(p^r) \cup_f C(P^3(p^r))$ in dimensions 2 and 4, respectively.

It is not hard to see that, if we have a factorization

$$f : P^3(p^r) \to S^3 \xrightarrow{F} S^2 \to P^3(p^r),$$

then $\overline{H}(f) = H(F)$ mod p^r. So $K_3(p^r)$ has at least p^r elements.

On the other hand, $K_3(p^r)$ is the epimorphic image of
$$[P^3(p^r), S^2]_* = \pi_3(S^2; \mathbb{Z}/p^r\mathbb{Z}) \cong \mathbb{Z}/p^r\mathbb{Z}$$
and hence $K_3(p^r) \cong \mathbb{Z}/p^r\mathbb{Z}$. In other words, maps $P^3(p^r) \to P^3(p^r)$ which vanish in mod p^r homology are detected by the mod p^r Hopf invariant in the mod p^r cohomology of the mapping cone.

For example, consider the map $p^r : P^3(p^r) \to P^3(p^r)$. We wish to determine whether this map is null homotopic. It certainly induces 0 in mod p^r homology. We consider the cofibration sequence
$$S^1 \xrightarrow{p^r} S^1 \to P^2(p^r)$$
and its smash with $P^2(p^r)$
$$S^1 \wedge P^2(p^r) \xrightarrow{p^r \wedge 1} S^1 \wedge P^2(p^r) \to P^2(p^r) \wedge P^2(p^r).$$

The mod p^r cohomology of $P^2(p^r)$ is the free $\mathbb{Z}/p^r\mathbb{Z}$ module generated by classes $1, e_1, e_2$ of dimensions $0, 1, 2$, respectively. Furthermore, we know from Steenrod and Epstein that
$$e_1 \cup e_1 = \frac{p^r(p^r+1)}{2} e_2.$$
Thus the Künneth theorem shows that the mod p^r cohomology of $P^2(p^r) \wedge P^2(p^r)$ is the free $\mathbb{Z}/p^r\mathbb{Z}$ module
$$\langle 1, e_1, e_2 \rangle \otimes \langle 1, e_1, e_2 \rangle.$$
Setting $u_2 = e_1 \otimes e_1, u_4 = e_2 \otimes e_2$, we compute that
$$u_2 \cup u_2 = -\left\{\frac{p^r(p^r+1)}{2}\right\}^2 u_4$$
and thus the mod p^r Hopf invariant of $p^r : P^3(p^r) \to P^3(p^r)$ is nontrivial only in the single case $p^r = 2$. Therefore we have the exponent result

Proposition 6.1.6. *The maps $p^r : P^3(p^r) \to P^3(p^r)$ are null homotopic if $p^r > 2$. If $p^r = 2$, the map is essential.*

It follows that the short exact sequence
$$0 \to K_3(p^r) \to \pi_3(P^3(p^r); \mathbb{Z}p^r\mathbb{Z}) \xrightarrow{\phi} \mathbb{Z}/p^r\mathbb{Z} \to 0$$
is split if $p^r > 2$ and not split if $p^r = 2$.

This completes the proof of Propositions 6.1.4 and 6.1.5 in the case $m = 3$ and p is an odd prime.

If $m \geq 4$ only some minor modifications are needed in the above argument.

First of all, $\pi_m(S^{m-1}) \cong \mathbb{Z}/2\mathbb{Z}, m \geq 4$, generated by the Hopf map $\eta: S^m \to S^{m-1}$. Furthermore, η is detected by the cohomology operation Sq^2 in the mod 2 cohomology of the mapping cone $\Sigma^{m-3}(CP^2)$.

If p is an odd prime, then localization shows that any factorization

$$P^m(p^r) \to S^m \to S^{m-1} \to P^m(p^r)$$

vanishes. Hence, $K_m(p^r) = 0$ if p is odd and $m \geq 4$.

If $p = 2$ and $f: P^m(2^r) \to P^m(2^r)$ is trivial in mod 2^r homology with $m \geq 4$, then f is essential if and only if Sq^2 is nontrivial in the mod 2 cohomology of the mapping cone C_f, that is, $Sq^2: H^{m-1} \cong H^{m+1}$.

The mod 2 cohomology of $P^m(2^r)$ has a $\mathbb{Z}/2\mathbb{Z}$ basis $1, e_{m-1}, e_m$ with $Sq^1(e_{m-1}) = e_m$ if $r = 1$ and $Sq^1(e_{m-1}) = 0$ if $r \geq 2$. Thus, in the mod 2 cohomology of $P^2(2^r) \wedge P^{m-1}(2^r)$, the mapping cone of $[2^r]: P^m(2^r) \to P^m(2^r)$, the Cartan formula shows that

$$Sq^2(e_1 \otimes e_{m-2}) = Sq^1(e_1) \otimes Sq^1(e_{m-2}) = \begin{cases} e_2 \otimes e_{m-1} & \text{if } r = 1 \\ 0 & \text{if } r \geq 2. \end{cases}$$

Hence

Proposition 6.1.7. *If $m \geq 3$ the maps $p^r: P^m(p^r) \to P^m(p^r)$ are null homotopic if $p^r > 2$. (Hence, all the homotopy groups $\pi_m(X; \mathbb{Z}/p^r\mathbb{Z})$ are annihilated by p^r if $p^r > 2$ and $m \geq 3$.) If $p^r = 2$, the map is essential.*

This completes the proof of Propositions 6.1.4 and 6.1.5 in all cases. □

Exercises

(1) Consider the mod k homology Serre spectral sequence of the principal bundle sequence $\Omega S^m \xrightarrow{\iota} F^m\{k\} \to P^m(k)$. It is a module over the mod k homology algebra $H_*(\Omega S^m)$.

   (a) Show that
   $$E^2_{*,*} \cong H_*(P^m(k)) \otimes H_*(\Omega S^m)$$
   $$\cong \mathbb{Z}/k\mathbb{Z}\langle 1, e_{m-1}, e_m \rangle \otimes T(u_{m-1})$$
   with zero differentials until $d^m(1) = 0, d^m(e_{m-1}) = 0, d^m(e_m) = u_{m-1}$.

   ($\mathbb{Z}/k\mathbb{Z}\langle 1, e_{m-1}, e_m \rangle$ is the vector space generated by $1, e_{m-1}, e_m$.)

   (b) Show that $E^{m+1}_{*,*} \cong \mathbb{Z}/k\mathbb{Z}\langle e_{m-1} \rangle \otimes T(u_{m-1})$ on the vertical line $* = m - 1$ and 0 otherwise except for the origin $(*, *) = (0, 0)$.

(c) Use b) above and the integral homology Serre spectral sequence in the proof of 6.1.2 to give a new proof of 6.1.1, that is, that $H_*(F^m\{k\}$ is either infinite cyclic or 0.

(2) Consider the diagram

$$\begin{array}{ccc} S^n & \xrightarrow{F} & S^{n-1} \\ \uparrow & & \uparrow = \\ P^n(p^r) & \to & S^{n-1} \\ \downarrow = & & \downarrow \\ P^n(p^r) & \xrightarrow{f} & P^n(p^r). \end{array}$$

(a) If $n = 3$ show that $H(F) = H(f) \bmod p^r$.

(b) If $n \geq 4$ and $p = 2$, show that Sq^2 is nontrivial in the mod 2 cohomology of the mapping cone of F if and only if it is nontrivial in the mod 2 cohomology of the mapping cone of f.

6.2 Existence of the smash decomposition

Let

$$X \xrightarrow{f} Y \xrightarrow{\iota} C_f \xrightarrow{j} \Sigma X \xrightarrow{\Sigma f} \Sigma Y$$

be a cofibration sequence with X connected and Y simply connected. The following lemma is an exercise. (The connectivity hypotheses are used to check that a homology equivalence in (c) is a homotopy equivalence.)

Lemma 6.2.1. *The following are equivalent:*

(a) f *is null homotopic.*

(b) *There is a retraction* $r : C_f \to Y$, *that is,* $r \circ \iota \simeq 1_Y$.

(c) *There is a homotopy equivalence* $C_f \to \Sigma X \vee C_f \xrightarrow{1 \vee r} \Sigma X \vee Y$ *where the first map is the standard coaction. (r is the retraction in b.)*

(d) *There is a section* $s : \Sigma X \to C_f$, *that is,* $j \circ s \simeq 1_{\Sigma X}$.

(e) *There is a homotopy equivalence* $s \vee \iota : \Sigma X \vee Y \to C_f$. *($s$ is the section in d.)*

Since $P^m(p^r) \wedge P^n(p^r)$ is the mapping cone of

$$S^{m-1} \wedge P^n(p^r) \xrightarrow{p^r \wedge 1} S^{m-1} \wedge P^n(p^r),$$

the above lemma and 6.1.7 imply:

6.3 Samelson and Whitehead products   167

Proposition 6.2.2. *If $m \geq 2, n \geq 2$ and $p^r \geq 3$, there is a homotopy equivalence*
$$\iota \vee s : P^{n+m-1}(p^r) \vee P^{n+m}(p^r) \to P^m(p^r) \wedge P^n(p^r).$$

Definition 6.2.3. We shall denote the section s by
$$\Delta_{m,n} : P^{m+n}(p^r) \to P^m(p^r) \wedge P^n(p^r).$$

We now describe the effect of $\Delta_{m,n}$ in mod p^r homology.

Let the reduced mod p^r homology $\overline{H}_*(P^m(p^r); \mathbb{Z}/p^r)$ be generated by e_{m-1} in dimension $m-1$ and e_m in dimension m. Associated to the exact coefficient sequence
$$0 \to \mathbb{Z}/p^r\mathbb{Z} \to \mathbb{Z}/p^{2r} \to \mathbb{Z}/p^r\mathbb{Z} \to 0$$
we have the Bockstein connection $\beta e_m = e_{m-1}$.

Since $\Delta_{m,n}$ is a section, we must have
$$\Delta_{m,n_*}(e_{m+n}) = \phi(\Delta_{m,n}) = e_m \otimes e_n$$
where ϕ is the mod p^r Hurewicz homomorphism.

Since the Bockstein is a natural derivation, we have
$$\begin{aligned}\Delta_{m,n_*}(e_{m+n-1}) &= \beta(e_m \otimes e_n) \\ &= \beta(e_m) \otimes e_n + (-1)^m e_m \otimes \beta(e_n) \\ &= e_{m-1} \otimes e_n + (-1)^m e_m \otimes e_{n-1}.\end{aligned}$$

6.3 Samelson and Whitehead products

Recall the definition of the general Samelson product for mappings into a group-like space.

Let G be a group-like space, that is, G is a homotopy associative H-space with multiplication $\mu : G \times G \to G$ and homotopy inverse $\iota : G \to G$. We write
$$\mu(x,y) = xy \quad \text{and} \quad \iota(x) = x^{-1}.$$

Then we write the commutator map $[\ ,\ ] : G \times G \to G$ as
$$[x,y] = \mu(\mu(\mu(x,y), \iota(x)), \iota(y)) = ((xy)x^{-1})y^{-1}$$
or, if we ignore the homotopy associativity, as
$$[x,y] = xyx^{-1}y^{-1}.$$

Since it is null homotopic on the bouquet, the commutator map factors as follows:
$$G \times G \to G \wedge G \xrightarrow{\overline{[\ ,\ ]}} G.$$

If $f : X \to G$ and $g : Y \to G$ are maps, then the commutator
$$C(f, g) = [\ ,\ ] \circ (f \times g) = fgf^{-1}g^{-1} : X \times Y \to G \times G \to G$$
factors up to homotopy through the map
$$[f, g] = \overline{[\ .\ ]} \circ (f \wedge g) : X \wedge Y \to G \wedge G \to G.$$

Definition 6.3.1. *The map* $[f, g] : X \wedge Y \to G$ *is called the Samelson product of* $f : X \to G$ *and* $g : Y \to G$.

It is well defined up to homotopy since the sequence of cofibrations
$$X \vee Y \to X \times Y \to X \wedge Y \to \Sigma X \vee \Sigma Y \to \Sigma(X \times Y)$$
is such that the last map has a retraction.

Observe that:

Proposition 6.3.2. *The Samelson product vanishes if the range G is homotopy commutative, for example, if G is the loops on an H-space.*

Of course, Samelson products are natural with respect to maps $f_1 : X_1 \to X$, $g_1 : Y_1 \to Y$, and morphisms of group-like spaces $\psi : G \to H$, that is,
$$[\psi \circ f \circ f_1, \psi \circ g \circ g_1] \simeq \psi \circ [f, g] \circ (f_1 \wedge g_1).$$

Given maps $\overline{f} : \Sigma X \to \mathbb{Z}$, $\overline{g} : \Sigma Y \to \mathbb{Z}$ with respective adjoints
$$f : X \xrightarrow{\Sigma} \Omega\Sigma X \xrightarrow{\Omega\overline{f}} \Omega\mathbb{Z}, \qquad g : X \xrightarrow{\Sigma} \Omega\Sigma X \xrightarrow{\Omega\overline{g}} \Omega\mathbb{Z}$$
we define the Whitehead product $[\overline{f}, \overline{g}]_w$ to be the adjoint of the Samelson product $[f, g]$, namely,

Definition 6.3.3. *The Whitehead product $[\overline{f}, \overline{g}]_w$ is the composition*
$$\Sigma(X \wedge Y) \xrightarrow{\Sigma[f, g]} \Sigma\Omega\mathbb{Z} \xrightarrow{e} \mathbb{Z}$$
where e is the evaluation map (= *the adjoint of the identity*).

As with Samelson products, Whitehead products are natural with respect to maps. Given
$$f_1 : X_1 \to X, \quad g_1 : Y_2 \to Y, \quad h : \mathbb{Z} \to \mathbb{Z}_1$$
we get
$$[h \circ \overline{f} \circ \Sigma f_1, h \circ \overline{g} \circ \Sigma g_1]_w \simeq h \circ [\overline{f}, \overline{g}]_w \circ \Sigma(f_1 \wedge g_1).$$

6.3 Samelson and Whitehead products

Proposition 6.3.2 can be reinterpreted as:

Proposition 6.3.4. *The Whitehead product vanishes when the range is an H-space.*

We claim:

Proposition 6.3.5. *For $f: X \to \Omega \mathbb{Z}$, $g: Y \to \Omega \mathbb{Z}$, the Whitehead product*
$$[\overline{f}, \overline{g}]_w : \Sigma(X \wedge Y) \to \mathbb{Z}$$
induces the 0 map in homology with any coefficients.

Proof: For any space $\mathbb{Z}$, the homology suspension σ_* is defined as the composition
$$\overline{H}_*(\Omega \mathbb{Z}) \xleftarrow{\partial \cong} H_{*+1}(P\mathbb{Z}, \Omega \mathbb{Z}) \to H_{*+1}(\mathbb{Z}, *).$$

The following commutative diagram appears in George Whitehead's book [134]. (The cohomology version appears in John Harper's book [50].)

$$\begin{array}{ccc} \overline{H}_*(\Omega\mathbb{Z}) & \xrightarrow{\sigma_*} & H_{*+1}(\mathbb{Z}, *) \\ \downarrow \sigma \cong & & \downarrow = \\ \overline{H}_{*+1}(\Sigma\Omega\mathbb{Z}) & \xrightarrow{e_*} & \overline{H}_{*+1}(\mathbb{Z}). \end{array}$$

In this diagram, $\sigma: \overline{H}_*(\Omega \mathbb{Z}) \to \overline{H}_{*+1}(\Sigma\Omega\mathbb{Z})$ is the standard suspension isomorphism.

Since σ_* is essentially e_*, the image of σ_* is contained in the primitive elements and, in the 1961 Cartan Seminar [95], Moore shows the following lemma.

Lemma 6.3.6. *The homology suspension σ_* vanishes on decomposable elements of the algebra $H_*(\Omega\mathbb{Z})$ and the image of σ_* is contained in the module of primitive elements of $H_{*+1}(\mathbb{Z})$.*

The following lemma is an exercise.

Lemma 6.3.7.

If $\alpha \varepsilon \overline{H}_*(X), \beta \varepsilon \overline{H}_*(Y)$, then $[f,g]_*(\alpha \otimes \beta)$ is a decomposable element in $H_*(\Omega \mathbb{Z})$. If α and β are primitive elements, then
$$[f,g]_*(\alpha \otimes \beta) = f_*(\alpha)g_*(\beta) - (-1)^{\deg(\alpha)\deg(\beta)} g_*(\beta)f_*(\alpha) = [f_*(\alpha), g_*(\beta)].$$

Now Proposition 6.3.5 follows from Whitehead's diagram and Moore's lemma. □

Recall that the multiplicative extension of a map $h: A \to \Omega W$ is defined to be the composition
$$\Omega \Sigma A \xrightarrow{\Omega \Sigma h} \Omega \Sigma \Omega W \to \Omega W.$$

Definition 6.3.8.

(a) Given a space W and a cyclic coefficient group G, define the Whitehead subgroup $Wh_*(W;G) \subseteq \pi_*(W;G)$ to be the subgroup generated by all compositions $P^*(G) \xrightarrow{f} \bigvee \Sigma(X \wedge Y) \to W$ where f is any map and $\bigvee \Sigma(X \wedge Y) \to W$ is any bouquet of Whitehead products.

(b) Similarly we define the Samelson subgroup $\text{Sam}_*(\Omega W; G) \subseteq \pi_*(\Omega W; G)$ to be the subgroup generated by all compositions $P^*(G) \to \Omega\Sigma(\bigvee X \wedge Y) \to \Omega W$ of an abitrary map with the multiplicative extension of a bouquet of Samelson products.

We leave as an exercise that $\text{Sam}_*(\Omega W; G) \cong Wh_{*+1}(W;G)$ under the usual isomorphism.

Proposition 6.3.5 asserts that $Wh_*(W;G)$ is contained in the kernel of the Hurewicz map.

If A_* is a graded abelian group, we shall say that a homomorphism $\psi: \pi_*(W;G) \to A_*$ is an isomorphism modulo Whitehead products if:

(1) ϕ is an actual epimorphism,

(2) $\ker(\psi) \subseteq Wh_*(W;G)$.

Similarly, a homomorphism $\xi: \pi_*(\Omega W;G) \to A_*$ is an isomorphism modulo Samelson products if:

(1) ξ is an actual epimorphism,

(2) $\ker(\xi) \subseteq \text{Sam}_*(\Omega W)$.

In other words, ξ is an isomorphism modulo Samelson products if and only if the corresponding map $\pi_{*+1}(W;G) \cong \pi_*(\Omega W; G) \to A_*$ is an isomorphism modulo Whitehead products.

The Hilton–Milnor theorem implies that

$$\pi_*(\Sigma X \vee \Sigma Y) \to \pi_*(\Sigma X \times \Sigma Y)$$

is an isomorphism modulo Whitehead products.

Exercises

(1) (a) If $\iota: G \to G$ is the homotopy inverse map of an group-like space G, that is, $\iota(x) = x^{-1}$, and α is a primitive element in $H_*(G)$, show that $\iota_*(\alpha) = -\alpha$.

(b) Prove 6.3.7 by using the definition of the commutator $C(f,g)$ as the composition

$$X \times Y \xrightarrow{f \times g} G \times G \xrightarrow{\Delta \times \Delta} G \times G \times G \times G$$
$$\xrightarrow{1 \times T \times 1} G \times G \times G \times G$$
$$\xrightarrow{1 \times 1 \times \iota \times \iota} G \times G \times G \times G \xrightarrow{\text{mult}} G \ .$$

(Δ is the diagonal and T is the twist map.)

(2) Show that $\text{Sam}_*(\Omega W; G) \cong \text{Wh}_{*+1}(W; G)$ under the usual isomorphism.

(3) If X is a pointed space, recall that the infinite symmetric product $SP^\infty(X)$ is the free abelian monoid generated by the points of X subject to the single relation that the basepoint is the unit. The result of Dold–Thom says that the natural inclusion $\iota : X \to SP^\infty(X)$ induces the integral Hurewicz map $\pi_*(X) \to \pi_*(SP^\infty(X)) = \overline{H}_*(X)$. Use this result to show that Whitehead products always induce the zero map in integral homology.

(4) **The classical definition of the Whitehead product:**

(a) Let $\iota_{n+1} : S^{n+1} \to S^{n+1} \vee S^{m+1}$ and $\iota_{m+1} : S^{m+1} \to S^{n+1} \vee S^{m+1}$ be the standard inclusion maps. Use the Hilton–Milnor theorem and the definition of the Whitehead product as the adjoint of the Samelson product to show that the Whitehead product $[\iota_{n+1}, \iota_{m+1}]_w : S^{n+m+1} \to S^{n+1} \vee S^{m+1}$ is a generator of the kernel of

$$\pi_{m+n+1}(S^{n+1} \vee S^{m+1}) \to \pi_{n+m+1}(S^{n+1} \times S^{m+1}).$$

(b) Conclude that

$$S^{n+1} \times S^{m+1} = (S^{n+1} \vee S^{m+1}) \cup_{[\iota_{n+1}, \iota_{m+1}]_w} e^{m+n+2}.$$

(c) Let $f : S^{n+1} \to X$ and $g : S^{m+1} \to X$ be maps. Use the naturality of the Whitehead product to show that $[f, g]_w : S^{n+m+1} \to X$ is the obstruction to an extension problem in the sense:

$$[f, g]_w = 0$$

if and only if $f \vee g : S^{n+1} \vee S^{m+1} \to X$ extends to a map

$$S^{n+1} \times S^{m+1} \to X.$$

(d) Use the characterization in c) to show that $[f, g] = 0$ if X is an H-space.

6.4 Uniqueness of the smash decomposition

The following spaces play an important role in this book.

172   Samelson products

Definition 6.4.1. Let $k: S^m \to S^m$ be a map of degree k on a sphere and let $S^m\{k\}$ be the homotopy theoretic fibre of k so that we have a fibration sequence

$$\dots \Omega S^m \xrightarrow{\Omega k} \Omega S^m \xrightarrow{\partial} S^m\{k\} \to S^m \xrightarrow{k} S^m.$$

We specialize to the case where $m = 2n+1$ is odd and where $k = p^r = $ a power of an odd prime. Then

$$\pi_q(S^{2n+1}\{p^r\}; \mathbb{Z}/p^r\mathbb{Z}) = \begin{cases} \mathbb{Z}/p^r\mathbb{Z} & \text{if } q = 2n \text{ or } q = 2n+1 \\ 0 & \text{if } 2n+1 < q < 2n + 2p - 3. \end{cases}$$

Let $P^{2n+1}(p^r) \to S^{2n+1}\{p^r\}$ be the map which induces an isomorphism of mod p^r homotopy groups in dimensions $2n$ and $2n+1$. The domain is the cofibre of a degree p^r map on an sphere and the range is the homotopy theoretic fibre. In the stable category, these are the same thing. The next result is an unstable version of this equality. The spaces $S^{2n+1}\{p^r\}$ are excellent approximations to the Moore spaces modulo Whitehead products, namely:

Proposition 6.4.2. Let p be an odd prime and let $n \geq 1$.

(a) *The map $P^{2n+1}(p^r) \to S^{2n+1}\{p^r\}$ induces an isomorphism on mod p^s homotopy groups modulo Whitehead products in dimensions $\leq 2n + 2p - 4$.*

(b) *If p is an odd prime, then there is a map*

$$\Omega P^{2n+2}(p^r) \to S^{2n+1}\{p^r\}$$

which induces an isomorphism on mod p^s homotopy groups modulo Samelson products. This is valid in all dimensions.

Part (a) of the above is shown in Section 9.2 and part (b) is shown in Section 11.8. The proofs of these facts require only the existence of Samelson products, not their uniqueness.

Actually, the fibre of the map $P^{2n+1}(p^r) \to S^{2n+1}\{p^r\}$ is $2pn - 3$ connected modulo Whitehead products. Hence, the connection is actually stronger than Proposition 6.4.2 indicates. But our definition means that we must be careful not to lose the epimorphic criterion for an isomorphism modulo Whitehead products.

The following results are immediate corollaries of the above.

Proposition 6.4.3. Let p be an odd prime.

(a) *For all $n \geq 1$, the Hurewicz map*

$$\phi: \pi_*(P^{2n+2}(p^r); \mathbb{Z}/p^s\mathbb{Z}) \to \overline{H}_*(P^{2n+2}(p^r); \mathbb{Z}/p^s\mathbb{Z})$$

6.4 Uniqueness of the smash decomposition

is an isomorphism modulo Whitehead products in dimensions $\leq 2n + 2p - 3$.

(b) *For all $n \geq 1$, the Hurewicz map*

$$\phi : \pi_*(P^{2n+1}(p^r); \mathbb{Z}/p^s\mathbb{Z})) \to \overline{H}_*(P^{2n+1}(p^r); \mathbb{Z}/p^s\mathbb{Z})$$

is an isomorphism modulo Whitehead products in dimensions $\leq 2n + 2p - 4$.

These results say that, except for Whitehead products, the first nontrivial homotopy group of the Moore space is a consequence of the first nontrivial higher homotopy group

$$\pi_{2n+2p-3}(S^{2n+1}\{p^r\}) \cong \pi_{2n+2p-2}(S^{2n+1}; \mathbb{Z}/p^r\mathbb{Z}) \cong \mathbb{Z}/p\mathbb{Z}.$$

The bouquet decompositions

$$P^m(p^r) \wedge P^n(p^r) \simeq P^{m+n}(p^r) \vee P^{m+n-1}(p^r),$$

$$P^m(p^r) \wedge P^n(p^r) \wedge P^q(p^r)$$
$$\simeq P^{m+n+q}(p^r) \vee P^{m+n+q-1}(p^r) \vee P^{m+n+q-1}(p^r) \vee P^{m+n+q-2}(p^r),$$

and Proposition 6.4.3 imply that the mod p^r Hurewicz map is a faithful representation in the following sense.

Corollary 6.4.4. *Let p be an odd prime.*

(a) *If $m \geq 3$, then*

$$\phi : \pi_*(P^m(p^r); \mathbb{Z}/p^s\mathbb{Z}) \to \phi : H_*(P^m(p^r); \mathbb{Z}/p^s\mathbb{Z})$$

is an isomorphism modulo Whitehead products in positive dimensions $\leq m$.

(b) *If $m \geq 2$ and $n \geq 2$, then*

$$\phi : \pi_*(P^m(p^r) \wedge P^n(p^r); \mathbb{Z}/p^s\mathbb{Z}) \to H_s(P^m(p^r) \wedge P^n(p^r); \mathbb{Z}/p^s\mathbb{Z})$$

is an isomorphism modulo Whitehead products in positive dimensions $\leq m + n$.

(c) *If p is a prime greater than 3, and $m \geq 2, n \geq 2, q \geq 2$, then*

$$\phi : \pi_*(P^m(p^r) \wedge P^n(p^r) \wedge P^q(p^r); \mathbb{Z}/p^s\mathbb{Z})$$
$$\to H_*(P^m(p^r) \wedge P^n(p^r) \wedge P^q(p^r); \mathbb{Z}/p^s\mathbb{Z})$$

is an isomorphism modulo Whitehead products in positive dimensions $\leq m + n + q$.

Thus we have

Corollary 6.4.5. *Let p be an odd prime.*

(a) *The map $\Delta_{m,n} : P^{m+n}(p^r) \to P^m(p^r) \wedge P^n(p^r)$ is characterized uniquely up to compositions with Whitehead products by its Hurewicz image, $\Delta_{m,n_*}(e_{m+n}) = e_m \otimes e_n$.*

(b) *The maps $\Delta_{m,n}$ are cocommutative up to compositions with Whitehead products, that is, if T is the twist map, the following diagram is commutative up to compositions with Whitehead products:*

$$\begin{array}{ccc} P^{m+n}(p^r) & \xrightarrow{\Delta_{m,n}} & P^m(p^r) \wedge P^n(p^r) \\ \downarrow (-1)^{mn} & & \downarrow T \\ P^{m+n}(p^r) & \xrightarrow{\Delta_{n,m}} & P^n(p^r) \wedge P^m(p^r). \end{array}$$

(c) *If p is a prime greater than 3, the maps $\Delta_{m,n}$ are coassociative up to compositions with Whitehead products; that is, the following diagram is commutative up to compositions with Whitehead products:*

$$\begin{array}{ccc} P^{m+n+q}(p^r) & \xrightarrow{\Delta_{m,n+q}} & P^m(p^r) \wedge P^{n+q}(p^r) \\ \downarrow \Delta_{m+n,q} & & \downarrow 1 \wedge \Delta_{n,q} \\ P^{m+n}(p^r) \wedge P^q(p^r) & \xrightarrow{\Delta_{m,n} \wedge 1} & P^m(p^r) \wedge P^n(p^r) \wedge P^q(p^r). \end{array}$$

Hence, we have the following results which are fundamental for our subsequent development of Samelson products in mod p^r homotopy groups.

Proposition 6.4.6.

(a) *If p is an odd prime and if $f : P^m(p^r) \wedge P^n(p^r) \to G$ is any map into an H-space, then*

$$f \circ \Delta_{m,n} \circ (-1)^{mn} \simeq f \circ T \circ \Delta_{n,m}.$$

(b) *If p is a prime greater than 3 and if $f : P^m(p^r) \wedge P^n(p^r) \wedge P^q(p^r) \to G$ is any map into an H-space, then*

$$f \circ (\Delta_{m,n} \wedge 1) \circ \Delta_{m+n,q} \simeq f \circ (1 \wedge \Delta_{n,q}) \circ \Delta_{m,n+q}.$$

Recall the maps $\rho : P^m(p^r) \to P^m(p^{r+s})$ and $\eta : P^m(p^{r+s}) \to P^m(p^s)$ which form the Bockstein cofibration sequence

$$P^m(p^r) \xrightarrow{\rho} P^m(p^{r+s}) \xrightarrow{\eta} P^m(p^s).$$

On the chain level, these are characterized on generators by $\rho(e_m) = e_m$ and $\eta(e_m) = p^r e_m$

6.4 Uniqueness of the smash decomposition

It is an easy exercise to check that the following diagrams commute modulo Whitehead products

$$
\begin{array}{ccc}
P^{m+n}(p^r) & \xrightarrow{\Delta_{m,n}} & P^m(p^r) \wedge P^n(p^r) \\
\downarrow \rho & & \downarrow \rho \wedge \rho \\
P^{m+n}(p^{r+s}) & \xrightarrow{\Delta_{m,n}} & P^m(p^{r+s}) \wedge P^n(p^{r+s}) \\
\downarrow p^r \eta & & \downarrow \eta \wedge \eta \\
P^{m+n}(p^s) & \xrightarrow{\Delta_{m,n}} & P^m(p^s) \wedge P^n(p^s)
\end{array}
$$

Exercises

(1) Show that the Hurewicz maps in parts (a),(b),(c) of Corollary 6.4.4 are actual isomorphisms if

(a) $m \geq 4$.

(b) $m + n \geq 5$.

(c) $m + n + q \geq 8$.

(2) Give a new proof of Exercise 1 by showing that

(a) If $m \geq 4$ and p is an odd prime, then there is a map

$$P^m(p^r) \to K(\mathbb{Z}/p^r\mathbb{Z}, m-1)$$

which is an isomorphism on homotopy groups in dimensions $\leq m$ and an epimorphism in dimension $m + 1$.

(b) If $m + n \geq 5$ and p is an odd prime, then there is a map

$$P^m(p^r) \wedge P^n(p^r) \to K(\mathbb{Z}/p^r\mathbb{Z}, m+n-2) \times K(\mathbb{Z}/p^r\mathbb{Z}, m+n-1),$$

which is an isomorphism on homotopy groups in dimensions $\leq m + n$ and an epimorphism in dimension $m + n + 1$.

(c) If $m + n + q \geq 8$ and p is a prime greater than 3, then there is a map

$$P^m(p^r) \wedge P^n(p^r) \wedge P^q(p^r)$$
$$\to K(\mathbb{Z}/p^r\mathbb{Z}, m+n-3) \times K(\mathbb{Z}/p^r\mathbb{Z}, m+n-2)$$
$$\times K(\mathbb{Z}/p^r\mathbb{Z}, m+n-2) \times K(\mathbb{Z}/p^r\mathbb{Z}, m+n-1),$$

which is an isomorphism on homotopy groups in dimensions $\leq m + n + q$ and an epimorphism in dimension $m + n + q + 1$.

(3) Let p be a prime and k a positive integer. The symmetric group Σ_k acts on the k-fold smash product

$$P = P^2(p^r) \wedge P^2(p^r) \wedge \cdots \wedge P^2(p^r)$$

via
$$\sigma(x_1 \wedge x_2 \wedge \cdots \wedge x_k) = y_1 \wedge y_2 \wedge \cdots \wedge y_k,$$
where
$$y_j = x_{\sigma^{-1}j}$$
for $\sigma \varepsilon \Sigma_k$. Thus the reduced mod p^r homology $M = \overline{H}_*(P)$ is a left module over the group ring $R = \mathbb{Z}/p^r\mathbb{Z}[\Sigma_k]$. Suppose $p > k$ and consider the trace element
$$e = \frac{1}{k!} \sum_{\sigma \varepsilon \Sigma_k} \sigma.$$

(a) Show that e is a central element, that is, $ex = xe$ for all $x \varepsilon R$.

(b) Show that, for any $R-$ module N, $eN =$ the submodule of invariant elements $\{x \varepsilon N : \sigma x = x \quad \text{for all} \quad \sigma \varepsilon \Sigma_k\}$.

(c) Show that e is an idempotent, that is, $e^2 = e$.

(d) Show that $eM =$ the image in mod p^r homology of the iterated map
$$\Delta : P^{2k}(p^r) \to P.$$

To be specific, Δ is the composition
$$P^{2k}(p^r) \xrightarrow{\Delta_{2,2k-2}} P^2(p^r) \wedge P^{2k-2}(p^r)$$
$$\xrightarrow{1 \wedge \Delta_{2,2k-4}} P^2(p^r) \wedge P^2(p^r) \wedge P^{2k-4}(p^r) \to \cdots$$
$$\cdots \to P^2(p^r) \wedge P^2(p^r) \wedge \cdots \wedge P^2(p^r) \wedge P^4(p^r)$$
$$\xrightarrow{1 \wedge \cdots \wedge 1 \wedge \Delta_{2,2}} P^2(p^r) \wedge P^2(p^r) \wedge \cdots \wedge P^2(p^r) = P.$$

Now show that $eM = 0$ in dimensions less than $2k - 1$ and $eM = \Delta_*(H_*(P^2(p^r); \mathbb{Z}/p^r\mathbb{Z}))$ in dimensions $2k - 1$ and $2k$.

(4) Localize at the prime 3.

 (a) If $n \geq 1$ show that
$$\alpha : S^{2n+3} \xrightarrow{\alpha_1} \Omega S^{2n+1} \to S^{2n+1}\{3^r\}$$
generates $\pi_{2n+3}(S^{2n+1}\{3^r\}) \cong \mathbb{Z}/3\mathbb{Z}$.

 (b) If $m \geq 3$, show that a composition
$$P^{m+2}(3^s) \to S^{m+2} \to S^{m-1} \subseteq P^m(3^r)$$
generates $\pi_{m+2}(P^m(3^r); \mathbb{Z}/3^s\mathbb{Z})$ modulo Whitehead products and actually generates it if $m > 3$.

(c) If $m, n, q \geq 2$ show that the kernel of the mod p^s Hurewicz map

$$\varphi : \pi_*(P^m(3^r) \wedge P^n(3^r) \wedge P^q(3^r); \mathbb{Z}/p^s\mathbb{Z})$$
$$\to H_*(P^m(3^r) \wedge P^n(3^r) \wedge P^q(3^r); \mathbb{Z}/p^s\mathbb{Z})$$

is generated in dimension $m + n + q$ by Whitehead products and a map

$$P^{m+n+q}(3^s) \to S^{m+n+q-3} \to P^m(3^r) \wedge P^n(3^r) \wedge P^q(3^r)$$

of order 3.

6.5 Lie identities in groups

George Whitehead [133] had the fundamental insight that the Lie identities for Samelson products are a consequence of certain analogous Lie identities for groups. In this section we present these Lie identities for groups. The treatment here is heavily influenced by that found in Serre's book [119].

Let x, y be elements of a group G and define

(1) The congugate homomorphisms are $x^y = y^{-1}xy$. Recall that $(xy)^z = x^z y^z$ and $(x^y)^{y^{-1}} = x$.

(2) The commutators are $[x, y] = xyx^{-1}y^{-1}$. Thus, $[x, y]^z = [x^z, y^z]$.

The Lie identities in groups are the following formulas.

Proposition 6.5.1. *For elements x, y, z in a group G,*

(1) *exponentiation modulo a commutator:*

$$x^y = x[x^{-1}, y^{-1}].$$

(2) *inverse of a commutator:*

$$[x, y]^{-1} = [y, x], \qquad [x^{-1}, y] = [y, x]^x.$$

(3) *commutativity modulo commutators:*

$$xy = [x, y]yx.$$

(4) *bilinearity modulo commutators:*

$$[x, yz] = [x, y] \, [x, z]^{(y^{-1})}, \qquad [xy, z] = [y, z]^{(x^{-1})} \, [x, z].$$

(5) *Jacobi identity modulo commutators*

$$[x^{(y^{-1})}, [z, y]] \, [y^{(z^{-1})}, [x, z]] \, [z^{(x^{-1})}, [y, x]] = 1.$$

It may be difficult to discover some of the above formulas but there can be no doubt that they are straightforward to prove. Merely write them in the form $c = 1$ and reduce the word c to the identity via successive applications of

(1) $wdw^{-1} = 1$ if and only if $d = 1$.

(2) $ww^{-1} = 1$.

This amounts to reducing a word in a free group to the identity.

Let A and B be two subgroups of a group G. Denote by $A \cdot B$ the subgroup generated by all products ab where $a \varepsilon A$ and $b \varepsilon B$. In other words, $A \cdot B$ is the subgroup generated by the union $A \cup B$.

Similarly, the commutator $[A, B]$ is the group generated by all commutators $[a, b]$.

Definition 6.5.2. The descending central series of a group G is the descending sequence of subgroups

$$G = G_1 \supseteq G_2 \supseteq G_3 \supseteq \cdots \supseteq G_i \supseteq G_{i+1} \supseteq \cdots$$

defined by

(1) $G_1 = G$

(2) $G_{i+1} = [G, G_i] =$ the subgroup generated by all commutators $[x, y]$ with $x \varepsilon G$, $y \varepsilon G_i$.

Remarks 6.5.3.

(1) It follows by induction and the Lie identities, in particular (1) and (4), that G_i is generated by commutators $[x_1, [x_2, \ldots [x_{i-1}, x_i] \ldots]]$ of length i in elements $x_1, x_2, \ldots, x_i$ of G. Thus $G_{i+1} \subseteq G_i$ and every G_i is a normal subgroup of G.

(2) It follows by induction on j, the Jacobi identity (5), and bilinearity (4) that $[G_i, G_j] \subseteq G_{i+j}$. In more detail, the case $j = 1$ is true by definition. For $j > 1$, assume that $[G_i, G_{j-1}] \subseteq G_{i+j-1}$ and note that

$$[G_i, G_j] = [G_i, [G, G_{j-1}]] \subseteq [[G_{j-1}, G_i], G] \cdot [[G_i, G], G_{j-1}]$$
$$\subseteq [G_{i+j-1}, G] \cdot [G_{i+1}, G_{j-1}] \subseteq G_{i+j} \cdot G_{i+j} \subseteq G_{i+j}.$$

We remark that G is called nilpotent of length $\leq n$ if $G_n = \{*\} =$ the trivial group. This is the same as requiring that all commutators of length n are trivial. In particular, nilpotent of length ≤ 2 is the same as abelian.

If we define $L_i = G_i/G_{i+1}$, then the multiplication in G, $(x, y) \mapsto xy$ induces an abelian operation in each L_i, written additively as $(x, y) \mapsto x + y \equiv xy$. Thus, each L_i is an abelian group with $x + y \equiv xy$ and $-x \equiv x^{-1}$. The commutator in G, $(x, y) \mapsto [x, y]$ induces a bilinear pairings $L_i \times L_j \to L_{i+j}$, written as $(x, y) \mapsto [x, y]$ which satisfy the classic (ungraded) Lie identities:

(1) $[x, x] = 0$

(2) $[y, x] = -[x, y]$

(3) $[x, [z, y]] + [y, [x, z]] + [z, [y, x]] = 0$.

We usually write the Jacobi identity in the equivalent derivation form:

$$[x, [y, z]] = [[x, y], z] + [y, [x, z]]$$

We wish to emphasize that the associated graded object L_* is not what is called a graded Lie algebra, the definition of which requires some signs. It is a classical Lie algebra which happens to have a grading.

Exercises

(1) Verify the Lie identities in Proposition 6.5.1.

(2) (a) If A, B are subgroups of a group G, prove that

$$[A, B] = [B, A].$$

(b) If A, B, C are normal subgroups of a group G, prove that

$$[A, [B, C]] \subseteq [[A, B], C] \cdot [B, [A, C]].$$

6.6 External Samelson products

Let G be a grouplike space and let $f : X \to G$ and $g : Y \to G$ be maps. In this section, we introduce hypotheses which enable us to prove the Lie identities for Samelson products $[f, g] : X \wedge Y \to G$. We are primarily interested in the case where X and Y are Moore spaces. In this case $X \wedge Y$ will usually not be a Moore space and so we call these Samelson products by the name external Samelson products in order to distinguish them from the internal Samelson products which are introduced in the next section.

The fundamental concept is:

Definition 6.6.1. A space X is conilpotent of length $\leq n$ if the composition

$$\overline{\Delta} : X \xrightarrow{\Delta} \prod X \xrightarrow{q} \bigwedge X$$

is null homotopic where Δ is the diagonal into the n-fold product and q is the standard projection.

It is obvious that, if X is conilpotent of length $\leq n$ and Y is any space, then the smash product $X \wedge Y$ is also conilpotent of length $\leq n$.

Being conilpotent of length $\leq n$ is a generalization of being of category $\leq n$.

Definition 6.6.2. A space X is of category $\leq n$ if the iterated diagonal $\Delta : X \to \prod X$ into the n-fold product factors up to homotopy through the fat wedge $\{(x_1, \ldots, x_n) : \text{at least one } x_i = * = \text{the basepoint}\}$.

Since any n-fold commutator map

$$G \xrightarrow{\Delta} \prod G \xrightarrow{C} G$$

factors as

$$G \xrightarrow{\overline{\Delta}} \bigwedge G \xrightarrow{\overline{C}} G$$

it follows that

Proposition 6.6.2. *If X is conilpotent of length $\leq n$, then the pointed mapping group $[X, G]_*$ is nilpotent of length $\leq n$.*

In particular, X is called coabelian if $\overline{\Delta} : X \to X \wedge X$ is null homotopic. In this case, the group $[X, G]_*$ is abelian. In Section 1.10 we showed that, if k is odd, then $P^2(k)$ is a coabelian space and thus $\pi_2(G; \mathbb{Z}/k\mathbb{Z})$ is an abelian group for all grouplike G. Note that $P^2(k)$ is not a space of category ≤ 2, that is, the diagonal is not homotopic to a composition $P^2(k) \to P^2(k) \vee P^2(k) \to P^2(k) \times P^2(k)$. This would imply a factorization of fundamental groups through the free product

$$\mathbb{Z}/k\mathbb{Z} \to (\mathbb{Z}/k\mathbb{Z}) * (\mathbb{Z}/k\mathbb{Z}) \to (\mathbb{Z}/k\mathbb{Z}) \times (\mathbb{Z}/k\mathbb{Z}).$$

This is impossible since the free product has no elements of finite order except for the two copies of $\mathbb{Z}/k\mathbb{Z}$.

Let $P_1, P_2, \ldots, P_n$ be pointed spaces and consider the following increasing filtration

$$* = F_0 \subseteq F_1 \subseteq \cdots \subseteq F_n = P_1 \times P_2 \times \cdots \times P_n$$

of the product by increasingly fatter wedges:

$$F_k = \{(x_1, x_2, \ldots, x_n) : x_i \neq * \text{ for at most } k \text{ values of } i\}$$
$$= \bigcup_{A \subseteq \{1,2,\ldots,n\}, |A|=k} \prod_{\alpha \in A} P_\alpha.$$

For example,

$$F_1 = P_1 \vee P_2 \vee \cdots \vee P_n$$

and

$$F_2 = P_1 \times P_2 \cup \cdots \cup P_{n-1} \times P_n.$$

Thus
$$\Sigma F_k = \bigvee_{A\subseteq\{1,2,\ldots,n\}, |A|\leq k} \Sigma \bigwedge_{\alpha \varepsilon A} P_\alpha =$$
a bouquet of smash products of $\leq k$ of the spaces $P_1, \ldots, P_n$.

The cofibration sequences
$$F_{k-1} \to F_k \to Q_k = \bigvee_{A\subseteq\{1,2,\ldots,n\}, |A|=k} \bigwedge_{\alpha \varepsilon A} P_\alpha$$
have cofibres equal to bouquets of smashes of exactly k of the spaces $P_1, \ldots, P_n$. The suspensions of these cofibration sequences are cofibration sequences which are split by sections
$$\Sigma Q_k \to \Sigma F_k.$$
Hence, there is an equivalence
$$\Sigma Q_k \vee \Sigma F_{k-1} \to \Sigma F_k.$$
The long cofibration sequences
$$F_{k-1} \to F_k \to Q_k \to \Sigma F_{k-1} \to \Sigma F_k \to \Sigma Q_k$$
show that

(1) $[\Sigma F_{k-1}, G]_* \leftarrow [\Sigma F_k, G]_*$ is an epimorphism,

(2) $[Q_k, G]_* \leftarrow [\Sigma F_{k-1}, G]_*$ is the trivial map, and

(3) $[F_k, G]_* \leftarrow [Q_k, G]_*$ is a monomorphism.

As a special case, we have

Proposition 6.6.3. *The natural maps $q : P_1 \times \cdots \times P_n \to P_1 \wedge \cdots \wedge P_n$ induce monomorphisms*
$$q^* : [P_1 \wedge \cdots \wedge P_n, G]_* \to [P_1 \times \cdots \times P_n, G]_*.$$

Thus
$$[P_1 \wedge \cdots \wedge P_n, G]_* = [(P_1 \times \cdots \times P_n, F_{n-1}), (G, *)] \subseteq [P_1 \times \cdots \times P_n, G]_*$$
may be regarded as a subgroup.

If $f_i : P_i \to G$ is any pointed map and $p_i : P_1 \times \cdots \times P_n \to P_i$ is the i-th projection map, then the composition
$$\overline{f_i} = f_i \circ p_i : P = P_1 \times \cdots \times P_n \xrightarrow{p_i} P_i$$

represents what is called a special element of the group $[P_1 \times \cdots \times P_n, G]_*$. Thus special elements are exactly those classes which are represented by a group valued function of one variable.

Definition 6.6.4. If $f_i : P_i \to G$ are pointed maps for $1 \leq i \leq n$ and $C : G \times \cdots \times G \to G$ is any commutator map of length n, then the composition

$$P_1 \times \cdots \times P_n \xrightarrow{f_1 \times \cdots \times f_n} G \times \cdots \times G \xrightarrow{C} G$$

factors up to homotopy through a homotopically unique map

$$P_1 \wedge \cdots \wedge P_n \xrightarrow{f_1 \wedge \cdots \wedge f_n} G \wedge \cdots \wedge G \xrightarrow{\overline{C}} G$$

and represents the external Samelson product $C(f_1, \ldots, f_n)$ in the subgroup $[P_1 \wedge \ldots P_n, G]_*$.

The homotopical uniqueness of the factorization has the consequence that the external Samelson products are well defined.

In other words, a commutator map C induces a well defined map

$$C : [P, G]_* \times \cdots \times [P, G]_* \to [P_1 \wedge \cdots \wedge P_n, G]_*$$

and, on special elements $\overline{f_i} : P_i \to G$, $1 \leq i \leq n$, we have

$$C(\overline{f_1}, \ldots, \overline{f_n}) = C(f_1, \ldots, f_n) = \overline{C} \cdot (f_1 \wedge \cdots \wedge f_n).$$

In order to make this clear, consider the case $n = 2$ with maps $f_1 : P_1 \to G$, $f_2 : P_2 \to G$. The external Samelson product $[f_1, f_2]$ is the map represented by

$$[\overline{f_1}, \overline{f_2}] = \overline{f_1}\,\overline{f_2}\,\overline{f_1}^{-1}\,\overline{f_2}^{-1}$$

or by the composition

$$[\ ,\ ] \circ (f_1 \wedge f_2)$$

in the subgroup

$$[P_1 \wedge P_2, G]_* \subseteq [P_1 \times P_2, G]_*.$$

It is uniquely characterized by the homotopy commutative diagram

$$\begin{array}{ccc} P_1 \times P_2 & \xrightarrow{\overline{f_1 f_2}\,(\overline{f_1})^{-1}(\overline{f_2})^{-1}} & G \\ \downarrow & & \uparrow [f_1, f_2] \\ P_1 \wedge P_2 & = & P_1 \wedge P_2 \end{array}$$

The map $[f_1, f_2] : P_1 \wedge P_2 \to G$ is represented, up to the canonical homotopies for units, inverses, and associativity, on elements by

$$x \wedge y \mapsto [f_1(x), f_2(y)].$$

Or consider the case $n = 3$ with maps $f_1 : P_1 \to G, f_2 : P_2 \to G, f_3 : P_3 \to G$. The external Samelson product $[f_1, [f_1, f_2]]$ is represented by

$$\overline{[f_1, [\overline{f_2, f_3}]]}$$

or by the composition

$$\overline{[\ , [\ ,\ ]\ ]} \circ (f_1 \wedge f_2 \wedge f_3)$$

in the subgroup

$$[P_1 \wedge P_2 \wedge P_3, G]_* \subseteq [P_1 \times P_2 \times P_3, G]_*.$$

It is uniquely characterized by the homotopy commutative diagram

$$\begin{array}{ccc} P_1 \times P_2 \times P_3 & \xrightarrow{\overline{[f_1, [\overline{f_2, f_3}]]}} & G \\ \downarrow & & \uparrow [f_1, [f_2, f_3]] \\ P_1 \wedge P_2 \wedge P_3 & = & P_1 \wedge P_2 \wedge P_3 \end{array}$$

The map $[f_1, [f_2, f_3]] : P_1 \wedge P_2 \wedge P_3 \to G$ is represented up to canonical homotopies on elements by

$$x \wedge y \wedge z \mapsto [f_1(x), [f_2(y), f_3(z)]].$$

A possible ambiguity is resolved by the following consequence of the homotopical uniqueness of the factorization:

Lemma 6.6.5. *If $C = [\ ,\ ] \circ (C_i \times C_j)$ is an iterated commutator of length $i + j$, then*

$$C(f_1, \ldots, f_i, f_{i+1}, \ldots, f_{i+j}) = [C_i(f_1, \ldots, f_i), C_j(f_{i+1}, \ldots, f_{i+j})].$$

For example, $[f, [g, h]]$ may be regarded as a triple Samelson product

$$P_1 \wedge P_2 \wedge P_3 \xrightarrow{f \wedge g \wedge h} G \wedge G \wedge G \xrightarrow{\overline{[\ , [\ ,\ ]\ ]}} G$$

or, equivalently, as the composition

$$P_1 \wedge (P_2 \wedge P_3) \xrightarrow{f \wedge [g,h]} G \wedge G \xrightarrow{\overline{[\ ,\ ]}} G.$$

Proposition 6.6.6. *Let $f_{\alpha(i)} : P_{\alpha(i)} \to G$, $1 \leq i \leq A$ be a collection of functions where the indices $\alpha(i)$ are in nondecreasing order (that is, $\alpha(i) \leq \alpha(i+1)$) and satisfy*

(1) $1 \leq \alpha(i) \leq n$ *for all i*

(2) *the $\alpha(i)$ include all of $\{1, \ldots, n\}$ with possible repetitions.*

(3) *the cardinality of the set of i is $A \geq n$.*

Let $C : G \times \cdots \times G \to G$ be any commutator map of length A. Then $C(\overline{f_{\alpha(1)}}, \ldots, \overline{f_{\alpha(A)}})$ is represented by a map

$$P_1 \wedge \cdots \wedge P_n \to G.$$

Furthermore, if the spaces $P_1, \ldots, P_n$ are all coabelian and $A > n$, this map is null homotopic.

Proof: $C(\overline{f_{\alpha(1)}}, \ldots, \overline{f_{\alpha(A)}})$ is represented by a map

$$P_{\alpha(1)} \wedge \cdots \wedge P_{\alpha(A)} \xrightarrow{f_{\alpha(1)} \wedge \cdots \wedge f_{\alpha(A)}} G \wedge \cdots \wedge G \xrightarrow{\overline{C}} G.$$

The standard map

$$\delta : P_1 \wedge \cdots \wedge P_n \to P_{\alpha(1)} \wedge \cdots \wedge P_{\alpha(A)}$$

is the identity map if $A = n$ and includes some diagonal maps if $A > n$. Thus, δ is null homotopic if $A > n$.

Then $C(\overline{f_{\alpha(1)}}, \ldots, \overline{f_{\alpha(A)}})$ is represented by the composition

$$P_1 \wedge \cdots \wedge P_n \xrightarrow{\delta} P_{\alpha(1)} \wedge \cdots \wedge P_{\alpha(A)} \xrightarrow{f_{\alpha(1)} \wedge \cdots \wedge f_{\alpha(A)}} G \wedge \cdots \wedge G \xrightarrow{\overline{C}} G.$$

$\square$

Proposition 6.6.6 says that, any commutator of one variable functions vanishes if it is defined on a smash product of lesser length (than the length of the commutator) involving coabelian spaces.

The Lie identities for groups now translate into Lie identities for external Samelson products. Roughly speaking, Proposition 6.6.6 allows the erasure of any extra commutators in the group version 6.5.1 of the Lie identities.

Proposition 6.6.7. *Let G be a grouplike space and let P_1, P_2, P_3 be coabelian spaces. If $f, f_1 : P_1 \to G, g, g_1 : P_2 \to G, h : P_3 \to G$ are maps then the external Samelson products $[f, g] : P_1 \wedge P_2 \to G$ and $[f, [g, h]] : P_1 \wedge P_2 \wedge P_3 \to G$ satisfy the following Lie identities:*

(1) *Anti-commutativity:*

$$[f, g] = -[g, f] \circ \tau$$

where

$$\tau = T_{(1,2)} : P_1 \wedge P_2 \to P_2 \wedge P_1$$

is the standard transposition.

(2) *Bilinearity:*

$$[f + f_1, g] = [f, g] + [f_1, g], \quad [f, g + g_1] = [f, g] + [f, g_1].$$

(3) *Jacobi identity:*
$$[f,[g,h]] + [h,[f,g]] \circ \sigma + [g,[h,f]] \circ \sigma^2 = 0$$
where
$$\sigma = T_{(1,2,3)} : P_1 \wedge P_2 \wedge P_3 \to P_3 \wedge P_1 \wedge P_2$$
and
$$\sigma^2 = T_{(1,3,2)} : P_1 \wedge P_2 \wedge P_3 \to P_2 \wedge P_3 \wedge P_1$$
are the standard cyclic permutations.

Proof:

(1) In $[P_1 \times P_2, G]_*$, the composition $-[g,f] \circ \tau$ is represented by $[\overline{g}, \overline{f}]^{-1} = [\overline{f}, \overline{g}]$ and this also represents the Samelson product $[f,g]$.

(2) The Lie identities in Proposition 6.5.1 assert that $[f + f_1, g]$ is represented in $[P_1 \times P_2, G]_*$, by
$$[\overline{ff_1}, \overline{g}] = [\overline{f_1}, \overline{g}]^{\overline{f}^{-1}} [\overline{f}, \overline{g}] = ([\overline{f_1}, \overline{g}][[\overline{g}, \overline{f_1}], \overline{f}])[\overline{f}, \overline{g}].$$
Because of the length of the commutators (6.6.6), in the abelian group $[P_1 \wedge P_2, G]_*$ this is representative of $[f_1, g] + [f, g] = [f, g] + [f_1, g]$.

Similarly, $[f, g + g_1] = [f, g] + [f, g_1]$.

(3) In the group $[P_1 \times P_2 \times P_3, G]_*$ we have
$$[\overline{f}^{\overline{g}^{-1}}, [\overline{h}, \overline{g}]] = [\overline{f}[\overline{f}^{-1}, \overline{g}], [\overline{h}, \overline{g}]] = ([[\overline{f}^{-1}, \overline{g}], [\overline{h}, \overline{g}]]^{\overline{f}^{-1}})[\overline{f}, [\overline{h}, \overline{g}]].$$
Because of the length of commutators, this represents $[f, [h, g]]$ in $[P_1 \wedge P_2 \wedge P_3, G]_*$.

Similarly, $[\overline{g}^{\overline{h}^{-1}}, [\overline{f}, \overline{h}]]$ and $[\overline{h}^{\overline{f}^{-1}}, [\overline{g}, \overline{f}]]$ represent $[g, [f, h]] \circ \sigma$ and $[h, [g, f]] \circ \sigma^2$, respectively.

Since
$$[\overline{f}^{\overline{g}^{-1}}, [\overline{h}, \overline{g}]] \cdot [\overline{g}^{\overline{h}^{-1}}, [\overline{f}, \overline{h}]] \cdot [\overline{h}^{\overline{f}^{-1}}, [\overline{g}, \overline{f}]] = 1$$
in the group $[P_1 \times P_2 \times P_3, G]_*$, it follows that
$$[f, [h, g]] + [g, [f, h]] \circ \sigma + [h, [g, f]] \circ \sigma^2 = 0$$
in the abelian subgroup $[P_1 \wedge P_2 \wedge P_3, G]_*$. □

Remark. It is tempting to think that external Samelson products satisfy $[f, f] = 0$ since this is true for commutators in groups. This is wrong since

186   Samelson products

$[f, f]$ is represented by the map $P_1 \wedge P_1 \to G$ which, up to homotopy, is

$$x \wedge y \mapsto [f(x), f(y)] \neq 1.$$

Exercise

(1) A commutator map of length 1 is just a map $C_1 : G \to G$. Inductively, if $C_i : \prod G \to G$ and $C_j : \prod G \to G$ are commutator maps of lengths i and j, respectively, then the map $C_{i+j} : (\prod G) \times (\prod G) \xrightarrow{C_i \times C_j} G \times G \xrightarrow{[\ ,\ ]} G$ is called a commutator map of length $i + j$. Show that any commutator map $C : \prod G \to G$ factors up to homotopy as

$$\prod G \to \wedge G \xrightarrow{\overline{C}} G.$$

6.7 Internal Samelson products

Let G be a grouplike space and let p be a prime. In the previous section, we showed that two maps $f : P^m(p^r) \to G$ and $g : P^n(p^r) \to G$ define an external Samelson product, here denoted by

$$[f, g]_e : P^m(p^r) \wedge P^n(p^r) \to G.$$

We now define the internal Samelson product $[f, g] : P^{m+n}(p^r) \to G$ by using the maps $\Delta_{m,n} : P^{m+n}(p^r) \to P^m(p^r) \wedge P^n(p^r)$. If p^r is greater than 2, these maps exist. If p is an odd prime, these maps are characterized modulo Whitehead products by their induced maps in mod p^r homology:

$$\Delta_{m,n*}(e_{m+n}) = e_m \otimes e_n.$$

Definition 6.7.1. Let $p^r > 2$. For $m, n \geq 2$, the internal Samelson product

$$[\ ,\ ] : \pi_m(G; \mathbb{Z}/p^r\mathbb{Z}) \times \pi_n(G; \mathbb{Z}/p^r\mathbb{Z}) \to \pi_{m+n}(G; \mathbb{Z}/p^r\mathbb{Z})$$

is represented by the composition

$$[f, g] = [f, g]_e \circ \Delta_{m,n} : P^{n+m}(p^r) \to P^m(p^r) \wedge P^n(p^r) \to G.$$

Since Samelson products map into an H-space, the above internal Samelson product is well defined if p is an odd prime. If $\tau : P^m(p^r) \wedge P^n(p^r) \to P^n(p^r) \wedge P^m(p^r)$ is the standard transposition, then

$$\begin{array}{ccc} P^{m+n}(p^r) & \xrightarrow{\Delta_{m,n}} & P^m(p^r) \wedge P^n(p^r) \\ \downarrow (-1)^{mn} & & \downarrow \tau \\ P^{m+n}(p^r) & \xrightarrow{\Delta_{n,m}} & P^n(p^r) \wedge P^m(p^r) \end{array}$$

commutes modulo Whitehead products when p is an odd prime.

6.7 Internal Samelson products

Let $\sigma : P^m(p^r) \wedge P^n(p^r) \wedge P^q(p^r) \to P^q(p^r) \wedge P^m(p^r) \wedge P^n(p^r)$ be the standard permutation. If p is a prime greater than 3, the following diagrams commute modulo Whitehead products

$$\begin{array}{ccccc}
P^{m+n+q}(p^r) & \xrightarrow{\Delta_{m,n+q}} & P^m(p^r) \wedge P^{n+q}(p^r) & \xrightarrow{1 \wedge \Delta_{n,q}} & P^m(p^r) \wedge P^n(p^r) \wedge P^q(p^r) \\
\downarrow (-1)^{q(m+n)} & & & & \downarrow \sigma \\
P^{m+n+q}(p^r) & \xrightarrow{\Delta_{q,m+n}} & P^q(p^r) \wedge P^{m+n}(p^r) & \xrightarrow{1 \wedge \Delta_{m,n}} & P^q(p^r) \wedge P^m(p^r) \wedge P^n(p^r) \\
P^{m+n+q}(p^r) & \xrightarrow{\Delta_{m,n+q}} & P^m(p^r) \wedge P^{n+q}(p^r) & \xrightarrow{1 \wedge \Delta_{n,q}} & P^m(p^r) \wedge P^n(p^r) \wedge P^q(p^r) \\
\downarrow (-1)^{m(n+q)} & & & & \downarrow \sigma^2 \\
P^{m+n+q}(p^r) & \xrightarrow{\Delta_{n,m+q}} & P^n(p^r) \wedge P^{m+q}(p^r) & \xrightarrow{1 \wedge \Delta_{q,m}} & P^n(p^r) \wedge P^q(p^r) \wedge P^m(p^r).
\end{array}$$

Therefore, Proposition 6.6.7 for external Samelson products translates into the following result for internal Samelson products

Proposition 6.7.2. *Let p be an odd prime. For $m, n, q \geq 2$, let f, $f_1 \varepsilon \pi_m(G; \mathbb{Z}/p^r\mathbb{Z})$, $g, g_1 \varepsilon \pi_n(G; \mathbb{Z}/p^r\mathbb{Z})$, and $h \varepsilon \pi_q(G; \mathbb{Z}/p^r\mathbb{Z})$.*

(a) *Bilinearity:*

$$[f + f_1, g] = [f, g] + [f_1, g], \quad [f, g + g_1] = [f, g] + [f, g_1].$$

(b) *Anti-commutativity:*

$$[f, g] = -(-1)^{mn}[g, f].$$

(c) *If p is greater than 3, the Jacobi identity:*

$$[f, [g, h]] + (-1)^{q(m+n)}[h, [f, g]] + (-1)^{m(n+q)}[g, [h, f]] = 0.$$

Remark. We usually prefer the Jacobi identity to be written in the equivalent derivation form which we believe is easier to remember:

$$[f, [g, h]] = [[f, g], h] + (-1)^{mn}[g, [f, h]].$$

Remark. If $p = 3$, then it follows from Exercise 4 in Section 6.4 that the permutation diagrams which preceed 6.6.7 commute modulo Whitehead products and an element of order 3. Hence the Jacobi element

$$[f, [g, h]] + (-1)^{q(m+n)}[h, [f, g]] + (-1)^{m(n+q)}[g, [h, f]]$$

is an element of order 3. In other words, the Jacobi identity holds modulo elements of order 3.

From Lemma 6.3.7, we get

Proposition 6.7.3. *If $p > 2$, then the mod p^r Hurewicz map $\varphi : \pi_*(G; \mathbb{Z}/p^r\mathbb{Z}) \to H_*(G; \mathbb{Z}/p^r\mathbb{Z})$ is compatible with internal Samelson products in the sense that*

$$\varphi[f, g] = [\varphi(f), \varphi(g)] = \varphi(f)\varphi(g) - (-1)^{\deg(f)\deg(g)}\varphi(g)\varphi(f).$$

Remark. If $p^r > 2$, the above is true in dimensions ≥ 3.

Other Samelson products arise from the equivalences
$$S^{m+n} \xrightarrow{=} S^m \wedge S^n$$
$$P^{m+n}(p^r) \xrightarrow{=} S^m \wedge P^n(p^r)$$
$$P^{m+n}(p^r) \xrightarrow{=} P^m(p^r) \wedge S^n.$$

If we add the definition
$$\pi_1(G; \mathbb{Z}/p r^r \mathbb{Z}) = \pi_1(G) \otimes \mathbb{Z}/p^r \mathbb{Z}$$
we get the full result

Propositon 6.7.4. *Let p be an odd prime and suppose $m, n, q \geq 1$. There are bilinear pairings*

$$[\ ,\ ] : \pi_m(G) \otimes \pi_n(G) \to \pi_{m+n}(G)$$

$$[\ ,\ ] : \pi_m(G) \otimes \pi_n(G; \mathbb{Z}/p^r \mathbb{Z}) \to \pi_{m+n}(G; \mathbb{Z}/p^r \mathbb{Z})$$

$$[\ ,\ ] : \pi_m(G; \mathbb{Z}/p^r \mathbb{Z}) \otimes \pi_n(G) \to \pi_{m+n}(G; \mathbb{Z}/p^r \mathbb{Z})$$

$$[\ ,\ ] : \pi_m(G; \mathbb{Z}/p^r \mathbb{Z}) \otimes \pi_n(G; \mathbb{Z}/p^r \mathbb{Z}) \to \pi_{m+n}(G; \mathbb{Z}/p^r \mathbb{Z})$$

which satisfy

(1) *bilinearity,*

(2) *anti-commutativity,*

(3) *and, if p is a prime greater than 3, the Jacobi identity.*

Remark. If at least one of the classes involved is in the homotopy group with integral coefficients, then the Jacobi identity is valid even if we do not require the prime to be greater than 3, but merely that it is odd.

Furthermore, these Samelson products are all compatible with the Hurewicz maps, for example, if $f \varepsilon \pi_m(G)$ and $g \varepsilon \pi_n(G; \mathbb{Z}/p^r \mathbb{Z})$, then the diagram below commutes

Lemma 6.7.5.

$$\begin{array}{ccc} \pi_m(G) \otimes \pi_n(G; \mathbb{Z}/p^r \mathbb{Z}) & \xrightarrow{[\ ,\ ]} & \pi_{m+n}(G; \mathbb{Z}/p^r \mathbb{Z})) \\ \downarrow \varphi \otimes \varphi & & \downarrow \varphi \\ H_m(G) \otimes H_n(G; \mathbb{Z}/p^r \mathbb{Z}) & \xrightarrow{[\ ,\ ]} & H_{m+n}(G; \mathbb{Z}/p^r \mathbb{Z})). \end{array}$$

Recall that the pinch map $\bar{p} : P^m(p^r) \to S^n$ induces the mod p^r reduction map
$$\rho = \bar{p}^* : \pi_m(G) \to \pi_m(G; \mathbb{Z}/p^r \mathbb{Z}).$$

And the inclusion map of the bottom cell $\bar{\beta}: S^{m-1} \to P^m(p^r)$ induces the Bockstein map

$$\beta: \pi_m(G; \mathbb{Z}/p^r\mathbb{Z}) \to \pi_{m-1}(G).$$

The reduction maps are homomorphisms and the Bockstein maps are derivations in the following sense.

Proposition 6.7.6. *If p is an odd prime, then*

(a) *if $f, g \varepsilon \pi_*(G)$,*

$$\rho[f, g] = [\rho f, \rho g].$$

(b) *if $f \varepsilon \pi_*(G; \mathbb{Z}/p^r\mathbb{Z}), g \varepsilon \pi_*(G)$,*

$$[f, g] = [f, \rho g].$$

(c) *if $f \varepsilon \pi_*(G; \mathbb{Z}/p^r\mathbb{Z}), g \varepsilon \pi_*(G; \mathbb{Z}/p^r\mathbb{Z})$,*

$$\rho\beta[f, g] = [\beta f, g] + (-1)^{\deg(f)}[f, \beta g].$$

(d) *if $f \varepsilon \pi_*(G), g \varepsilon \pi_*(G; \mathbb{Z}/p^r\mathbb{Z})$,*

$$\beta[f, g] = (-1)^{\deg(f)}[f, \beta g].$$

Part (a) uses the commutativity of

$$\begin{array}{ccc} P^{m+n}(p^r) & \xrightarrow{\Delta_{m,n}} & P^m(p^r) \wedge P^n(p^r) \\ \downarrow \bar{\rho} & & \downarrow \bar{\rho} \\ S^{m+n} & \xrightarrow{=} & S^m \wedge S^n. \end{array}$$

Part (c) uses the commutativity modulo Whitehead products of

$$\begin{array}{ccc} P^{m+n}(p^r) & \xrightarrow{\Delta_{m,n}} & P^m(p^r) \wedge P^n(p^r) \\ \uparrow \bar{\beta} & & \uparrow \\ S^{m+n-1} & & | \\ \uparrow \bar{\rho} & & | \\ P^{m+n-1}(p^r) & \xrightarrow{1+(-1)^m} & (S^{m-1} \wedge P^n(p^r)) \vee (P^m(p^r) \wedge S^{n-1}). \end{array}$$

Parts (a) and (d) are even easier.

Definition 6.7.7. Let R be a commutative ring in which 2 is a unit. A graded Lie algebra $L = L_*$ over R is a graded R-module together with bilinear pairings

$$[\ ,\ ]: L_m \otimes L_n \to L_{m+n}$$

such that

(1)
$$[x,y] = -(-1)^{\deg(x)\deg(y)}[y,x].$$

(2) if the degree of x is even,
$$[x,x] = 0.$$

(3) if the degree of x is odd,
$$[x,[x,x]] = 0.$$

(4)
$$[x,[y,z]] = [[x,y],z] + (-1)^{\deg(x)\deg(y)}[y,[x,z]].$$

If 2 is a unit in the ring, it is clear that (1) implies (2) since:
$$[x,x] = -[x,x]$$
when x has even degree.

If 3 is a unit in the ring, then (1) and (4) imply (3) since:
$$[x,[x,x]] = [[x,x],x] - [x,[x,x]] = -2[x[x,x]]$$
when x has odd degree.

So it is clear that, if G is a connected grouplike space, then $\pi_*(G)$ is a graded Lie algebra over the integers localized away from 2 and 3, and, if p is a prime greater than 3, then $\pi_*(G;\mathbb{Z}/p^r\mathbb{Z})$ is a graded Lie algebra.

Definition 6.7.8. A differential graded Lie algebra is a graded Lie algebra L together with a linear map $d: L_m \to L_{m-1}$ which is a differential, $d^2 = 0$, and a derivation,
$$d[x,y] = [dx,y] + (-1)^{\deg(x)}[x,dy].$$

If p is a prime greater than 3 and G is a connected grouplike space, then the composition $d = \rho \circ \beta : \pi_m(G;\mathbb{Z}/p^r\mathbb{Z}) \to \pi_{m-1}(G) \to \pi_{m-1}(G;\mathbb{Z}/p^r\mathbb{Z})$ makes $\pi_*(G;\mathbb{Z}/p^r\mathbb{Z})$ into a graded Lie algebra.

Exercises

(1) Verify Proposition 6.7.4, Lemma 6.7.5, and Proposition 6.7.6 in detail.

6.8 Group models for loop spaces

In this section we discuss group models for loop spaces. We begin by defining a general notion of principle bundle:

6.8 Group models for loop spaces

Definition 6.8.1. Let $F \to E \xrightarrow{p} B$ be a fibration sequence and let $F \vee E \to B$ and $F \times E \to B$ be the maps which both send a point (f, e) to $p(e)$. We say that the fibration has a left action if the natural map over the base B of the bouquet

$$F \vee E \to E$$

extends to a map over B of the product

$$F \times E \to E.$$

Any principal bundle $E \to B$ with group G has a left action $G \times E \to E$ in this sense.

If we consider the multiplication in the usual path space PX, then the usual left action $\Omega X \times PX \to PX$ is fibre homotopic to the inclusion on the bouquet $\Omega X \vee PX \to PX$. Hence, we can alter the usual left action by a fibre homotopy to get a left action $\Omega X \times PX \to X$ in the above sense.

Proposition 6.8.2. *Let $p : E \to B$ and $q : E' \to B'$ be fibrations. Let $f : B \to B'$ be any map and suppose that E is contractible. Then*

(a) *there exists a fibre map*

$$\begin{array}{ccc} E & \xrightarrow{\Phi} & E' \\ \downarrow \pi & & \downarrow q \\ B & \xrightarrow{f} & B' \end{array}$$

(b) *and any two choices of Φ are fibre homotopic.*

Proof:

(a) Pick basepoints $b_0 \varepsilon B$, $b_0' \varepsilon B'$, $e_0' \varepsilon E'$ such that $f(b_0) = b_0'$, $q(e_0') = b_0'$. Let $F : E \times I \to B'$ be a homotopy from the constant map at b_0' to the composition $E \xrightarrow{p} B \xrightarrow{f} B'$. Let $G : E \times 0 \to E'$ be the constant map to e_0'. The maps F and G being compatible, the homotopy lifting property gives a map $H : E \times I \to E'$ which extends G and lifts F. Then the restriction $H_1 : E = E \times 1 \to E'$ is the required fibre map Φ.

(b) Given another choice of a fibre map $\Psi : E \to E'$ which covers $f \circ p$, we observe that Ψ is homotopic to a map $E \to (p')^{-1}(b_0')$ via a homotopy which covers F. Since E is contractible, in fact, Ψ is homotopic to the constant map to e_0' via a homotopy which covers the composition K of F with a constant homotopy.

We note the the same is true for Φ, that is, it is homotopic to the constant map to e_0 via a homotopy which covers K.

Thus, we have compatible maps
$$E \times I \times I \to E \times I \xrightarrow{K} B'$$
by projecting to the first two factors and following with K and
$$(E \times I \times \{0, 1\}) \cup (E \times 0 \times I) \to E'$$
by using the homotopies for Φ and Ψ on the first piece and by using the constant map to e'_0 on the second piece.

Hence we get a map $L : E \times I \times I \to E'$ which lifts the first and extends the second. The restriction of L to $E \times 1 \times I$ is the required fibre homotopy between Φ and Ψ. □

Now let $F \to E \to B$ and $F' \to E' \to B'$ be fibration sequences as in 6.8.2 and suppose that there are left actions in the sense of Definition 6.8.1. Then:

Proposition 6.8.3. *Suppose E is contractible and B' is simply connected. Suppose also that the pair (E, F) is a CW pair. If $\Phi : E \to E'$ is a fibre map covering $f : B \to B'$, the diagram of left actions is fibre homotopy commutative*

$$\begin{array}{ccc} F \times E & \xrightarrow{\nu} & E \\ \downarrow \Phi \times \Phi & & \downarrow \Phi \\ F' \times E' & \xrightarrow{\nu'} & E'. \end{array}$$

Proof: The diagram is exactly commutative when restricted to the bouquet $F \vee E$. Since B' is simply connected and $F \wedge E$ is contractible, there are no local coefficients and the obstructions to fibre homotopy commutativity lie in the vanishing groups groups
$$H^*(F \times E, F \vee E; \pi_*(F')) = H^*(F \wedge E; \pi_*(F')) = 0.$$
□

Corollary 6.8.4. *Under the hypotheses of Proposition 6.8.3, the restriction $\Phi : F \to F'$ is an H-map.*

We see that, at certain points when obstruction theory must be used, it is helpful to be dealing with CW-complexes. Fortunately, Milnor's geometric realization of simplicial sets provides a simple device for doing this.

Recall the singular complex $S(X)$ of a topological space X. It is a functor from the category of topological spaces to the category of simplicial sets. It preserves products, $S(X \times Y) = S(X) \times S(Y)$.

Recall also the geometric realization $|K|$ of a simplical set. It is a functor from the category of simplicial sets to the category of compactly generated topological spaces. In fact, each geometric realization K is a CW-complex, and it preserves

6.8 Group models for loop spaces

products in the compactly generated category, $|K \times L| = |K| \times |L|$. In order to have this convenient property, it is important in this section to work in the category of compactly generated spaces and to take the product topology in that category.

Furthermore, these are adjoint functors, that is,

$$\mathrm{map}(|K|, X) \cong \mathrm{map}(K, S(X))$$

for simplicial sets K and topological spaces X. The resulting adjoints of the identity maps $S(X) \to S(X)$ and $|K| \to |K|$ are the natural transformations

$$\alpha : |S(X)| \to X, \quad \beta : K \to S|K|$$

and satisfy:

(1) α is a weak equivalence for all spaces X.

(2) β is a homotopy equivalence for all Kan complexes K.

Of the next two facts the first is elementary and the second is due to Quillen:

(1) If $F \to E \to B$ is a Serre fibration sequence of topological spaces, then the singular complexes, $S(F) \to S(E) \to S(B)$, yield a Kan fibration sequence.

(2) If $A \to B \to C$ is a Kan fibration sequence, then the geometric realizations, $|A| \to |B| \to |C|$, yield a Serre fibration sequence.

Definition 6.8.5. A group model for a loop space ΩX is a topological group G together with a map of the geometric realization of the singular complex

$$\Phi : |S(\Omega X)| \to G$$

such that Φ is both a weak homotopy equivalence and an H-map.

Notice that we have weak homotopy equivalences and H-maps:

$$\alpha : |S(\Omega X)| \to \Omega X.$$

If X has the homotopy type of a CW-complex, then a result of Milnor says that ΩX also has the homotopy type of a CW-complex. In this case, it follows that a group model is the same as a map $\Psi : \Omega X \to G$, which is a weak homotopy equivalence and an H-map.

If X is a space, then we know the path space fibration $\pi : PX \to X$, $\omega \mapsto \omega(1)$, is a Serre fibration and can be regarded, up to homotopy, as a principal bundle with a left action

$$\mu : \Omega X \times PX \to PX, \quad (\gamma, \omega) \mapsto \gamma * \omega.$$

The same is true for the Serre fibration $|S(\pi)| : |S(PX)| \to |S(X)|$ and we have commutative diagrams of action maps

$$\begin{array}{ccc} |S(\Omega X)| \times |S(PX)| & \xrightarrow{\mu} & |S(PX)| \\ \downarrow \alpha \times \alpha & & \downarrow \alpha \\ \Omega X \times PX & \xrightarrow{\mu} & PX. \end{array}$$

The following is an immediate corollary of Proposition 6.8.3 and Corollary 6.8.4.

Proposition 6.8.6. *Let B be a simply connected space and let $p : E \to B$ be a principal bundle with structural group G and left action*

$$\nu : G \times E \to E.$$

Suppose $f : X \to B$ is any map and E is contractible. Then

(a) *There exists a fibre map*

$$\begin{array}{ccc} |S(PX)| & \xrightarrow{\Phi} & E \\ \downarrow \pi & & \downarrow p \\ |S(X)| & \xrightarrow{f \circ \alpha} & B. \end{array}$$

(b) *Any two choices of Φ are fibre homotopic.*

(c) *Up to homotopy, Φ is a map of principle bundles in the sense that the following diagram is homotopy commutative*

$$\begin{array}{ccc} |S(\Omega X)| \times |S(PX)| & \xrightarrow{\Phi \times \Phi} & G \times E \\ \downarrow \mu & & \downarrow \nu \\ |S(PX)| & \xrightarrow{\Phi} & E. \end{array}$$

(d) *The map $\Phi : |S(\Omega X)| \to G$ is an H-map.*

The group models for loop spaces are natural in the following sense.

Corollary 6.8.7. *If X is a simply connected space and $E \to X$ is a principal G bundle with E contractible, then there are weak equivalences which are H-maps:*

$$G \leftarrow |S(\Omega X)| \to \Omega X$$

and, up to homotopy, these equivalences are natural.

The first construction of a group model for a loop space was given by Milnor [85]. Let X be a simplicial complex with basepoint $*$. Define the space of piecewise linear paths $E_n(X)$ of length n to be all sequences of points

$$(x_0, x_1, x_2, \ldots, x_n)$$

with $x_0 = *$ and such that any two successive points x_i, x_{i+1} belong to a common simplex. Let $E(X)$ be the disjoint union of all the $E_n(X)$ modulo the identifications:

$$(x_0, x_1, \ldots, x_{i-1}, x_i, x_{i+1}, \ldots, x_n) \sim (x_0, x_1, \ldots, x_{i-1}, x_{i+1}, \ldots, x_n)$$

if $x_{i-1} = x_i$ or if $x_{i-1} = x_{i+1}$. Define the map $p : E(X) \to X$ by $p(x_0, x_1, \ldots, x_n) = x_n$. If $G(X) = p^{-1}(*)$, then it is easy to see that $G(X)$ is a topological group and Milnor shows that:

Proposition 6.8.8. *If X is a simplicial complex, the map $p : E(X) \to X$ is a principal $G(X)$ bundle with left action $G(X) \times E(X) \to E(X)$ given by juxtaposition of sequences and $E(X)$ is contractible. Furthermore, both $E(X)$ and $G(X)$ are CW complexes.*

Thus, if X is a simply connected simplicial complex then Milnor's $G(X)$ is a group model for the loop space. It remains the most plausible model but it has the disadvantage that it is functorial only on the category of simplicial complexes and simplicial maps. Hence, it is useful to develop Kan's simplicial group model which is completely functorial in our context of connected spaces.

As a general principle, we note that the singular complex functor $S(X)$ and the geometric realization functor $|K|$ preserve the usual notions of homotopy theory. There are no restrictions if we start with any spaces in the topological category but, if we start in the simplicial category, we should restrict to Kan complexes, that is, those simplicial sets which satisfy the extension condition. Then, for example, a space X is n-connected if and only if its singular complex $S(X)$ is. And, a Kan complex K is n-connected if and only if its geometric realization $|K|$ is. We also have that a map of spaces $X \to Y$ is a weak homotopy equivalence if and only if the map of singular complexes $S(X) \to S(Y)$ is. And so on: weak equivalences of Kan complexes give weak equivalences of geometric realizations. We have already seen that Serre fibrations correspond to Kan fibrations. What this means is that every result that we need concerning simplicial sets can be translated to a result about spaces where, presumably, the definitions are more familiar.

First of all, let K be a simplicial set with basepoint. Recall that the n-th Eilenberg subcomplex $E_n K$ is the subcomplex of K consisting of all simplices σ with all faces of dimension less than n equal to the basepoint.

Fact (1): If K is a Kan complex, then so are all its Eilenberg subcomplexes.

Fact (2): If K is an 0-connected Kan complex, then the inclusion $E_1 K \to K$ is a weak homotopy equivalence.

We say that K is reduced if $E_1 K = K$, that is, if K has only one vertex.

Fact (3): If $A \to B \to C$ is a Kan fibration sequence with the fibre A 0-connected, then $E_1 A \to E_1 B \to E_1 C$ is a Kan fibration sequence.

We recall without proof Dan Kan's result:[69, 81]

Proposition 6.8.9. *If K is a reduced Kan complex, there exists a functorial simplicial group $G(K)$ and a functorial simplicial principal $G(K)$ bundle $p: T(K) \to K$ with the following properties:*

(a) *$G(K)$ is a Kan complex and $T(K)$ is a contractible Kan complex.*

(b) *If $f: K \to L$ is an surjection of reduced simplicial sets, then $G(f): G(K) \to G(L)$ is also a surjection.*

To be specific, let K be a reduced simplicial set with face and degeneracy operators $d_i : K_n \to K_{n-1}$ and $s_i : K_n \to K_{n+1}$ for $0 \le i \le n$. Then $G(K)$ is the simplicial group defined by:

(1) $G(K)_n$ is the free group with one generator $\bar{x}$ for each $x \varepsilon K_{n+1}$ and one relation $\overline{s_0 x} = e$ for each $x \varepsilon K_n$.

(2) The face and degeneracy operators of $G(K)$ are given by

$$d_0 \bar{x} = \overline{(d_1 x)}\, \overline{(d_0 x)}^{-1},$$
$$d_i \bar{x} = \overline{d_{i+1} x} \quad \text{for} \quad i > 0,$$
$$s_i \bar{x} = \overline{s_{i+1} x} \quad \text{for} \quad i \ge 0.$$

If X is a connected space, we get a functorial topological group $G_X = |G(E_1 S(X))|$ and a contractible principal G_X bundle $T_X = |T(E_1 S(X))| \to |E_1 S(X)|$. Thus, if X is simply connected G_X is a group model for $\Omega(|E_1 S(X)|)$. Since $|E_1 S(X)| \to |S(X)| \to X$ are weak equivalences, G_X is a group model for ΩX and it is clearly functorial.

The functorial group model G_X leads to a useful group model for the loops on fibrations.

Let $F \xrightarrow{\iota} E \xrightarrow{p} B$ be a Serre fibration of simply connected spaces. Let K be the kernel of the induced map $G(p): G(E_1 S(E)) \to G(E_1 S(B))$ of simplicial groups. Since $G(p)$ is a surjection of simplicial groups, it is a Kan fibration with fibre K and the geometric realization $|K| \to G_E \to G_B$ is a Serre fibration of topological groups. In fact, it is a short exact sequence of topological groups with $|K|$ the kernel and therefore a normal subgroup.

Proposition 6.8.10. *If $F \xrightarrow{\iota} E \xrightarrow{p} B$ is a Serre fibration of simply connected spaces, then*

$$|K| \to G_E \to G_B$$

is a group model for the fibration sequence of loop spaces

$$\Omega F \xrightarrow{\Omega \iota} \Omega E \xrightarrow{\Omega p} \Omega B.$$

Proof: Write $\overline{X} = |E_1 S(X)|$ and note that there is a weak equivalence $\overline{X} \to X$ when X is connected.

Consider the commutative diagram

$$\begin{array}{ccccc} G_F & \to & G_E & \to & G_B \\ \downarrow & & \downarrow & & \downarrow \\ T_F & \to & T_E & \to & T_B \\ \downarrow & & \downarrow & & \downarrow \\ \overline{F} & \to & \overline{E} & \to & \overline{B}. \end{array}$$

Clearly, the top row is a sequence of group models for $\Omega F \to \Omega E \to \Omega B$ and G_F maps to $|K|$ by a homomorphism. If we embed the bottom right-hand square in a totally fibred square, we get from Proposition 3.2.3 that the fibre $|K|$ and the fibre G_F are homotopy equivalent, more precisely, the map $G_F \to |K|$ is a homomorphism and a homotopy equivalence.

Loosely speaking, G_F is the fibre of the left-hand vertical column and $|K|$ is the fibre of the top row. By Proposition 3.2.3 they are homotopy equivalent. □

Exercises

(1) Let G be a simplicial group and let B be a simplicial set. Suppose we are given a twisting function $\tau : B_n \to G_{n-1}$ which satisfies the identities:

(A) $d_0 \tau(b) = \tau(d_1 b) [\tau(d_0 b)]^{-1}$.

(B) $d_i \tau(b) = \tau(d_{i+1} b)$ for $i > 0$.

(C) $s_i \tau(b) = \tau(s_{i+1} b)$ for $i \geq 0$.

(D) $\tau(s_0 b) = e_n =$ the unit in G_n if $b \varepsilon B_n$.

Define the twisted cartesian product $E(\tau) = G \times_\tau B$ by:

(a) $E(\tau)_n = G_n \times B_n$.

with face and degeneracy operators

(b)
$$d_i(g, b) = (d_i g, d_i b), \quad i > 0$$
$$d_0(g, b) = ((d_0 g)\tau(b), d_0 b)$$
$$s_i(g, b) = (s_i g, s_i b), \quad i \geq 0.$$

Show that $E(\tau)$ is a simplicial G bundle.

(2) Let K be a simplicial set and let $G(K)$ be Kan's loop group as in Proposition 6.8.9.

(a) Show that $G(K)$ is a simplicial group.

(b) Show that the twisting function $\tau : K_n \to G(K)_{n-1}$, $\tau(x) = \bar{x}$, satisfies the identities A,B,C,D in Exercise 1. (Remark: The twisted cartesian product $E(\tau) = G(K) \times_\tau K$ is the principle bundle $T(K)$ in Proposition 6.8.5.)

(3) Prove Facts 1, 2, 3 in this section, at least for simplicial sets which are singular complexes.

6.9 Relative Samelson products

In this section, we use the group models of the previous section to define relative Samelson products.

Let G be a topological group and let H be a normal subgroup. Let P_1 and P_2 be pointed topological spaces.

If $f : P_1 \to G$ and $g : P_2 \to H$ are continuous pointed maps, the commutator map

$$[f,g] : P_1 \times P_2 \xrightarrow{f \times g} G \times H \xrightarrow{[\ ,\ ]} H$$

maps the bouquet $P_1 \vee P_2$ to the unit. Therefore, it factors through a map of the smash

$$[f,g]_e : P_1 \wedge P_2 \to H.$$

This map is called a relative external Samelson product and is defined without the use of any homotopies. Of course, $[g,f]_e : P_2 \wedge P_1 \to H$ is also defined and we have a strictly commutative diagram

$$\begin{array}{ccc} P_1 \wedge P_2 & \xrightarrow{[f,g]_e} & H \\ \downarrow \tau & & \downarrow (\ )^{-1} \\ P_2 \wedge P_1 & \xrightarrow{[g,f]_e} & H. \end{array}$$

In fact, the Lie identities for groups, Proposition 6.5.1, says that, if

$$\tau = T_{(1,2)} : P_1 \wedge P_2 \to P_2 \wedge P_1$$
$$\sigma = T_{(1,2,3)} : P_1 \wedge P_2 \wedge P_3 \to P_3 \wedge P_1 \wedge P_2$$
$$\sigma^2 = T_{(1,3,2)} : P_1 \wedge P_2 \wedge P_3 \to P_2 \wedge P_3 \wedge P_1$$

are the standard cyclic permutations, then we have the following identities

(a)
$$[f,g]_e = [g,f]_e^{-1} \circ \tau : P_1 \wedge P_2 \to H$$

(b)
$$[ff_1,g]_e = [f_1,g]_e^{(f^{-1})} [f,g]_e : P_1 \wedge P_2 \to H$$

$$[f,gg_1]_e = [f,g]_e [f,g_1]_e^{(g^{-1})} : P_1 \wedge P_2 \to H$$

(c)
$$([f^{(g^{-1})},[h,g]_e)([g^{(h^{-1})},[f,h]_e \circ \sigma)([h^{(f^{-1})},[g,f]_e]_e \circ \sigma^2)$$
$$= 1 : P_1 \wedge P_2 \wedge P_3 \to H.$$

So far, no homotopies have been used. Now assume that the spaces P_i are coabelian. The analog of Proposition 6.6.6 remains true; that is, given any commutator of one variable functions defined on a lesser length smash product of coabelian spaces, that commutator is null homotopic. It follows that we get the analog of Proposition 6.6.7 for relative external Samelson products.

Proposition 6.9.1. *Let P_1, P_2, P_3 be coabelian spaces. If*

$$f, f_1 : P_1 \to H, g, g_1 : P_2 \to G, h : P_3 \to G$$

are maps then the external Samelson products $[f,g]_e : P_1 \wedge P_2 \to H$ and $[f,[g,h]_e]_e : P_1 \wedge P_2 \wedge P_3 \to H$ satisfy the following Lie identities:

(1) *Anti-commutativity:*
$$[f,g]_e = -[g,f]_e \circ \tau$$

where
$$\tau = T_{(1,2)} : P_1 \wedge P_2 \to P_2 \wedge P_1$$

is the standard transposition.

(2) *Bilinearity:*
$$[f+f_1,g]_e = [f,g]_e + [f_1,g]_e, \quad [f,g+g_1]_e = [f,g]_e + [f,g_1]_e.$$

(3) *Jacobi identity:*
$$[f,[g,h]_e]_e + [h,[f,g]_e]_e \circ \sigma + [g,[h,f]_e]_e \circ \sigma^2 = 0.$$

Introduce relative internal Samelson products as expected. Let p be an odd prime and let $\Delta_{m,n} : P^{m+n}(p^r) \to P^n(p^r) \wedge P^n(p^r)$ be any map such that $\Delta_{m,n_*}(e_{m+n}) = e_m \otimes e_n$. Relative internal Samelson products are defined as

$$[f,g] = [f,g]_e \circ \Delta_{m,n} : P^{m+n}(p^r) \to H.$$

This leads to the following definition.

Definition 6.9.2. Let $L' \to L$ be a morphism of graded Lie algebras. We call L' an extended ideal of L if there are two bilinear pairings (called Lie brackets):

$$[\ ,\ ] : L' \times L \to L'$$

$$[\ ,\ ] : L \times L' \to L',$$

such that

(1) the diagram of Lie brackets commutes

$$\begin{array}{ccc} L' \times L' & \xrightarrow{[\ ,\ ]} & L' \\ \downarrow & & \downarrow \\ L' \times L & \xrightarrow{[\ ,\ ]} & L' \\ \downarrow & & \downarrow \\ L \times L & \xrightarrow{[\ ,\ ]} & L. \end{array}$$

(2) for all x, y, and z in the union of L' and L,

$$[x, y] = -(-1)^{\deg(x) \cdot \deg(y)}[y, x]$$

$$[x, [y, z]] = [[x, y], z] + (-1)^{\deg(x) \cdot \deg(y)}[y, [x, z]].$$

Definition 6.9.3. An extended differential ideal $L' \to L$ is a morphism of differential graded Lie algebras which is an extended ideal and such that the differential d is a derivation in the sense that, for all x and y in the union of L' and L,

$$d[x, y] = [dx, y] + (-1)^{\deg(x)}[x, dy].$$

Now the same proofs as before show:

Proposition 6.9.4. *Let H and G be topological groups and let H be a normal subgroup of G.*

(a) *If we localize away from 2 and 3, the map of integral homotopy groups $\pi_*(H) \to \pi_*(G)$ is an extended ideal.*

(b) *If p is a prime greater than 3, the map of homotopy groups with coefficients $\pi_*(H; \mathbb{Z}/p^r\mathbb{Z}) \to \pi_*(G; \mathbb{Z}/p^r\mathbb{Z})$ is an extended differential ideal with the Bockstein differential $d = \rho \circ \beta$..*

Finally, we consider the case of a fibration of simply connected spaces $F \to E \to B$. In the previous section, we introduced the short exact sequence of topological

groups
$$|K| \to G_E \to G_B$$
which is a group model for the fibration sequence of loop spaces
$$\Omega F \to \Omega E \to \Omega B.$$
Since $H = |K|$ is a normal subgroup of $G = G_E$, we get relative Samelson products
$$[\ ,\ ] : \pi_*(\Omega F) \otimes \pi_*(\Omega E) \to \pi_*(\Omega F)$$
with various coefficients.

Corollary 6.9.5.

(a) *If we localize away from 2 and 3, the map $\pi_*(\Omega F) \to \pi_*(\Omega E)$ is an extended ideal.*

(b) *If p is a prime greater than 3, the map $\pi_*(\Omega F; \mathbb{Z}/p^r\mathbb{Z}) \to \pi_*(\Omega E; \mathbb{Z}/p^r\mathbb{Z})$ is an extended differential ideal.*

Exercises

(1) Let p be an odd prime. Let H be a normal subgroup of a topological group G.

   (a) Construct two bilinear pairings
   $$[\ ,\ ] : \pi_*(H) \times \pi_*(G; \mathbb{Z}/p^r\mathbb{Z}) \to \pi_*(H; \mathbb{Z}/p^r\mathbb{Z})$$
   $$[\ ,\ ] : \pi_*(H; \mathbb{Z}/p^r\mathbb{Z}) \times \pi_*(G)) \to \pi_*(H; \mathbb{Z}/p^r\mathbb{Z}).$$

   (b) Show that these pairings satisfy anti-commutativity and the Jacobi identity, even if $p = 3$.

   (c) Show that these pairings, together with the reduction maps $\eta : \pi_*(\ ) \to \pi_*(\ , \mathbb{Z}/p^r\mathbb{Z})$ and the Bocksteins $\beta : \pi_*(\ ; \mathbb{Z}/p^r\mathbb{Z}) \to \pi_{*-1}(\ )$ satisfy the analogs of those in Proposition 6.7.6.

(2) Let H be a normal subgroup of a topological group G.

   (a) Show that the commutator $[\ ,\ ] : H \times G \to H$ induces a bilinear bracket pairing
   $$[\ ,\ ] : H_*(H) \times H_*(G) \to H_*(H),$$
   which is anti-commutative and satisfies the Jacobi identity. (The coefficients can be $\mathbb{Z}$, $\mathbb{Z}/p^r\mathbb{Z}$, or some combination of the two.)

   (b) Show that these pairings, together with the reduction maps and the Bocksteins, satisfy the analogs of those in exercise 1c.

(c) Show that the Hurewicz maps are compatible with these pairings, that is, show that the diagram below commutes:

$$\begin{array}{ccc} \pi_m(H) \otimes \pi_n(G; \mathbb{Z}/p^r\mathbb{Z}) & \xrightarrow{[\ ,\ ]} & \pi_{m+n}(H; \mathbb{Z}/p^r\mathbb{Z})) \\ \downarrow \varphi \otimes \varphi & & \downarrow \varphi \\ H_m(H) \otimes H_n(G; \mathbb{Z}/p^r\mathbb{Z}) & \xrightarrow{[\ ,\ ]} & H_{m+n}(H; \mathbb{Z}/p^r\mathbb{Z})). \end{array}$$

6.10 Universal models for relative Samelson products

Let p be an odd prime. In this section, we assume that we have a theory of (internal) Samelson products for mod p^r homotopy groups, **a theory which is defined and functorial on the category of loop spaces ΩE and loop maps.** That is, we assume that we have bilinear maps

$$[\ ,\ ]: \pi_*(\Omega E; \mathbb{Z}/p^r\mathbb{Z}) \otimes \pi_*(\Omega E; \mathbb{Z}/p^r\mathbb{Z}) \to \pi_*(\Omega E; \mathbb{Z}/p^r\mathbb{Z}).$$

This theory satisfies anti-commutativity, the derivation formula for the Bockstein, and, if p is a prime greater than 3, the Jacobi identity.

We then show that this implies that there is a theory of relative Samelson products which is defined and functorial on loops of fibration sequences

$$\Omega F \to \Omega E \to \Omega B$$

and loop maps between them. That is, we have bilinear maps

$$[\ ,\ ]: \pi_*(\Omega F; \mathbb{Z}/p^r\mathbb{Z}) \otimes \pi_*(\Omega E; \mathbb{Z}/p^r\mathbb{Z}) \to \pi_*(\Omega F; \mathbb{Z}/p^r\mathbb{Z}) \quad \text{and}$$
$$[\ ,\ ]: \pi*(\Omega E; \mathbb{Z}/p^r\mathbb{Z}) \otimes \pi*(\Omega F; \mathbb{Z}/p^r\mathbb{Z}) \to \pi*(\Omega F; \mathbb{Z}/p^r\mathbb{Z}).$$

The identities such as bilinearity, anti-commutativity, and the Jacobi identity (if $p > 3$) are also valid.

We shall use the technique of universal models to do this. This technique is based on certain fibration sequences which are split as a consequence of the Hilton–Milnor theorem.

Definition 6.10.1. Let p be an odd prime and let $m, n > 2$. The basic two variable universal model for mod p^r homotopy groups in dimensions m and n is the space

$$\Sigma P^{m,n} = \Sigma P^m(p^r) \vee \Sigma P^n(p^r).$$

Remark. If $P^m = P^m(p^r)$ and $P^n = P^n(p^r)$, there are the standard inclusions

$$1_m: \Sigma P^m \to \Sigma P^{m,n} \quad \text{and} \quad 1_n: \Sigma P^n \to \Sigma P^{m,n}.$$

6.10 Universal models for relative Samelson products

Any mod p^r homotopy classes

$$x : \Sigma P^m \to E \quad \text{and} \quad y : \Sigma P^n \to E$$

define a unique homotopy class

$$x \vee y : \Sigma P^{m,n} \to E$$

which restricts to $x = (x \vee y) \cdot 1_m$ and $y = (x \vee y) \cdot 1_n$, respectively. Let

$$\overline{x} : P^m \to \Omega E \quad \text{and} \quad \overline{y} : P^n \to \Omega E$$

be the respective adjoints.

Thus, we get the loop map

$$\Omega \Sigma P^{m,n} \to \Omega E$$

which is the multiplicative extension $\mu(\overline{x} \vee \overline{y})$ of the adjoint $\overline{x} \vee \overline{y} : P^{m,n} \to \Omega E$.

Remark. If $\iota_m : P^m \to \Omega P^{m,n}$ and $\iota_n : P^n \to \Omega E$ are the adjoints of the standard inclusions, then naturality asserts that the internal Samelson product $[\overline{x}, \overline{y}] : P^{m+n} \to \Omega E$ is given by the formula

$$[\overline{x}, \overline{y}] = \mu(\overline{x} \vee \overline{y}) \cdot [\iota_m, \iota_n].$$

Thus all internal Samelson products are determined by Samelson products in the universal models.

Definition 6.10.2. The two variable universal model for relative Samelson products in mod p^r homotopy groups is the standard projection map $\pi : \Sigma P^{m,n} \to \Sigma P^m$.

Remarks. If $\tau : E \to B$ is a fibration with fibre F and $\overline{x} : P^m \to \Omega E$ and $\overline{y} : P^n \to \Omega F$ are homotopy classes, then we have a commutative diagram

$$\begin{array}{ccc}
\Sigma P^n & \xrightarrow{y} & F \\
\downarrow & & \downarrow \\
\Sigma P^{m,n} & \xrightarrow{x \vee y} & E \\
\downarrow & & \downarrow \tau \\
\Sigma P^m & \xrightarrow{\tau \cdot x} & B.
\end{array}$$

If $F^{m,n} \to E^{m,n} \xrightarrow{\pi} \Sigma P^m$ is the fibration sequence which is the result of the standard replacement of the map $P^{m,n} \to P^m$ by a fibration, Exercise 1 below gives an extension to a commutative diagram

$$\begin{array}{ccccc} \Sigma P^n & \to & F^{m,n} & \xrightarrow{x\tilde{\vee} y} & F \\ \downarrow & & \downarrow & & \downarrow \\ \Sigma P^{m,n} & \xrightarrow{\simeq} & E^{m,n} & \xrightarrow{x\tilde{\vee} y} & E \\ \downarrow & & \downarrow \pi & & \downarrow \tau \\ \Sigma P^m & \xrightarrow{=} & \Sigma P^m & \xrightarrow{\tau \cdot x} & B. \end{array}$$

Exercise 1 also asserts that the right-hand map of fibration sequences can be chosen to be unique up to fibre homotopy.

It follows from the Hilton–Milnor theorem that the fibration sequence

$$\Omega F^{m,n} \to \Omega E^{m,n} \to \Omega \Sigma P^m$$

is split, that is, it has a section. Since the Samelson product $[\iota_m, \iota_n]$ maps to a null map in the base ΣP^m, it has a homotopy unique representative $[\iota_m, \iota_n]_r$ in the fibre $\Omega F^{m,n}$. Thus, the following definition is well defined.

Definition 6.10.3. If $\overline{x} \in \pi_*(\Omega E)$ and $\overline{y} \in \pi_*(\Omega F)$, the relative Samelson product $[\overline{x}, \overline{y}]_r \in \pi_*(\Omega F)$ is defined by the formula

$$[\overline{x}, \overline{y}]_r = \Omega(x \tilde{\vee} y)_* [\iota_m, \iota_n]_r.$$

Similarly, using the universal model $P^{n,m} \to P^m$, the relative Samelson product $[\overline{y}, \overline{x}]_r$ is defined by the formula

$$[\overline{y}, \overline{x}]_r = \Omega(y \tilde{\vee} x)_* [\iota_n, \iota_m]_r.$$

Remarks. The relative Samelson product is characterized by two properties. One, it is natural with respect to loop maps of fibration sequences and, two, it maps to the internal Samelson product in the total space. Hence, it is identical with the relative Samelson product as previously defined via group models. In special case of the universal model, the relative Samelson product is uniquely characterized by the fact that its image in the total space is the usual internal Samelson product. In addition, naturality and the commutative diagram of vertical fibration sequences

$$\begin{array}{ccccc} F & \to & F & \to & E \\ \downarrow & & \downarrow & & \downarrow = \\ F & \to & E & \to & E \\ \downarrow & & \downarrow & & \downarrow \\ * & \to & B & \to & * \end{array}$$

6.10 Universal models for relative Samelson products

show that internal and relative Samelson products are compatible. In other words, the following diagram commutes

$$\begin{array}{ccc} \pi * (\Omega F; \mathbb{Z}/p^r\mathbb{Z}) \otimes \pi * (\Omega F; \mathbb{Z}/p^r\mathbb{Z}) & \to & \pi * (\Omega F; \mathbb{Z}/p^r\mathbb{Z}) \\ \downarrow & & \downarrow = \\ \pi * (\Omega F; \mathbb{Z}/p^r\mathbb{Z}) \otimes \pi * (\Omega E; \mathbb{Z}/p^r\mathbb{Z}) & \to & \pi * (\Omega F; \mathbb{Z}/p^r\mathbb{Z}) \\ \downarrow & & \downarrow \\ \pi * (\Omega E; \mathbb{Z}/p^r\mathbb{Z}) \otimes \pi * (\Omega F; \mathbb{Z}/p^r\mathbb{Z}) & \to & \pi * (\Omega E; \mathbb{Z}/p^r\mathbb{Z}). \end{array}$$

Note that the twist map $T : P^{n,m} \to P^{m,n}$ leaves the internal Samelson product invariant,

$$T_*[\iota_n, \iota_m] = [\iota_n, \iota_m].$$

Since $[\iota_n, \iota_m] = -(-1)^{mn}[\iota_m, \iota_n]$ for the internal Samelson product, it follows that

$$[\bar{y}, \bar{x}]_r = \Omega(y \,\tilde{\vee}\, x)_*[\iota_n, \iota_m]_r = \Omega(x \,\tilde{\vee}\, y)_* T[\iota_n, \iota_m]_r = \Omega(x \,\tilde{\vee}\, y)_*[\iota_n, \iota_m]_r$$
$$= -(-1)^{nm}\Omega(x \,\tilde{\vee}\, y)_*[\iota_m, \iota_n]_r = -(-1)^{nm}[\bar{x}, \bar{y}]_r.$$

That is,

Lemma 6.10.4. $[\bar{y}, \bar{x}]_r = -(-1)^{mn}[\bar{x}, \bar{y}]_r$ for the relative Samelson product.

The next lemma determines the Bockstein of a relative Samelson product.

Lemma 6.10.5. If $\beta : \pi_*(X; \mathbb{Z}/p^r\mathbb{Z}) \to \pi_{*-1}(X; \mathbb{Z}/p^r\mathbb{Z})$ is the Bockstein, then the following derivation formula is valid for relative Samelson products,

$$\beta[\bar{x}, \bar{y}]_r = [\beta\bar{x}, \bar{y}]_r + (-1)^m [\bar{x}, \beta\bar{y}]_r.$$

Proof: First of all,

$$\beta[\iota_m, \iota_n]_r = [\beta\iota_m, \iota_n]_r + (-1)^m [\iota_m, \beta\iota_n]_r$$

since the formula is valid for internal Samelson products and the homotopy of the fibre injects into that of the total space. The naturality of relative Samelson products converts this formula into the desired formula. □

In order to prove identities for relative Samelson products which involve three variables, it is necessary to introduce:

Definition 6.10.6. If $P^{m,n,q} = P^m \vee P^n \vee P^q$, then the basic three variable model for the mod p^r homotopy groups is the space $\Sigma P^{m,n,q}$ and the three variable model for relative Samelson products in mod p^r homotopy groups is the standard projection map

$$\Sigma P^{m,n,q} \to \Sigma P^{m,n}.$$

Remarks. Similar to the two variable case, we have the following mapping properties. If $\tau : E \to B$ is a fibration with fibre F and $\overline{x} : P^m \to \Omega E, \overline{y} : P^n \to \Omega E$, and $\overline{z} : P^q \to \Omega F$ are homotopy classes, then we have a commutative diagram

$$\begin{array}{ccc} \Sigma P^q & \xrightarrow{y} & F \\ \downarrow & & \downarrow \\ \Sigma P^{m,n,q} & \xrightarrow{x \vee y \vee z} & E \\ \downarrow & & \downarrow \tau \\ \Sigma P^{m,n} & \xrightarrow{\tau \cdot (x \vee y)} & B. \end{array}$$

If $F^{m,n,q} \to E^{m,n,q} \xrightarrow{\pi} \Sigma P^m$ is the fibration sequence which is the result of the standard replacement of the map $P^{m,n,q} \to P^{m,n}$ by a fibration, we have an extension to a commutative diagram

$$\begin{array}{ccccc} \Sigma P^q & \to & F^{m,n,q} & \xrightarrow{x \vee \tilde{y} \vee z} & F \\ \downarrow & & \downarrow & & \downarrow \\ \Sigma P^{m,n,q} & \xrightarrow{\simeq} & E^{m,n,q} & \xrightarrow{x \vee \tilde{y} \vee z} & E \\ \downarrow & & \downarrow \pi & & \downarrow \tau \\ \Sigma P^{m,n} & \xrightarrow{=} & \Sigma P^{m,n} & \xrightarrow{\tau \cdot (x \vee y)} & B. \end{array}$$

As before, it follows from the Hilton–Milnor theorem that the fibration sequence

$$\Omega F^{m,n,q} \to \Omega E^{m,n,q} \to \Omega \Sigma P^{m,n}$$

is split, that is, it has a section. Since the iterated Samelson product $[\iota_m, \iota_n, \iota_q] = [\iota_m, [\iota_n, \iota_q]]$ maps to a null map in the base $\Sigma P^{m,n}$, it has a homotopy unique representative $[\iota_m, \iota_n, \iota_q]_r$ in the fibre $\Omega F^{m,n,q}$. Thus, the following definition is well defined.

Definition 6.10.7. If $\overline{x} \in \pi_*(\Omega E), \overline{y} \in \pi_*(\Omega E)$, and $\overline{z} \in \pi_*(\Omega F)$ the relative Samelson product

$$[\overline{x}, \overline{y}, \overline{z}]_r \in \pi_*(\Omega F)$$

is defined by the formula

$$[\overline{x}, \overline{y}, \overline{z}]_r = \Omega(x \vee \tilde{y} \vee z)_*[\iota_m, \iota_n, \iota_q]_r.$$

Similarly, using the model $P^{m,n,q} \to P^{m,n}$,

$$[[\overline{x}, \overline{y}, \overline{z}]]_r \in \pi_*(\Omega F)$$

is defined by the formula

$$[[\overline{x}, \overline{y}, \overline{z}]]_r = \Omega(x \vee \tilde{y} \vee z)_*[[\iota_m, \iota_n, \iota_q]]_r$$

where $[[\iota_m, \iota_n, \iota_q]]_r$ is the homotopy class in the fibre which maps to the Samelson product $[[\iota_m, \iota_n], \iota_q]$.

6.10 Universal models for relative Samelson products

Finally, using the model $P^{n,m,q} \to P^{m,n}$,
$$[[[\bar{y}, \bar{x}, \bar{z}]]]_r \in \pi_*(\Omega F)$$
is defined by the formula
$$[[[\bar{y}, \bar{x}, \bar{z}]]]_r = \Omega(y \vee \tilde{x} \vee z)_*[[[\iota_n, \iota_m, \iota_q]]]_r$$
where $[[[\iota_n, \iota_m, \iota_q]]]_r$ is the homotopy class in the fibre which maps to the Samelson product $[\iota_n, [\iota_m, \iota_q]]$.

We need to relate three variable products to two variable products.

Lemma 6.10.8.
$$[\bar{x}, \bar{y}, \bar{z}]_r = [\bar{x}, [\bar{y}, \bar{z}]_r]_r$$
$$[[\bar{x}, \bar{y}, \bar{z}]]_r = [[\bar{x}, \bar{y}], \bar{z}]_r$$
$$[[[\bar{y}, \bar{x}, \bar{z}]]]_r = [\bar{y}, [\bar{x}, \bar{z}]_r]_r$$

Proof: For example, the first of the above formulas is an immediate consequence of the commutative diagram

$$\begin{array}{ccccc} \Omega F^{m,n+q} & \to & \Omega F^{m,n,q} & \to & \Omega F \\ \downarrow & & \downarrow & & \\ \Omega\Sigma P^{m,n+q} & \to & \Omega\Sigma P^{m,n,q} & \to & \Omega E \\ \downarrow & & \downarrow & & \downarrow \\ \Omega\Sigma P^m & \to & \Omega\Sigma P^{m,n} & \to & \Omega B. \end{array}$$

The remaining two formulas have similar proofs. □

Let $p > 3$. Since the internal Samelson products are invariant under the map induced by the natural map $S: P^{m,n,q} \to P^{n,m,q}$, the Jacobi identity for internal Samelson products
$$[\iota_m, [\iota_n, \iota_q]] = [[\iota_m, \iota_n], \iota_q] + (-1)^{mn}[\iota_n, [\iota_m, \iota_q]]$$
yields the Jacobi identity for relative Samelson products:

Lemma 6.10.9. *If p is a prime greater than 3 and if $\bar{x} \in \pi_m(\Omega E; \mathbb{Z}/p^r\mathbb{Z})$, $\bar{y} \in \pi_n(\Omega E; \mathbb{Z}/p^r\mathbb{Z})$, and $\bar{z} \in \pi_q(\Omega F; \mathbb{Z}/p^r\mathbb{Z})$, then*
$$[\bar{x}, [\bar{y}, \bar{z}]_r]_r = [[\bar{x}, \bar{y}], \bar{z}]_r + (-1)^{mn}[\bar{y}, [\bar{x}, \bar{z}]_r]_r.$$

Proof: The equation
$$[\iota_m, [\iota_n, \iota_q]] = [[\iota_m, \iota_n], \iota_q] + (-1)^{mn}[\iota_n, [\iota_m, \iota_q]]$$
can be rewritten as
$$[\iota_m, \iota_n, \iota_q]_r = [[\iota_m, \iota_n, \iota_q]]_r + (-1)^{mn}(\Omega\Sigma S)_*[[[\iota_n, \iota_m, \iota_q]]]_r.$$

Since $\Omega(x \vee y \vee z) \cdot \Omega\Sigma S = \Omega(y \vee x \vee z)$, the required Jacobi identity is the image under the map $\Omega(x \vee y \vee z)$. □

In order to prove the bilinearity of relative Samelson products, we consider two universal examples

$$\Sigma P^{m,n,q} \to \Sigma P^{m,n}$$
$$\Sigma P^{m,n,q} \to \Sigma P^{m}.$$

In the first case, we consider the relative Samelson product $[\iota_m + \iota_n, \iota_q]_r$ where $m = n$. Clearly,

$$[\iota_m + \iota_n, \iota_q] = [\iota_m, \iota_q] + [\iota_n, \iota_q]$$

and hence the same formula for relative Samelson products. Now naturality gives Lemma 6.10.10a below.

In the second case, we consider the relative Samelson product $[\iota_m, \iota_n + \iota_q]_r$ where $n = q$. Clearly,

$$[\iota_m, \iota_n + \iota_q] = [\iota_m, \iota_n] + [\iota_m, \iota_q]$$

and hence the same formula for relative Samelson products. Now naturality gives 6.10.10b below.

Lemma 6.10.10.

(a) *If* $\bar{x}, \bar{y} \in \pi_m(\Omega E; \mathbb{Z}/p^r\mathbb{Z})$ *and* $\bar{z}, \in \pi_q(\Omega F; \mathbb{Z}/p^r\mathbb{Z})$, *then*

$$[\bar{x} + \bar{y}, \bar{z}]_r = [\bar{x}, \bar{z}]_r + [\bar{y}, \bar{z}]_r.$$

(b) *If* $\bar{x}, \in \pi_m(\Omega E; \mathbb{Z}/p^r\mathbb{Z})$ *and* $\bar{y}, \bar{z}, \in \pi_q(\Omega F; \mathbb{Z}/p^r\mathbb{Z})$, *then*

$$[\bar{x}, \bar{y} + \bar{z}]_r = [\bar{x}, \bar{y}]_r + [\bar{x}, \bar{z}]_r.$$

Corollary 6.10.11. $\bar{x}, \in \pi_m(\Omega E; \mathbb{Z}/p^r\mathbb{Z})$ *and* $\bar{z}, \in \pi_q(\Omega F; \mathbb{Z}/p^r\mathbb{Z})$ *If α is a scalar, then*

$$[\alpha\bar{x}, \bar{z}]_r = [\bar{x}, \alpha\bar{z}]_r = \alpha[\bar{x}, \bar{z}]_r.$$

6.10 Universal models for relative Samelson products

Of course, Corollary 6.10.11 is a corollary of Lemma 6.10.10 but it can also be proved directly.

Exercises

(1) Let $f : X \to Y$ be any continuous map and let
$$X \xrightarrow{\iota_f} E_f \xrightarrow{\pi_f} Y$$
be the standard factorization of f into the composite of a homotopy equivalent inclusion ι_f and a fibration π_f, that is,
$$E_f = \{(x,\omega)| x \epsilon X, \omega : I \to Y, \omega(0) = f(x)\}$$
$$\pi_f(x,\omega) = \omega(1), \quad \iota_f(x) = (x, \omega_x),$$
where ω_x is the constant path at x.

(a) If $\tau : E \to B$ is a fibration, show that any commutative diagram
$$\begin{array}{ccc} X & \xrightarrow{h} & E \\ \downarrow f & & \downarrow \tau \\ Y & \xrightarrow{g} & B \end{array}$$
can be embedded in a commutative diagram
$$\begin{array}{ccccc} X & \xrightarrow{\iota_f} & E_f & \xrightarrow{\overline{h}} & E \\ \downarrow f & & \downarrow \pi_f & & \downarrow \tau \\ Y & = & Y & \xrightarrow{g} & B, \end{array}$$
where $h = \overline{h} \cdot \iota_f$.

(b) Show that this larger diagram can be chosen to be unique up to fibre homotopy of the right-hand map of fibrations.

(c) Prove the following parametrized version of (a): Any commutative diagram
$$\begin{array}{ccc} X & \xrightarrow{h_t} & E \\ \downarrow f & & \downarrow \tau \\ Y & \xrightarrow{g_t} & B \end{array}$$
where g_t and h_t are homotopies can be embedded in a commutative diagram
$$\begin{array}{ccccc} X & \xrightarrow{\iota_f} & E_f & \xrightarrow{\overline{h_t}} & E \\ \downarrow f & & \downarrow \pi_f & & \downarrow \tau \\ Y & = & Y & \xrightarrow{g_t} & B. \end{array}$$
where $h_t = \overline{h_t} \cdot \iota_f$.

(2) Give a direct proof of Lemma 6.10.11.

(3) Show that there is a relative Samelson product defined on mod p homotopy Bockstein spectral sequences (see Chapter 7)
$$E^r(\Omega F) \otimes E^r(\Omega E) \to E^r(\Omega F).$$

6.11 Samelson products over the loops on an H-space

Let p be an odd prime. We shall show that there is a theory of so-called H-based relative Samelson products. This theory is defined and functorial on loops of fibration sequences
$$\Omega F \to \Omega E \to \Omega B$$
where B is a homotopy commutative H-space and where the maps of fibration sequences are loop maps with the maps on the base being loops of H-maps. In this case, we have bilinear maps
$$[\ ,\ ] : \pi_*(\Omega E; \mathbb{Z}/p^r\mathbb{Z}) \otimes \pi_*(\Omega E; \mathbb{Z}/p^r\mathbb{Z}) \to \pi_*(\Omega F; \mathbb{Z}/p^r\mathbb{Z}).$$

The important thing is that the usual identities are also valid.

We use the method of universal models with more care. Loosely speaking, the universal two variable models for the mod p^r homotopy in dimensions m and n of the fibrations $E \to B$ with B an H-space are the maps $\Sigma P^{m,n} \to \Sigma P^m \times \Sigma P^n$. But this model does not have a base which is an H-space so it is desirable to change the category for which this is a universal model.

Consider the category of two variable models in which the objects are ordered triples consisting of:

(1) a map $\tau : E \to B$.

(2) maps $x : \Sigma P^m \to E$ and $y : \Sigma P^n \to E$.

(3) a commutative diagram

$$\begin{array}{ccc} \Sigma P^{m,n} & \xrightarrow{x \vee y} & E \\ \downarrow & & \downarrow \tau \\ \Sigma P^m \times \Sigma P^n & \xrightarrow{\Phi} & B, \end{array}$$

where the left-hand vertical map is the standard inclusion of the bouquet into the product. Thus, Φ is an extension of a map from the bouquet to the product. (Of course, we are aware that (3) includes all the data for a model.)

Morphisms in this category of models are maps from $E \to B$ to another map $E' \to B'$ which preserve the structures in (1), (2), (3).

6.11 Samelson products over the loops on an H-space 211

A universal model is just an initial object in this category.

Definition 6.11.1. The universal model is the object:

(1) The standard inclusion of the bouquet into the product
$$\Sigma P^{m,n} \to \Sigma P^m \times \Sigma P^n.$$

(2) The standard inclusions $1_n : \Sigma P^m \to \Sigma P^m \times \Sigma P^n$ and $1_n : \Sigma P^n \to \Sigma P^m \times \Sigma P^n$.

(3) The commutative diagram
$$\begin{array}{ccc} \Sigma P^{m,n} & \xrightarrow{1_m \vee 1_n} & \Sigma P^{m,n} \\ \downarrow & & \downarrow \\ \Sigma P^m \times \Sigma P^n & \xrightarrow{1} & \Sigma P^m \times \Sigma P^n \end{array}$$

It is customary to abuse the terminology and simply refer to the map
$$\Sigma P^{m,n} \to \Sigma P^m \times \Sigma P^n$$
as the universal two variable model. With this simplified terminology, the universal mapping property is given by the unique map
$$\begin{array}{ccc} \Sigma P^{m,n} & \xrightarrow{x \vee y} & E \\ \downarrow & & \downarrow \tau \\ \Sigma P^m \times \Sigma P^n & \xrightarrow{\Phi} & B. \end{array}$$

In other words, the universal mapping property is the model itself.

Now suppose that $F \xrightarrow{\iota} E \xrightarrow{\tau} B$ is a fibration sequence and let $F^{m,n} \xrightarrow{\iota} E^{m,n} \xrightarrow{\pi} \Sigma P^m \times \Sigma P^n$ be the fibration sequence which results from the standard replacement of $\Sigma P^{m,n} \to \Sigma P^m \times \Sigma P^n$ by a fibration. Exercise 1 in the previous section gives an extension of the map in the previous paragraph to a commutative diagram

$$\begin{array}{ccccc} & & F^{m,n} & \xrightarrow{\overline{\overline{\Phi}}} & F \\ & & \downarrow \iota & & \downarrow \iota \\ \Sigma P^{m,n} & \xrightarrow{\simeq} & E^{m,n} & \xrightarrow{\overline{\Phi}} & E \\ \downarrow & & \downarrow \pi & & \downarrow \tau \\ \Sigma P^m \times \Sigma P^n & \xrightarrow{=} & \Sigma P^m \times \Sigma P^n & \xrightarrow{\Phi} & B. \end{array}$$

The choice of the lift $\overline{\Phi} : E^{m,n} \to E$ is uniquely determined up to fibre homotopy and thus the choice of the map $\overline{\overline{\Phi}} : F^{m,n} \to F$ is uniquely determined up to homotopy.

If we loop the above diagram, the Hilton–Milnor theorem implies that the middle vertical fibration sequence
$$\Omega F^{m,n} \xrightarrow{\Omega \iota} \Omega E^{m,n} \xrightarrow{\Omega \pi} \Omega(\Sigma P^m \times \Sigma P^n)$$

is split. If $\iota_m : P^m \to \Omega P^{m,n}$ and $\iota_n : P^n \to \Omega P^{m,n}$ are the standard inclusions, the internal Samelson product $[\iota_m, \iota_n]$ maps via $\Omega\pi$ to zero in the base $\Omega(\Sigma P^m \times \Sigma P^n)$ and hence there is a unique mod p^r homotopy class $[\iota_m, \iota_n]_1 : P^{m+n} \to \Omega F^{m,n}$ which maps to $[\iota_m, \iota_n]$ via $\Omega\iota$. That is, $(\Omega\iota)_*[\iota_m, \iota_n]_1 = [\iota_m, \iota_n]$.

Definition 6.11.2. If $\overline{x} : P^n \to \Omega E$ and $\overline{y} : P^n \to \Omega E$ are the respective adjoints of x and y, then the Φ-based Samelson product is the class

$$[\overline{x}, \overline{y}]_\Phi : P^{m+n} \to \Omega F$$

uniquely defined by

$$[\overline{x}, \overline{y}]_\Phi = (\Omega\overline{\overline{\Phi}})_*[\iota_m, \iota_n]_1.$$

Remarks. The Φ-based Samelson products are functorial or natural on the category. That is, given a commutative diagram

$$\begin{array}{ccccccc}
 & & F^{m,n} & \xrightarrow{\overline{\overline{\Phi}}} & F & \xrightarrow{h} & F_1 \\
 & & \downarrow \iota & & \downarrow \iota & & \downarrow \iota \\
\Sigma P^{m,n} & \xrightarrow{\simeq} & E^{m,n} & \xrightarrow{\overline{\Phi}} & E & \xrightarrow{g} & E_1 \\
\downarrow & & \downarrow \pi & & \downarrow \tau & & \downarrow \tau_1 \\
\Sigma P^m \times \Sigma P^n & \xrightarrow{=} & \Sigma P^m \times \Sigma P^n & \xrightarrow{\Phi} & B & \xrightarrow{f} & B_1,
\end{array}$$

we have the naturality formula

$$(\Omega h)_*[\overline{x}, \overline{y}]_\Psi = [g_*\overline{x}, g_*\overline{y}]_{f \cdot \Psi}.$$

In particular, $[\iota_m, \iota_n]_1$ is the 1-based Samelson product. Thus, this Samelson product is universal among Φ-based Samelson products and it is uniquely characterized by the fact that its image in the total space of the universal model is the usual internal Samelson product.

Twisted anti-commutativity

Consider the twist map T and its extension to the replacement fibrations, that is, consider the diagram

$$\begin{array}{ccccccc}
\Sigma P^{n,m} & \xrightarrow{T} & \Sigma P^{m,n} & \xrightarrow{x \vee y} & E \\
\downarrow & & \downarrow & & \downarrow \tau \\
\Sigma P^n \times \Sigma P^m & \xrightarrow{T} & \Sigma P^m \times \Sigma P^n & \xrightarrow{\Phi} & B
\end{array}$$

6.11 Samelson products over the loops on an H-space

and its extension to

$$
\begin{array}{ccccc}
F^{n,m} & \xrightarrow{\overline{\overline{T}}} & F^{m,n} & \xrightarrow{\overline{\overline{\Phi}}} & F \\
\downarrow & & \downarrow & & \downarrow \\
E^{n,m} & \xrightarrow{\overline{T}} & E^{m,n} & \xrightarrow{\overline{\Phi}} & E \\
\downarrow & & \downarrow & & \downarrow \tau \\
\Sigma P^n \times \Sigma P^m & \xrightarrow{T} & \Sigma P^m \times \Sigma P^n & \xrightarrow{\Phi} & B.
\end{array}
$$

Since

$$(\Omega T)_*[\iota_n, \iota_m] = [\iota_n, \iota_m] = -(-1)^{nm}[\iota_m, \iota_n]$$

that is,

$$(\Omega \overline{T})_*[\iota_n, \iota_m] = [\iota_n, \iota_m] = -(-1)^{nm}[\iota_m, \iota_n].$$

The injection of the homotopy of the fibre into the homotopy of the total space gives

$$(\Omega \overline{\overline{T}})_*[\iota_n, \iota_m]_1 = -(-1)^{nm}[\iota_m, \iota_n]_1$$

in the universal models. Hence, the natural definition implies the twisted anti-commutativity:

Lemma 6.11.3.

$$[\overline{y}, \overline{x}]_{\Phi \cdot T} = -(-1)^{nm}[\overline{x}, \overline{y}]_\Phi.$$

Relative Samelson products

The relative Samelson product $[\overline{x}, \overline{y}]_r$ is a special case of the Φ-based product. Suppose $x : \Sigma P^m \to E$ and $y : \Sigma P^m \to F$. Consider the diagram

$$
\begin{array}{ccccc}
\Sigma P^{m,n} & \xrightarrow{=} & \Sigma P^{m,n} & \xrightarrow{x \vee y} & E \\
\downarrow & & \downarrow & & \downarrow \tau \\
\Sigma P^m \times \Sigma P^n & \xrightarrow{\pi_1} & \Sigma P^m & \xrightarrow{x} & B,
\end{array}
$$

where π_1 is the projection on the first factor. It follows that

Lemma 6.11.4. *If $x : \Sigma P^m \to E$ and $y : \Sigma P^m \to F$, then*

$$[\overline{x}, \overline{y}]_r = [\overline{x}, \overline{y}]_{x \cdot \pi_1}$$

Compatibility of Samelson products in the loops on a fibration sequence

The above lemma is part of the general compatibility of Φ-based, relative, and internal Samelson products. That is, we have a commutative diagram

$$\begin{array}{ccc}
\pi_*(\Omega E;\mathbb{Z}/p^r\mathbb{Z}) \otimes \pi_*(\Omega F;\mathbb{Z}/p^r\mathbb{Z}) & \xrightarrow{[\ ,\ ]_r} & \pi_*(\Omega F;\mathbb{Z}/p^r\mathbb{Z}) \\
\downarrow & & \downarrow = \\
\pi_*(\Omega E;\mathbb{Z}/p^r\mathbb{Z}) \otimes \pi_*(\Omega E;\mathbb{Z}/p^r\mathbb{Z}) & \xrightarrow{[\ ,\ ]_\Phi} & \pi_*(\Omega F;\mathbb{Z}/p^r\mathbb{Z}) \\
\downarrow = & & \downarrow \\
\pi_*(\Omega E;\mathbb{Z}/p^r\mathbb{Z}) \otimes \pi_*(\Omega E;\mathbb{Z}/p^r\mathbb{Z}) & \xrightarrow{[\ ,\ ]} & \pi_*(\Omega E;\mathbb{Z}/p^r\mathbb{Z}).
\end{array}$$

The compatibility of the Φ-based product with the internal product is a consequence of the next lemma. Consider the diagram

$$\begin{array}{ccc}
F^{m,n} & \xrightarrow{\overline{\overline{\Phi}}} & F \\
\downarrow \iota & & \downarrow \iota \\
E^{m,n} & \xrightarrow{\overline{\Phi}} & E \\
\downarrow & & \downarrow \tau \\
\Sigma P^m \times \Sigma P^n & \xrightarrow{\Phi} & B.
\end{array}$$

Hence,

$$(\Omega\iota)_*[\overline{x},\overline{y}]_\Phi = (\Omega\iota) * (\Omega\overline{\overline{\Phi}})_*[\iota_m,\iota_n]_1$$
$$= (\Omega\overline{\Phi})_*(\Omega\iota)_*[\iota_m,\iota_n]_1 = (\Omega\overline{\Phi})_*[\iota_m,\iota_n] = [\overline{x},\overline{y}],$$

that is,

Lemma 6.11.4. *If $x : \Sigma P^m \to E$ and $y : \Sigma P^n \to E$, then*

$$(\Omega\iota)_*[\overline{x},\overline{y}]_\Phi = [\overline{x},\overline{y}].$$

Bocksteins and Φ-based Samelson products

Recall that the Bockstein βx of $x : \Sigma P^m \to E$ is defined by the composition

$$\Sigma P^{m-1} \xrightarrow{\overline{\beta}} \Sigma P^m \xrightarrow{x} E.$$

Consider the two diagrams

$$\begin{array}{ccc}
F^{m-1,n} & \xrightarrow{(\overline{\beta}\times 1)} & F^{m,n} & \xrightarrow{\overline{\overline{\Phi}}} & F \\
\downarrow & & \downarrow \iota & & \downarrow \iota \\
E^{m-1,n} & \xrightarrow{(\overline{\beta}\times 1)} & E^{m,n} & \xrightarrow{\overline{\Phi}} & E \\
\downarrow & & \downarrow & & \downarrow \tau \\
\Sigma P^{m-1} \times \Sigma P^n & \xrightarrow{\overline{\beta}\times 1} & \Sigma P^m \times \Sigma P^n & \xrightarrow{\Phi} & B
\end{array}$$

$$
\begin{array}{ccccc}
F^{m,n-1} & \xrightarrow{\overline{\overline{(1\times\overline{\beta})}}} & F^{m,n} & \xrightarrow{\overline{\overline{\Phi}}} & F \\
\downarrow & & \downarrow \iota & & \downarrow \iota \\
E^{m,n-1} & \xrightarrow{\overline{(1\times\overline{\beta})}} & E^{m,n} & \xrightarrow{\overline{\Phi}} & E \\
\downarrow & & \downarrow & & \downarrow \tau \\
\Sigma P^m \times \Sigma P^{n-1} & \xrightarrow{1\times\overline{\beta}} & \Sigma P^m \times \Sigma P^n & \xrightarrow{\Phi} & B.
\end{array}
$$

In $\Omega E^{m,n}$, we have

$$\beta[\iota_m, \iota_n] = [\beta\iota_m, \iota_n] + (-1)^m [\iota_m, \beta\iota_n].$$

Since the homotopy of the fibre injects into that of the total space, in $\Omega F^{m,n}$, we have

$$\beta[\iota_m, \iota_n]_1 = [\beta\iota_m, \iota_n]_{\overline{\beta}\times 1} + (-1)^m [\iota_m, \beta\iota_n]_{1\times\overline{\beta}}.$$

Naturality gives the derivation formula for the Bockstein

Lemma 6.11.5. *If $x : \Sigma P^m \to E$ and $y : \Sigma P^n \to E$, then*

$$\beta[\overline{x}, \overline{y}]_\Phi = [\beta\overline{x}, \overline{y}]_{\Phi \cdot (\overline{\beta}\times 1)} + (-1)^m [\overline{x}, \beta\overline{y}]_{\Phi \cdot (1\times\overline{\beta})}.$$

Three-variable models

Let $x : \Sigma P^m \to E$, $y : \Sigma P^n \to E$, and $z : \Sigma P^q \to E$. A three-variable model is a commutative diagram

$$
\begin{array}{ccc}
\Sigma P^{m,n,q} & \xrightarrow{x\vee y\vee z} & E \\
\downarrow & & \downarrow \tau \\
\Sigma P^m \times \Sigma P^n \times \Sigma P^q & \xrightarrow{\Psi} & B.
\end{array}
$$

As in the two-variable case, we extend the maps to maps of homotopy equivalent fibration sequences

$$
\begin{array}{ccccc}
 & & F^{m,n,q} & \xrightarrow{\overline{\overline{\Psi}}} & F \\
 & & \downarrow \iota & & \downarrow \iota \\
\Sigma P^{m,n,q} & \xrightarrow{\simeq} & E^{m,n,q} & \xrightarrow{\overline{\Psi}} & E \\
\downarrow & & \downarrow & & \downarrow \tau \\
\Sigma P^m \times \Sigma P^n \times \Sigma P^q & \xrightarrow{=} & \Sigma P^m \times \Sigma P^n \times \Sigma P^q & \xrightarrow{\Psi} & B.
\end{array}
$$

We have the universal three-variable model given by the identity map of

$$\Sigma P^{m,n,q} \to \Sigma P^m \times \Sigma P^n \times \Sigma P^q.$$

Bilinearity

For the consideration of bilinearity, assume that $m = n$ in the three-variable model and consider the three diagrams

$$\begin{array}{ccccccc}
\Sigma P^{m,q} & \xrightarrow{\nabla \vee 1} & \Sigma P^{m,n,q} & \xrightarrow{x \vee y \vee z} & E \\
\downarrow & & \downarrow & & \downarrow \tau \\
\Sigma P^m \times \Sigma P^q & \xrightarrow{\Delta \times 1} & \Sigma P^m \times \Sigma P^n \times \Sigma P^q & \xrightarrow{\Psi} & B
\end{array}$$

where $\Delta : \Sigma P^m \to \Sigma P^m \times \Sigma P^n$ is the diagonal, $\nabla : \Sigma P^m \to \Sigma P^{m,n}$ is the coproduct which defines addition, and

$$\begin{array}{ccccccc}
\Sigma P^{m,q} & \xrightarrow{1_m \vee 1} & \Sigma P^{m,n,q} & \xrightarrow{x \vee y \vee z} & E \\
\downarrow & & \downarrow & & \downarrow \tau \\
\Sigma P^m \times \Sigma P^q & \xrightarrow{1_m \times 1} & \Sigma P^m \times \Sigma P^n \times \Sigma P^q & \xrightarrow{\Psi} & B
\end{array}$$

where $1_m : \Sigma P^m \to \Sigma P^{m,n} \to \Sigma P^m \times \Sigma P^n$ is the inclusion of one summand, and

$$\begin{array}{ccccccc}
\Sigma P^{n,q} & \xrightarrow{1_n \vee 1} & \Sigma P^{m,n,q} & \xrightarrow{x \vee y \vee z} & E \\
\downarrow & & \downarrow & & \downarrow \tau \\
\Sigma P^n \times \Sigma P^q & \xrightarrow{1_n \times 1} & \Sigma P^m \times \Sigma P^n \times \Sigma P^q & \xrightarrow{\Psi} & B
\end{array}$$

where $1_n : \Sigma P^n \to \Sigma P^{m,n} \to \Sigma P^m \times \Sigma P^n$ is the inclusion of one summand. In $\Omega \Sigma P^{m,n,q}$ and hence also in $\Omega E^{m,n,q}$,

$$[\iota_m + \iota_n, \iota_q] = [\iota_m, \iota_q] + [\iota_n, \iota_q].$$

Since the homotopy of the fibre injects into that of the total space, it follows that

$$[\iota_m + \iota_n, \iota_q]_{(\Delta \times 1)} = [\iota_m, \iota_q]_{(1_m \times 1)} + [\iota_n, \iota_q]_{(1_n \times 1)}$$

and, by naturality,

Lemma 6.11.6.

$$[\bar{x} + \bar{y}, \bar{z}]_{\Psi \cdot (\Delta \times 1)} = [\bar{x}, \bar{z}]_{\Psi \cdot (1_m \times 1)} + [\bar{y}, \bar{z}]_{\Psi \cdot (1_n \times 1)}$$

The Jacobi identity

In the three-variable model as above, consider the iterated Samelson product

$$[\iota_m, \iota_n, \iota_q] = [\iota_m, [\iota_n, \iota_q]]$$

in $\Omega \Sigma P^{m,n,q}$ or, equivalently, in $\Omega E^{m,n,q}$. Since the fibration sequence

$$\Omega F^{m,n,q} \xrightarrow{\iota} \Omega E^{m,n,q} \xrightarrow{\pi} \Sigma P^m \times \Sigma P^n \times \Sigma P^q$$

6.11 Samelson products over the loops on an H-space   217

is split, there is a unique mod p^r homotopy class $[\iota_m, \iota_n, \iota_q]_1$ in $\Omega F^{m,n,q}$ such that $\iota_*[\iota_m, \iota_n, \iota_q]_1 = [\iota_m, \iota_n, \iota_q]$.

Definition 6.11.6. The Ψ-based Samelson product is defined by

$$[\overline{x}, \overline{y}, \overline{z}]_\Psi = \Omega(\overline{\overline{\Psi}})_*[\iota_m, \iota_n, \iota_q]_1.$$

There are three obvious restrictions of

$$\Psi : \Sigma P^m \times \Sigma P^n \times \Sigma P^q \to B$$

to maps

$$\Phi = \Phi_m : \Sigma P^n \times \Sigma P^q \to B$$
$$\Phi = \Phi_n : \Sigma P^m \times \Sigma P^q \to B$$
$$\Phi = \Phi_q : \Sigma P^m \times \Sigma P^n \to B.$$

The following lemma which relates three-variable products to two variable products is a trivial application of the injectivity in the universal model.

Lemma 6.11.7.

$$[\iota_m, \iota_n, \iota_q]_1 = [\iota_m, [\iota_n, \iota_q]_1]_r$$

and naturality gives

$$[\overline{x}, \overline{y}, \overline{z}]_\Psi = [\overline{x}, [\overline{y}, \overline{z}]_{\Phi_m}]_r.$$

If $R : \Sigma P^q \times \Sigma P^m \times \Sigma P^n \to \Sigma P^m \times \Sigma P^n \times \Sigma P^q$ and $S : \Sigma P^n \times \Sigma P^m \times \Sigma P^q \to \Sigma P^m \times \Sigma P^n \times \Sigma P^q$ are the permutations, then the Jacobi identity becomes

$$[\overline{x}, \overline{y}, \overline{z}]_\Psi = (-1)^{q(m+n)+1}[\overline{z}, \overline{x}, \overline{y}]_{\Psi \cdot R} + (-1)^{mn}[\overline{y}, \overline{x}, \overline{z}]_{\Psi \cdot S}$$

which is a rewrite of

$$[\overline{x}, [\overline{y}, \overline{z}]_{\Phi_m}]_r = (-1)^{q(m+n)+1}[\overline{z}, [\overline{x}, \overline{y}]_{\Phi_q}]_r + (-1)^{mn}[\overline{y}, [\overline{x}, \overline{z}]_{\Phi_n}]_r.$$

In the total space of the universal example $\Omega E^{m,n,q}$, this is equivalent to

$$[\iota_m, [\iota_n, \iota_q]] = (-1)^{q(m+n)+1}(\Omega R)_*[\iota_q, [\iota_m, \iota_n]] + (-1)^{mn}(\Omega S)_*[\iota_n, [\iota_m, \iota_q]].$$

Since $R : \Sigma P^{q,m,n} \to \Sigma P^{m,n,q}$ and $S : \Sigma P^{n,m,q} \to \Sigma P^{m,n,q}$ induce the formal identity on these Samelson products, this identity is the same as the usual Jacobi identity

$$[\iota_m, [\iota_n, \iota_q]] = (-1)^{q(m+n)+1}[\iota_q, [\iota_m, \iota_n]] + (-1)^{mn}[\iota_n, [\iota_m, \iota_q]].$$

Hence naturality gives

Lemma 6.11.8. *The Jacobi identity is valid if p is a prime greater than 3, that is,*

$$[\overline{x},\overline{y},\overline{z}]_\Psi = (-1)^{q(m+n)+1}[\overline{z},\overline{x},\overline{y}]_{\Psi \cdot R} + (-1)^{mn}[\overline{y},\overline{x},\overline{z}]_{\Psi \cdot S}$$

or the more familiar

$$[\overline{x},[\overline{y},\overline{z}]_{\Phi_m}]_r = (-1)^{q(m+n)+1}[\overline{z},[\overline{x},\overline{y}]_{\Phi_q}]_r + (-1)^{mn}[\overline{y},[\overline{x},\overline{z}]_{\Phi_n}]_r.$$

Finally, suppose that $F \xrightarrow{\iota} E \xrightarrow{\pi} B$ is a fibration sequence with B an H-space with multiplication $\mu : B \times B \to B$. We can assume that the multiplication has a strict unit. In this case we have a canonical choice for the map $\Phi : \Sigma P^m \times \Sigma P^n \to B$, namely, we set Φ to be the composition

$$\Sigma P^m \times \Sigma P^n \xrightarrow{x \times y} B \times B \xrightarrow{\mu} B.$$

Definition 6.11.9. The H-based Samelson product $[\overline{x},\overline{y}]_\mu$ is defined to be the Φ-based Samelson product $[\overline{x},\overline{y}]_\Phi$.

In addition, if we choose Ψ to be the composition

$$\Sigma P^m \times \Sigma P^n \times \Sigma P^q \xrightarrow{x \times y \times z} B \times B \times B \xrightarrow{1 \times \mu} B \times B \xrightarrow{\mu} B,$$

then the three restrictions to two factors are all equal to the above map, that is,

$$\Phi_m = \Phi_n = \Phi_q = \Phi.$$

Thus, we get

(1) Twisted anti-commutativity:
$$[\overline{y},\overline{x}]_{\mu \cdot T} = -(-1)^{mn}[\overline{x},\overline{y}]_\mu$$

(2) Derivation formula for the Bockstein:
$$\beta[\overline{x},\overline{y}]_\mu = [\beta\overline{x},\overline{y}]_\mu + (-1)^m[\overline{x},\beta\overline{y}]_\mu$$

(3) Bilinearity:
$$[\overline{x}+\overline{y},\overline{z}]_\mu = [\overline{x},\overline{z}]_\mu + [\overline{y},\overline{z}]_\mu$$

(4) The Jacobi identity if $p > 3$:
$$[\overline{x},[\overline{y},\overline{z}]_\mu]_r = (-1)^{q(m+n)+1}[\overline{z},[\overline{x},\overline{y}]_\mu]_r + (-1)^{mn}[\overline{y},[\overline{x},\overline{z}]_\mu]_r.$$

If, in addition, the multiplication is homotopy commutative, that is, if $\mu \cdot T \simeq \mu$, we get true anti-commutativity

$$[\overline{y},\overline{x}]_\mu = -(-1)^{mn}[\overline{x},\overline{y}]_\mu.$$

Remarks. The verification of true anti-commutativity in the case of homotopy commutativity uses the fact that the commuting homotopy can be chosen to be stationary on the wedge. This fact, known to Frank Adams and to Michael Barratt,

6.11 Samelson products over the loops on an H-space 219

will be proved in Chapter 11. We use the fact that the commuting homotopy on $\Sigma P^m \times \Sigma P^n$ is stationary on the wedge $\Sigma P^m \vee \Sigma P^n$ in order to lift this homotopy on the base to a fibre homotopy on $E^{m,n}$, giving a homotopy on the fibre $F^{m,n}$ and hence also on its loop space $\Omega F^{m,n}$ which is where the H-based Samelson product lives.

Remarks. Since the prime p is odd, the restriction to a homotopy commutative base is really no restriction at all. If we localize at p, the base space B can be made homotopy commutative by replacing the original multiplication μ by the homotopy commutative multiplication

$$\nu(\alpha,\beta) = \frac{1}{2}(\mu(\mu(\alpha,\beta),\mu(\beta,\alpha))) = \frac{1}{2}(\mu(\alpha,\beta) + \mu(\beta,\alpha))$$

where $+$ is the addition (that is, the multiplication) in the H-space and $\frac{1}{2}: B \to B$ is the homotopy inverse to the squaring map $2: B \to B$, $\alpha \mapsto \alpha^2$.

Remarks. Consider the standard maps $S^{n-1} \to P^n(p^r)$, $P^n(p^r) \to S^n$, $P^n(p^r) \to P^n(p^{r+s})$, $P^n(p^{r+s}) \to P^n(p^s)$, and similarly with m replacing n. The obvious variations of the above apply when $P^n(p^r)$ and $P^m(p^r)$ are replaced by various combinations of the domains and ranges of these maps. Hence, identities which involve Bocksteins and reductions of maps are valid for H-based Samelson products if they are valid for the usual Samelson products.

Remarks. When the base B is a homotopy commutative H-space, it is clear that the morphism of mod p^r homotopy groups $\pi(\Omega F; \mathbb{Z}/p^r\mathbb{Z}) \to \pi(\Omega E; \mathbb{Z}/p^r\mathbb{Z})$ satisfies the following two definitions where p is a prime greater than 3 and the differential d in Definition 6.10.2 is the Bockstein differential $d = \rho \circ \beta = \beta$:

Definition 6.11.10. Let $L' \to L$ be a morphism of graded Lie algebras. We call L' a strong extended ideal of L if there is a bilinear pairing (called a Lie bracket):

$$[\ ,\ ]: L \times L \to L'$$

such that

(1) the diagram of Lie brackets commutes

$$\begin{array}{ccc} L' \times L' & \xrightarrow{[\ ,\ ]} & L' \\ \downarrow & & \downarrow \\ L \times L & \xrightarrow{[\ ,\ ]} & L' \\ \downarrow & & \downarrow \\ L \times L & \xrightarrow{[\ ,\ ]} & L. \end{array}$$

(2) for all $x, y,$ and z in the union of L' and L,

$$[x,y] = -(-1)^{\deg(x)\cdot\deg(y)}[y,x]$$

$$[x,[y,z]] = [[x,y],z] + (-1)^{\deg(x)\cdot\deg(y)}[y,[x,z]].$$

Definition 6.11.11. A strong extended differential ideal $L' \to L$ is a morphism of differential graded Lie algebras which is a strong extended ideal and such that the differential d is a derivation in the sense that, for all x and y in the union of L' and L,

$$d[x, y] = [dx, y] + (-1)^{\deg(x)}[x, dy].$$

Exercises

(1) Let

$$\begin{array}{ccccc} F_1 & \to & E_1 & \to & B_1 \\ \downarrow h & & \downarrow g & & \downarrow f \\ F_2 & \to & E_2 & \to & B_2 \end{array}$$

be a map of fibration sequences with $f : B_1 \to B_2$ being an H-map of H-spaces, then the map on the loop spaces of the fibres sends H-based Samelson products to H-based Samelson products, that is,

$$(\Omega h)_*[\overline{x}, \overline{y}]_\mu = [(\Omega h)_*\overline{x}, (\Omega h)_*\overline{y}]_\mu.$$

(2) Given that, in a homotopy commutative H-space B, a commuting homotopy

$$h_t : B \times B \to B, \quad h_0 = \mu, \quad h_1 = \mu \cdot T$$

can be made to be stationary on the wedge, show that the H-based Samelson product is anti-commutative:

$$[\overline{y}, \overline{x}]_\mu = -(-1)^{mn}[\overline{x}, \overline{y}]_\mu.$$

7 Bockstein spectral sequences

This chapter begins with an exposition of exact couples [78, 79] which follows the presentation in MacLane's Homology [77]. The important feature of MacLane's exposition is that it stresses the explicit identification of the r-th term of the spectral sequence. We specialize to the case of the homotopy Bockstein exact couple and blend this treatment of spectral sequences by exact couples with a treatment which is a simplification of that of Cartan–Eilenberg via spectral systems or $H(p,q)$ systems [23]. Cartan–Eilenberg gives an alternative description of the r-th term of the Bockstein spectral sequence. In particular, this alternate description is well suited for the introduction of Samelson products into the spectral sequence.

We determine the convergence of the mod p homotopy Bockstein spectral sequence, at least in the case where the classical homotopy groups have p-torsion of bounded order. From general principles it is clear that the E^∞ term should be a function of the homotopy groups of the p-completion, but an example shows that there are significant difficulties in the case when the p-torsion is of unbounded order.

But the principal application of the Bockstein spectral sequence is not to compute the E^∞ term. There are usually better ways to do that. Application of the Bockstein spectral sequence comes from its differentials. The differentials determine the torsion in the integral homotopy groups. In addition, there is a strong connection between differentials and the problem of extending maps which originate on a Moore space.

Homology and cohomology versions of Bockstein spectral sequences first appear in the work of William Browder on H-spaces [19]. We give a brief treatment of these spectral sequences here and then apply them, as Browder did, to prove some classical theorems on finite H-spaces. These remain some of the most attractive and powerful applications of Bockstein spectral sequences.

Also important to us is the Samelson product structure in the homotopy Bockstein spectral sequence and the Hurewicz representation of this in the Pontrjagin structure of of the homology Bockstein spectral sequence.

In a subsequent chapter (Section 9.6) we apply these Samelson products in the homotopy Bockstein spectral sequence to establish the existence of higher order torsion in the homotopy groups of odd primary Moore spaces. It is also vital in the product decompositions of Chapter 11 which lead to the odd primary exponent theorem for spheres.

7.1 Exact couples

We base our definitions of Bockstein spectral sequences on William Massey's notion of exact couple: [78, 79]

Definition 7.1.1. An exact couple C consists of two graded modules A, E and an exact triangle of homomorphisms

$$\begin{array}{ccc} A & \xrightarrow{\iota} & A \\ & \partial \nwarrow \swarrow j & \\ & E & \end{array}$$

where $\iota : A \to A$ has degree 0 and $\partial : E \to A$ has degree ± 1.

Remark. If the degree of ∂ is -1, the exact couple is called a homology exact couple. If the degree of ∂ is $+1$, the exact couple is called a cohomology exact couple.

We note that, if we define $d = j \circ \partial : E \to E$, then d is a differential, $d \circ d = 0$, and we can define a homology group

$$H(E, d) = \mathbb{Z}(E, d)/B(E, d) = \ker(d)/\mathrm{im}(d).$$

Exact couples yield new exact couples by a process of derivation defined in the following manner.

Definition 7.1.2. Let C be an exact couple as above. The derived couple C' has the graded modules $A' = \iota A = \mathrm{im}(\iota), E' = H(E, d)$ and the triangle of maps

$$\begin{array}{ccc} A' & \xrightarrow{\iota'} & A' \\ & \partial' \nwarrow \swarrow j' & \\ & E' & \end{array}$$

where

$$\iota'(\iota a) = \iota^2(a),$$
$$j'(\iota a) = ja + \mathrm{im}(d), \ a \varepsilon A,$$
$$\partial'(e + \mathrm{im}(d)) = \partial e, \ e \varepsilon E.$$

We leave as exercises two key facts:

(1) the maps ι', j', ∂' are well defined, that is, they are independent of choices of representatives and they land in the appropriate groups.

(2) the derived couple is exact.

The sequence of successive derived exact couples $C, C', C'', C''', \ldots$ defines a spectral sequence via

$$E^1 = E, \ E^2 = E', \ E^3 = E'', \ldots$$

with differentials $d^r : E^r \to E^r$

$$d^1 = d = j \circ \partial, \ d^2 = d' = j' \circ \partial', \ d^3 = d'' = j'' \circ \partial'', \ d^4 = d''' = j''' \circ \partial''', \ldots$$

and we have that

$$E^{r+1} = H(E^r, d^r).$$

It is sometimes convenient to define the successive derived couples in one step as follows:

Let C be an exact couple. Define couples C^r as follows:

$$A^r = \mathrm{im}(\iota^{r-1} : A \to A) = \iota^{r-1} A,$$
$$E^r = \mathbb{Z}^r / B^r = \partial^{-1}(\mathrm{im}(\iota^{r-1})) / j(\ker(\iota^{r-1}))$$

and maps

$$\iota_r = \iota : A^r \to A^r, \qquad \iota_r(\iota^{r-1} a) = \iota^r a,$$
$$j_r = j \circ \iota^{-(r-1)} : A^r \to E^r \qquad j_r(\iota^{r-1} a) = ja + j(\mathrm{im}(\iota^{r-1})),$$
$$\partial_r = \partial : E^r \to A^r, \qquad \partial_r(e + j(\mathrm{im}(\iota^{r-1}))) = \partial e.$$

It is another exercise to check that the maps ι_r, j_r, ∂_r are well defined.

Thus, $C^1 = C$ and we need:

Lemma 7.1.3. *The couples C^r are all exact and C^{r+1} is the derived couple of C^r.*

Proof: We assume that C^r is exact. Then it is sufficient to show that C^{r+1} is the derived couple of C^r.

Suppose $\bar{e} = e + B^r$ is a coset in E^r, that is,

$$B^r = j(\ker(\iota^{r-1})),$$

and

$$e \varepsilon \mathbb{Z}^r = \partial^{-1}(\mathrm{im}(\iota^{r-1})).$$

The r-th differential is described as follows:
$$d^r(\bar{e}) = ja + j(\ker(\iota^{r-1})) = \overline{ja}, \quad \text{where} \quad \partial e = \iota^{r-1}a, \ a \varepsilon A.$$
First we determine the group of boundaries $\mathrm{im}(d^r) \subseteq E^r$:
$\iota^r a = \iota \partial e = 0$ and $ja\ \varepsilon j(\ker(\iota^r)) = B^{r+1}$. Thus,
$$\mathrm{im}(d^r) = B^{r+1} + B^r.$$
Next we determine the group of cycles $\ker(d^r)$:
Observe that $d^r \bar{e} = \bar{0}$ if and only if $ja\ \varepsilon B^r$. That is,
$$ja = jb, \quad \iota^{r-1}b = 0$$
$$\therefore a - b = \iota c,$$
$$\partial e = \iota^{r-1} a = \iota^{r-1}(b + \iota c) = \iota^r c.$$
Thus,
$$\ker(d^r) = \mathbb{Z}^{r+1} + B^r.$$
And,
$$\therefore H(E^r, d^r) = E^{r+1}.$$

$\square$

Note that d^r is defined by the relation
$$\begin{array}{ccccc} E^1 & \supseteq & \mathbb{Z}^r & \to & E^r \\ \downarrow \partial & & & & \\ A^1 & & & & \bigg| d^r \\ \uparrow \iota^{r-1} & & & & \downarrow \\ A^1 & \xrightarrow{j} & \mathbb{Z}^r & \to & E^r. \end{array}$$

It is customary to display an exact couple as follows:
$$\begin{array}{ccccccccc} & & \downarrow \iota & & & & \downarrow \iota & & \\ \xrightarrow{j} & E & \xrightarrow{\partial} & A & \xrightarrow{j} & E & \xrightarrow{\partial} & A & \xrightarrow{\iota} \\ & & \downarrow \iota & & & & \downarrow \iota & & \\ \xrightarrow{j} & E & \xrightarrow{\partial} & A & \xrightarrow{j} & E & \xrightarrow{\partial} & A & \xrightarrow{\iota} \\ & & \downarrow \iota & & & & \downarrow \iota & & \\ \xrightarrow{j} & E & \xrightarrow{\partial} & A & \xrightarrow{j} & E & \xrightarrow{\partial} & A & \xrightarrow{\iota} \\ & & \downarrow \iota & & & & \downarrow \iota & & \end{array}$$

In this picture, the r-th differential $d^r = j \circ \iota^{1-r} \circ \partial$ is the relation which starts at E, moves one step to the right to A, followed by $r - 1$ steps up to A, and ends with one step to the right to E. Observe that $\mathbb{Z}^r$ is the domain of definition of d^r, B^r is the range of d^{r-1}, and $\mathbb{Z}^{r+1}$ is the kernel of d^r.

7.2 Mod p homotopy Bockstein spectral sequences 225

Exercises

(1) Show that the derived couple of an exact couple is well defined and exact.
(2) Show that the maps the ι_r, j_r, ∂_r in the definition of the couple C^r are well defined.

7.2 Mod p homotopy Bockstein spectral sequences

Let p be a prime and let X be a space which is either simply connected or a connected H-space. In the first case, we note that $\pi_2(X; \mathbb{Z}/p^r\mathbb{Z}) = \pi_2(X) \otimes \mathbb{Z}/p^r\mathbb{Z}$. In the second case, we define $\pi_1(X; \mathbb{Z}/p^r\mathbb{Z}) = \pi_1(X) \otimes \mathbb{Z}/p^r\mathbb{Z}$. In either case, the homotopy groups with coefficients $\pi_m(X; \mathbb{Z}/p^r\mathbb{Z})$ are defined and are groups (possibly zero) for all $m \geq 1$. We showed in Chapter 1 that the Bockstein exact sequences are valid.

This leads to the following definition:

Definition 7.2.1. The mod p homotopy Bockstein exact couple is

$$\pi_*(X) \xrightarrow{p} \pi_*(X)$$
$$\beta \nwarrow \qquad \swarrow \rho$$
$$E$$

where $p: \pi_*(X) \to \pi_*(X)$ is multiplication by p, $\rho: \pi_*(X) \to \pi_*(X; \mathbb{Z}/p\mathbb{Z})$ is the reduction map, and $\beta: \pi_*(X; \mathbb{Z}/p\mathbb{Z}) \to \pi_{*-1}(X)$ is the Bockstein associated to the exact sequence of coefficient groups $0 \to \mathbb{Z} \xrightarrow{p} \mathbb{Z} \xrightarrow{\rho} \mathbb{Z}/p\mathbb{Z} \to 0$.

We denote the successive derived homotopy Bockstein couples by

$$A_\pi^r(X)_* = A_*^r = \mathrm{im}(\pi_*(X) \xrightarrow{p^{r-1}} \pi_*(X))$$
$$E_\pi^r(X)_* = \mathbb{Z}_\pi^r(X)_*/B_\pi^r(X)_*.$$

Note that $E_\pi^1(X)_* = \pi_*(X; \mathbb{Z}/p\mathbb{Z})$ and the first differential β^1 is just the usual Bockstein associated to the exact coefficient sequence $\mathbb{Z}/p\mathbb{Z} \to \mathbb{Z}/p^2\mathbb{Z} \to \mathbb{Z}/p\mathbb{Z}$.

We use the long exact Bockstein sequences to determine $\mathbb{Z}_\pi^r(X)_*, B_\pi^r(X)_*$, and $E_\pi^r(X)_*$.

Let p be an odd prime. Recall the maps from Sections 1.3 and 1.5 which give the Bockstein long exact sequences associated to coefficient groups:

$$\rho: \pi_*(X) \to \pi_*(X; \mathbb{Z}/p^r\mathbb{Z}),$$
$$\rho: \pi_*(X; \mathbb{Z}/p^{r+s}\mathbb{Z}) \to \pi_*(X; \mathbb{Z}/p^r\mathbb{Z}), \quad r, s \geq 0$$
$$\eta: \pi_*(X; \mathbb{Z}/p^s\mathbb{Z}) \to \pi_*(X; \mathbb{Z}/p^{r+s}\mathbb{Z}), \quad r, s \geq 0$$
$$\beta: \pi_*(X; \mathbb{Z}/p^r\mathbb{Z}) \to \pi_{*-1}(X),$$
$$\overline{\beta} = \rho \circ \beta: \pi_*(X; \mathbb{Z}/p^r\mathbb{Z}) \to \pi_{*-1}(X) \to \pi_{*-1}(X; \mathbb{Z}/p^s)$$

We have, in dimensions ≥ 4:

$$\eta \circ \rho = p^{r-1} : \pi_*(X; \mathbb{Z}/p^r\mathbb{Z}) \to \pi_*(X; \mathbb{Z}/p\mathbb{Z}) \to \pi_*(X; \mathbb{Z}/p^r\mathbb{Z}).$$

$$\begin{aligned}
Z^r_\pi(X)_* &= \beta^{-1}(\mathrm{im}(\pi_{*-1}(X) \xrightarrow{p^{r-1}} \pi_{*-1}(X))) \\
&= \beta^{-1}(\ker(\pi_{*-1}(X) \xrightarrow{\rho} \pi_{*-1}(X; \mathbb{Z}/p^{r-1}\mathbb{Z}))) \\
&= \ker(\pi_*(X; \mathbb{Z}/p\mathbb{Z}) \xrightarrow{\beta} \pi_{*-1}(X; \mathbb{Z}/p^{r-1}\mathbb{Z})) \\
&= \mathrm{im}(\rho : \pi_*(X; \mathbb{Z}/p^r\mathbb{Z}) \to \pi_*(X; \mathbb{Z}/p\mathbb{Z}))
\end{aligned}$$

$$\begin{aligned}
B^r_\pi(X)_* &= \rho(\ker(\pi_*(X) \xrightarrow{p^{r-1}} \pi_*(X))) \\
&= \rho(\mathrm{im}(\pi_{*+1}(X; \mathbb{Z}/p^{r-1}\mathbb{Z}) \xrightarrow{\beta} \pi_*(X))) \\
&= \mathrm{im}(\beta : \pi_{*+1}(X; \mathbb{Z}/p^{r-1}\mathbb{Z}) \to \pi_*(X; \mathbb{Z}/p\mathbb{Z}))
\end{aligned}$$

We need the following easy lemma of Cartan–Eilenberg [23]:

Lemma 7.2.2. *Suppose that there is a commutative diagram with the bottom row exact:*

$$\begin{array}{ccccc}
 & & W & & \\
 & \nearrow & \downarrow \gamma & \searrow \epsilon & \\
X & \xrightarrow{\alpha} & Y & \xrightarrow{\beta} & Z
\end{array}$$

Then β induces an isomorphism $\mathrm{im}(\gamma)/\mathrm{im}(\alpha) \xrightarrow{\cong} \mathrm{im}(\epsilon)$.

Applying Lemma 7.2.2 to

$$\begin{array}{ccccc}
 & & \pi_*(X; \mathbb{Z}/p^r\mathbb{Z}) & & \\
 & \nearrow & \downarrow \rho & \searrow p^{r-1} & \\
\pi_{*+1}(X; \mathbb{Z}/p^{r-1}\mathbb{Z}) & \xrightarrow{\beta} & \pi_*(X; \mathbb{Z}/p\mathbb{Z}) & \xrightarrow{\eta} & \pi_*(X; \mathbb{Z}/p^r\mathbb{Z})
\end{array}$$

yields the first part of

Proposition 7.2.3. *In dimensions ≥ 4:*

$$E^r_\pi(X)_* = \mathrm{im}\ p^{r-1} : \pi_*(X; \mathbb{Z}/p^r\mathbb{Z}) \to \pi_*(X; \mathbb{Z}/p^r\mathbb{Z})$$

and the r-th differential $\beta^r : E^r_\pi(X)_* \to E^r_\pi(X)_{*-1}$ is induced by the Bocksteins $\beta : \pi_*(X; \mathbb{Z}/p^r\mathbb{Z}) \to \pi_{*-1}(X) \to \pi_{*-1}(X; \mathbb{Z}/p^r\mathbb{Z})$ on the domain and range of p^{r-1}.

7.2 Mod p homotopy Bockstein spectral sequences

The identification of the differential comes from the commutative diagram

$$\begin{array}{ccccc}
\pi_*(X; \mathbb{Z}/p^r\mathbb{Z}) & \xrightarrow{\rho} & \pi_*(X; \mathbb{Z}/p\mathbb{Z}) & \xrightarrow{\eta} & \pi_*(X; \mathbb{Z}/p^r\mathbb{Z}) \\
\downarrow \beta & & \downarrow \beta & & \downarrow \beta \\
\pi_{*-1}(X) & \xrightarrow{p^{r-1}} & \pi_{*-1}(X) & = & \pi_{*-1}(X) \\
\downarrow \rho & & \downarrow \rho & & \downarrow \rho \\
\pi_{*-1}(X; \mathbb{Z}/p^r\mathbb{Z}) & \xrightarrow{p^{r-1}} & \pi_{*-1}(X; \mathbb{Z}/p^r\mathbb{Z}) & = & \pi_{*-1}(X; \mathbb{Z}/p^r\mathbb{Z})
\end{array}$$

Remark. It follows that $pE_\pi^r(X)_* = 0$ if p is an odd prime.

We now prove a sort of universal coefficient exact sequence for $E_\pi^r(X)_*$.

Proposition 7.2.4. *There is a short exact sequence*

$$0 \to p^{r-1}(\pi_*(X) \otimes \mathbb{Z}/p^r\mathbb{Z}) \xrightarrow{\rho} E_\pi^r(X)_* \xrightarrow{\beta} p^{r-1}\mathrm{Tor}^\mathbb{Z}(\pi_{*-1}(X), \mathbb{Z}/p^r\mathbb{Z}) \to 0$$

and the r-th differential β^r is the composition

$$E_\pi^r(X)_* \xrightarrow{\beta} p^{r-1}\mathrm{Tor}^\mathbb{Z}(\pi_{*-1}(X), \mathbb{Z}/p^r\mathbb{Z})$$
$$\to p^{r-1}(\pi_{*-1}(X) \otimes \mathbb{Z}/p^r\mathbb{Z}) \xrightarrow{\rho} E_\pi^r(X)_{*-1}$$

where the middle map is induced by the inclusion $\mathrm{Tor}^\mathbb{Z}(\pi_{*-1}(X), \mathbb{Z}/p^r\mathbb{Z}) \subseteq \pi_{*-1}(X)$.

Proof: Consider the diagram of known universal coefficient sequences

$$\begin{array}{ccccccc}
0 \to & \pi_*(X) \otimes \mathbb{Z}/p^r\mathbb{Z} & \to & \pi_*(X; \mathbb{Z}/p^r\mathbb{Z}) & \to & \mathrm{Tor}^\mathbb{Z}(\pi_{*-1}(X), \mathbb{Z}/p^r\mathbb{Z}) & \to 0 \\
& \downarrow 1 \otimes \rho & & \downarrow \rho & & \downarrow p^{r-1} & \\
0 \to & \pi_*(X) \otimes \mathbb{Z}/p\mathbb{Z} & \to & \pi_*(X; \mathbb{Z}/p\mathbb{Z}) & \to & \mathrm{Tor}^\mathbb{Z}(\pi_{*-1}(X), \mathbb{Z}/p\mathbb{Z}) & \to 0 \\
& \downarrow 1 \otimes \eta & & \downarrow \eta & & \downarrow \mathrm{include} & \\
0 \to & \pi_*(X) \otimes \mathbb{Z}/p^r\mathbb{Z} & \to & \pi_*(X; \mathbb{Z}/p^r\mathbb{Z}) & \to & \mathrm{Tor}^\mathbb{Z}(\pi_{*-1}(X), \mathbb{Z}/p^r\mathbb{Z}) & \to 0
\end{array}$$

The universal coefficient sequence that we desire says that the images of the 3 columns form a short exact sequence:

$$0 \to \mathrm{im}_1 \to \mathrm{im}_2 \to \mathrm{im}_3 \to 0.$$

The only nontrivial part is the exactness in the middle. This is an easy consequence of the facts that the upper left hand corner map $1 \otimes \rho$ is an epimorphism and that the lower right hand corner map include is a monomorphism.

The description of β^r is just the fact that it is induced on the image by the usual Bockstein. $\square$

Remark. The above universal coefficient sequence is split if p is an odd prime since it is a short exact sequence of vector spaces. To be more specific, assume that $\pi_*(X)$ is finitely generated in each degree and has a decomposition into cyclic summands with a set of generators $\{x_i, y_j, z_k\}_{i,j,k}$

with order$(x_i) = \infty$, order$(y_j) = p^{r_j}$, and order$(z_k) = q_k$ with q_k relatively prime to p. Then $E_\pi^1(X)_* = \pi_*(X; \mathbb{Z}/p\mathbb{Z})$ contains the following elements which generate it (and are a basis if p is odd):

(1) $\overline{x_i}, \overline{y_j} \varepsilon \pi_*(X; \mathbb{Z}/p\mathbb{Z})$ such that $x_i \otimes 1 = \overline{x_i}, y_j \otimes 1 = \overline{y_j}$ via the reduction map.

(2) and $\sigma(y_j) \varepsilon \pi_{*+1}(X; \mathbb{Z}/p\mathbb{Z})$ such that

$$\beta(\sigma(y_j)) = p^{r_j-1} y_j \varepsilon \text{Tor}^\mathbb{Z}(\pi_*(X), \mathbb{Z}/p\mathbb{Z}) \subseteq \pi_*(X).$$

We have the following description of the differentials:

(1) $\beta^s(\overline{x_i}) = \beta^s(\overline{y_j}) = 0$ for all $1 \leq s < \infty$.
(2) $\beta^s(\sigma(y_j)) = 0$ for all $1 \leq s < r_j$ and $\beta^{r_j}(\sigma(y_j)) = \overline{y_j}$.
(3) if p is odd, $E_\pi^r(X)_*$ has a vector space basis $\overline{x_i}, \overline{y_j}, \sigma(y_j)$, with $r_j \geq r$.
(4) $E_\pi^\infty(X)_* = E_\pi^r(X)_*$ for r sufficiently large and has a basis $\overline{x_i}$.

This translation of Proposition 7.2.4 is the sense in which differentials in the mod p homotopy Bockstein spectral sequence determine the p-primary torsion in the homotopy groups $\pi_*(X)$.

In Section 7.4, we shall give a more natural and general description of $E_\pi^\infty(X)_*$.

Exercise

(1) Suppose that X is either a simply connected space or a connected loop space.

  (a) Show that there are isomorphisms of the mod p homotopy Bockstein spectral sequence of X, of its localization and of its completion, that is,

  $$E_\pi^r(X)_* \xrightarrow{\cong} E_\pi^r(X_{(p)})_* \xrightarrow{\cong} E_\pi^r(\hat{X}_p)_*$$

  (b) Show directly from the defining exact couple that, if $\pi_*(X)$ consists entirely of p-torsion of order bounded by p^r, then

  $$E_\pi^{r+1}(X)_* = 0.$$

  (c) Show directly from the defining exact couple that, if the p-torsion in $\pi_*(X)$ is entirely of order bounded by p^r, then

  $$E_\pi^{r+1}(X)_* = E_\pi^{r+2}(X)_* = E_\pi^{r+3}(X)_* = \cdots.$$

7.3 Reduction maps and extensions

Let p be an odd prime. In this section, we make explicit the general connection between Bockstein differentials and the usual Bocksteins. These ideas are important in the applications of the Bockstein spectral sequence.

There are two natural reduction maps into the r-th term of the mod p homotopy Bockstein spectral sequence, namely:

Definition 7.3.1. From the identification $E^r_\pi(X)_* = p^{r-1}\pi_*(X; \mathbb{Z}/p^r\mathbb{Z})$, we have the natural map
$$\varrho = p^{r-1} : \pi_*(X; \mathbb{Z}/p^r\mathbb{Z}) \to E^r_\pi(X).$$

We also have the natural map
$$\varrho = p^{r-1} \circ \rho : \pi_*(X) \to \pi_*(X; \mathbb{Z}/p^r\mathbb{Z}) \to E^r_\pi(X)_*.$$

The following diagram commutes

$$\begin{array}{ccc} \pi_*(X) & \xrightarrow{\varrho} & E^r_\pi(X)_* \\ \downarrow \rho & & \downarrow = \\ \pi_*(X; \mathbb{Z}/p^r\mathbb{Z}) & \xrightarrow{\varrho} & E^r_\pi(X)_* \\ \downarrow \beta & & \downarrow \beta^r \\ \pi_{*-1}(X; \mathbb{Z}/p^r\mathbb{Z}) & \xrightarrow{\varrho} & E^r_\pi(X)_{*-1} \end{array}$$

That is,
$$\beta^r \circ \varrho = \varrho \circ \beta.$$

Meaning of a nonzero Bockstein differential

When a nonzero differential occurs in the Bockstein spectral sequence, it can be interpreted as follows:

Suppose
$$x \in E^r_\pi(X)_*, \quad y \in E^r_\pi(X)_{*-1},$$
and
$$\beta^r x = y.$$

Then Proposition 7.2.4 gives:

there exists $\overline{y} \in \pi_{*-1}(X)$ such that $\varrho(\overline{y}) = y, \quad p^r \overline{y} = 0$.

This implies that

there exists $z : P^*(p^r) \to X$ such that $\beta z = \overline{y} : S^{*-1} \to X$.

And
$$\varrho z \; \varepsilon E_\pi^r(X)_*, \quad \beta^r \varrho z = \varrho \beta z = y.$$

Meaning of a zero Bockstein

Suppose there is a zero Bockstein associated to the short exact coefficient sequence $\mathbb{Z}/p\mathbb{Z} \to \mathbb{Z}/p^{2r}\mathbb{Z} \to \mathbb{Z}/p^r\mathbb{Z}$. This implies a zero r-th Bockstein differential β^r and an extension of a map out of a mod p^r Moore space as follows:

Suppose $x : P^*(p^r) \to X$ is such that $\beta x : P^{*-1}(p) \to X$ is 0. Then the standard cofibration sequence implies that there exists an extension $y : P^*(p^{r+1}) \to X$ which reduces to x, that is, a factorization

$$x : P^*(p^r) \xrightarrow{\rho} P^*(p^{r+1}) \xrightarrow{y} X, \quad x = \rho y.$$

And
$$\beta^r \varrho x = 0, \quad \beta^{r+1} \varrho y = \varrho \beta y.$$

7.4 Convergence

Let p be an odd prime. We determine convergence of the mod p homotopy Bockstein spectral sequence when the integral homotopy has p-torsion of bounded order in each degree.

Lemma 7.4.1. *If $\pi_*(X)$ has p-torsion of bounded order in each degree, then for r sufficiently large,*

$$\mathbb{Z}_\pi^r(X)_* \cong \pi_*(X) \otimes \mathbb{Z}/p\mathbb{Z}$$

$$E_\pi^r(X)_* \cong \mathrm{im}\ \pi_*(X) \otimes \mathbb{Z}/p\mathbb{Z} \xrightarrow{1 \otimes \eta} \pi_*(X) \otimes \mathbb{Z}/p^r\mathbb{Z}.$$

Proof: Consider the commutative diagram

$$\begin{array}{ccc}
0 & & 0 \\
\downarrow & & \downarrow \\
\pi_*(X) \otimes \mathbb{Z}/p^r\mathbb{Z} & \xrightarrow{\rho} & \pi_*(X) \otimes \mathbb{Z}/p\mathbb{Z} \\
\downarrow & & \downarrow \\
\pi_*(X; \mathbb{Z}/p^r\mathbb{Z}) & \xrightarrow{\rho} & \pi_*(X; \mathbb{Z}/p\mathbb{Z}) \\
\downarrow & & \downarrow \\
\mathrm{Tor}^\mathbb{Z}(\pi_{*-1}(X), \mathbb{Z}/p^r\mathbb{Z}) & \xrightarrow{p^{r-1}} & \mathrm{Tor}^\mathbb{Z}(\pi_{*-1}(X), \mathbb{Z}/p\mathbb{Z}) \\
\downarrow & & \downarrow \\
0 & & 0
\end{array}$$

For r sufficiently large, the bottom map is 0. Hence the middle map factors through $\pi_*(X) \otimes \mathbb{Z}/p\mathbb{Z}$. Since the top map is an epimorphism, $\mathbb{Z}_\pi^r(X)_*$, which is the image of the middle map, is just $\pi_*(X) \otimes \mathbb{Z}/p\mathbb{Z}$.

Hence,
$$E_\pi^r(X)_* = \text{image } \mathbb{Z}_\pi^r(X)_* \to \pi_*(X;\mathbb{Z}/p\mathbb{Z}) \xrightarrow{\eta} \pi_*(X;\mathbb{Z}/p^r\mathbb{Z})$$
is as stated. $\square$

It follows that, if $\pi_*(X)$ has p-torsion of bounded order in each degree, then for r sufficiently large, there are maps
$$E_\pi^r(X)_* \to E_\pi^{r+1}(X)_* \to E_\pi^{r+2}(X)_* \to \cdots.$$
We define
$$E_\pi^\infty(X)_* = \lim_{s \to \infty} E_\pi^{r+s}(X)_*$$
to be the direct limit.

Corollary 7.4.2. *If $\pi_*(X)$ has p-torsion of bounded order, then*
$$E_\pi^\infty(X)_* = \lim_{\to} E_\pi^r(X)_*$$
$$= \text{im } (\pi_*(X) \otimes \mathbb{Z}/p\mathbb{Z}) \to \pi_*(X) \otimes \mathbb{Z}(p^\infty)$$
$$\cong (\pi_*(X)/\text{torsion}) \otimes \mathbb{Z}/p\mathbb{Z}.$$

Proof: The corollary follows immediately from Definition 7.3.1 and the general result:
$$\text{image } (A \to A \otimes \mathbb{Z}(p^\infty)) \cong (A/\text{torsion}) \otimes \mathbb{Z}/p\mathbb{Z}.$$
We see this as follows. The map factors as
$$A \to A \otimes \mathbb{Z}/p^r A \xrightarrow{p^{r-1}} A \otimes \mathbb{Z}/p^r \mathbb{Z} \to A \otimes \mathbb{Z}(p^\infty).$$
Thus, the torsion subgroup T of A maps to 0.

If $B = A/T$ we can factor
$$\begin{array}{ccc} A & \to & A \otimes \mathbb{Z}(p^\infty) \\ \downarrow & \nearrow & \downarrow \\ B & \to & B \otimes \mathbb{Z}(p^\infty) \end{array}.$$

Clearly, pA and pB map to 0 in $A \otimes \mathbb{Z}(p^\infty)$.

We claim that the kernel of $B \to B \otimes \mathbb{Z}(p^\infty)$ is exactly pB. But, if $b \varepsilon B$ and $b \mapsto 0$ in $B \otimes \mathbb{Z}(p^\infty)$, then $b \mapsto 0$ in some $B \otimes \mathbb{Z}/p^s\mathbb{Z}$. Hence, there exists $b_1 \varepsilon B$ with $p^{r-1}b = p^r b_1$. Since B is torsion free, $b = pb_1$, that is, this kernel is exactly pB. $\square$

Exercise

(1) This exercise illustrates a problem with convergence of the mod p homotopy Bockstein spectral sequence:

(a) Show that, for finite r,
$$E_\pi^r\left(\prod_\alpha X_\alpha\right) = \prod_\alpha E_\pi^r(X_\alpha).$$

(b) Show that
$$E_\pi^r(K(\mathbb{Z}/p^s\mathbb{Z}, k))_* = 0$$
for $r \geq s$.

(c) If
$$A = \prod_{n \geq 1} \mathbb{Z}/p^n\mathbb{Z}$$
show that A is p-complete and
$$(A/\text{torsion}) \otimes \mathbb{Z}/p\mathbb{Z} \neq 0.$$

(d) Since it is reasonable to expect that $E_\pi^\infty(X)_*$, if it exists, should commute with products, there seems to be no reasonable value for $E_\pi^\infty(K(A,k))_*$ even though $K(A,n)$ is p-complete.

7.5 Samelson products in the Bockstein spectral sequence

Let p be an odd prime. Given spaces A, B, C, a pairing of mod p homotopy Bockstein spectral sequences
$$E_\pi^r(A)_* \otimes E_\pi^r(B)_* \to E_\pi^r(C)_*$$
is a collection of bilinear pairings
$$\langle\,,\,\rangle : \pi_*(A; \mathbb{Z}/p^r\mathbb{Z}) \otimes \pi_*(B; \mathbb{Z}/p^r\mathbb{Z}) \to \pi_*(C; \mathbb{Z}/p^r\mathbb{Z}), \quad x \otimes y \mapsto \langle x, y\rangle$$
which are compatible with reduction maps,
$$\rho : \pi_*(\ ; \mathbb{Z}/p^{r+s}\mathbb{Z}) \to \pi_*(\ ; \mathbb{Z}/p^r\mathbb{Z}), \quad \rho\langle a,b\rangle = \langle\rho a, \rho b\rangle,$$
and which satisfy the Bockstein derivation property
$$\beta[a,b] = [\beta a, b] + (-1)^{\deg(a)}[a, \beta b]$$
where $\beta : \pi_*(\ ; \mathbb{Z}/p^r\mathbb{Z}) \to \pi_{*-1}(\ ; \mathbb{Z}/p^r\mathbb{Z})$ is the Bockstein associated to the exact coefficient sequence $\mathbb{Z}/p^r\mathbb{Z} \to \mathbb{Z}/p^{2r}\mathbb{Z} \to \mathbb{Z}/p^r\mathbb{Z}$.

Given such, it defines pairings of E^r terms,
$$E_\pi^r(A)_* \otimes E_\pi^r(B)_* \xrightarrow{\langle\,,\,\rangle_r} E_\pi^r(C)_*,$$
via:

7.5 Samelson products in the Bockstein spectral sequence

if
$$a = p^{r-1}a_1 \ \varepsilon E^r \pi(A)_* = p^{r-1}\pi_*(A; \mathbb{Z}/p^r\mathbb{Z}),$$
$$b = p^{r-1}b_1 \ \varepsilon E^r \pi(B)_* = p^{r-1}\pi_*(B; \mathbb{Z}/p^r\mathbb{Z}),$$

then
$$\langle a, b \rangle_r = p^{r-1}\langle a_1, b_1 \rangle.$$

Warning: The r in the notation of the above pairing $\langle \ , \ \rangle_r$ is not to be confused with the r used in Sections 6.10 and 6.11 for relative Samelson products. Just as we have dropped the use of that r, we shall eventually drop this r and write just $\langle a, b \rangle = \langle a, b \rangle_r$. But, until the end of this section, doing this would cause great confusion.

Bilinearity of the original pairings implies that this is a well defined bilinear pairing of E^r terms. One easily checks the derivation property for the r-th Bockstein differentials

$$\beta^r \langle a, b \rangle_r = \langle \beta^r a, b \rangle_r + (-1)^{\deg(a)} \langle a, \beta^r b \rangle_r.$$

Thus, the bilinear pairing of E^r terms induces in the usual way a pairing of the E^{r+1} terms and this is the same as defined above.

Of course, principal examples of such pairings are the internal Samelson products, both absolute and relative:

$$[\ ,\ ] : \pi_*(G; \mathbb{Z}/p^r\mathbb{Z}) \otimes \pi_*(G; \mathbb{Z}/p^r\mathbb{Z}) \to \pi_*(G; \mathbb{Z}/p^r\mathbb{Z})$$

where G is a grouplike space and

$$[\ ,\ ] : \pi_*(\Omega E; \mathbb{Z}/p^r\mathbb{Z}) \otimes \pi_*(\Omega F; \mathbb{Z}/p^r\mathbb{Z}) \to \pi_*(\Omega F; \mathbb{Z}/p^r\mathbb{Z})$$

where $F \to E \to B$ is a fibration sequence of simply connected spaces.

And, if $F \to E \to B$ is a fibration sequence of simply connected spaces with B an H-space, we also have the stronger relative pairings

$$[\ ,\ ] : \pi_*(\Omega E; \mathbb{Z}/p^r\mathbb{Z}) \otimes \pi_*(\Omega E; \mathbb{Z}/p^r\mathbb{Z}) \to \pi_*(\Omega F; \mathbb{Z}/p^r\mathbb{Z}).$$

We obviously have

Proposition 7.5.1. *If p is a prime greater than 3, then the Samelson product pairings $[\ ,\ ]_r$ make $E_\pi^r(G)$ into a spectral sequence of differential graded Lie algebras for all $r \geq 1$. And $E^r(\Omega F) \to E^r(\Omega E)$ is a spectral sequence of extended Lie ideals (and of strong extended Lie ideals if B is an H-space).*

If $p = 3$, the pairings exist for all $r \geq 1$ and they are bilinear and anti-commutative. Furthermore, we have the surprising but easy result that the mod 3 homotopy Bockstein spectral sequence corrects bad phenomena as we progress to higher terms. More precisely, if we shall say that bilinear pairings with anti-commutativity and the Jacobi identity constitute a graded quasi Lie algebra, then we have

Proposition 7.5.2. *If G is a grouplike space, the mod 3 homotopy Bockstein spectral sequence $E^r \pi(G)_*$ is a spectral sequence of differential graded quasi Lie algebras for all $r \geq 2$. If $r \geq 3$, it is a spectral sequence of differential graded Lie algebras for $r \geq 3$. And, if $F \to E \to B$ is a fibration sequence of simply connected spaces, the analogous statements for relative products and products over the loops on an H-space are true.*

Proof: Let $f = 3^{r-1} f_1, g = 3^{r-1} g_1, h = 3^{r-1} h_1$ be elements of $E^r \pi(G)_*$. Note that

$$[g,h]_r = 3^{r-1}[g_1, h_1]$$

implies that

$$[f, [g,h]_r]_r = [3^{r-1} f_1, 3^{r-1}[g_1, h_1]]_r = 3^{r-1}[f_1, [g_1, h_1]].$$

Thus, if $r \geq 2$, the Jacobi element in E^r = the deviation from the truth of the Jacobi identity =

$$[f, [g,h]_r]_r + (-1)^{q(m+n)}[h, [f,g]_r]_r + (-1)^{m(n+q)}[g, [h,f]_r]_r$$
$$= 3^{r-1}([f_1, [g_1, h_1]] + (-1)^{q(m+n)}[h_1, [f_1, g_1]] + (-1)^{m(n+q)}[g_1, [h_1, f_1]])$$

equals 0 since all Jacobi elements have order 3. (See the Remark in Section 6.7.) Thus, the Jacobi identity is valid in E^r for $r \geq 2$.

Lemma 7.5.3. *If x has odd degree, anti-commutativity and the Jacobi identity imply $3[x, [x, x]] = 0$.*

Recall that the proof of this is:

$$[x, [x, x]] = [[x, x]x] - [x, [x, x]] = -2[x, [x, x]].$$

Now suppose $f = 3^{r-1} f_1$ has odd degree in E^r.

$$[f, [f, f]_r]_r = 3^{r-1}[f_1, [f_1, f_1]].$$

But, since the deviation from the Jacobi identity has order 3,

$$3[f_1, [f_1, f_1]]$$

has order 3. Thus, if $r \geq 3$,

$$[f, [f, f]_r]_r = 0.$$

□

7.6 Mod p homology Bockstein spectral sequences

The mod p homology Bockstein spectral sequence may be defined in the same way that the homotopy Bockstein spectral sequence is defined, via an exact couple coming from the exact coefficient sequence

$$\mathbb{Z} \xrightarrow{p} \mathbb{Z} \xrightarrow{\rho} \mathbb{Z}/p\mathbb{Z}.$$

The homology Bockstein exact couple is

$$\begin{array}{ccc} H_*(X) & \xrightarrow{p} & H_*(X) \\ \partial \nwarrow & & \swarrow \rho \\ & H_*(X;\mathbb{Z}/p\mathbb{Z}) & \end{array}.$$

Derivation of this exact couple leads to the mod p homology Bockstein spectral sequence

$$\begin{aligned} E_H^r(X)_* &= \mathbb{Z}_H^r(X)_*/B_H^r(X)_* \\ &= \frac{\partial^{-1}(\mathrm{im}(p^{r-1} : H_*(X) \to H_*(X)))}{\rho(\ker(p^{r-1} : H_*(X) \to H_*(X)))} \\ &= \mathrm{im}\ (p^{r-1} : H_*(X;\mathbb{Z}/p^r\mathbb{Z}) \to H_*(X;\mathbb{Z}/p^r\mathbb{Z})). \end{aligned}$$

Note that $E_H^1(X)_* = H_*(X;\mathbb{Z}/p\mathbb{Z})$ and that the first differential β^1 is just the usual Bockstein, which is exactly analogous to the homotopy Bockstein spectral sequence.

The differential $\beta^r : E_H^r(X)_* \to E_H^r(X)_{*-1}$ is defined either by the relation

$$\beta^r = j \circ p^{1-r} \circ \partial$$

or by the Bockstein associated to the exact sequence $\mathbb{Z}/p^r\mathbb{Z} \xrightarrow{\eta} \mathbb{Z}/p^{2r}\mathbb{Z} \xrightarrow{\rho} \mathbb{Z}/p^r\mathbb{Z}$, that is,

$$\beta^r(p^{r-1}x) = p^{r-1}\beta(x).$$

The first description of the differential leads to a description in terms of chain representatives of homology classes,

$$\beta^r(c) = \rho\left(\frac{\partial(c)}{p^r}\right).$$

Just as before, we have the universal coefficient exact sequence

$$0 \to p^{r-1}H_*(X) \otimes \mathbb{Z}/p^r\mathbb{Z} \xrightarrow{\rho} E_H^r(X)_* \xrightarrow{\beta} p^{r-1}\mathrm{Tor}^{\mathbb{Z}}(H_{*-1}(X),\mathbb{Z}/p^r\mathbb{Z}) \to 0.$$

Thus, in the case where the integral homology is finitely generated in each degree, we have the analogous description of the homology Bockstein differentials in terms of the decomposition of the integral homology into a direct sum of cyclic groups.

In the case where $H_*(X)$ has p-torsion of bounded order in each degree, we have convergence and
$$E_H^\infty(X)_* \cong (H_*(X)/\text{torsion}) \otimes \mathbb{Z}/p\mathbb{Z}.$$
We have the usual reduction maps
$$\varrho : H_*(X; \mathbb{Z}/p^r\mathbb{Z}) \to E_H^r(X)_*, \quad \varrho(x) = p^{r-1}x, \quad \beta^r \circ \varrho = \varrho \circ \beta$$
where β is the Bockstein associated to the short exact sequence $\mathbb{Z}/p^r\mathbb{Z} \to \mathbb{Z}/p^{2r}\mathbb{Z} \to \mathbb{Z}/p^r\mathbb{Z}$.

And we have
$$\varrho = \varrho \circ \rho : H_*(X) \to H_*(X; \mathbb{Z}/p^r\mathbb{Z}) \to E_H^r(X)_*, \quad \varrho(x) = p^{r-1}\rho(x),$$
$$\beta^r \circ \varrho = 0.$$

The various Hurewicz maps
$$\varphi : \pi_*(X) \to H_*(X)$$
$$\varphi : \pi_*(X; \mathbb{Z}/p^r\mathbb{Z}) \to H_*(X; p^r\mathbb{Z})$$
commute with the Bocksteins β associated to exact coefficient sequences, reduction maps ρ associated to coefficient maps $\mathbb{Z}/p^{r+s}\mathbb{Z} \to \mathbb{Z}/p^r\mathbb{Z}$, and expansion maps η associated to coefficient maps $\mathbb{Z}/p^s\mathbb{Z} \to \mathbb{Z}/p^{r+s}\mathbb{Z}$.

Hence we have an induced Hurewicz homomorphism of mod p Bockstein spectral sequences
$$\varphi : E_\pi^r(X)_* \to E_H^r(X)_*$$
compatible of course with the structure maps
$$\beta^r \circ \varphi = \varphi \circ \beta^r, \quad \varrho \circ \varphi = \varphi \circ \varrho.$$
The mod p homology Bockstein spectral sequence has a differential coalgebra structure. This is based on the Eilenberg–Zilber maps
$$\nabla : S_*(X; \mathbb{Z}/p^r\mathbb{Z}) \otimes S_*(Y; \mathbb{Z}/p^r\mathbb{Z}) \to S_*(X \times Y; \mathbb{Z}/p^r\mathbb{Z})$$
which induce cross product pairings
$$H_*(X; \mathbb{Z}/p^r\mathbb{Z}) \otimes H_*(Y; \mathbb{Z}/p^r\mathbb{Z}) \to H_*(S_*(X; \mathbb{Z}/p^r\mathbb{Z}) \otimes S_*(Y; \mathbb{Z}/p^r\mathbb{Z}))$$
$$\xrightarrow{\nabla_*} H_*(X \times Y; \mathbb{Z}/p^r\mathbb{Z})$$
defined by
$$[x] \otimes [y] \mapsto [x \otimes y] \mapsto \nabla_*([x \otimes y]) = x \times y.$$
One checks that this is compatible with reduction maps and that the Bockstein differentials have the derivation property. Hence, there is cross product pairing of

7.6 Mod p homology Bockstein spectral sequences

spectral sequences
$$E_H^r(X)_* \otimes E_H^r(Y)_* \to E_H^r(X \times Y)_*, \quad x \otimes y \mapsto x \times y$$
such that
$$\beta^r(x \times y) = \beta^r(x) \times y + (-1)^{\deg(x)} x \times \beta^r(y).$$
Clearly, the pairing is an isomorphism for $r = 1$ and thus for all $r \geq 1$.

Definition 7.6.1. The differential coalgebra structure on $E_H^r(X)_*$ is induced by the diagonal $\Delta : X \to X \times X$:
$$E_H^r(X)_* \xrightarrow{\Delta_*} E_H^r(X \times X) \xleftarrow{\cong} E_H^r(X) \otimes E_H^r(X)_*.$$

Now suppose G is an H-space with multiplication $\mu : G \times G \to G$.

Definition 7.6.2. The Pontrjagin algebra structure on $E_H^r(G)_*$ is induced by the multiplication:
$$E_H^r(X)_* \otimes E_H^r(X)_* \xrightarrow{\cong} E_H^r(X \times X)_* \xrightarrow{\mu_*} E_H^r(X)_*,$$
$$x \otimes y \mapsto xy = \mu_*(x \times y) = \mu_*(x \otimes y)$$

Thus, $E_H^r(G)$ is a spectral sequence of differential Hopf algebras.

The Hurewicz map is compatible with Lie structures:

Proposition 7.6.3. *If p is a prime greater than 3, the Hurewicz map is a morphism of differential graded Lie algebras*
$$\varphi : E_\pi^r(X)_* \to E_H^r(X)_*,$$
$$\varphi[x, y] = [\varphi x, \varphi y] = (\varphi x)(\varphi y) - (-1)^{\deg(x)\deg(y)}(\varphi y)(\varphi x)$$

Remark. If $p = 3$, we need $r \geq 2$ to get a quasi-Lie algebra structure on the mod p homotopy Bockstein spectral sequence and $r \geq 3$ to get a Lie algebra structure. But the Hurewicz map is still a homomorphism of bracket structures in all dimensions.

Exercises

(1) Use the universal example $P^m(p^r)$ to show that the image of the mod p Hurewicz map of Bockstein spectral sequences
$$\varphi : E_\pi^r(X)_* \to E_H^r(X)_*$$
is contained in the module of primitives $PE_H^r(X)_*$ for all $r \geq 1$ if p is an odd prime or if $* \geq 3$.

(2) Let $\Omega F \to \Omega E \to \Omega B$ be a fibration sequence of connected loop spaces.

(a) Show that the commutator
$$[\ ,\ ]: \Omega E \times \Omega F \to \Omega F$$
induces a bracket pairing
$$H_*(\Omega E; \mathbb{Z}/p^r\mathbb{Z}) \otimes H_*(\Omega F; \mathbb{Z}/p^r\mathbb{Z}) \to H_*(\Omega F; \mathbb{Z}/p^r\mathbb{Z})$$

(b) Show that there is a bracket pairing of spectral sequences
$$E_H^r(\Omega E)_* \otimes E_H^r(\Omega F)_* \to E_H^r(\Omega F)_*.$$

(c) If p is an odd prime, show that the Hurewicz map induces a map of extended ideals (of Lie algebras if $p > 3$ or $r \geq 3$)
$$\begin{array}{ccc} E_\pi^r(\Omega F)_* & \to & E_H^r(\Omega F)_* \\ \downarrow \varphi & & \downarrow \varphi \\ E_\pi^r(\Omega E)_* & \to & E_H^r(\Omega E)_* \end{array}$$

(3) Suppose X is a space with integral homology $H_*(X)$ finitely generated in each degree.

(a) If $H_*(X)$ is torsion free, show that $H_*(\Omega\Sigma X)$ is torsion free.

(b) If the reduced homology $\overline{H}_*(X)$ consists entirely of p-torsion of order bounded by p^r, show that the same is true of the reduced homology $\overline{H}_*(\Omega\Sigma X)$.

7.7 Mod p cohomology Bockstein spectral sequences

The short exact sequence $\mathbb{Z} \xrightarrow{p} \mathbb{Z} \xrightarrow{\rho} \mathbb{Z}/p\mathbb{Z}$ leads to the mod p cohomology Bockstein exact couple

$$\begin{array}{ccc} H^*(X;\mathbb{Z}) & \xrightarrow{p} & H^*(X;\mathbb{Z}) \\ {}_{\overline{\beta}}\nwarrow & & \swarrow_\rho \\ & H^*(X;\mathbb{Z}/p\mathbb{Z}) & \end{array}$$

where the cohomology Bockstein $\overline{\beta}$ has degree $+1$.

Deriving this couple leads to the mod p cohomology Bockstein spectral sequence
$$E_r^H(X)^*, \beta_r$$
with
$$E_1^H(X)^* = H^*(X;\mathbb{Z}/p\mathbb{Z}).$$

The first Bockstein differential β_1 is the usual Bockstein associated to the coefficient sequence $\mathbb{Z}/p\mathbb{Z} \to \mathbb{Z}/p^2\mathbb{Z} \to \mathbb{Z}/p\mathbb{Z}$. The differentials
$$\beta_r : E_r^H(X)^* \to E_r^H(X)^{*+1}$$

7.7 Mod p cohomology Bockstein spectral sequences

are of degree $+1$ and give
$$H(E_r^H(X)^*, \beta_r) \cong E_{r+1}^H(X)^*.$$
On cochains, the differential is represented by $\beta_r(c) = \rho(\frac{\delta c}{p^r})$.

For a map $f : A \to B$ of abelian groups, we set
$$\operatorname{coim}(f) = A/\ker(f).$$
Of course, this is isomorphic to the image of f but it can be more convenient in some contexts.

For example, in order to fit nicely with duality, we use
$$E_r^H(X) = \operatorname{coim}\,(H^*(X; \mathbb{Z}/p^r\mathbb{Z}) \xrightarrow{p^{r-1}} H^*(X; \mathbb{Z}/p^r\mathbb{Z}))$$
and the differential β_r is induced by the cohomology Bockstein $\overline{\beta}$ associated to the short exact coefficient sequence $\mathbb{Z}/p^r\mathbb{Z} \to \mathbb{Z}/p^{2r}\mathbb{Z} \to \mathbb{Z}/p^r\mathbb{Z}$. That is,
$$\beta_r \overline{a} = \overline{\overline{\beta a}}, \quad \text{where} \quad a \,\varepsilon\, H^*(X; \mathbb{Z}/p^r\mathbb{Z})$$
and $\overline{a}$ and $\overline{\overline{\beta a}}$ denote the classes in the coimages.

Note that there is a unique embedding $\mathbb{Z}/p^r\mathbb{Z} \subseteq \mathbb{Q}/\mathbb{Z}$ into the elements of order p^r where $\mathbb{Q}/\mathbb{Z}$ is a divisible abelian group and therefore injective. Thus
$$\begin{aligned} H^*(X; \mathbb{Z}/p^r\mathbb{Z}) &= H^*\operatorname{Hom}(S_*(X), \mathbb{Z}/p^r\mathbb{Z}) \\ &= H^*\operatorname{Hom}(S_*(X; \mathbb{Z}/p^r\mathbb{Z}), \mathbb{Z}/p^r\mathbb{Z}) \\ &= H^*\operatorname{Hom}(S_*(X; \mathbb{Z}/p^r\mathbb{Z}), \mathbb{Q}/\mathbb{Z}) \\ &= \operatorname{Hom}(H_*(X; \mathbb{Z}/p^r\mathbb{Z}), \mathbb{Q}/\mathbb{Z}) \\ &= \operatorname{Hom}(H_*(X; \mathbb{Z}/p^r\mathbb{Z}), \mathbb{Z}/p^r\mathbb{Z}) \end{aligned}$$
the penultimate equation holding since $\mathbb{Q}/\mathbb{Z}$ is injective.

Let D be an injective module and set
$$A^* = \operatorname{Hom}(A, D).$$
We leave the proof of the following to the exercises.

Lemma 7.7.1. *If $f : A \to B$ is a homomorphism and $f^* : B^* \to A^*$ is the dual, then*
$$(\operatorname{im} f)^* \cong \operatorname{coim} f^*$$
via
$$\overline{b^*}(fa) = b^*(fa) = (f^*b^*)(a), \quad b^* \,\varepsilon\, B^*, \quad a \,\varepsilon\, A.$$

If we let f be the map $p^{r-1} : H_*(X; \mathbb{Z}/p^r\mathbb{Z}) \to H_*(X; \mathbb{Z}/p^r\mathbb{Z})$, we get that the mod p homology and cohomology Bockstein spectral sequences are dual in the following way:

Corollary 7.7.2.
$$E_r^H(X)^* \cong \text{Hom}(E_H^r(X)_*, Q/\mathbb{Z})$$
$$\cong \text{Hom}(E_H^r(X)_*, \mathbb{Z}/p\mathbb{Z})$$

and

$$\beta_r \cong \text{Hom}(\beta^r, Q/\mathbb{Z}) \cong \text{Hom}(\beta^r, \mathbb{Z}/p\mathbb{Z}).$$

Recall the Alexander–Whitney maps

$$H^*(X; \mathbb{Z}/p^r\mathbb{Z}) \otimes H^*(Y; \mathbb{Z}/p^r\mathbb{Z}) \to H^*(X \times Y; \mathbb{Z}/p^r\mathbb{Z}).$$

These give a pairing of Bockstein spectral sequences

$$E_r^H(X)^* \otimes E_r^H(Y)^* \to E_r^H(X \times Y)^*$$

compatible with the differentials. If the mod p homologies are finitely generated in each degree, these maps are isomorphisms.

Hence we have the cup product pairings

$$E_r^H(X)^* \otimes E_r^H(X)^* \to E_r^H(X \times X)^* \xrightarrow{\Delta^*} E_r^H(X)^*$$

making $E_r^H(X)^*$ a spectral sequence of graded commutative algebras. This algebra structure in $E_r^H(X)^*$ is dual to the coalgebra structure in $E_H^r(X)_*$ via

$$(x \cup y)(z) = (x \otimes y)(\Delta_* z), \quad x, y \: \varepsilon E_r^H(X)^*, \quad z \: \varepsilon E_H^r(X)_*.$$

We often write $xy = x \cup y = \Delta^*(x \otimes y)$.

Let G be an H-space with multiplication $\mu : G \times G \to G$. If G has finitely generated mod p homology, we get a spectral sequence of Hopf algebras via the diagonals

$$\mu^* : E_r^H(G)^* \xrightarrow{\mu^*} E_r^H(G \times G) \xleftarrow{\cong} E_r^H(G)^* \otimes E_r^H(G)^*.$$

These Hopf algebras are dual to the Hopf algebras $E_H^r(X)_*$.

Exercises

(1) Show that $\mathbb{Z}/p^r\mathbb{Z}$ is an injective module in the category of $\mathbb{Z}/p^r\mathbb{Z}$ modules.

(2) Prove 7.7.1 for injective modules D.

(3) (a) Show that we have a universal coefficient exact sequence

$$0 \to p^{r-1} H^*(X) \otimes \mathbb{Z}/p^r \mathbb{Z} \xrightarrow{\rho} E_r^H(X)^*$$
$$\xrightarrow{\beta} p^{r-1} \operatorname{Tor}^{\mathbb{Z}}(H^{*+1}(X), \mathbb{Z}/p^r \mathbb{Z}) \to 0.$$

(b) In the case where the integral cohomology is finitely generated in each degree, describe the cohomology Bockstein differentials in terms of the decomposition of the integral cohomology into a direct sum of cyclic groups.

(c) In the case where $H^*(X)$ has p–torsion of bounded order in each degree, show that we have convergence and that

$$E_\infty^H(X)_* \cong (H^*(X)/\text{torsion}) \otimes \mathbb{Z}/p\mathbb{Z}.$$

7.8 Torsion in H-spaces

In this section present some applications of the homology and cohomology Bockstein spectral sequences and of the theory of Hopf algebras to the study of finite H-spaces, that is, to H-spaces which are finite cell complexes. These applications are due to William Browder [19].

Let p be a prime. Begin by recalling without proof the following result from Milnor and Moore [90].

Let A be a connected Hopf algebra over a perfect field of characteristic p (for example, over the field $\mathbb{Z}/p\mathbb{Z}$).

Let $\xi : A \to A$ be the Frobenius map (restricted to even degrees if p is odd), $\xi(a) = a^p$ for $a \, \varepsilon A$, and let ξA be the subHopf algebra of p-th powers (of even degree elements if p is odd).

As usual, $P(A)$ denotes the module of primitive elements, $D(A) = \overline{A} \cdot \overline{A}$ denotes the module of decomposable elements, and $Q(A) = \overline{A}/D(A)$ denotes the module of indecomposable elements.

Proposition 7.8.1. *If A is a connected Hopf algebra over a perfect field and A has a (graded) commutative and associative multiplication, then the sequence*

$$0 \to P(\xi A) \to P(A) \to Q(A)$$

is exact, that is,

$$P(A) \cap D(A) = P(\xi A).$$

Corollary 7.8.2. *Let A be a differential Hopf algebra with a differential d of degree ± 1. If A is isomorphic as an algebra to an exterior algebra generated by odd degree elements, then $d = 0$.*

Proof: Let n be the first degree in which d is nonzero. Then n must be odd. But it is easy to see that
$$d(A_n) \subseteq P(A)$$
and, of course,
$$d(A_n) \subseteq A_{n-1} \subseteq D(A)$$
since this is in an even degree. Hence,
$$d(A_n) \subseteq P(\xi A) = 0.$$
Thus, $d = 0$ on all of A. □

Since $E_1^H(G)^* = E_\infty^H(G)^*$ in the mod p cohomology spectral sequence, Corollary 7.8.2 and Exercise 1 below give

Corollary 7.8.3. *Let p be a prime and G a finite H-space. Then the integral cohomology $H^*(G)$ has no p-torsion if and only if the mod p cohomology algebra $H^*(G; \mathbb{Z}/p\mathbb{Z})$ is an exterior algebra generated by elements of odd degree.*

Our next result uses the fact that, for an H-space with integral homology of finite type, the homology and cohomology Bockstein spectral sequences are dual Hopf algebras.

We adopt the following notation: Let
$$x \ \varepsilon E_H^r(G)_*$$
and
$$\bar{x} \ \varepsilon E_r^H(G)^*.$$
Write
$$\bar{x}(x) = \langle \bar{x}, x \rangle \ \varepsilon \mathbb{Z}/p\mathbb{Z}$$
to indicate the duality
$$E_r^H(G)^* = \operatorname{Hom}(E_H^r(G)_*; \mathbb{Z}/p\mathbb{Z}).$$

If $x \ \varepsilon E_H^r(G)_*$ is a β^r cycle and $\bar{x} \ \varepsilon E_r^H(G)^*$ is a β_r cocycle, we shall indicate by $[x] \ \varepsilon E_H^{r+1}(G)^*$ and $[\bar{x}] \ \varepsilon E_{r+1}^H(G)^*$ the respective homology and cohomology classes.

The following is an easy and well known exercise:

7.8 Torsion in H-spaces

Lemma 7.8.4. *The pairing is well defined on homology and cohomology classes*

$$[\overline{x}]([x]) = \overline{x}(x).$$

Just in case of nonassociativity, define $x^n = x^{n-1} \cdot x$. The main result of this section is:

Browder's implication theorem 7.8.5. *Let G be an H-space with integral homology of finite type. If there exists $y \; \varepsilon E_H^r(G)_{2n+1}$ such that*

$$\beta^r y = x \neq 0 \quad \text{and} \quad x \; \varepsilon PE_H^r(G)_{2n} =$$

the module of primitive elements, then one of the following must hold:

(1)
$$x^p \neq 0$$

and, of course,

$$x^p \; \varepsilon PE_H^r(G)_{2np} \text{ and } \beta^r(x^{p-1}y) = x^p$$

or

(2)
$$x^p = 0 \text{ and there exists } z \neq 0,$$

$$z \; \varepsilon PE_H^{r+1}(G)_{2np} \text{ such that } \beta^{r+1}([x^{p-1}y]) = z.$$

Remark. In case 2 above we observe that $\beta^r(x^{p-1}y) = x^p = 0$ so that $[x^{p-1}y]$ exists in $E_H^{r+1}(G)_{2np+1}$.

The remarkable thing about this theorem is that it clearly can be iterated to imply the so-called infinite implication:

For all $k \geq 0$ the integral homology $H_{2np^k}(G)$ contains nonzero p-torsion of order $\geq p^r$.

Since this can never happen in a finite H-space, we get

Corollary 7.8.6. *If G is a finite H-space, then the image of the reduction map*

$$\rho : H_*(G) \to H_*(G; \mathbb{Z}/p\mathbb{Z})$$

contains no nonzero primitive elements of even degree.

Proof: Suppose α is an even degree nonzero primitive element in the image of ρ. Then α represents a permanent cycle in the Bockstein spectral sequence. If there exists y such that $\beta^r y = x$, we have a contradiction by means of infinite implication. If this never happens, then x survives to give a nonzero even dimensional primitive element in the homology algebra $E_H^\infty(G)_*$.

Since $P(A)^* = Q(A^*)$ for any finite type Hopf algebra, this contradicts Exercise 1 below which says that $E_\infty^H(G)^*$ is an exterior algebra generated by odd degree elements. □

Since spherical classes are primitive, we get

Corollary 7.8.7. *Let G be a finite H-space and let*

$$\rho \circ \varphi : \pi_*(G) \to H_*(G) \to H_*(G; \mathbb{Z}/p\mathbb{Z})$$

be the mod p reduction of the integral Hurewicz homomorphism. Then the image of this map is zero in even degrees.

The following is a generalization of a theorem of E. Cartan.

Corollary 7.8.8. *If G is a finite H-space, then the first nonvanishing homotopy group $\pi_*(G)$, if any, occurs in an odd degree.*

Proof: The first time $\pi_* G$ is nonzero must be the first time that $H_*(G) \otimes \mathbb{Z}/p\mathbb{Z} = H_*(G; \mathbb{Z}/p\mathbb{Z})$ is nonzero for some prime p. This implies that there is a nonzero primitive element in that degree. □

Cartan's result was that $\pi_2(G) = 0$ when G is a Lie group.

The remainder of this section is devoted to the proof of Browder's implication theorem 7.8.5.

We can assume that we are in case (2) of Theorem 7.8.5:

$$y = \beta^r x, \quad \beta^r(x^{p-1}y) = x^p = 0.$$

We begin with an algebraic lemma with several parts:

Let x and y be as above and let $\bar{x} \in E_r^H(G)^{2n}$, $\bar{y} = \beta_r \bar{x}$.

Lemma 7.8.9.

(a)
$$\Delta(x^k) = \sum_{i+j=k} (i,j) x^i \otimes x^j.$$

(b)
$$(\bar{x}^k)(x^k) = k!\{\bar{x}(x)\}^k.$$

(c) *If the characteristic is p and*
$$\bar{x}(x) \neq 0,$$
then
$$(\bar{x}^{p-1}\bar{y})(x^{p-1}y) \neq 0.$$

(d) *If the characteristic is p, then $[x^{p-1}y]$ is primitive in $E_H^{r+1}(G)_{2np+1}$.*

Proof:

(a) This part is proved in the usual way by induction on k.

(b) Use induction and

$$\begin{aligned}(\overline{x}^k)(x^k) &= (\Delta^*(\overline{x}^{k-1} \otimes \overline{x}))(x^k)\\ &= (\overline{x}^{k-1} \otimes \overline{x})(\Delta(x^k))\\ &= (\overline{x}^{k-1} \otimes \overline{x})(\Sigma_{i+j=k}(i,j)x^i \otimes x^j)\\ &= (k-1,1)((k-1)!\{\overline{x}(x)\}^k) = k!\{\overline{x}(x)\}^k.\end{aligned}$$

(c) Note that

$$\overline{y}(y) = (\beta_r \overline{x})(y) = \overline{x}(x) \neq 0.$$

Let

$$\Delta(y) = y \otimes 1 + 1 \otimes y + \sum y_\alpha \otimes y_\beta.$$

Then

$$\begin{aligned}(\overline{x}^{p-1}\overline{y})&(x^{p-1}y)\\ &= \Delta^*(\overline{x}^{p-1} \otimes \overline{y})(x^{p-1}y)\\ &= (\overline{x}^{p-1} \otimes \overline{y})(\Delta(x^{p-1}y))\\ &= (\overline{x}^{p-1} \otimes \overline{y})(\Delta(x^{p-1}) \cdot \Delta(y))\\ &= (\overline{x}^{p-1} \otimes \overline{y})(x^{p-1} \otimes y + \Sigma_{\deg(y_\beta)=1}(p-2,1)x^{p-2}y_\alpha \otimes xy_\beta)\\ &= \{(\overline{x}^{p-1})(x^{p-1})\}\{\overline{y}(y)\}\\ &\quad + \Sigma_{\deg(y_\beta)=1}(p-2,1)\{\overline{x}^{p-1}(x^{p-2}y_\alpha)\}\{\overline{y}(xy_\beta)\}\end{aligned}$$

and so it suffices to show

$$\overline{y}(xy_\beta) = (\beta_r\overline{x})(xy_\beta) = \overline{x}(\beta^r(xy_\beta)) = \overline{x}(x \cdot \beta^r y_\beta) = \overline{x}(0) = 0$$

by Exercise 3(b).

(d) Let

$$a = \sum_{i+j=p-2}(-1)^j x^i y \otimes x^j y.$$

Then

$$\begin{aligned}\beta^r a &= \Sigma_{i+j=p-1}(-1)^j\{x^i \otimes x^j y + x^i y \otimes x^j\} - \{x^{p-1}y \otimes 1 + 1 \otimes x^{p-1}y\}\\ &= \Delta(x^{p-1}) \cdot \Delta(y) - \{x^{p-1}y \otimes 1 + 1 \otimes x^{p-1}y\}\end{aligned}$$

using the fact that
$$\binom{p-1}{j} = (-1)^j \mod p.$$

Observe that $\beta^r \{\Delta(y) - y \otimes 1 - 1 \otimes y\} = 0$ since $\beta^r y = x$ is primitive. Thus, if
$$b = \Delta(x^{p-2}y)\{\Delta(y) - y \otimes 1 - 1 \otimes y\},$$
we get
$$\beta^r b = \beta^r \{\Delta(x^{p-1})y\}\{\Delta(y) - y \otimes 1 - 1 \otimes y\}$$
$$= \Delta(x^{p-1}) \cdot \{\Delta(y) - y \otimes 1 - 1 \otimes y\}$$

Therefore,
$$\Delta(x^{p-1}y) = \Delta(x^{p-1}) \cdot \{\Delta(y)\}$$
$$= \Delta(x^{p-1}) \cdot \{\Delta(y) - y \otimes 1 - 1 \otimes y\} + \Delta(x^{p-1})(y \otimes 1 + y \otimes 1)$$
$$= \beta^r b + \beta^r a + \{x^{p-1}y \otimes 1 + 1 \otimes x^{p-1}y\}.$$

$\square$

We need the following lemma on the action of the mod 2 Steenrod algebra in the cohomology of an H-space:

Lemma 7.8.10. *Let x and y be mod 2 homology classes of dimensions $2n$ and $2n+1$, respectively, and let $\bar{z}$ be a cohomology class of dimension $2n+1$. If x is primitive, then*
$$(Sq^{2n}\bar{z})(xy) = 0.$$

Proof: Consider $\bar{a} \otimes \bar{b}$ where $\bar{a}$ and $\bar{b}$ are cohomology classes with $\deg(\bar{a}) + \deg(\bar{b}) = 2n+1$. The Cartan formula and the fact that Steenrod operations vanish on classes of smaller degree imply that
$$Sq^{2n}(\bar{a} \otimes \bar{b}) = \bar{a}^2 \otimes (Sq^{2n-\deg(\bar{a})}\bar{b}) + (Sq^{2n-\deg(\bar{b})}\bar{a}) \otimes \bar{b}^2.$$

Now
$$Sq^{2n}(\bar{a} \otimes \bar{b})(x \otimes y)$$
$$= \{\bar{a}^2(x)\}\{(Sq^{2n-\deg(\bar{a})}\bar{b})(y)\} + \{(Sq^{2n-\deg(\bar{b})}\bar{a})(x)\}\{\bar{b}^2(y)\}$$
$$= 0 + 0 = 0$$

since:
$$\bar{a}^2(x) = \Delta^*(\bar{a} \otimes \bar{a})(x) = (\bar{a} \otimes \bar{a})(\Delta_*(x)) = 0$$

for primitive x, and
$$\bar{b}^2(y) = 0$$
for odd degree y.

Hence
$$(Sq^{2n}\bar{z})(xy) = (Sq^{2n}\bar{z})\mu_*(x \otimes y) = (Sq^{2n}(\mu^*\bar{z})(x \otimes y) = 0.$$

□

Browder's implication theorem is based on H. Cartan's computations [22] of the mod p cohomology of Eilenberg–MacLane spaces. These results can be found on page 142 of Harper's book [50] where they are referred to as Browder's theorem: Let
$$\iota_{2n} \ \varepsilon H^{2n}(K(\mathbb{Z}/p^r\mathbb{Z}, 2n); \mathbb{Z}/p\mathbb{Z}) = E_r^H(K(\mathbb{Z}/p^r\mathbb{Z}, 2n))^{2n}$$
be a generator and let $\eta = \beta_r \iota$.

If p is an odd prime:
$$\beta_r(\iota^p) = 0, \ \beta_r(\iota^{p-1}\eta) = 0$$
and the classes in E_{r+1} satisfy
$$\beta_{r+1}([\iota^p]) = [\iota^{p-1}\eta] \neq 0.$$

If $p = 2$ and $r > 1$, the situation is similar:
$$\beta_r(\iota^2) = 0, \ \beta_r(\iota\eta) = \eta^2 = Sq^{2n+1}(\eta) = Sq^1 Sq^{2n}(\eta) = \beta_1 Sq^{2n}(\eta) = 0$$
in E_r and the classes in E_{r+1} satisfy
$$\beta_{r+1}([\iota^2]) = [\iota\eta] \neq 0.$$

If $p = 2$ and $r = 1$, a variation occurs:
$$\beta_1(\iota^2) = 0, \ \beta_1(\iota\eta) = \eta^2 = Sq^{2n+1}(\eta) = Sq^1 Sq^{2n}(\eta) = \beta_1 Sq^{2n}(\eta)$$
in E_r and now the classes in E_{r+1} satisfy
$$\beta_2([\iota^2]) = [\iota\eta + Sq^{2n}\eta] \neq 0.$$

Since Eilenberg–MacLane spaces constitute the universal examples for mod p cohomology, we get the same formulas as above when we replace ι and η by $\bar{x} \ \varepsilon E_r^H(G)^{2n}$ and $\bar{y} = \beta_r x \ \varepsilon E_r^H(G)^{2n+1}$, respectively.

Return to the homology Bockstein spectral sequence E^{r+1} and let $z = \beta^{r+1}([x^{p-1}y])$. Then z is primitive since $[x^{p-1}y]$ is.

If p is odd or $p = 2, r > 1$, then
$$[\bar{x}^p](z) = [\bar{x}^p](\beta^{r+1}[x^{p-1}y]) = (\beta_{r+1}[\bar{x}^p])([x^{p-1}y]) = [\bar{x}^{p-1}\bar{y}]([x^{p-1}y]) \neq 0$$

by Lemma 7.8.9(c).

If $p = 2, r = 1$, then

$$[\overline{x}^2](z) = [\overline{x}^2](\beta^2[xy]) = (\beta_2[\overline{x}^p])([xy]) = [\overline{xy} + Sq^{2n}\overline{y}]([x^{p-1}y]) \neq 0$$

by Lemma 7.8.9(c) and Lemma 7.8.10.

Thus

$$\beta^{r+1}[x^{p-1}y] = z \neq 0 \; \varepsilon PE_H^{r+1}(G)_{2pn}.$$

This completes the proof of the Browder implication theorem. $\square$

We point out two extreme examples of the Browder implication theorem phenomena:

(1) Let $G = \Omega P^{2n+2}(p^r)$, $n \geq 1$. Then

$$\beta^r v = u \neq 0 \; \varepsilon H_{2n}(\Omega P^{2n+2}(p^r); \mathbb{Z}/p\mathbb{Z})$$

and, since p-th powers never vanish, we get infinitely much homology torsion of order exactly p^r:

$$\beta^r(u^{p^r - 1}v) = u^{p^r} \text{ for all } r \geq 0.$$

Of course, we already knew this since $\Omega P^{2n+2}(p^r)$ is the loops on a suspension and the whole homology Bockstein spectral sequence is very simple.

(2) Let $G = K(\mathbb{Z}/p^r\mathbb{Z}, 2n)$, $n \geq 1$. Then there exists $\beta^r y = x \neq 0 \; \varepsilon H_{2n}(K(\mathbb{Z}/p^r\mathbb{Z}); \mathbb{Z}/p\mathbb{Z})$.

If p is an odd prime, the mod p cohomology is a polynomial tensor exterior algebra primitively generated by Steenrod operations on ι. Hence, the mod p homology consists of the tensor product of divided power algebras and exterior algebras. All p-th powers in mod p homology are 0.

If $p = 2$, the mod 2 cohomology is entirely polynomial and the mod 2 homology is entirely divided power. Again, all squares are 0.

In both cases, we have infinitely many elements

$$\beta^{r+s} w_s = z_s \neq 0 \; \varepsilon H_{2np^s}(K(\mathbb{Z}/p^r\mathbb{Z}); \mathbb{Z}/p\mathbb{Z})$$

for $s \geq 0$. Hence, there is torsion of order p^{r+s} in the integral homology $H_{2np^s}(K(\mathbb{Z}/p^r\mathbb{Z}, 2n))$.

This concludes this section except for the following items.

7.8 Torsion in H-spaces

In order to do the exercises below, recall without proof the following two classic theorems on Hopf algebras.

Theorem of Hopf-Borel. *Let A be a commutative and associative connected Hopf algebra over a field of characteristic 0. Then A is isomorphic as an algebra to the tensor product of an exterior algebra on elements of odd degree and a polynomial algebra on elements of even degree.*

Hopf proved that A is an exterior algebra if it is finite dimensional and the result was extended as above by Borel who also extended it as below to Hopf algebras over fields of finite characteristic.

Theorem of Borel. *Let A be a commutative and associative connected Hopf algebra over a perfect field of finite characteristic p and assume that A has a commutative and associative multiplication.*

If p is odd, then A is isomorphic as an algebra to the tensor product of an exterior algebra on elements of odd degree, a polynomial algebra on elements of even degree, and polynomial algebras on even degree elements truncated at powers of p.

If p is 2, then A is isomorphic as an algebra to the tensor product of a polynomial algebra and polynomial algebras on elements truncated at powers of 2.

Exercises

(1) Let G be a connected H-space and assume that the integral cohomology $H^*(G)$ is finitely generated in each degree and that the rational cohomology $H^*(G;Q)$ is finite dimensional.

  (a) Show that, in the mod p cohomology Bockstein spectral sequence, the E_∞ term $E_\infty^H(G)*$ is an exterior algebra on odd degree generators. (Hint: Use the fact that there are algebra maps
  $$H*(G;Q) \supseteq H^*(G)/\text{torsion} \to E_\infty^H(G)^*$$
  to observe that $H^*(G;Q)$ and $E_\infty^H(G)^*$ are Hopf algebras with identical Poincare polynomials. Then show by considering the roots of the polynomials that this is possible only if $E_\infty^H(G)^*$ has no even dimensional generators.)

  (b) Show that $H^*(G;Q)$ and $E_\infty^H(G)^*$ are exterior algebras generated by elements in the same odd degrees.

(2) Use
$$(x+1)^p = x^p + 1$$
$$= (x+1)(x^{p-1} - x^{p-2} + x^{p-3} - \cdots + (-1)^i x^i + \cdots + 1)$$

to show the following formula for binomial coefficients mod p:
$$\binom{p-1}{i} = (-1)^i \text{ modulo } p.$$

(3) Let A be a differential Hopf algebra with differential d of degree ± 1.

   (a) Show that $d(1) = 0$.

   (b) If A is connected and dx is in dimension 0, show that $dx = 0$.

8 Lie algebras and universal enveloping algebras

In this chapter we present the theory of graded Lie algebras. Although this theory is usually restricted to the case where 2 is a unit in the ground ring, this restriction is unnecessary and is removed here. The price one pays is the introduction of the additional structure of a squaring operation on odd degree classes. When 2 is a unit in the ground ring, this squaring map is a consequence of the Lie bracket structure. With this modification, the definition of graded Lie algebra satisfies a suitable version of the Poincare–Birkhoff–Witt theorem. The author suspects that this fact was known many years ago, at least to John Moore and to Frank Adams.

The Poincare–Birkhoff–Witt theorem has the immediate consequence that graded Lie algebras embed in their universal enveloping algebras. It also implies that the universal enveloping algebra of an ambient graded Lie algebra is a free module over the universal enveloping algebra of a sub graded Lie algebra.

The universal enveloping algebra has a Hopf algebra structure in which the graded Lie algebra is primitive. In fact, in prime characteristic p, the module of primitives is generated by the Lie elements and p^k-th powers of even degree Lie elements.

The free graded Lie algebras are characterized by the property that their universal enveloping algebras are tensor algebras. The fact that tensor algebras have global dimension one, when combined with the Poincare–Birkhoff–Witt theorem, yields the important result that subalgebras of free graded Lie algebras are themselves free graded Lie algebras.

It is important to determine the generators of subalgebras of free graded Lie algebras. In important cases, the module of generators is a free module over a tensor algebra via a Lie bracket action. When this is combined with an argument involving Euler-Poincare series, a complete determination of the module of generators results. This leads to algebraic analogues of the Hilton–Milnor theorem and of Serre's decomposition of the loops on an even dimensional sphere localized away from 2.

Some topological applications of this chapter will be given in the next chapter.

8.1 Universal enveloping algebras of graded Lie algebras

The definition of a graded Lie algebra has a complication when 2 is not a unit in the ground ring. It is necessary to add another operation to the usual bilinear Lie bracket operation. Namely, one must add a quadratic operation $x \mapsto x^2$ which is defined on odd degree classes. If 2 is a unit in the ground ring, then this squaring operation may be omitted since $x^2 = \frac{1}{2}[x,x]$ for odd degree x. More precisely,

Definition 8.1.1. A graded Lie algebra L is a graded R module together with two operations:

(1) bilinear pairings called Lie brackets
$$[\ ,\ ] : L_m \otimes L_n \to L_{m+n}, \quad x \otimes y \mapsto [x,y]$$
(2) and a quadratic operation called squaring defined on odd degree classes
$$(\ )^2 : L_k \to L_{2k}, \quad x \mapsto x^2$$
with k odd. The quadratic requirement is expressed in the identities
$$(ax)^2 = a^2 x^2, \quad (x+y)^2 = x^2 + y^2 + [x,y] \quad \text{for all scalars } a$$
and all x and y of equal odd degree.

These operations must satisfy the identities

(1) anti-symmetry:
$$[x,y] = -(-1)^{\deg(x)\deg(y)}[y,x] \quad \text{for all } x, y,$$

(2) Jacobi identity:
$$[x,[y,z]] = [[x,y],z] + (-1)^{\deg(x)\deg(y)}[y,[x,z]] \quad \text{for all } x,y,z,$$

(3)
$$[x,x] = 0 \quad \text{for all } x \text{ of even degree,}$$

(4)
$$2x^2 = [x,x], \quad [x,x^2] = 0 \quad \text{for all } x \text{ of odd degree.}$$

(5)
$$[y,x^2] = [[y,x],x]$$
for all y and for all x of odd degree.

Remark. If 2 is a unit in the ground ring, then the squaring operation may be defined in terms of the Lie bracket, that is, $x^2 = \frac{1}{2}[x,x]$ for x of odd degree. In the axioms for a graded Lie algebra, we may omit all reference to the squaring operation and add the requirement that $[x,[x,x]] = 0$ for all x of odd degree.

8.1 Universal enveloping algebras of graded Lie algebras

Remark. The first example of a graded Lie algebra is a graded associative algebra A with the Lie bracket

$$[a, b] = ab - (-1)^{\deg(a)\deg(b)} ba$$

and the squaring operation c^2 for c of odd degree. The above identities are all valid.

Remark. A homomorphism of graded Lie algebras $f : L \to L'$ is a linear map which preserves the two operations, that is,

$$f[x, y] = [fx, fy]$$

for all x, y and

$$f(x^2) = (fx)^2$$

for all odd degree x.

There is some related terminology:

(1) If all of the above identities are satisfied except for

$$[x, x^2] = 0 \quad \text{for all } x \text{ of odd degree}$$

we shall say that we have a quasi graded Lie algebra. Localized away from 2, the Samelson product makes the integral homotopy of a group like space into a quasi graded Lie algebra.

(2) In the one case when R is a field of characteristic 2 and L satisfies anti-commutativity, the Jacobi identity, and

$$[x, x] = 0 \quad \text{for all } x,$$

then we shall say that we have a Lie algebra with a grading (but not a graded Lie algebra.) Of course, the signs are unimportant here and graded associative algebras over fields of characteristic 2 provide examples of Lie algebras with gradings.

Definition 8.1.2. If L is a graded Lie algebra, the universal enveloping algebra $U(L)$ is the graded associative algebra (with unit) uniquely characterized up to isomorphism by the following universal property:

(1) there is a homomorphism of graded Lie algebras

$$\iota : L \to U(L)$$

(2) and, for any graded associative algebra A and any homomorphism

$$f : L \to A$$

of graded Lie algebras, there is a unique homomorphism of graded algebras
$$\overline{f} : U(L) \to A \quad \text{such that} \quad \overline{f} \circ \iota = f.$$
That is, $\overline{f}$ is the unique extension of f to an algebra homomorphism.

The uniqueness of $U(L)$ is proven by the standard categorical argument. The existence of $U(L)$ is proved by the following construction:

Let
$$T(L) = R \oplus L \oplus (L \otimes L) \oplus (L \otimes L \otimes L) \oplus \cdots$$
be the tensor algebra. The tensor algebra is the free associative algebra generated by L, that is, any linear map $f : L \to A$ has a unique extension to an algebra homomorphism $g : T(L) \to A$.

Then g vanishes on $I =$ the 2-sided ideal of $T(L)$ generated by all
$$x \otimes y - (-1)^{\deg(x)\deg(y)} y \otimes x - [x,y] \quad \text{for all} \quad x, y \ \varepsilon L$$
and by all
$$x \otimes x - x^2 \quad \text{for all } x \text{ of odd degree}.$$

If
$$L \xrightarrow{\iota} U(L) = T(L)/I$$
is the natural map, then ι has the required universal property.

Some examples of universal enveloping algebras are the following:

Example 8.1.3. Universal enveloping algebra of a direct sum

The universal enveloping algebra of a direct sum is the tensor product. That is, let L and M be graded Lie algebras and consider
$$L \oplus M \xrightarrow{\iota} U(L) \otimes U(M), \quad \iota(x, y) = \iota_L x \otimes 1 + 1 \otimes \iota_M y.$$
Given a Lie homomorphism $\phi : L \oplus M \to A$ into an associative algebra, the fact that
$$[\phi L, \phi M] = 0$$
implies that there is a unique extension of ϕ to an algebra homomorphism $\overline{\phi} : U(L) \otimes U(M) \to A$ given in terms of the extensions to UL and UM by
$$\overline{\phi}(a \otimes b) = (\overline{\phi}a) \cdot (\overline{\phi}b), \quad a \ \varepsilon U(L), \quad b \ \varepsilon U(M).$$

Example 8.1.4. Universal enveloping algebra of an abelian Lie algebra

Let V be a graded module which is free in every dimension. Let $S(V)$ be the free graded commutative algebra generated by V, that is, $\iota : V \to S(V)$ has the

8.1 Universal enveloping algebras of graded Lie algebras

universal extension property for maps $V \to A$ into a graded commutative algebra in which squares of odd degree classes are zero. Then

$$S(V) = E(V_{\text{odd}}) \otimes P(V_{\text{even}}) =$$

the tensor product of the exterior algebra generated by elements of odd degree and of the polynomial algebra generated by elements of even degree. If L is an abelian Lie algebra, that is, if $[L, L] = 0$ and $L^2_{\text{odd}} = 0$, then

$$U(L) = S(L).$$

Example 8.1.5. Universal enveloping algebra of a free Lie algebra

Let V be a graded module which is free in every dimension. The free graded Lie algebra $L(V)$ generated by V is characterized by the map $\iota : V \to L(V)$ having the universal extension property for maps $V \to L$ into an graded Lie algebra. Then

$$U(L(V)) = T(V) =$$

the tensor algebra (or free associative algebra) generated by V.

The case where $V = \langle x \rangle$ and $L(x) = L(V)$ is free on one basis element x is illustrative of the difference between graded and ungraded Lie algebras.

(1) If x has even degree, then $L(x) = \langle x \rangle$ of rank 1 is abelian and

$$UL(x) = T(x) = S(x) = P(x)$$

is a tensor algebra or polynomial algebra.

(2) If x has odd degree, then $L(x) = \langle x, x^2 \rangle$ of rank 2 is not abelian and

$$UL(x) = T(x)$$

is a tensor algebra and not an exterior algebra.

Example 8.1.6. Hopf algebra structure on the universal enveloping algebra

Given a universal enveloping algebra $\iota : L \to U(L)$, a Hopf algebra structure can be defined on $U(L)$ by making $\iota(L)$ primitive, that is, the diagonal or coalgebra structure is defined as the unique extension to an algebra map of the diagonal

$$\Delta : L \to L \oplus L \xrightarrow{\iota} U(L) \otimes U(L), \quad \Delta x = (x, x) \equiv x \otimes 1 + 1 \otimes x.$$

The counit or augmentation is the map $\epsilon : UL \to R$ to the ground ring which sends 1 to 1, is 0 on $\iota(L)$, and is a homomorphism of algebras.

As usual, if $\epsilon : A \to R$ is an augmentation (= algebra map to the ground ring), the kernel of ϵ is called the augmentation ideal and denoted by

$$\ker(\epsilon) = I(A) = \overline{A}.$$

The decomposables of A are defined as

$$D(A) = \overline{A} \cdot \overline{A} = \text{image } (\overline{A} \otimes \overline{A} \subseteq A \otimes A \xrightarrow{\text{mult}} A).$$

The indecomposables are

$$Q(A) = A/D(A) = A/\overline{A} \cdot \overline{A}.$$

Definition 8.1.7. If L is a graded Lie algebra, the abelianization of L is the abelian Lie algebra

$$Ab(L) = L/\{[L,L] + L_{\text{odd}}^2\}.$$

Lemma 8.1.8. *If L is a graded Lie algebra, then the abelianization of L is the same as the indecomposables of $U(L)$, that is,*

$$Ab(L) \cong Q(U(L))).$$

Proof: Consider the natural epimorphism $\pi : L \to Ab(L)$ and the commutative diagram

$$\begin{array}{ccccc} L & \xrightarrow{\iota} & U(L) & \xrightarrow{\overline{\pi}} & U(Ab(L)) = S(Ab(L)) \\ \downarrow \pi & & \downarrow & & \downarrow \\ Ab(L) & \to & QU(L) & \to & QU\,Ab(L) = Ab(L). \end{array}$$

The bottom composition is the identity. Since $L \to QU(L)$ is an epimorphism, the bottom right arrow is an epimorphism. Therefore, both bottom arrows are isomorphisms. $\square$

Remark. Sometimes, $[L,L] + L_{\text{odd}}^2$ is called the decomposables of L and $Ab(L)$ is called the indecomposables of L.

Exercises

(1) Show that the image of the graded Lie algebra $\iota(L)$ generates the universal enveloping algebra $U(L)$ as an algebra.

(2) Give the justifications for

(a) Example 8.1.3:

$$U(L \oplus M) \cong UL \otimes UM.$$

(b) Example 8.1.4: L abelian implies that $U(L) = S(L) = $ the free graded commutative algebra.

(c) Example 8.1.5: V a free graded module implies that

$$U(L(V)) = T(V).$$

8.2 The graded Poincaré–Birkhoff–Witt theorem

The graded version of the Poincaré–Birkhoff–Witt theorem shows that Lie algebras embed into their universal enveloping algebras and also provide a complete determination of the coalgebra structure of the universal enveloping algebra. This key theorem is essential to the application of universal enveloping algebras. When 2 is a unit in the ground ring it was first proved by Bruce Jordan [68]. His proof was modeled on Cartan–Eilenberg's proof of the ungraded version [23]. We give an alternate proof which is modeled on Nathan Jacobson's proof of the ungraded version [65].

Let V be a graded R module and let $v_1, v_2, v_3, \ldots, v_n$ be elements of V. A monomial of length k is a tensor product

$$v_1 \otimes v_2 \otimes v_3 \otimes \cdots \otimes v_n.$$

The increasing length filtration $F_n T(V)$ on the tensor algebra $T(V)$ is the span of all monomial tensors of length $\leq n$.

Let L be a graded Lie algebra over a commutative ring R. Let $\iota : L \to U(L)$ be the map into the universal enveloping algebra. The Lie filtration $F_n U(L)$ is the increasing filtration on $U(L)$ which is the epimorphic image of the length filtration on the tensor algebra $T(L)$. Thus, the n-th filtration $F_n U(L)$ is the submodule generated by all products of Lie elements (= elements of $\iota(L)$) of length $\leq n$. In other words,

Definition 8.2.1. The Lie filtration on the universal enveloping algebra is the increasing filtration defined as follows:

(1)
$$F_0 U(L) = R,$$

(2)
$$F_n U(L) = F_{n-1} U(L) + \text{image}(\ L \otimes F_{n-1} U(L)$$
$$\xrightarrow{\iota \otimes \text{incl}} U(L) \otimes U(L) \xrightarrow{\text{mult}} U(L))$$

The multiplication is compatible with the Lie filtration:
$$F_m U(L) \cdot F_n U(L) \subseteq F_{m+n} U(L).$$

The coproduct is compatible with the Lie filtration:
$$\Delta(F_n U(L)) \subseteq \bigoplus_{i+j=n} F_i U(L) \otimes F_j U(L).$$

It follows that the associated graded object,
$$E_*^0(U(L)) = F_* U(L) / F_{*-1} U(L),$$

is a Hopf algebra. As an algebra, $E_*^0(U(L))$ is generated by filtration 1, that is, by $\iota(L)$. Since

$$xy - (-1)^{\deg(x)\deg(y)} yx = [x,y] \text{ for } x, y \; \varepsilon L$$

and

$$xx = x^2$$

for odd degree x, it follows that $E_*^0(U(L))$ has a graded commutative multiplication with odd degree elements having square zero.

The graded Poincare–Birkhoff–Witt theorem describes the coalgebra $U(L)$ completely. Define a Hopf algebra structure in the free commutative algebra $S(L)$ by requiring L to be primitive and give $S(L)$ the length filtration.

Graded Poincare–Birkhoff–Witt theorem 8.2.2: *If L is a graded Lie algebra which is a free module in each degree, then there is an isomorphism of coalgebras*

$$\Psi : S(L) \to U(L)$$

which is length preserving, that is, $\Psi(F_n S(L)) \subseteq F_n U(L)$ for all n.

Proof: Let x_α be a basis for L and choose a well ordering $\leq$ of the indices of this basis set.

A monomial of length n in the free commutative algebra $S(L)$ is a product of basis elements

$$x_{\alpha_1} \cdot x_{\alpha_2} \cdot x_{\alpha_3} \cdots x_{\alpha_n}.$$

Monomials of length $\leq n$ span the length filtration $F_n S(L)$.

Define the index of the monomial as the number of pairs α_i, α_j with $i < j$ but $\alpha_i > \alpha_j$, that is, the index is the number of variables that are out of order in the monomial. If the index is 0, the monomial is said to be in proper order.

A monomial is said to be reduced if $\alpha_i = \alpha_{i+1}$ implies that the degree of x_{α_i} is even. It is clear that a basis of $S(L)$ is given by the reduced monomials in proper order and we define a linear map

$$\Psi : S(L) \to U(L)$$

by

$$\Psi(x_{\alpha_1} \cdot x_{\alpha_2} \cdot x_{\alpha_3} \cdots x_{\alpha_n}) = x_{\alpha_1} \cdot x_{\alpha_2} \cdot x_{\alpha_3} \cdots x_{\alpha_n}$$

for reduced monomials in proper order. We claim that

Lemma 8.2.3. Ψ *is a map of coalgebras.*

8.2 The graded Poincare–Birkhoff–Witt theorem

Proof: After collapsing any repetitions of the same element into a single power, a reduced monomial in $S(L)$ lies in some subcoalgebra

$$S(x_{\beta_1}) \otimes S(x_{\beta_2}) \otimes \cdots \otimes S(x_{\beta_k})$$

where the x_{β_i} are basis elements in strict proper order. Restricted to this coalgebra, the map Ψ is the composition of coalgebra maps

$$S(x_{\beta_1}) \otimes S(x_{\beta_2}) \otimes \cdots \otimes S(x_{\beta_k}) \to U(L) \otimes U(L) \otimes \cdots \otimes U(L) \xrightarrow{\text{mult}} U(L).$$

Thus, Ψ is a coalgebra map. $\square$

We need to check that Ψ is an isomorphism.

We first show that $\Psi : F_n S(L) \to F_n U(L)$ is an epimorphism for all n. We prove this by induction using

$$x \cdot y - (-1)^{\deg(x)\deg(y)} y \cdot x = [x, y]$$

and, if x has odd degree,

$$x \cdot x = x^2.$$

If M is any monomial in $F_n U(L)$, then M is congruent modulo $F_{n-1} U(L)$ to a monomial N in proper order. If N is reduced, it is clearly in the image of Ψ and, if N is not reduced, then it is congruent to 0 modulo $F_{n-1} U(L)$ and this is of course in the image of Ψ.

The fact that Ψ is a monomorphism is harder.

Let P be the free module which has a basis consisting of the reduced monomials in proper order. Note that P has a natural increasing length filtration $F_n P$. We shall show

Lemma 8.2.4. *There is a linear map $J : T(L) \to P$ which is length filtration preserving and such that*

(1)

$$J(x_{\alpha_1} \otimes x_{\alpha_2} \otimes \cdots \otimes x_{\alpha_n}) = x_{\alpha_1} \otimes x_{\alpha_2} \otimes \cdots \otimes x_{\alpha_n}$$

for reduced monomials in proper order,

(2) *the switching identity*

$$J(x_{\alpha_1} \otimes \cdots \otimes x_{\alpha_i} \otimes x_{\alpha_{i+1}} \otimes \cdots \otimes x_{\alpha_n})$$
$$- (-1)^{\deg(x_{\alpha_i})\deg(\alpha_{i+1})} J(x_{\alpha_1} \otimes \cdots \otimes x_{\alpha_{i+1}} \otimes x_{\alpha_i} \otimes \cdots \otimes x_{\alpha_n})$$
$$= J(x_{\alpha_1} \otimes \cdots \otimes [x_{\alpha_i}, x_{\alpha_{i+1}}] \otimes \cdots \otimes x_{\alpha_n})$$

is valid for all monomials and,

(3) *the contracting identity*

$$J(x_{\alpha_1} \otimes \cdots \otimes x_{\alpha_i} \otimes x_{\alpha_{i+1}} \otimes \cdots \otimes x_{\alpha_n})$$
$$= J(x_{\alpha_1} \otimes \cdots \otimes x_{\alpha_i}^2 \otimes \cdots \otimes x_{\alpha_n})$$

is valid whenever $x_{\alpha_i} = x_{\alpha_{i+1}}$ has odd degree.

Remark. The bilinearity of the Lie bracket implies that, if the switching identities hold for elements in a basis, then the switching identities hold for all elements of the graded Lie algebra. Similarly, the quadratic property of squaring odd elements implies that, in the presence of the switching identities, if the contracting identities hold for odd degree elements in a basis, then the contracting identities hold for all odd degree elements.

Thus $J : T(L) \to P$ defines a quotient map $\overline{J} : U(L) \to P$ compatible with the Lie filtration on $U(L)$ and the composition

$$S(L) \xrightarrow{\Psi} U(L) \xrightarrow{\overline{J}} P$$

is an isomorphism. Hence, Ψ is an isomorphism, which was to be demonstrated.

It remains to prove Lemma 8.2.4.

There are three main points:

(A) Define J recursively in terms of the length and index ordering of monomials, that is, define J in terms of the previous definition on monomials of lesser length or of the same length but lesser index.

(B) Show that the definition of J is unambiguous, that is, it does not depend on how one reduces the length or index. Show this for all possible ways of reducing the length or index and you will have automatically accomplished step C below.

(C) Define J so that the identities 1), 2), and 3) are satisfied.

Suppose that J has been defined for all monomial tensors of length $< n$ and all monomial tensors of length $= n$ and index $< i$ in such a way that 1) and 2) are satisfied for all tensors in this space. Let $M = x_{\alpha_1} \otimes x_{\alpha_2} \otimes \cdots \otimes x_{\alpha_n}$ be a monomial of length n and index i.

Case 1: index $i = 0$: When the index is 0, the monomial M is in proper order. If it is reduced, define

$$J(M) = x_{\alpha_1} \otimes x_{\alpha_2} \otimes \cdots \otimes x_{\alpha_n}.$$

If it is not reduced, then $x_{\alpha_i} = x_{\alpha_{i+1}}$ for some element of odd degree. Define

$$J(M) = J(x_{\alpha_1} \otimes \cdots \otimes x_{\alpha_i}^2 \otimes \cdots \otimes x_{\alpha_n}).$$

8.2 The graded Poincaré–Birkhoff–Witt theorem

We shall call this a contraction. Writing $u = x_{\alpha_i}$, this definition of J also ensures the required switching identity that

$$J(\cdots \otimes u \otimes u \otimes \cdots) + J(\cdots \otimes u \otimes u \otimes \cdots) = J(\cdots \otimes [u,u] \otimes \cdots).$$

Suppose there is another way to reduce the length, that is, we have one of the following two situations

(1)
$$J(\cdots \otimes u \otimes u \otimes \cdots \otimes v \otimes v \otimes \cdots)$$

or

(2)
$$J(\cdots \otimes u \otimes u \otimes u \otimes \cdots)$$

with u and v of odd degree.

In the first case (1), the definition is

$$J(\cdots \otimes u^2 \otimes \cdots \otimes v^2 \otimes \cdots),$$

no matter which of the two possible contractions one does first. Thus, the definition of J is unambiguous in this case and satisfies 3), 2), and 1).

In the second case (2), the one contraction leads to

$$\begin{aligned} J(M) &= J(\cdots \otimes u^2 \otimes u \otimes \cdots) \\ &= J(\cdots \otimes u \otimes u^2 \otimes \cdots) + J(\cdots \otimes [u^2, u] \otimes \cdots) \\ &= J(\cdots \otimes u \otimes u^2 \otimes \cdots) \end{aligned}$$

since $[u^2, u] = 0$. The other contraction leads directly to

$$J(M) = J(\cdots \otimes u \otimes u^2 \otimes \cdots).$$

Thus, the definition is unambiguous in this case and the identities are satisfied.

Case 2: index $i > 0$: When the index is positive, there must a pair $x_{\alpha_i} = u, x_{\alpha_{i+1}} = v$ with $\alpha_i > \alpha_{i+1}$. In this case, we can define

$$J(\cdots \otimes u \otimes v \otimes \cdots) = (-1)^{\deg(u)\deg(v)} J(\cdots \otimes v \otimes u \otimes \cdots)$$
$$+ J(\cdots \otimes [u,v] \otimes \cdots).$$

We note that J is now defined since the index or length has decreased. We shall call this a switch. Note that switches guarantee that the identities (1) and (2) are satisfied. This is true even for the reverse switch where u is less than v. It is also possible that J could be defined by a contraction as in Case 1.

We need to check that the definition is unambiguous when there are different ways of lowering length or index.

There are the following cases:

(1)
$$J(\cdots \otimes u \otimes v \otimes \cdots \otimes w \otimes z \otimes \cdots)$$
with $u > v$ and $w > z$.

(2)
$$J(\cdots \otimes u \otimes v \otimes w \otimes \cdots)$$
with $u > v > w$.

(3)
$$J(\cdots \otimes u \otimes v \otimes \cdots \otimes w \otimes w \otimes \cdots)$$
with $u > v$ and w of odd degree.

(4)
$$J(\cdots \otimes u \otimes u \otimes \cdots \otimes w \otimes z \otimes \cdots)$$
with u of odd degree and $w > z$.

(5)
$$J(\cdots \otimes u \otimes u \otimes v \cdots)$$
with u of odd degree and $u > v$.

(6)
$$J(\cdots \otimes u \otimes v \otimes v \otimes \cdots)$$
with $u > v$ and v of odd degree.

(7)
$$J(\cdots \otimes u \otimes u \otimes \cdots \otimes v \otimes v \otimes \cdots)$$
with u and v of odd degree.

(8)
$$J(\cdots \otimes u \otimes u \otimes u \otimes \cdots)$$
with u of odd degree.

Cases (7) and (8) have already been considered in the index 0 situation of Case 1.

8.2 The graded Poincare–Birkhoff–Witt theorem

We shall use the obvious shorthand of writing u for $\deg(u)$ when no confusion is possible.

In case (1), performing the two switches, no matter which is first, leads to

$$(-1)^{uv+wz} J(\cdots \otimes u \otimes v \otimes \cdots \otimes w \otimes z \otimes \cdots)$$
$$+ (-1)^{vu} J(\cdots \otimes v \otimes u \otimes \cdots \otimes [w,z] \otimes \cdots)$$
$$+ (-1)^{wz} J(\cdots \otimes [u,v] \otimes \cdots \otimes z \otimes w \otimes \cdots)$$
$$+ J(\cdots \otimes [u,v] \otimes \cdots \otimes [w,z] \otimes \cdots).$$

Thus, the definition is unambiguous in this case.

In case (2), performing the two switches leads to either of the following:

$$A = (-1)^{uv+uw+wz} J(\cdots \otimes w \otimes v \otimes u \otimes \cdots)$$
$$+ (-1)^{uv+uw} J(\cdots \otimes [v,w] \otimes u \otimes \cdots)$$
$$+ (-1)^{uv} J(\cdots \otimes v \otimes [u,w] \otimes \cdots)$$
$$+ J(\cdots \otimes [u,v] \otimes w \otimes \cdots).$$

$$B = (-1)^{uv+uw+wz} J(\cdots \otimes w \otimes v \otimes u \otimes \cdots)$$
$$+ (-1)^{vw+uw} J(\cdots \otimes w \otimes [u,v] \otimes \cdots)$$
$$+ (-1)^{vw} J(\cdots \otimes [u,w] \otimes v \otimes \cdots)$$
$$+ J(\cdots \otimes u \otimes [v,w] \otimes \cdots).$$

Thus, $A - B = J(\cdots \otimes ((-1)^{uv+uw}[[v,w],u] + (-1)^{uv}[v,[u,w]] + [[u,v],w]) \otimes \cdots) = J(0) = 0$ by the Jacobi identity. Thus, the definition is unambiguous in this case and the identities hold.

The cases (3) and (4) are straightforward and left to the exercises.

In case (5), performing the contraction leads to

$$J(\cdots \otimes u^2 \otimes v \otimes \cdots).$$

On the other hand, performing two switches and a contraction leads to

$$J(\cdots \otimes v \otimes u^2 \otimes \cdots) + J(\cdots \otimes \{(-1)^{uv}[u,v] \otimes u + u \otimes [u,v]\} \otimes \cdots).$$

Now another switch and the identities

$$J(\cdots \otimes \{(-1)^{uv}[u,v] \otimes u + u \otimes [u,v]\} \otimes \cdots)$$
$$= J(\cdots \otimes [u,[u,v]] \otimes \cdots)[v,u^2] + [u,[u,v]] = 0$$

prove that this second procedure yields the same answer as the first.

Case (6) is left to the exercises.

Since we have checked that all possible ways of defining J by recursion are equal, it follows that the definition is invariant under all switches, whether they raise or lower the index. Hence the identities (2) and (3) are satisfied for the extension of the definition to higher index and length.

This completes the proof of Lemma 8.2.4 and of the graded Poincare–Birkhoff–Witt theorem. □

Remark. Let L be a classical Lie algebra over any ring. Suppose L has a grading and is a free module. The classical Poincare–Birkhoff–Witt theorem asserts that there is a coalgebra isomorphism

$$\Psi : P(L) \to U(L)$$

where $P(L)$ is the primitively generated polynomial algebra generated by L. The proof is similar to the graded version but simpler.

There is a version of the graded Poincare–Birkhoff–Witt theorem which is only a little bit weaker and which does not depend on choosing an ordering of a basis for L, namely:

Graded Poincare–Birkhoff–Witt theorem, second version 8.2.4. *If L is a free module, then the natural map*

$$L \to F_1 U(L)/F_0 U(L) \to E_*^0 U(L)$$

extends to a Hopf algebra isomorphism

$$\Phi : S(L) \to E_*^0 U(L).$$

Proof: The coalgebra isomorphism $\Psi : S(L) \to U(L)$ is filtration preserving and induces the Hopf algebra isomorphism on the associated graded objects

$$\Phi = E^0 \Psi : E_*^0 S(L) = S(L) \to E_*^0 U(L). \qquad \square$$

Exercises

(1) Verify cases (3), (4), and (6) in the proof of Lemma 8.2.3.

(2) Show that, if L is a connected graded Lie algebra, that is, if $L_0 = 0$, then the fact that Φ is an isomorphism (8.2.4) implies that Ψ is an isomorphism (8.2.2). In other words, the second version of the Poincare–Birkhoff–Witt theorem implies the first.

8.3 Consequences of the graded Poincare–Birkhoff–Witt theorem

A first consequence of the graded Poincare–Birkhoff–Witt theorem is that graded Lie algebras embed into their universal enveloping algebras, more precisely, the basis of reduced monomials in proper order (8.2.2) shows that

8.3 Consequences of the graded Poincaré–Birkhoff–Witt theorem

Proposition 8.3.1. *Suppose L is a graded Lie algebra which is a free R module in every dimension. Then the universal enveloping algebra $U(L)$ is a free R module in every dimension and $\iota: L \to U(L)$ is an injection onto a summand.*

Remark. The meaning of Proposition 8.3.1 is that the identities for a graded Lie algebra are exactly right. In particular, the triple identity,

$$[x, x^2] = 0 \quad \text{for } x \text{ of odd degree},$$

is clearly essential for a graded Lie algebra to embed in its universal enveloping algebra. Adding it to the usual identities is a sufficient criterion for embedding when the graded Lie algebra is free as a module.

The coalgebra isomorphism $S(L) \xrightarrow{\cong} U(L)$ in the graded Poincaré–Birkhoff–Witt theorem determines the module of primitive elements in the Hopf algebra $U(L)$:

Proposition 8.3.2. *Let L be a graded Lie algebra which is a free R module with a basis $\{x_\alpha\}$.*

(a) *If the ground ring R is an integral domain of characteristic 0, then x_α is a basis of $PU(L)$, that is, $PU(L) = L$.*

(b) *If the ground ring R is a field of finite characteristic p, then $PU(L)$ has a basis*

$$\{x_\alpha : \text{ for all } \alpha\} \cup \{x_\alpha^p, x_\alpha^{p^2}, \ldots, x_\alpha^{p^k}, \ldots : x_\alpha \text{ of even degree}\}.$$

The following freeness result is fundamental.

Proposition 8.3.3. *Suppose L' is a subalgebra of the graded Lie algebra L such that L' has a basis $\{x_\alpha\}$ which extends to a basis $\{x_\alpha\} \cup \{y_\beta\}$ of L. Order the basis sets so that the x_α precede the y_β. Then $U(L)$ is a free left $U(L')$ with a basis consisting of all reduced monomials in $\{y_\beta\}$ which are in proper order. This includes the empty monomial 1.*

Proof: It follows from the graded Poincaré-Birkhoff-Witt theorem that a basis for $U(L)$ is given by all $M \cdot N$ where M is a reduced monomial in proper order involving the variables $\{x_\alpha\}$ and N is a reduced monomial in proper order involving the variables $\{y_\beta\}$. Thus,

$$U(L) = \bigoplus_N U(L') \cdot N. \qquad \square$$

Lemma 8.3.4. *Suppose $L' \subseteq L$ is an ideal in a graded Lie algebra, that is,*

$$[L, L'] \subseteq L'.$$

If $\overline{U(L')}$ is the augmentation ideal, then

$$U(L) \cdot \overline{U(L')} \subseteq \overline{U(L')} \cdot U(L).$$

Proof: Consider
$$xy = (-1)yx + [x, y].$$
Using this, it follows by induction on the length of monomials that, if $x \, \varepsilon L$,
$$x \cdot \overline{U(L')} \subseteq \overline{U(L')} \cdot x + \overline{U(L')}.$$
Hence,
$$U(L) \cdot \overline{U(L')} = \overline{U(L')} \cdot U(L). \qquad \square$$
In other words, if $L' \subseteq L$ is an ideal, then $\overline{U(L')}$ generates a two-sided ideal
$$I = U(L) \cdot \overline{U(L')} = \overline{U(L')} \cdot U(L)$$
and hence the quotient (see Definition 8.4.2)
$$U(L) \otimes_{U(L')} R = U(L)/I = U(L)/U(L) \cdot \overline{U(L')}$$
has a natural multiplication. It is easy to check that I is a Hopf ideal in the sense that
$$\Delta(I) \subseteq I \otimes U(L) + U(L) \otimes I.$$
Hence, the quotient
$$U(L) \otimes_{U(L')} R$$
has a natural Hopf algebra structure.

Proposition 8.3.5. *Suppose*
$$0 \to L' \xrightarrow{i} L \xrightarrow{j} L'' \to 0$$
is a short exact sequence of graded Lie algebras which are free modules (hence, the sequence is split as modules over the ground ring). Then there is a bijection
$$U(L') \otimes U(L'') \to U(L)$$
which is simultaneously an isomorphism of left $U(L')$ modules and of coalgebras.

Proof: The reduced monomials in proper order formed from a basis of L'' provide a section
$$\pi : U(L'') \to U(L), \quad U(j) \circ \pi = 1_{U(L'')}$$
which is a map of coalgebras. It follows from 8.3.3 that the composition
$$U(L') \otimes U(L'') \xrightarrow{U(i) \otimes \pi} U(L) \otimes U(L) \xrightarrow{\text{mult}} U(L)$$
is an isomorphism of left $U(L')$ modules. It is clearly a morphism of coalgebras.
$\square$

For example, let x be an odd degree element and consider the exact sequence of graded Lie algebras

$$0 \to \langle x^2 \rangle \to L(x) \to \langle x \rangle \to 0$$

where $\langle x^2 \rangle$ and $\langle x \rangle$ are abelian Lie algebras on one generator. Then there is a bijection

$$UL(x) = T(x) \cong P(x^2) \otimes E(x)$$

which is simultaneously an isomorphism of left $P(x^2)$ modules and of coalgebras.

8.4 Nakayama's lemma

Graded modules over a connected algebra share many of the properties of modules over a local ring. In particular, a strong form of Nakayama's lemma is true.

Recall the definitions. An augmented algebra A over a commutative ring R is called connected if the augmentation $\epsilon : A \to R$ is an isomorphism in degree 0. The kernel of the augmentation is called the augmentation ideal

$$I(A) = \overline{A} = \text{kernel}\,(\epsilon : A \to R).$$

Graded modules are assumed to be concentrated in nonnegative degrees.

Nakayama's lemma 8.4.1. *If A is a connected algebra over a commutative ring R and M is a graded left module over A, then the following equivalent statements are valid:*

(a)
$$M = I(A) \cdot M$$
implies that $M = 0$.

(b)
$$M/I(A) \cdot M = 0$$
implies that $M = 0$.

(c)
$$R \otimes_A M = 0$$
implies that $M = 0$.

Proof: (a) is true since A connected implies that there can be no first nonzero degree of M. Certainly, (a) and (b) are the same.

The exact sequence
$$0 \to I(A) \to A \xrightarrow{\epsilon} R \to 0$$
yields the exact sequence
$$I(A) \otimes_A M \to A \otimes_A M \to R \otimes_A M \to 0$$
and thus
$$M/I(A) \cdot M \cong R \otimes_A M.$$

$\square$

This leads to the following definition.

Definition 8.4.2. The module of generators of M is
$$M/I(A) \cdot M \cong R \otimes_A M.$$

Nakayama's lemma gives the following lemma on epimorphisms.

Lemma 8.4.3. *Suppose $f : M \to N$ is a morphism of graded modules over a connected algebra A. Then f is an epimorphism if and only if the map on the module of generators*
$$1 \otimes_A f : R \otimes_A M \to R \otimes_A N$$
is an epimorphism.

Proof: Since tensor product is right exact, we need only show that, if $1 \otimes_A f$ is an epimorphism, then f is also.

But, let C be the cokernel of f, so that
$$M \xrightarrow{f} N \to C \to 0$$
is exact. Hence
$$R \otimes_A M \to R \otimes_A N \to R \otimes_A C \to 0$$
is exact. If $1 \otimes_A f$ is an epimorphism, then $R \otimes_A C = 0$ and Nakayama's lemma implies that $C = 0$ and f is an epimorphism. $\square$

Definition 8.4.4. If A is a connected algebra with augmentation ideal $I(A)$, the module of indecomposables is
$$Q(A) = I(A)/I(A) \cdot I(A).$$

The following lemmas are left as exercises.

Lemma 8.4.5. *If A is a connected algebra, then $I(A) = 0$ if and only if $Q(A) = 0$.*

8.4 Nakayama's lemma

Lemma 8.4.6. *Let $f : A \to B$ be a morphism of connected algebras. Then f is an epimorphism if and only if the induced map on indecomposables $Q(f) : Q(A) \to Q(B)$ is an epimorphism.*

Definition 8.4.7. For an augmented algebra A, the filtration by powers of the augmentation ideal is the decreasing filtration
$$F^0 A = A, \quad F^1 A = I(A), \quad F^{n+1} A = I(A) \cdot F^n A, \quad n \geq 0.$$

Since $F^j A \cdot F^k A \subseteq F^{j+k} A$, the associated graded object
$$E_0^n(A) = F^n(A)/F^{n+1}(A)$$
of the filtration by powers of the augmentation ideal inherits an algebra structure, that is, it has a multiplication
$$E_0^n(A) \otimes E_0^m(A) \xrightarrow{\text{mult}} E_0^{n+m}(A).$$

Observe that
$$E_0^0(A) = A/I(A), \quad , E_0^1(A) = Q(A).$$

Furthermore, the first grading $E_0^1(A) = Q(A)$ generates the algebra $E_0^*(A)$.

The following lemmas are also left as exercises.

Lemma 8.4.8. *If A is a connected algebra, then the filtration by powers of the augmentation ideal is finite in each degree.*

Lemma 8.4.9. *Suppose that $g : A \to B$ is a morphism of connected algebras. Then*

(a) *g is an epimorphism if and only if $E_0^*(g) : E_0^*(A) \to E_0^*(B)$ is an epimorphism.*

(b) *g is an isomorphism if and only if $E_0^*(g) : E_0^*(A) \to E_0^*(B)$ is an isomorphism.*

(c) *If $E_0^*(g) : E_0^*(A) \to E_0^*(B)$ is an monomorphism, then g is a monomorphism.*

Exercises

(1) Prove Lemmas 8.4.5 and 8.4.6.

(2) Let $f : L \to K$ be a morphism of connected graded Lie algebras, that is, $L_0 = K_0 = 0$. Then f is an epimorphism if and only if the induced map on abelianizations $Ab(f) : Ab(L) \to Ab(K)$ is an epimorphism.

(3) Prove Lemmas 8.4.8 and 8.4.9.

(4) Give an example to show that the converse to Lemma 8.4.9(c) is not true.

(5) Suppose $f : A \to B$ is a map of connected tensor algebras. Show that, if the induced map on the module of indecomposables $Q(f) : Q(A) \to Q(B)$ is an isomorphism, then f is an isomorphism.

8.5 Free graded Lie algebras

If V is a graded R-module, recall that the free graded Lie algebra generated by V is characterized by the universal mapping property: Every linear map $V \to L$ into a graded Lie algebra has a unique extension to a map $L(V) \to L$ of graded Lie algebras. We also have that $UL(V) = T(V)$. In order to know that $L(V)$ embeds in its universal enveloping algebra, we need to know that $L(V)$ is a free R-module. The proof below is based on an argument in Serre's book on Lie algebras [119]. Namely, suppose you are given an algebraic construction defined over a commutative ring. Suppose that, in the finitely generated case, there is a formula for the dimension of that construction which is independent of the characteristic of the ground field. Then, even over an arbitrary ring in the possibly nonfinitely generated case, the given construction must be a free module over its ground ring. Details follow.

Proposition 8.5.1. *If V is a graded module which is a free over R, then the free graded Lie algebra $L(V)$ is a free R-module in each degree.*

Lemma 8.5.2. *If R is a field and V is finite dimensional in each degree, then*

$$L(V) = \bigoplus L(V)_n$$

where $L(V)_n$ is generated by Lie monomials of length n, each $L(V)_n$ is finite dimensional in each degree, and the dimension of $L(V)_n$ in each degree is independent of the characteristic.

Proof: If M is a bigraded object, define its Euler-Poincare series by

$$\Xi(M) = \Sigma m_{i,j} s^i t^j$$

where $m_{i,j}$ = dimension of $M_{i,j}$.

Let

$$\xi = \chi(V) = \sum_{i=0}^{\infty} d_i t^i$$

be the usual Euler–Poincare series of V, that is, d_i = the dimension of V_i. If we introduce an extra grading by giving each element of V length 1, we get a two-variable Euler–Poincare series

$$\Xi(V) = s\xi.$$

Since R is a field, the graded Lie algebra $L(V)$ is free in each degree and embeds as a summand in its universal enveloping algebra $UL(V) = T(V)$. Bigrade $T(V)$ by giving each element $v \varepsilon V_i$ the bigrading $(1, i)$, that is, $T(V)$ is bigraded by length and degree.

The two-variable Euler–Poincare series of

$$T(V) = R \oplus V \oplus (V \otimes V) \otimes \cdots$$

is

$$\Xi(T(V)) = 1 + s\xi + s^2\xi^2 + s^3\xi^3 + \cdots = \frac{1}{1 - s\xi}.$$

On the other hand, note that

$$L(V) = \bigoplus L(V)_n$$

inherits the length-degree bigrading and each bigrading $L(V)_n$ is finite dimensional. (This is why we have to use bigradings.)

Write the two-variable Euler–Poincare series

$$\Xi(L(V)) = \bigoplus_{i=0}^{\infty} \alpha_{i,j} s^i t^j.$$

Hence, the 2-variable Euler-Poincare series of $T(V) = U(L) \cong S(L_{\text{even}}) \otimes S(L_{\text{odd}})$ is

$$\Xi(U(L)) = \prod_{k=0}^{\infty} \frac{1}{(1 - s^i t^{2k})^{\alpha_{i,2k}}} \cdot \prod_{k=0}^{\infty} (1 + s^i t^{2k+1})^{\alpha_{i,2k+1}}.$$

Applying the logarithm function yields

$$-\log(1 - s\xi) = -\Sigma_{i,k} \alpha_{i,2k} \cdot \log(1 - s^i t^{2k}) + \Sigma_{i,k} \alpha_{i,2k+1} \cdot \log(1 + s^i t^{2k+1}).$$

If we choose a fixed pair i, j and reduce mod $s^{i+1} t^{j+1}$, then the expansion

$$\log(1 - x) = \sum_{i=1}^{\infty} \frac{1}{i} x^i$$

yields

$$-\log(1 - s\xi) \equiv -\sum_{a \leq i, b \leq j} (-1)^{b+1} \alpha_{a,b} \cdot \log(1 + (-1)^{b+1} s^a t^b).$$

On the right-hand side, the coefficient of $s^i t^j$ is

$$(-1)^{j+1} \alpha_{i,j} + \text{terms involving } \alpha_{a,b}, \ a < i, b < j.$$

So this can be solved to yield a recursive expression

$$\alpha_{i,j} = F(\alpha_{a,b};\ a < i, b < j)$$

which depends on $d_0, d_1, d_2, \ldots$ but is clearly independent of the characteristic. □

Now suppose that $R = \mathbb{Z}$ and suppose that V is a free $\mathbb{Z}$ module which is finitely generated in each degree. Let $L(V)$ be the free graded Lie algebra generated by V. It is clearly finitely generated in each length-degree bidegree. If p is a prime, then

$$L(V) \otimes \mathbb{Z}/p\mathbb{Z} \cong L(V \otimes \mathbb{Z}/p\mathbb{Z})$$

since it has the required universal mapping property. Since the dimension mod p of each bidegree is independent of the prime p, it follows from the basis theorem for finitely generated abelian groups that each bidegree is a free $\mathbb{Z}$ module. Thus $L(V)$ is a free $\mathbb{Z}$ module in each degree when V is a finite dimensional free $\mathbb{Z}$-module. In this case, $L(V)$ embeds in the tensor algebra $UL(V) = T(V)$.

Suppose V is a free $\mathbb{Z}$ module which is not finitely generated in each degree. Then

$$V = \varinjlim V_\alpha$$

where each V_α is finitely generated free in each degree. Therefore,

$$L(V) = \varinjlim L(V_\alpha)$$

embeds in the tensor algebra $\varinjlim T(V_\alpha) = T(V)$. Since $T(V)$ is a free $\mathbb{Z}$ module in each degree and $\mathbb{Z}$ is a principal ideal domain, it follows that, if V is free over $\mathbb{Z}$ in each degree, then $L(V)$ is a free $\mathbb{Z}$ module in each degree.

Now let R be any commutative ring and suppose that V is a graded $\mathbb{Z}$ module which is free in each degree. Then $V \otimes R$ is the general module which is free in each degree over R. Since

$$L(V) \otimes R \cong L(V \otimes R),$$

it follows that $L(V \otimes R)$ is a free R-module in each degree. □

For connected Lie algebras with a free abelianization, free Lie algebras can be characterized by their universal enveloping algebras. To be precise:

Proposition 8.5.3. *Suppose L is a graded connected Lie algebra such that the universal enveloping algebra $U(L)$ is isomorphic to a tensor algebra $T(V)$ with V a free module. Then L is isomorphic to the free graded Lie algebra $L(V)$.*

Proof: Since Lemma 8.1.8 implies that $Ab(L) \cong QT(V) \cong V$ and V is a free module, there is a linear map $V \to L$ such that the composition $V \to L \to Ab(L)$

8.5 Free graded Lie algebras

is an isomorphism. Consider the extensions

$$\begin{array}{ccc} L(V) & \xrightarrow{\iota} & UL(V) \\ \downarrow f & & \downarrow Uf \\ L & \xrightarrow{\iota} & UL. \end{array}$$

For connected Lie algebras the fact that $Ab(f)$ is an epimorphism implies that $f : L(V) \to L$ is an epimorphism. On the other hand, the fact that $L(V)$ is a free module implies that $\iota : L(V) \to UL(V)$ is a monomorphism. Therefore, $f : L(V) \to L$ is an isomorphism if $Uf : UL(V) \to UL$ is a monomorphism.

Since Uf is a map between tensor algebras and is an isomorphism on indecomposables, it is an isomorphism by Exercise 5 of Section 8.4. □

With mild restrictions, Proposition 8.5.3 says that a connected graded Lie algebra is free if and only if its universal enveloping algebra is a tensor algebra. It becomes important to be able to recognize tensor algebras. First of all, if $T(V)$ is a tensor algebra over a commutative ring R, then

$$0 \to T(V) \otimes V \xrightarrow{\text{mult}} T(V) \xrightarrow{\epsilon} R \to 0$$

is a free $T(V)$ resolution of R. Hence, for any $T(V)$ module M, we have that

$$\mathrm{Tor}_2^{T(V)}(M, R) = 0.$$

Conversely, we have

Proposition 8.5.4. *Suppose A is a connected associative algebra with a free module of indecomposables $Q(A) = W$ and suppose that*

$$\mathrm{Tor}_2^A(R, R) = 0.$$

Then A is isomorphic to the tensor algebra $T(W)$.

Proof: Since $W = Q(A)$ is free, the map $I(A) \to Q(A)$ has a section $W \to I(A)$. Consider the composition which is a map of A-modules

$$f : A \otimes W \to A \otimes I(A) \xrightarrow{\text{mult}} I(A).$$

We claim that f is an isomorphism.

Note that f is a epimorphism since the map on generators

$$1 \otimes f : R \otimes_A (A \otimes W) \xrightarrow{\cong} R \otimes_A I(A)$$

is an epimorphism.

Let N be the kernel of f and consider the short exact sequence

$$0 \to N \to A \otimes W \to I(A) \to 0.$$

In the standard way, this turns into a long exact sequence
$$\cdots \to \operatorname{Tor}_1^A(R, I(A)) \to R \otimes_A N \to R \otimes_A A \otimes W \to R \otimes_A I(A) \to 0.$$
Since
$$R \otimes_A A \otimes W = W \xrightarrow{\cong} R \otimes_A I(A) = Q(A)$$
is an isomorphism, it follows that
$$\operatorname{Tor}_1^A(R, I(A)) \to R \otimes_A N$$
is an epimorphism.

But the short exact sequence
$$0 \to I(A) \to A \to R \to 0$$
implies that
$$0 = \operatorname{Tor}_2^A(R, R) \cong \operatorname{Tor}_1^A(R, I(A))$$
and hence
$$R \otimes_A N = 0.$$
Nakayama's lemma implies that $N = 0$ and $f : A \otimes W \to I(A)$ is an isomorphism. A standard inductive argument implies that $A \cong T(W)$ as algebras. □

Exercises

(1) Show that, if A is a connected associative algebra with an isomorphism given by the composition
$$A \otimes W \to A \otimes I(A) \xrightarrow{\text{mult}} I(A),$$
then there is an isomorphism of algebras $T(W) \to A$.

8.6 The change of rings isomorphism

The change of rings theorem relates the homological algebra of a subalgebra to that of an ambiant algebra.

Suppose that
$$A' \xrightarrow{\iota} A \xrightarrow{j} A''$$
are maps of graded augmented algebras over a commutative ring R. Observe that
$$R \otimes_{A'} A \cong A/I(A') \cdot A$$
and that A'' is a right A module.

Assume that:

(a) ι makes A into a free left A' module.

(b) j induces an isomorphism of right A modules
$$R \otimes_{A'} A \cong A''$$

Change of rings theorem 8.6.1. *There is an isomorphism*
$$\operatorname{Tor}_*^{A'}(R, R) \xrightarrow{\cong} \operatorname{Tor}_*^A(A'', R).$$

Proof: Let
$$P_* \xrightarrow{\epsilon} R \to 0$$
be a free A resolution of R. Write
$$P_* = A \otimes_R \overline{P_*}.$$
Since P_* is also a free A' resolution of R,
$$\operatorname{Tor}_n^A(R, R) = H_n(R \otimes_{A'} P_*)$$
$$= H_n(R \otimes_{A'} A \otimes_R \overline{P_*}) = H_n(A'' \otimes_R \overline{P_*})$$
$$= H_n(A'' \otimes_A A \otimes_R \overline{P_*}) = \operatorname{Tor}_n(A'', R). \qquad \square$$

Under the hypotheses of the change of rings theorem, A'' is a 2-sided A module. Hence, the isomorphic objects
$$\operatorname{Tor}_*^{A'}(R, R) \cong \operatorname{Tor}_*^A(A'', R)$$
have the structure of a left A module. We now determine that structure in the case of the first derived functor $\operatorname{Tor}_1^{A'}(R, R)$.

First, note that the short exact sequence
$$0 \to I(A) \to A \to R \to 0$$
yields the exact sequence
$$0 = \operatorname{Tor}_1^A(R, A) \to \operatorname{Tor}_1^A(R, R) \to R \otimes I(A) \to R \otimes_A A \to R \otimes_A R \to 0.$$
Since $R \otimes_A A \cong R \cong R \otimes_A R$, it follows that:

Lemma 8.6.2.
$$\operatorname{Tor}_1^A(R, R) \cong R \otimes_A I(A) = Q(A).$$

Lemma 8.6.2 can also be proved by using the canonical bar resolution
$$B_*(A) \xrightarrow{\epsilon} R \to 0$$
with

(a)
$$B_n(A) = A \otimes I(A)^{\otimes n} =$$
the $n+1$ fold tensor product as indicated.

(b) $d_n : B_n(A) \to B_{n-1}(A)$ given by
$$d_n(a \otimes a_1 \otimes \cdots \otimes a_n) = aa_1 \otimes a_2 \otimes \cdots \otimes a_n$$
$$+ \sum_{i=1}^{n-1}(-1)^i a \otimes a_1 \otimes \cdots \otimes a_i a_{i+1} \otimes \cdots \otimes a_n.$$

Or we may use the classical notation
$$d_n(a[a_1|\cdots|a_n]) = aa_1[a_2|\cdots|a_n] + \sum_{i=1}^{n-1}(-1)^i a[a_1|\cdots|a_i a_{i+1}|\cdots|a_n].$$

This is the origin of the term bar resolution.

(c) $\epsilon : B_0(A) = A \to R$ is the augmentation.

Let
$$\overline{B}_*(A) = R \otimes_A B_*(A) = I(A)^{\otimes *}$$
be the so-called bar construction. One sees that
$$\operatorname{Tor}_n^A(R, R) = H_n(\overline{B}_*(A))$$
The first few terms of the bar resolution look like
$$\cdots \xrightarrow{d_3} A \otimes I(A) \otimes I(A) \xrightarrow{d_2} A \otimes I(A) \xrightarrow{d_1} A\epsilon R \to 0$$
with
$$d_1(a[a_1]) = aa_1[\,],$$
$$d_2(a[a_1|a_2]) = aa_1[a_2] - a[a_1 a_2].$$

Our first observation is that the nonzero elements of the bar construction $\overline{B}_n(A) = R \otimes_A B_n(A)$ are represented by tensors $a_1 \otimes \cdots \otimes a_n = [a_1|\cdots|a_n]$ with the differential
$$d_n([a_1|\cdots|a_n]) = \sum_{i=1}^{n-1}(-1)^i[a_1|\cdots|a_i a_{i+1}|\cdots|a_n].$$

Now it is easy to see that
$$\operatorname{Tor}_0^A(R, R) = R$$
$$\operatorname{Tor}_1^A(R, R) = Q(A).$$

8.6 The change of rings isomorphism

The change of rings isomorphism can be described as follows:

The map $A' \to A$ gives a natural map of bar resolutions $B_*(A') \to B_*(A)$ and the change of rings isomorphism is induced by the following map

$$R \otimes_{A'} B_*(A') \to A'' \otimes_A B_*(A), \quad [a'] \mapsto 1[a']$$

(We observe that the algebra map $A' \to A$ makes $R \to A''$ and $R \to R$ into maps of modules.)

So any element in $Q(A') \cong \operatorname{Tor}_1^{A'}(R, R) \xrightarrow{\cong} \operatorname{Tor}_1^A(A'', R)$ is represented by an element $a' \, \varepsilon A'$ via the map $[a'] \mapsto 1[a']$.

The left action of an element $a \, \varepsilon A$ on $\operatorname{Tor}_1^A(A'', R)$ is represented by $a \cdot [a'] = j(a)[a']$.

But

$$d_2(1[a|a'] - (-1)^{\deg(a)\deg(a')} 1[a'|a])$$
$$= 1 \cdot a[a'] - 1[a \cdot a'] - (-1)^{\deg(a)\deg(a')} \{1 \cdot a'[a] - 1[a' \cdot a]\}$$
$$= j(a)[a'] - 1[[a, a']]$$

since $j(a') = 0$. Thus, $j(a)[a']$ is homologous to $1[[a, a']]$.

Specialize to the special case of universal enveloping algebras as in Proposition 8.3.5. In that case, start with a short exact sequence of graded Lie algebras

$$0 \to L' \to L \to L'' \to 0$$

and the sequence of universal enveloping algebras

$$U(L') \to U(L) \to U(L'')$$

is a sequence

$$A' \to A \to A''$$

which satisfies the hypotheses of the change of rings theorem. In this Lie algebra case, it is clear that $[a, a'] \, \varepsilon L'$ for $a' \, \varepsilon L'$, $a \, \varepsilon L$. Hence we have shown

Lemma 8.6.3. *Under the change of rings isomorphism, the left action of $L \subseteq U(L)$ on the image of L' in $QU(L')$ is given by the Lie bracket*

$$a \cdot [a'] = [[a, a']].$$

8.7 Subalgebras of free graded Lie algebras

Let R be a commutative ring in which 2 is a unit and let R have the property that projective modules over R are free over R. For example, R could be a principal ideal domain or, by a theorem of Kaplansky, a local ring [71].

Proposition 8.7.1. *Let V be a free connected R module, that is, $V_0 = 0$. Let $L = L(V)$ be the free graded Lie algebra generated by V and suppose that $L' \subseteq L$ is a subalgebra which, as an R module, is a summand of L. Then L' is a free graded Lie algebra, that is, there is a free R module W such that $L' = L(W)$.*

Proof: Proposition 8.3.3 asserts that $UL(V) = T(V)$ is a free module over $U(L')$. Hence, the free $T(V)$ resolution

$$0 \to T(V) \otimes V \xrightarrow{d_1 = \text{mult}} T(V) \to R \to 0$$

is also a free $U(L')$ resolution.

Therefore the second derived functor

$$\text{Tor}_2^{UL'}(R, R) = 0.$$

In fact, applying $R \otimes_{UL'} (\ )$ to the resolution yields the complex

$$0 \to R \otimes_{UL'} T(V) \otimes V \xrightarrow{\phi = 1 \otimes d_1} R \otimes_{UL'} T(V) \to 0$$

with

$$\text{coker}(\phi) = \text{Tor}_0^{UL'}(R, R) = R,$$
$$\ker(\phi) = \text{Tor}_1^{UL'}(R, R) = QUL' = Ab(L') = W.$$

Clearly, the fact that R is projective over R implies the complex is split and hence that W is projective over R. We are assuming that projective R modules are free R modules so Proposition 8.5.4 and the vanishing of Tor_2 implies that $UL' \cong T(W)$ is a tensor algebra.

Now by Proposition 8.5.3, the fact that the universal enveloping algebra is a tensor algebra implies that $L' = L(W)$ is a free Lie algebra. □

Recall the circumstances in which the change of rings theorem is valid:

We are given a sequence of morphisms augmented algebras

$$A' \xrightarrow{\iota} A \xrightarrow{j} A''$$

with A a free module over A', and

$$A'' \cong R \otimes_{A'} A = A/I(A') \cdot A.$$

Then we have the following important freeness proposition:

8.7 Subalgebras of free graded Lie algebras

Proposition 8.7.2. *Suppose that A is a connected tensor algebra $T(V)$ with V free over R and suppose that A'' is a free module over some tensor subalgebra $T(W) \subseteq A''$ with W free over R. Then $Q(A')$ is a free $T(W)$ module via the Lie bracket action in the change of rings theorem.*

Proof: Recall that the change of rings isomorphism Proposition 8.6.1 asserts that

$$Q(A') \cong \mathrm{Tor}_1^{A'}(R, R) \cong \mathrm{Tor}_1^A(A'', R)$$

and that the actions of A'' and $T(W)$ are a consequence of the first variable of Tor.

Hence $Q(A')$ as an A'' module and as a $T(W)$ module is isomorphic to the kernel of ϕ:

$$0 \to A'' \otimes_{T(V)} T(V) \otimes V \xrightarrow{\phi = 1 \otimes \mathrm{mult}} A'' \otimes_{T(V)} T(V) \to 0.$$

But this the same as

$$0 \to A'' \otimes V \xrightarrow{\phi} A'' \to 0$$

Notice that $\mathrm{image}(\phi) = I(A'')$.

Write A'' as a free $T(W)$ module

$$A'' = \{T(W) \cdot 1\} \bigoplus_\alpha \{T(W) \cdot x_\alpha\}$$

with 1 as the first basis element. Then

$$\mathrm{image}(\phi) = I(A'') = \{I(T(W))\} \bigoplus_\alpha \{T(W) \cdot x_\alpha\}.$$

Since

$$I(T(W)) = W \oplus (W \otimes W) \oplus (W \otimes W \otimes W) \oplus \cdots = T(W) \otimes W$$

is a free $T(W)$ module, it follows that $I(A'')$ is a free $T(W)$ module.

Therefore, $Q(A') \cong \mathrm{kernel}(\phi)$ is a projective $T(W)$ module. Since $T(W)$ is a connected algebra, this implies that $Q(A')$ is a free $T(W)$ module. See Exercise 1 below. □

The preceding proposition combines with the next proposition on Euler–Poincare series to provide a powerful method for determining the generating module of a subalgebra of a free Lie algebra.

Proposition 8.7.3. *Suppose that*

$$0 \to L' \to L \to L'' \to 0$$

is a short exact sequence of graded Lie algebras which is split as a sequence of R modules and suppose that $L = L(V)$ is a free graded Lie algebra with V a connected free R module. Then $L' = L(W)$ is a free graded Lie algebra with W a free R module and with the Euler–Poincare series satisfying the formula

$$\chi(W) = 1 + \chi(UL'')\{\chi(V) - 1\}.$$

Proof: We already know from Theorem 8.6.1 that $L' = L(W)$ so that it only remains to determine the Euler–Poincare series $\chi(W)$.

Recall that the Euler–Poincare series of the tensor algebra

$$T(V) = R \oplus V \oplus V^{\otimes 2} \oplus V^{\otimes 3} \oplus \cdots$$

is

$$\chi(T(V)) = 1 + \chi(V) + \chi(V)^2 + \chi(V)^3 + \cdots = \frac{1}{1 - \chi(V)}.$$

The tensor product decomposition

$$T(V) = UL(V) \cong UL(W) \otimes UL'' = T(W) \otimes UL''$$

implies that

$$\chi(T(V)) = \frac{1}{1 - \chi(V)} = \chi(UL(W)) \cdot \chi(UL'') = \frac{\chi(UL'')}{1 - \chi(W)}$$

and the formula follows by easy manipulation. $\square$

We now illustrate two named applications of Lemmas 8.6.2 and 8.6.3. The names indicate their future use in geometric applications.

Hilton–Milnor Example 8.7.4. Let K be the kernel of the natural epimorphism of free graded Lie algebras $L(V \oplus W) \to L(V)$. Since K is an R-split subalgebra of a free graded Lie algebra, it is itself a free graded Lie algebra $K = L(X)$. We need to determine the R free module X.

First of all,

$$\chi(X) = 1 + \frac{1}{1 - \chi(V)}\{\chi(V) + \chi(W) - 1\} = \frac{\chi(W)}{1 - \chi(V)}$$

$$= \chi(W) \cdot \{1 + \chi(V) + \chi(V)^2 + \chi(V)^3 + \cdots\}.$$

We recognize this Euler–Poincare series as that of the tensor product $W \otimes T(V)$.

Note that W is indecomposable in $L(V \oplus W)$ and $W \subseteq K$. So W is certainly indecomposable in K, that is, W injects into QUK, and in fact the image of W is even indecomposable in QUK with respect to the Lie bracket action of $T(V)$. Since QUK is a free $T(V)$ module with respect to this action, we know that QUK has an R split submodule isomorphic to

$$T(V) \otimes W = W \oplus (V \otimes W) \oplus (V \otimes V \otimes W) \oplus \cdots$$

8.7 Subalgebras of free graded Lie algebras

and, since action is via the Lie bracket, this provides a lifting to an embedding

$$W \oplus [V, W] \oplus [V, [V, W]] \oplus \cdots \subseteq K$$

where $[V, W]$ is the isomorphic image of $V \otimes W$ via

$$v \otimes w \mapsto [v, w] = vw - (-1)^{\deg(v)\deg(w)} wv,$$

$[V, [V, W]]$ is the isomorphic image of $V \otimes V \otimes W$ via

$$v_1 \otimes v_2 \otimes w \mapsto [v_1, [v_2, w]],$$

and so on.

Therefore, since there is a R–split embedding

$$W \oplus (V \otimes W) \oplus (V \otimes V \otimes W) \oplus \cdots \subseteq X$$

and since the two Euler–Poincare series are equal, it follows that

$$X = W \oplus [V, W] \oplus [V, [V, W]] \oplus \cdots = \bigoplus_{i \geq 0} \mathrm{ad}^i(V)(W)$$

via the Lie bracket.

In other words, the kernel of the epimorphism $L(V \oplus W) \to L(V)$ is the free Lie algebra

$$K = L\left(\bigoplus_{i \geq 0} \mathrm{ad}^i(V)(W)\right).$$

The following special case is often used: If K is the kernel of the natural map of free graded Lie algebras $L(x, x_\alpha)_\alpha \to L(x)$, then K is the free graded Lie algebra

$$L(\mathrm{ad}^i(x)(x_\alpha))_{i \geq 0, \alpha}.$$

There are several increasing length filtrations which can be introduced into $L(x, x_\alpha)$ and into K. We can filter by the length in any variable, that is, the length in x or the length in any one of the x_α. Then the generators of K all have length ≤ 1 in one of the x_α and therefore they are not decomposable elements in K. In fact, in a free graded Lie algebra, the relations of anti-symmetry and the Jacobi identity are homogeneous in these length filtrations and these filtrations are therefore in fact length gradations.

Note that, if x has even degree, then $L(x) = \langle x \rangle$, but, if x has odd degree, then $L(x) = \langle x, x^2 \rangle$. This naturally suggests the consideration of another case.

Serre Example 8.7.5. Suppose x has odd degree and let K be the kernel of the natural map $L(x, x_\alpha) \to \langle x \rangle$. Then K is the free graded Lie algebra

$$L(x_\alpha, x^2, [x, x_\alpha])_\alpha.$$

Let V be the span of the x_α and let the degree of x be $2n + 1$. Note that $K = L(X)$ and compute the Euler–Poincare series using 8.6.3:

$$\chi(X) = \chi(V) + t^{4n+2} + t^{2n+1}\chi(V).$$

The computation is completed by noting that the elements $x_\alpha, x^2, [x, x_\alpha]$ form a set of independent generators of K. The Euler–Poincare series tells us that this is a complete set.

Exercises

(1) Suppose that M is a graded module over a connected algebra A and that the ground ring R has the property that projective modules over R are free over R. Show that, if M is a projective A module, then M is a free A module. (Hint: Use the graded version of Nakayama's lemma.)

(2) Consider a sequential inverse limit

$$L = \varprojlim L_n$$

of connected free graded Lie algebras over a field.

(a) Show that the inverse limit L is locally free, that is, if $K \subseteq L$ is any finitely generated subalgebra, then K is a free graded Lie algebra.

(b) Show that any connected locally free graded Lie algebra over a field is a free graded Lie algebra. (Hint: Use Proposition 8.5.4.)

9 Applications of graded Lie algebras

Graded Lie algebras and universal enveloping algebras have topological applications, for example, to product decomposition theorems for loop spaces and, via Samelson products in the Bockstein spectral sequence, to the existence of higher order torsion in homotopy groups.

An early result of Serre [118] asserts that, localized away from 2, the loop space of an even dimensional sphere has a product decomposition

$$\Omega S^{2n+2} \simeq \Omega S^{4n+3} \times S^{2n+1},$$

thus reducing the homotopy groups of an even dimensional sphere to a product of the homotopy groups of two odd dimensional spheres. This can be proven by a general method which we outline as follows.

The homology of certain loop spaces can be identified with a universal enveloping algebra of a graded Lie algebra. Short exact sequences of graded Lie algebras, especially those related to abelianization, lead to tensor product decompositions of universal enveloping algebras. These tensor product decompositions can sometimes be realized by the Kunneth isomorphism applied to a geometric decomposition of a loop space into a product.

Serre's decomposition of the loop space on an even dimensional sphere is based on the abelianization of a free graded Lie algebra on a single generator of odd degree. The abelianization of a free differential graded Lie algebra on two generators connected by a Bockstein leads to a decomposition of the loop space of an even dimensional odd primary Moore space into a product

$$\Omega P^{2n+2}(p^r) \simeq \Omega \bigvee_{k=0}^{\infty} P^{4n+2kn+3}(p^r) \times S^{2n+1}\{p^r\}$$

where the second factor is the homotopy theoretic fibre of the degree p^r map.

In this sense, at odd primes, the homotopy theory both of even dimensional spheres and of even dimensional Moore spaces is reduced to the odd dimensional cases.

284   Applications of graded Lie algebras

The universal enveloping algebra methods also yield an algebraic proof of the Hilton-Milnor theorem. It reduces to a decomposition of a tensor algebra into a tensor product of tensor algebras corresponding to certain Lie brackets.

Higher order torsion exists in the homotopy groups of primary Moore spaces. We prove this, at least when the prime is greater than 3, by studying Samelson products in the homotopy Bockstein spectral sequence of the loop space and their representation via the Hurewicz map.

Although it is not necessary for the proof of the existence of higher order torsion, knowledge of the mod p Hurewicz image is enlightening. Therefore (and also for historical reasons) we present present a summary of the equivalent forms of the nonexistence of elements of mod p Hopf invariant one. Nonexistence can be phrased in the following equivalent forms: (1) p-th powers are not in the mod p Hurewicz image, (2) certain truncated polynomial rings cannot be realized as the mod p cohomology ring of a space, (3) two-cell complexes with nontrivial mod p Steenrod operations do not exist, and (4) a certain homology class in the double loop space of a sphere is not the image of a homotopy class of order p.

Since higher order torsion in homotopy groups are detected by the Bockstein homology of the homotopy differential graded Lie algebra of Samelson products, it is of interest to compute the homology of differential graded Lie algebras. We conclude this chapter with a complete computation of the homology of free graded Lie algebras generated by acyclic modules.

9.1 Serre's product decomposition

The simplest example of the use of universal enveloping algebras of graded Lie algebras is to prove a product decomposition theorem for the loop space of an even dimensional sphere. It requires localization away from 2 and is originally due to Serre. He did not explicitly mention graded Lie algebras in his argument but we feel that the graded Lie algebras add clarity and generality to the method.

Serre's decomposition theorem 9.1.1. *If 2 is inverted there is a homotopy equivalence*

$$\Omega S^{4n+3} \times S^{2n+1} \to \Omega S^{2n+2}.$$

Start with the fact that, with any coefficients,

$$H_*(\Omega S^{2n+2}) = T(x) = UL(x)$$

= a tensor algebra = the universal enveloping algebra of a free graded Lie algebra, where x is a generator of odd degree $2n+1$.

9.1 Serre's product decomposition

The map from the free graded Lie algebra to its abelianization yields the short exact sequence of graded Lie algebras

$$0 \to \langle x^2 \rangle \xrightarrow{i} L(x) \xrightarrow{\pi} \langle x \rangle \to 0.$$

Choosing a section s to the map $U\pi : UL(x) \to U\langle x \rangle$ and multiplying maps yields the isomorphism of coalgebras and of left $U\langle x^2 \rangle$ modules in the standard manner

$$U\langle x^2 \rangle \otimes U\langle x \rangle \xrightarrow{i \otimes s} UL(x) \otimes UL(x) \xrightarrow{\text{mult}} UL(x).$$

This isomorphism involving the tensor product is the algebraic form of the product decomposition.

In order to produce a geometric realization of this algebraic decomposition, we must use coefficients in some ring in which 2 is a unit, for example, the ring $\mathbb{Z}[\frac{1}{2}]$ of integers with 2 inverted.

We note that

$$H_*(S^{2n+1}) = E(x) = U\langle x \rangle, \quad H_*(\Omega S^{4n+3}) = T(x^2) = U\langle x^2 \rangle.$$

Thus the algebraic form of the product decomposition is consistent with the geometric product decomposition in 9.1.1. We must produce maps which realize this decomposition.

Since $x^2 = \frac{1}{2}[x, x]$, it follows that the odd degree square x^2 is in the image of the Hurewicz map

$$\varphi : \pi_*(\Omega S^{2n+2}) \otimes \mathbb{Z}\left[\frac{1}{2}\right] \to H_*\left(\Omega S^{2n+2}; \mathbb{Z}\left[\frac{1}{2}\right]\right).$$

Explicitly, if $\iota : S^{2n+1} \to \Omega S^{2n+2}$ represents a generator of the homotopy group, then the Hurewicz image of the Samelson product is

$$\varphi([\iota, \iota]) = [x, x] = 2x^2.$$

Now the map $\iota : S^{2n+1} \to \Omega S^{2n+2}$ induces in homology the section

$$s : U\langle x \rangle \to UL(x).$$

This is the first step in our geometric realization.

The Samelson product $[\iota, \iota] : S^{4n+2} \to \Omega S^{2n+2}$ has the multiplication extension to the map

$$\overline{[\iota, \iota]} : \Omega \Sigma(S^{4n+2}) \to \Omega S^{2n+2}$$

and, up to the unit $\frac{1}{2}$, this induces in homology the inclusion $U\langle x^2 \rangle \to UL(x)$. This is the second step in our geometric realization.

We conclude by multiplying the two maps in steps one and two. In other words, the map which arises from the loop multiplication,

$$\Omega \Sigma S^{4n+2} \times S^{2n+1} \to \Omega S^{2n+2} \times \Omega S^{2n+2} \xrightarrow{\text{mult}} \Omega S^{2n+2},$$

induces an isomorphism of homology with coefficients $\mathbb{Z}[\frac{1}{2}]$.

Hence, it is an equivalence of spaces with 2 inverted and we are done. □

9.2 Loops of odd primary even dimensional Moore spaces

The next result is a theorem which is for odd primary Moore spaces both an algebraic and a topological analog of Serre's localized product decomposition of the loops on an even dimensional sphere.

Let p be a prime and recall that the odd primary Moore space $P^m(p^r)$ is the cofibre of the degree p^r map $p^r : S^{m-1} \to S^{m-1}$. Recall also that $S^m\{p^r\}$ is the homotopy theoretic fibre of the map $p^r : S^m \to S^m$.

Proposition 9.2.1. *If p is an odd prime and $n \geq 1$, then there is a homotopy equivalence*

$$\Omega \bigvee_{k=0}^{\infty} P^{4n+2kn+3}(p^r) \times S^{2n+1}\{p^r\} \to \Omega P^{2n+2}(p^r).$$

Proof: The first step in proving this is to identify the homology of the loop space $\Omega P^{2n+2}(p^r)$ as the universal enveloping algebra of a graded Lie algebra. It is convenient to use $\mathbb{Z}/p^r\mathbb{Z}$ coefficients since the relevant mod p^r homologies are all free over $\mathbb{Z}/p^r\mathbb{Z}$ (although $\mathbb{Z}/p\mathbb{Z}$ coefficients may also be used).

With $\mathbb{Z}/p^r\mathbb{Z}$ coefficients, the reduced homology

$$\overline{H}_*(P^{2n+1}(p^r)) = \langle u, v \rangle$$

is a free $\mathbb{Z}/p^r\mathbb{Z}$ module on two generators, u of degree $2n$ and v of degree $2n+1$.

Furthermore, if β is the Bockstein associated to the exact coefficient sequence $\mathbb{Z}/p^r\mathbb{Z} \to \mathbb{Z}/p^{2r}\mathbb{Z} \to \mathbb{Z}/p^r\mathbb{Z}$, then $\beta v = u$ and $\beta u = 0$. The Bott–Samelson theorem asserts that the homology of the loop space is a tensor algebra

$$H_*(\Omega \Sigma P^{2n+1}(p^r)) = H_*(\Omega P^{2n+2}(p^r)) = T(u,v) = UL(u,v)$$

with $\beta v = u$.

Now introduce a short exact sequence of graded Lie algebras. Consider the abelianization short exact sequence

$$0 \to K \to L(u,v) \to \langle u, v \rangle \to 0.$$

9.2 Loops of odd primary even dimensional Moore spaces 287

Lemma 9.2.2. *The kernel K is a free graded Lie algebra with a countable set of generators*

$$K = L(\mathrm{ad}^k(u)(v^2), \mathrm{ad}^k(u)([u,v]))_{k \geq 0}.$$

Furthermore, $\beta \, \mathrm{ad}^k(u)(v^2) = \mathrm{ad}^k(v)([u,v])$, that is, the module of generators is acyclic with respect to the Bockstein.

Proof of the lemma: K is a subalgebra of a free graded Lie algebra and it is a free module since it is a split submodule. Hence, K is itself a free graded Lie algebra.

Notice that K can be obtained by taking successive kernels, that is, we have two exact sequences

$$0 \to K_1 \to L(u,v) \to \langle v \rangle \to 0$$
$$0 \to K \to K_1 \to \langle u \rangle \to 0.$$

We see from Section 8.7 that

$$K_1 = L(u, v^2, [u,v])$$

and another application of this section shows that

$$K = L(\mathrm{ad}^k(u)(v^2), \mathrm{ad}^k(v)([u,v]))_{k \geq 0}.$$

Furthermore, $\beta u = 0$, $\beta v = u$ and the derivation property of β implies that $\beta v^2 = [u,v]$ and $\beta \mathrm{ad}^k(u)(v^2) = \mathrm{ad}^k(u)[u,v]$. □

The algebraic form of the product decomposition is the standard tensor product isomorphism related to a subalgebra, that is,

$$UK \otimes U\langle u,v \rangle \xrightarrow{\cong} UL(u,v).$$

We must find spaces whose homologies realize the two factors and construct maps which we can multiply to give the geometric realization of this algebraic decomposition.

Geometric realization of UK

Consider the tensor algebra factor

$$UK = UL(\mathrm{ad}^k(u)(v^2), \mathrm{ad}^k(u)([u,v]))_{k \geq 0}$$
$$= T(\mathrm{ad}^k(u)(v^2), \mathrm{ad}^k(u)([u,v]))_{k \geq 0} = T.$$

The mod p^r homology classes u and v are the respective Hurewicz images of the mod p^r homotopy classes

$$\mu \; \varepsilon \; \pi_{2n}(P^{2n+1}(p^r); \mathbb{Z}/p^r\mathbb{Z}), \quad \nu \; \varepsilon \; \pi_{2n+1}(P^{2n+1}(p^r); \mathbb{Z}/p^r\mathbb{Z})$$

and we have
$$\beta\nu = \mu$$
in the mod p^r homotopy groups.

Use the suspension map $\Sigma : P^{2n+1}(p^r) \to \Omega P^{2n+2}(p^r)$ to identify $\mu = \Sigma_*\mu$ and $\nu = \Sigma_*\nu$ with their images in the mod p^r homotopy groups of the loop space. Then
$$\varphi\mu = u, \quad \varphi\nu = v,$$
implies that the Samelson products have Lie brackets as Hurewicz images, that is,
$$\varphi[\nu,\nu] = [v,v] = 2v^2, \quad \varphi[\mu,\nu] = [u,v]$$
and
$$\varphi \,\mathrm{ad}^k(\mu)([\nu,\nu]) = \mathrm{ad}^k(u)([v,v]), \quad \varphi \,\mathrm{ad}^k(\mu)([\mu,\nu]) = \mathrm{ad}^k(u)([u,v]).$$

Let e generate the top nonzero dimension $2n+1$ of the mod p^r homology of $P^{4n+2+2kn}(p^r)$ and note that βe generates dimension $2n$.

The Samelson products
$$\mathrm{ad}^k(\mu)([\nu,nu]) : P^{4n+2+2kn}(p^r) \to \Omega P^{2n+2}(p^r)$$
induce the following images in reduced homology
$$\mathrm{ad}^k(\mu)([\nu,\nu])_*(e) = \mathrm{ad}^k(u)([v,v]),$$
$$\mathrm{ad}^k(\mu)([\nu,\nu])_*(\beta e) = \beta\mathrm{ad}^k(u)([v,v]) = 2\mathrm{ad}^k(u)([u,v]).$$

In other words, if we add the Samelson products $\mathrm{ad}^k(\mu)([\nu,\nu])$ together, we get a map of a countable bouquet
$$\iota = \bigvee_{k=0}^{\infty} \mathrm{ad}^k(\mu)([\nu,\nu]) : P = \bigvee_{k=0}^{\infty} P^{4n+2+2kn}(p^r) \to \Omega P^{2n+2}(p^r)$$

and the image in reduced homology of this map is, since 2 is a unit in $\mathbb{Z}/p^r\mathbb{Z}$, precisely the module of generators of $K = L(\mathrm{ad}^k(u)(v^2), \mathrm{ad}^k(u)([u,v])_{k\geq 0} \subseteq T$.

Let
$$\overline{\iota} : \Omega\Sigma(P) \to \Omega P^{2n+2}(p^r)$$
be the multiplicative extension of ι. Then
$$\overline{\iota}_* : H_*(\Omega\Sigma P) \to H_*(\Omega P^{2n+2}(p^r))$$
is an injection onto the subalgebra $T = UK \subseteq UL(u,v)$.

This is the geometric realization of the first factor UK.

9.2 Loops of odd primary even dimensional Moore spaces

Geometric realization of the factor $U\langle u, v\rangle$

The free commutative factor

$$U\langle u, v\rangle = P(u) \otimes E(v)$$

is realized by the fibre $S^{2n+1}\{p^r\}$ of the degree p^r map, $p^r : S^{2n+1} \to S^{2n+1}$. Consider the map of fibration sequences

$$\begin{array}{ccccccc}
\Omega S^{2n+1} & \to & S^{2n+1}\{p^r\} & \to & S^{2n+1} & \xrightarrow{p^r} & S^{2n+1} \\
\downarrow & & \downarrow s & & \downarrow t & & \downarrow \text{inclusion} \\
\Omega P^{2n+2}(p^r) & \xrightarrow{=} & \Omega P^{2n+2}(p^r) & \to & PP^{2n+2}(p^r) & \to & P^{2n+2}(p^r)
\end{array}$$

where the bottom row is the path fibration and the map t is a lift of the null homotopic composition

$$S^{2n+1} \xrightarrow{p^r} S^{2n+1} \xrightarrow{\text{inclusion}} P^{2n+2}(p^r).$$

The mod p^r homology Serre spectral sequence of the principal fibration

$$\Omega S^{2n+1} \to S^{2n+1}\{p^r\} \to S^{2n+1},$$

together with the principal action of the fibre ΩS^{2n+1} on the total space $S^{2n+1}\{p^r\}$, shows that

$$H_*(S^{2n+1}\{p^r\}) = P(u) \otimes E(v)$$

with $\beta v = u$ and the powers of u arise from the principle action of ΩS^{2n+1}.

From this it is clear that the map $s : S^{2n+1}\{p^r\} \to \Omega P^{2n+2}(p^r)$ induces in homology a section to the map $UL(u, v) \to U\langle u, v\rangle$.

This is the geometric realization of the second factor.

Geometric realization via multiplying maps

We multiply the maps to realize the algebraic decomposition and produce a mod p^r homology equivalence

$$\Omega\Sigma P \times S^{2n+1}\{p^r\} \xrightarrow{\bar{\iota}\times s} \Omega P^{2n+2}(p^r) \times \Omega P^{2n+2}(p^r) \xrightarrow{\text{mult}} \Omega P^{2n+2}(p^r).$$

Since the spaces all have integral homology which is p-torsion, they are localized at p. Since the homology is of finite type, the mod p^r equivalence is actually a homotopy equivalence. $\square$

9.3 The Hilton–Milnor theorem

Graded Lie algebras give an algebraic proof of the Hilton–Milnor theorem.

Hilton–Milnor theorem 9.3.1. *Let X and Y be connected spaces. There is a homotopy equivalence*

$$\Psi : \Omega\Sigma\left(\bigvee_{k\geq 0} X^{\wedge k} \wedge Y\right) \times \Omega\Sigma X \to \Omega\Sigma(X \vee Y).$$

Definition of the equivalence Ψ

The map Ψ is defined using Samelson products as follows:

Let

$$\iota_X : X \to \Omega\Sigma(X \vee Y)$$
$$\iota_Y : Y \to \Omega\Sigma(X \vee Y)$$

be the two inclusions and form the Samelson products

$$S_k = \operatorname{ad}(\iota_X)(\iota_Y) : X^{\wedge k} \wedge Y \to \Omega\Sigma(X \vee Y).$$

Form the bouquet of maps

$$S = \bigvee S_k : \mathbb{Z} = \bigvee_{k\geq 0} X^{\wedge k} \wedge Y \to \Omega\Sigma(X \vee Y)$$

and then form the multiplicative extension

$$\overline{S} : \Omega\Sigma\mathbb{Z} \to \Omega\Sigma(X \vee Y).$$

This is the geometric realization of the first factor.

Let

$$\overline{\iota_X} : \Omega\Sigma X \to \Omega\Sigma(X \vee Y)$$

be the standard inclusion. This is the geometric realization of the second factor.

Now multiply maps. That is, let Ψ be the composition

$$\Omega\Sigma\mathbb{Z} \times \Omega\Sigma X \xrightarrow{\overline{S}\times\overline{\iota_X}} \Omega\Sigma(X \vee Y) \times \Omega\Sigma(X \vee Y) \xrightarrow{\text{mult}} \Omega\Sigma(X \vee Y).$$

As a first step, we claim that Ψ induces a homology isomorphism with coefficients in any field.

Filtering by powers of the augmentation ideal

For the moment, assume that the coefficients are some field. If X and Y were to have a trivial comultiplication in homology, for example, if they were both

9.3 The Hilton–Milnor theorem

suspensions, then the proof that Ψ is an equivalence would be a simple consequence of the tensor product decomposition of 8.7.4:

$$T(V \oplus W) \cong T\left(\sum_{i=0}^{\infty} V^{\otimes i} \otimes W\right) \otimes T(V)$$

$$UL(V \oplus W) \cong UL\left(\sum_{i=0}^{\infty} V^{\otimes i} \otimes W\right) \otimes UL(V).$$

The point is that Samelson products $X^{\wedge i} \wedge Y \to \Omega\Sigma(X \vee Y)$ induce Lie brackets in homology only if the homologies of X and Y are primitive, that is, the comultipication is trivial.

In the general case, we need to introduce the filtration by powers of the augmentation ideal.

In order to show that Ψ is a homotopy equivalence we need to improve the Hopf algebra by replacing it with a primitively generated Hopf algebra. To do this, introduce the decreasing filtration defined by the powers of the augmentation ideal.

If A is any connected Hopf algebra, let $I = IA = \overline{A}$ be the augmentation ideal and consider the filtration defined by the powers of I, that is,

$$F^k A = I^k.$$

Since the filtration is multiplicative, that is,

$$F^i A \cdot F^j A \subseteq F^{i+j} A,$$

and comultiplicative, that is,

$$\Delta(F^k A) \subseteq \sum_{i+j=k} F^i A \otimes F^j A,$$

there is an induced Hopf algebra structure on the associated graded object

$$E_0^* = F^* A / F^{*+1} A.$$

We remark that the inverse (conjugation) map $\iota : A \to A$ induces the inverse map $\iota = E_0^*(\iota) : E_0^* A \to E_0^* A$.

Note that the associated graded Hopf algebra is always primitively generated by grading 1: $E_0^1 A = Q(A)$.

What commutators do in homology

The commutator in a grouplike space corresponds to the Hopf algebra Lie bracket below:

Definition 9.3.2. The Hopf algebra Lie bracket is the composition

$$[\ ,\ ]_H : A \otimes A \xrightarrow{\Delta \otimes \Delta} A \otimes A \otimes A \otimes A \xrightarrow{1 \otimes T \otimes 1} A \otimes A \otimes A \otimes A$$
$$\xrightarrow{1 \otimes 1 \otimes \iota \otimes \iota} A \otimes A \otimes A \otimes A \xrightarrow{\text{mult}} A.$$

The Hopf algebra Lie bracket on A induces the Hopf algebra Lie bracket on the associated graded

$$E_0^* A.$$

Recall

Lemma 9.3.3. *If x and y are primitive classes, then the Hopf algebra Lie bracket is the commutator Lie bracket, that is,*

$$[x, y]_H = [x, y] = xy - (-1)^{\deg(x)\deg(y)} yx.$$

Since commutators of primitive classes are primitive, it follows that any iterated Hopf algebra Lie bracket of primitive classes is equal to the corresponding iterated commutator.

A filtration compatible with the powers of the augmentation ideal

We define a decreasing filtration on the homology of

$$\Omega \Sigma Z \times \Omega \Sigma X$$

so that Ψ_* is filtration preserving.

Let V be the reduced homology of X, W be the reduced homology of Y, and $M = \sum_{k \geq 0} V^{\otimes k} W =$ the reduced homology of $Z = \bigvee_{k \geq 0} X^{\wedge k} \wedge Y$.

On the homology $T(V)$ of $\Omega \Sigma X$, choose the filtration by powers of the augmentation ideal.

Define the filtration on M by

$$F^0 M = F^1 M = M, \quad F^k M = \sum_{k \geq i} V^{\otimes k} \otimes W.$$

Extend this filtration multiplicatively to $T(M)$ via:

$$F^0 T(M) = T(M), F^1 T(M) = IT(M) = \overline{T(M)},$$
$$F^k (M \otimes \cdots \otimes M) = \sum_{i_1 + \cdots + i_m = k} F^{i_1} M \otimes \cdots F^{i_m} M.$$

9.3 The Hilton–Milnor theorem

Finally, extend the filtration to the tensor product $T(M) \otimes T(V)$ by

$$F^k(T(M) \otimes T(V)) = \sum_{i+j=k} F^i T(M) \otimes F^j T(V).$$

Then the associated graded of this filtration is

$$E^0(T(M) \otimes T(V)) \cong E^0 T(M) \otimes E^0 T(V) \cong T(M) \otimes T(V).$$

The isomorphism Φ in the primitively generated case

Observe that Ψ is filtration preserving and, since $E_0^* T(V \oplus W)$ is primitively generated, the map $E^O \Psi = \Phi$ is the isomorphism from Example 8.7.4:

$$\Phi : T(A) \otimes T(V) \xrightarrow{\iota_1 \otimes \iota_2} T(V \oplus W) \otimes T(V \oplus W) \xrightarrow{\text{mult}} T(V \otimes W)$$

which is defined as follows:

$$V^{\otimes k} \otimes W \to T(V \oplus W)$$

is the iterated commutator

$$x_1 \otimes x_k \otimes y \mapsto \text{ad}(x_1) \ldots \text{ad}(x_k)(y).$$

$$M = \sum_{k \geq 0} V^{\otimes k} \otimes W \to T(V \oplus W)$$

is the sum of these maps and

$$T(M) \to T(V \oplus W)$$

is the multiplicative extension.

$$T(V) \to T(V \oplus W)$$

is the inclusion.

From Example 8.7.4, Φ is an isomorphism.

Conclusion of the proof of the Hilton–Milnor theorem

Since the filtration is finite in each degree, $E^0 \Psi = \Phi$ is an isomorphism implies that Ψ induces an isomorphism in homology with any field coefficients, for example, with coefficients in the field with a prime number of elements.

If we restrict X and Y to be finite type CW complexes, we get that Ψ is a homology isomorphism with coefficients the integers localized at any prime.

Since any CW complex is a direct limit of finite type cell complexes and since homology commutes with direct limits, we get that Ψ induces an isomorphism for all cell complexes with coefficients localized at any prime. In other words, Ψ is a equivalence localized at any prime.

Hence, Ψ is a homotopy equivalence for all connected CW complexes X and Y. □

Exercises

(1) Let $\iota : A \to A$ be the inverse map in a connected Hopf algebra over a ring R, that is, it is the unique map such that the composition

$$A \xrightarrow{\Delta} A \otimes A \xrightarrow{\iota \otimes 1} A \otimes A \xrightarrow{\text{mult}} A$$

is the composition

$$A \xrightarrow{\varepsilon} R \xrightarrow{\eta} A.$$

(a) Show that ι preserves the filtration by powers of the augmentation ideal.

(b) Show that, if x is primitive, then $\iota(x) = -x$.

9.4 Elements of mod p Hopf invariant one

This section is a summary of the equivalent odd primary forms of the Hopf invariant one problem and will be used in the forthcoming section on the existence of higher order torsion in the homotopy groups of odd primary Moore spaces.

Throughout this section, let p be an odd prime. The vanishing Theorem 2.12.2 is a very strong form of the nonexistence of elements of mod p Hopf invariant one:

Liulevicius–Shimada–Yamanoshita vanishing theorem. *Suppose p is an odd prime. If X is a space such that the degree one Bockstein β and the first Steenrod operation P^1 of degree $2p - 2$ both vanish in the mod p cohomology of X, then all Steenrod operations vanish.*

The above vanishing theorem seems to stronger than the mutually equivalent forms below. The following theorem lists some of the classical equivalent forms of what is called the existence or nonexistence of elements of mod p Hopf invariant one.

Proposition 9.4.1. *If p is an odd prime, the following are all equivalent:*

(a) *If $\iota \in H_{2n}(\Omega S^{2n+1}; \mathbb{Z}/p\mathbb{Z})$ is a generator, then the p-th power ι^p is in the image of the mod p Hurewicz map*

$$\varphi : \pi_{2np}(\Omega S^{2n+1}; \mathbb{Z}/p\mathbb{Z}) \to H_{2np}(\Omega S^{2n+1}; \mathbb{Z}/p\mathbb{Z}).$$

(b) *There exists a map $f : P^{2np}(p) \to \Omega S^{2n+1}$ such that the mapping cone C_f has a mod p cohomology ring which contains a truncated polynomial algebra generated by a class $\alpha \in H^{2n}(C_f; \mathbb{Z}/p\mathbb{Z})$ truncated at height exactly*

9.4 Elements of mod p Hopf invariant one

$p+1$, *that is,*

$$\alpha^p \neq 0, \quad \alpha^{p+1} = 0.$$

(c) *For any or all $s \geq 1$, there exists a 2-cell complex*

$$X_s = S^{2n+s} \cup_\gamma e^{2n+s+2n(p-1)}$$

for which the mod p Steenrod operation P^n is nonzero in mod p cohomology.

(d) *The Hurewicz map φ maps an element of order p in $\pi_{2pn-2}(\Omega^2 S^{2n+1})$ to a generator of the $\mathbb{Z}_{(p)}$ homology group $H_{2pn-2}(\Omega^2 S^{2n+1}; \mathbb{Z}_{(p)}) \cong \mathbb{Z}/p\mathbb{Z}$.*

(e) *Liulevicius–Shimada–Yamanoshita:*

$$n = 1.$$

We prove the proposition by proving

$$a) \implies b) \implies c) \implies e) \implies a)$$

and

$$a) \implies d) \implies a).$$

The equivalence of condition (e) is essentially the Liulevicius–Shimada–Yamanoshita Theorem 2.12.2 which we do not prove.

Proof that $(a) \implies (b)$**:**

Assume (a) and let $f : P^{2np}(p) \to \Omega S^{2n+1}$ be a map such that

$$\varphi[f] = \iota^p.$$

Recall the identification with the James construction

$$\Omega S^{2n+1} \simeq J(S^{2n})$$

and the increasing filtration $J_k(S^{2n})$ which captures the first k powers of ι. In particular the map f factors through the subspace $J_p(S^2 n)$. In the cofibration sequence

$$P^{2np}(p) \xrightarrow{f} J_p(S^{2n}) \xrightarrow{j} C_f,$$

collapse the cofibration sequence

$$* \to J_{p-1}(S^{2n}) \xrightarrow{=} J_{p-1}(S^{2n})$$

to a point. The result is the standard cofibration sequence

$$P^{2np}(p) \to S^{2np} \xrightarrow{p} S^{2np}.$$

The integral cohomology

$$H(\Omega S^{2n+1}) = \Gamma[\alpha] =$$

the divided power algebra on a generator α of degree $2n$. If

$$\xi \in H^{2np}(C_f)$$

is a generator, we see that

$$j^*\xi = p\gamma_p(\alpha) = p\left(\frac{\alpha^p}{p!}\right) = \frac{\alpha^p}{(p-1)!}.$$

Reducing mod p, we see that $H^*(C_f; \mathbb{Z}/p\mathbb{Z})$ contains a truncated polynomial algebra generated by α of height $p+1$.

Proof that $(b) \implies (c)$:

Let $C_f = J_p(S^{2n}) \cup_f C(P^{2np}(p))$ be the complex which validates (b) and let $s \geq 1$. Clearly the Steenrod operation P^n is nontrivial in the mod p cohomology of $\Sigma^s C_f$. Using the bouquet decomposition

$$\Sigma J_{p-1}(S^{2n}) \simeq \bigvee_{k=1}^{p-1} S^{2nk+1}$$

we can form the two-cell complex

$$X_s = \Sigma^s C_f / \bigvee_{k=2}^{p-1} S^{2nk+s}$$

for which the mod p Steenrod operation P^n is nonzero.

Proof that $(c) \implies (e)$:

The Liulevicius–Shimada–Yamanoshita theorem [50, 76, 121] implies that P^1 must be nonzero in the mod p cohomology of the two-cell complex

$$X_s = S^{2n+s} \cup_\gamma e^{2n+s+2n(p-1)}.$$

Hence, $n = 1$.

Proof that $(e) \implies (a)$:

Suppose $n = 1$.

Consider the integral homology Serre spectral sequence of the fibration sequence

$$S^1 \to \Omega S^3 \langle 3 \rangle \to \Omega S^3$$

where $S^3 \langle 3 \rangle$ is the three-connected cover. It is a spectral sequence of Hopf algebras with

$$E^2_{*,*} = H_*(\Omega S^3) \otimes H_*(S^1) = P[\iota] \otimes E[x]$$

9.4 Elements of mod p Hopf invariant one

where ι is a polynomial generator of degree 2, x is an exterior generator of degree 1, and $d^2\iota = x$. Thus,
$$d^2(\iota^p) = p\iota^{p-1}x$$
shows that
$$H_{2p-1}(\Omega S\langle 3\rangle) = \mathbb{Z}/p\mathbb{Z}.$$

Now the mod p homology spectral sequence shows that
$$H_*(\Omega S^3\langle 3\rangle) = P[\iota^p] \otimes E[\iota^{p-1}]$$
and we have the mod p Bockstein differential
$$\beta^1[\iota^p] = [\iota^{p-1}x].$$
Clearly, $[\iota^p]$ is in the mod p Hurewicz image in $\Omega S^3\langle 3\rangle$ and hence also in ΩS^3.

Proof that $(a) \implies (d) \implies (a)$:

First we perform a computation of the homology of $\Omega^2 S^{2n+1}$ through a small range of dimensions. (In a subsequent chapter, we will compute it all.)

Consider the path space fibration
$$\Omega^2 S^{2n+1} \to P\Omega S^{2n+1} \to \Omega S^{2n+1}.$$
In the Serre spectral sequence for $\mathbb{Z}_{(p)}$ homology we see immediately that
$$H_{2n-1}(\Omega^2 S^{2n+1}; \mathbb{Z}_{(p)}) \cong \mathbb{Z}_{(p)}$$
generated by x with
$$d^{2n}\iota = x.$$

The formula
$$d^{2n}(\iota^p) = p\iota^{p-1}x$$
tells us that
$$H_{2pn-2}(\Omega^2 S^{2n+1}; \mathbb{Z}_{(p)}) \cong \mathbb{Z}/p\mathbb{Z}$$
generated by σ with
$$d^{2n(p-1)}(\iota^{p-1}x) = \sigma.$$

Since the element $\iota\sigma$ must die in the spectral sequence, we get the computation that the reduced homology $\overline{H}_*(\Omega^2 S^{2n+1}; \mathbb{Z}_{(p)})$ is

(1)
$$\mathbb{Z} \text{ for } * = 2n+1 \text{ generated by } x \text{ with } d^{2n}\iota = x.$$
$$\mathbb{Z}/p\mathbb{Z} \text{ for } * = 2np-2 \text{ generated by } \sigma \text{ with } d^{2n(p-1)}(\iota^{p-1}x) = \sigma.$$
$$0 \text{ for } * < 2np+2n-3.$$

In the Serre spectral sequence for $\mathbb{Z}/p\mathbb{Z}$, only minor changes occur and we get that the reduced homology $\overline{H}_*(\Omega^2 S^{2n+1}; \mathbb{Z}/p\mathbb{Z})$ is

(2)
$$\mathbb{Z}/p\mathbb{Z} \text{ for } * = 2n+1 \text{ generated by } x \text{ with } d^{2n}\iota = x.$$
$$\mathbb{Z}/p\mathbb{Z} \text{ for } * = 2np-2 \text{ generated by } \sigma \text{ with } d^{2n(p-1)}(\iota^{p-1}x) = \sigma.$$
$$\mathbb{Z}/p\mathbb{Z} \text{ for } * = 2np-1 \text{ generated by } \tau \text{ with } d^{2np}(\iota^p) = \tau \text{ and } \beta^1 \tau = \sigma.$$
$$0 \text{ for } * < 2np+2n-3.$$

Note that, if trans = the transgression, then
$$\operatorname{trans}(\iota^p) = \tau.$$

We now prove the equivalence of (a) and (d):

$\iota^p \; \varepsilon \text{ image } \varphi : \pi_{2pn}(\Omega S^{2n+1}; \mathbb{Z}/p\mathbb{Z}) \to H_{2pn}(\Omega S^{2n+1}; \mathbb{Z}/p\mathbb{Z})$

if and only if:

$$\tau = \operatorname{trans}(\iota^p) \; \varepsilon \text{ image } \varphi : \pi_{2pn-1}(\Omega^2 S^{2n+1}; \mathbb{Z}/p\mathbb{Z})$$
$$\to H_{2pn-1}(\Omega^2 S^{2n+1}; \mathbb{Z}/p\mathbb{Z})$$

(write $\varphi(\tilde{\tau}) = \tau$.)

if and only if:

the integral Bockstein $\beta\tilde{\tau} = \tilde{\sigma}$ is an element of order p which maps onto the Bockstein $\beta\tau = \sigma$ which is a generator of $H_{2np-2}(\Omega^2 S^{2n+1}; \mathbb{Z}_{(p)})$. □

Let $C(n)$ be the homotopy theoretic fibre of the double suspension $\Sigma^2 : S^{2n-1} \to \Omega^2 S^{2n+1}$ localized at p. The Serre spectral sequence tells us that $C(n)$ is $2np-4$ connected and that

(1) $\overline{H}_*(C(n); \mathbb{Z}_{(p)}) = \mathbb{Z}/p\mathbb{Z}$ for $* = 2np-3$ and 0 otherwise for $* < 2pn + 2n - 4$.

(2) $\overline{H}_*(C(n); \mathbb{Z}/p\mathbb{Z}) = \mathbb{Z}/p\mathbb{Z}$ for $* = 2np-3, 2pn-4$ and 0 otherwise for $* < 2pn+2n-4$.

These results correspond under transgression to the above computations for the homology of $\Omega^2 S^{2n+1}$.

Now recall that

$$H_*(\Omega P^{2n+1}(p^r) : \mathbb{Z}/p\mathbb{Z}) = T(u,v) =$$

a tensor algebra generated by u of degree $2n-1$ and v of degree $2n$. We have the loops on the pinch map

$$\Omega q : \Omega P^{2n+1}(p^r) \to \Omega S^{2n+1}$$

and in mod p homology this induces

$$(\Omega q)_* : T(u,v) \to T(\iota), \quad u \mapsto 0, \quad v \mapsto \iota.$$

There are the following additional equivalent forms of the nonexistence of elements of mod p Hopf invariant one:

Proposition 9.4.2. *If $s \geq 1$ the following are equivalent:*

(a) *The p^s power v^{p^s} is in the image of the mod p Hurewicz homomorphism.*

(b) *The p^s power ι^{p^s} is in the image of the mod p Hurewicz homomorphism.*

(c) $n = 1$ *and* $s = 1$.

Proof: It is clear that (a) implies (b). Assume (b) and consider the p-th Hopf invariant maps

$$\Omega S^{2n+1} \xrightarrow{h_p} \Omega S^{2pn+1} \xrightarrow{h_p} \Omega S^{2p^2n+1} \xrightarrow{h_p} \cdots$$

By Proposition 5.2.2, $h_p(\iota_{2n}^{p^s}) = \nu \cdot \iota_{2np}^{p^{s-1}}$ where ν is a unit. It follows from 9.4.1 that $n = 1$ and $s = 1$.

That (c) implies (a) will be not proved here [29]. It is sometimes called desuspending the Adams map. $\square$

9.5 Cycles in differential graded Lie algebras

In this section we are going to exhibit some nontrivial homology classes which arise in differential graded Lie algebras. We begin with the definition of a differential graded Lie algebra.

Definition 9.5.1. A differential graded Lie algebra is a graded Lie algebra L together with a degree -1 linear map $d : L \to L$ such that

(1) d is a differential, that is, $d \circ d = 0$.

(2) d is a derivation, that is,

$$d[x,y] = [dx, y] + (-1)^{\deg(x)}[x, dy] \quad \text{for all } x, \ y \ \varepsilon L,$$

and
$$dx^2 = [dx, x] \quad \text{for all } x \text{ of odd degree.}$$

Remark. Notice that, if A is a differential graded associative algebra, then the graded Lie algebra A (given by the bracket operation and the squaring of odd dimensional classes) is a differential graded Lie algebra.

Remark. Of course, if 2 is a unit in the ground ring, then $2x^2 = [x, x]$ for odd degree x and $d[x, x] = [dx, x] - [x, dx] = 2[dx, x]$ imply that the formula $dx^2 = [dx, x]$ is a consequence of the derivation property for Lie brackets.

Recall that, in a graded algebra,
$$\text{ad}(x)(y) = [x, y] = xy - (-1)^{\deg(x)\deg(y)} yx$$
and we write
$$\text{ad}(x)^k(x)(y) = \text{ad}^k(x)(y)$$
for the iterated Lie bracket.

Lemma 9.5.2. *Let x be an even degree element in a graded associative differential algebra with differential d. Then*

(a)
$$dx^n = \sum_{j=0}^{n-1}(j, n-j)\text{ad}^{n-j-1}(x)(dx)x^j$$

(b) *and thus, if the characteristic of the ground field is a prime p,*
$$dx^{p^k} = \text{ad}^{p^k-1}(x)(dx).$$

We set $\tau_k(x) = \text{ad}^{p^k-1}(x)(dx)$ and note that
$$d(\tau_k(x)) = 0.$$

Proof: Part (a) follows by induction using the derivation formula
$$dx^n = d(x^{n-1} \cdot x) = (dx^{n-1})x + x^{n-1}(dx)$$
$$= (dx^{n-1})x + \text{ad}(x^{n-1})(x) + (dx)x^{n-1}.$$

Part (b) follows immediately from part (a). □

Hence, in characteristic p, we have found a nonzero Lie element $\tau_k(x)$ which is a cycle. We are going to find another one.

9.5 Cycles in differential graded Lie algebras

Lemma 9.5.3. *Let x be an even degree element in a differential Lie algebra. If the differential is d and $k \geq 1$, then*

$$d(\mathrm{ad}^{k-1}(x)(dx))$$

$$= \sum_{j=1}^{s}(j, k-j)[\mathrm{ad}^{j-1}(x)(dx), \mathrm{ad}^{k-1-j}(x)(dx)] \quad \textit{if } k = 2s+1 \textit{ is odd}$$

and

$$= \left\{ \sum_{j=1}^{s-1}(j, k-j)[\mathrm{ad}^{j-1}(x)(dx), \mathrm{ad}^{k-1-j}(x)(dx)] \right\}$$
$$+ (s, s)\{\mathrm{ad}^{s-1}(x)(dx)\}^2 \quad \textit{if } k = 2s \textit{ is even.}$$

Proof: It is simplest to proceed as follows:

If 2 is a unit in the ground ring, then the desired formula can be rewritten as

$$d(\mathrm{ad}^{k-1}(x)(dx)) = \frac{1}{2}\sum_{j=1}^{k-1}(j, k-j)[\mathrm{ad}^{j-1}(x)(dx), \mathrm{ad}^{k-1-j}(x)(dx)].$$

This form can be proven by induction using the well known formula for the sum of binomial coefficients, $(i-1, j) + (i, j-1) = (i, j)$, and the derivation formula

$$d(\mathrm{ad}^k(x)(dx)) = d[x, \mathrm{ad}^{k-1}(x)(dx)] = [dx, \mathrm{ad}^{k-1}(x)(dx)]$$
$$+ [x, d(\mathrm{ad}^{k-1}(x)(dx))].$$

It is easy to do. Assume that it has been done.

The truth of the general case of the formula then follows by the following considerations:

(1) With the ground ring the integers $\mathbb{Z}$, consider the differential free graded Lie algebra $L(x, dx)$ generated by two elements x and dx. This embeds in the free graded Lie algebra $L(x, dx) \otimes Q$ over the rationals Q where 2 is a unit. Thus, the formula is true in this integral case.

(2) Hence, it is true in the free graded Lie algebra $L(x, dx) \otimes R$ over any commutative ring R.

(3) It follows that it is true in any graded Lie algebra over any commutative ring. $\square$

We see in an explicit way that $d(\mathrm{ad}^{p^k-1}(x)(dx)) = d(\tau_k(x))$ has all of its coefficients divisible by p and we define

$$\sigma_k(x) = \frac{1}{p}d(\tau_k(x))$$

$$= \begin{cases} \sum_{j=1}^{\frac{p^k-1}{2}} \frac{(j, p^k-j)}{p}[\mathrm{ad}^{j-1}(x)(dx), \mathrm{ad}^{p^k-j-1}(x)(dx)] & \text{if } p \text{ is an odd prime} \\ \{\sum_{j=1}^{2^{k-1}} \frac{(j, 2^k-j)}{2}[\mathrm{ad}^{j-1}(x)(dx), \mathrm{ad}^{2^k-j-1}(x)(dx)]\} \\ + \frac{(2^{k-1}, 2^{k-1})}{2}\{\mathrm{ad}^{2^{k-1}-1}(x)(dx)\}^2 & \text{if } p = 2. \end{cases}$$

If 2 is a unit in the ground ring, then the definition assumes the more symmetric form

$$\sigma_k(x) = \frac{1}{2} \sum_{j=1}^{p^k-1} \frac{(j, p^k-j)}{p}[\mathrm{ad}^{j-1}(x)(dx), \mathrm{ad}^{p^k-1-j}(x)(dx)].$$

Over any commutative ring which has the property that all projective modules are free, let $L(x, dx)$ be the differential free graded Lie algebra generated by x and dx. The kernel L_0 of the natural map

$$L(x, dx) \to L(x), \quad x \mapsto x, \quad dx \mapsto 0$$

is the free Lie algebra generated by all $\mathrm{ad}^{k-1}(x)(dx)$. Among other things, it is a free module over its ground ring. And two-fold brackets of its distinct odd degree generators together with the squares of odd degree generators are linearly independent in it.

In particular, since $d(p\sigma_k(x)) = d(d\tau_k(x)) = 0$, it follows that

$$d(\sigma_k(x)) = 0, \quad \sigma_k(x) \neq 0$$

in L_0. Thus, we have

Lemma 9.5.4. *If x is an even degree element, then*

(1)
$$d\sigma_k(x) = 0.$$

(2)
$$[\sigma_k(x)] \neq 0$$

in $HL(x, dx)$.

(3) *If the ground field has characteristic p, then*

$$d\tau_k(x) = 0.$$

(4) *If the characteristic is p, then*

$$[\tau_k(x)] \neq 0$$

in $HL(x, dx)$.

Proof: Parts (1) and (3) are already done. Part (4) follows from looking at the bigrading that comes from counting the number of occurences of x and dx. More precisely, $\tau_k(x)$ has bidegree $(p^k - 1, 1)$ and $L(x, dx)$ is 0 in bidegree $(p^k, 0)$. Hence, $\tau_k(x)$ cannot be a boundary.

Part (2) follows from the fact that $\sigma_k(x)$ has bidegree $(p^k - 2, 2)$ and $L(x, dx)$ is generated by $\tau_k(x)$ in bidegree $(p^k - 1, 1)$. Since $d\tau_k(x)$ is divisible by p, it follows that $\sigma_k(x)$ is not a boundary.

Remark. Lemma 9.5.4 remains true in a differential graded quasi-Lie algebra. The proof does not require the use of the triple Jacobi identity $[x, [x, x]] = 0$ for odd degree elements.

Exercises

(1) If $L = L(x_\alpha)_\alpha$ is a free graded Lie algebra over a ring for which projective modules are free modules, show that the elements $[x_\alpha, x_\beta]$, $\alpha \neq \beta$ and x_γ^2 with x_γ of odd degree are linearly independent.

9.6 Higher order torsion in odd primary Moore spaces

Throughout this section, let p be an odd prime. We are going to prove that higher order torsion exists in the homotopy groups of a mod p^r Moore space. We need a Lie algebra structure on mod p Samelson products. Therefore, in order to prove some of these results, we assume either that p is greater than 3 or that $r \geq 2$. The results are true even if $p = 3$ and $r = 1$ but the proof is harder and will not be given in this book.

Let

$$\mu : P^{2n-1}(p) \to \Omega P^{2n+1}(p^r)$$

and

$$\nu : P^{2n}(p) \to \Omega P^{2n+1}(p^r)$$

represent generators of the mod p homotopy groups with the Bockstein differentials $\beta^r \nu = \mu, \beta^r \mu = 0$.

Consider the β^r cycles from Lemma 9.5.4:

$$\tau_k(\nu) = \mathrm{ad}^{p^k-1}(\nu)(\mu) \; \varepsilon \; E^r_\pi(\Omega P^{2n+1}(p^r))_{2np^k-1}$$

$$\sigma_k(\nu) = \frac{1}{2}\Sigma^{p^k-1}_{j=1}[\mathrm{ad}^{j-1}(\nu)(\mu), \mathrm{ad}^{p^k-j-1}(\nu)(\mu)] \; \varepsilon \; E^r_\pi(\Omega P^{2n+1}(p^r))_{2np^k-2}$$

If p is a prime greater than 3, then the mod p homotopy Bockstein spectral sequence is a spectral sequence of differential graded Lie algebras and these are always cycles. If $p = 3$, then the mod p homotopy Bockstein spectral sequence is a spectral sequence of quasi graded Lie algebras if $r \geq 2$ and, since the triple vanishing identity is not required for the proof, these are cycles if $r \geq 2$.

$$H_*(\Omega P^{2n+1}(p^r); \mathbb{Z}/p\mathbb{Z}) = T(u,v) = E^1_H(\Omega P^{2n+1}) = \cdots = E^r_H(\Omega P^{2n+1})$$

the tensor algebra generated by u of degree $2n-1$ and v of degree $2n$ with $\beta^r v = u$. On the other hand, all mod p homology information disappears at

$$E^{r+1}_H(\Omega P^{2n+1}(p^r)) = H_*(T(u,v), \beta^r) = \mathbb{Z}/p\mathbb{Z}.$$

Note that the free graded Lie algebra $L(u,v)$ is embedded as a Lie subalgebra of the tensor algebra $T(u,v)$. In fact,

$$L(u,v) \subseteq P(u,v) \subset T(u,v)$$

where $P(u,v)$ is the submodule of primitive elements. Furthermore,

$$P(u,v) = L(u,v) \cup \xi L(u,v) \cup \xi^2 L(u,v) \cup \cdots$$

where ξ is the Frobenius map which raises even degree elements to the p-th power and which is 0 on odd degree elements. In other words, the only difference between the module of all primitives and the module of all Lie elements is that the former has p^k-th powers of even dimensional Lie elements. The image of the mod p Hurewicz map contains all the Lie elements as images of Samelson products and, on the other hand, the mod p Hurewicz image is contained in the module of primitive elements.

Consider the Hurewicz map of mod p Bockstein spectral sequences

$$\varphi^r : E^r_\pi(\Omega P^{2n+1}(p^r)) \to E^r_H(\Omega P^{2n+1}(p^r)).$$

Differentials are zero and nothing happens in the homology Bockstein spectral sequence until we reach the r-th stage. Clearly,

$$L(u,v) \subseteq \mathrm{im}\, \varphi^r \subseteq P(u,v)$$

and the results on the nonexistence of elements of mod p Hopf invariant one place severe restrictions on the extent to which the Hurewicz image can be larger than the graded Lie algebra $L(u,v)$. In particular, the only time that v^{p^k} is in the Hurewicz image is when $n = 1, k = 1$.

9.6 Higher order torsion in odd primary Moore spaces

We detect higher order torsion in homotopy groups by means of the Hurewicz representation of the mod p homotopy Bockstein spectral sequence in the mod p homology Bockstein spectral sequence. We note that the Hurewicz image

$$\text{im } \varphi^r \subseteq PE_H^r(\Omega P^{2n+1}(p^r))$$

is a differential submodule of the module of primitives.

Definition 9.6.1. The factored Hurewicz map is

$$\overline{\varphi}^{r+1} : E_\pi^{r+1}(\Omega P^{2n+1}(p^r)) \to HPE_H^r(\Omega P^{2n+1}(p^r))$$

which factors the Hurewicz map of Bockstein spectral sequences as

$$E_\pi^{r+1}(\Omega P^{2n+1}(p^r)) \to H(\text{im}\varphi^r) \to HPE_H^r(\Omega P^{2n+1}(p^r))$$
$$\to E_H^{r+1}(\Omega P^{2n+1}(p^r)).$$

Thus, the range of the factored Hurewicz map is the β^r homology of the module of primitives. If we replace this range by the β^r homology of the Hurewicz image itself, we shall call it the strongly factored Hurewicz map.

Proposition 9.6.2. *With the restrictions on p and r as indicated in the first paragraph of this section,*

(a) *The elements $[\sigma_k(\nu)] \in E_\pi^{r+1}(\Omega P^{2n+1}(p^r))$ are nontrivial for all $k \geq 1$.*

(b) *The elements $[\tau_k(\nu)] \in E_\pi^{r+1}(\Omega P^{2n+1}(p^r))$ are nontrivial for all $k \geq 2$ or all $k \geq 1, n > 1$.*

This is a consequence of the fact that these classes have nontrivial image under the strongly factored Hurewicz map. The elements $\tau_k(\nu)$ have Hurewicz image $\beta^r v^{p^k} = \tau_k(v)$ but the nonexistence of elements of mod p Hopf invariant one implies that the powers v^{p^k} are not in the Hurewicz image except in one case, $r = 1, n = 1$.

Hence, we have proved that there exist elements of order $\geq p^{r+1}$ in $\pi_*(\Omega P^{2n+1}(p^r)) = \pi_{*+1}(P^{2n+1}(p^r))$. The exact order and dimension require that we know more, namely, that $\overline{\varphi}^{r+1} \circ \beta^{r+1}[\tau_k(\nu)] = [\sigma_k(v)] \neq 0$. If we knew this nontriviality of the $r+1$-st Bockstein differential, then the properties of the Bockstein spectral sequence would imply:

Proposition 9.6.3.

$$\pi_{2p^k n - 1}(P^{2n+1}(p^r))$$

contains a $\mathbb{Z}/p^{r+1}\mathbb{Z}$ summand.

The nontriviality of the homology class, that is, the fact that $[\sigma_k(v)] \neq 0$ in the homology of the primitives

$$HPE_H^r(\Omega P^{2n+1}(p^r)) \to E_H^{r+1}(\Omega P^{2n+1}(p^r))$$

follows from the identity

$$HPE^r_H(\Omega P^{2n+1}(p^r)) = HPUL(u,v) = HP(u,v)$$

and from Proposition 9.7.5 that $[\sigma_k(v)] \neq 0 \varepsilon HP(u,v)$.

The equation

$$\overline{\varphi}^{r+1} \circ \beta^{r+1}[\tau_k(\nu)] = [\sigma_k(u)]$$

will be proved in the computation of the mod p homology Bockstein spectral sequence for the so-called fibre of the pinch map $P^{2n+1}(p^r) \to S^{2n+1}$ in Proposition 11.5.2.

Exercises

(1) If p is a prime greater than 3, find a countable number of nonzero elements in the $r+1$-st term of the mod p homotopy Bockstein spectral sequence of the loop space of the even dimensional Moore space $\Omega P^{2n+2}(p^r)$. (Hint: Consider $[\nu,\nu]$ where ν generates $\pi_{2n+1}(\Omega P^{2n+2}(p^r); \mathbb{Z}/p\mathbb{Z})$.)

(2) Use Section 9.2 and this section to give another solution to exercise 1.

9.7 The homology of acyclic free differential graded Lie algebras

In this section the ground ring will be a field K of finite characteristic p. We shall show that, in a sense to be made precise, the operations $\tau_k(x)$ and $\sigma_k(x)$ generate all of the homology of a so-called acyclic differential graded Lie algebra.

Algebras generated by acyclic modules

Recall that an acyclic graded module V is a differential module with $H(V,d) = 0$. Given such V, we define three objects which are referred to as acyclic even if, in fact, they are not, they are merely generated by an acyclic module.

The acyclic tensor algebra is $T(V)$ with the usual differential. We note that it is in fact acyclic in the sense that $HT(V) = T(H(V)) = T(O) = K$.

The acyclic free graded Lie algebra is $L(V) \subseteq T(V)$. We have already seen that $\tau_k(x)$ and $\sigma_k(x)$ represent cycles which are nonhomologous to zero in the so-called acyclic free graded Lie algebra $L(x,dx)$ with the degree of x even. It is the purpose of this section to compute the homology $HL(V)$.

Consider also the related differential object, $P(V) = $ the module of primitives in $T(V)$. $P(V)$ may be called the acyclic free primitive module or the acyclic free restricted graded Lie algebra. We shall compute the homology $HL(V)$ first and then $HP(V)$ will be a corollary.

9.7 The homology of acyclic free differential graded Lie algebras

In the previous section, we saw that the factored Hurewicz map detects higher order torsion in the Bockstein spectral sequence. The homology of the graded Lie algebra $HL(V)$ corresponds to homotopy in the Bockstein spectral sequence. The map $HL(V) \to HP(V)$ is the algebraic version of the factored Hurewicz map.

Useful operations on even dimensional classes

Define three operations $\tau(x)$, $\sigma(x)$, $\eta(x)$ on an even dimensional class x as follows:

(1)
$$\tau(x) = \mathrm{ad}^{p-1}(x)(dx) = \tau_1(x) = d(x^p).$$

The purpose of $\tau(x)$ is to be killed by x^p.
Note that $d(x^{p^k}) = \tau_k(x) = \mathrm{ad}^{p^k-1}(x)(dx) = \tau(x^{p^k})$.

(2)
$$\sigma(x) = \sigma_1(x) = \frac{1}{p} d(\tau(x)).$$

We shall see that, up to a nonzero factor, $\sigma(x^{p^k})$ and $\sigma_k(x)$ are homologous cycles, that is,

$$[\sigma(x^{p^k})] = \lambda_k [\sigma_k(x)], \quad \lambda_k \neq 0 \; \varepsilon K.$$

(3)
$$\eta(x) = x^{p-1}(dx).$$

We shall see that there exists a noncommutative polynomial $\kappa(x)$ of degree $< p$ in the Lie elements from $L(x, dx)$ such that

$$d(\eta(x) + \kappa(x)) = \sigma(x).$$

In other words, the purpose of $\eta(x)$ is to kill $\sigma(x)$.

The homology of acyclic free graded Lie algebras is given by:

Proposition 9.7.1. *Write $L(V) = HL(V) \oplus K$ where K is an acyclic module. If K has an acyclic basis, that is, a basis*

$$\{x_\alpha, y_\alpha, z_\beta, w_\beta\}$$

with

$$dx_\alpha = y_\alpha, \quad \deg(x_\alpha) \text{ even},$$
$$dz_\beta = w_\beta, \quad \deg(z_\beta) \text{ odd},$$

then $HL(V)$ has a basis

$$\{\tau_k(x_\alpha), \sigma_k(x_\alpha)\}_{\alpha, k \geq 1}.$$

Remarks. Before we begin the proof of Proposition 9.7.1, we review some useful facts concerning connected Hopf algebras and universal enveloping algebras.

Recall the following result of Milnor and Moore: [90]

Proposition 9.7.2. *Let B be a connected Hopf algebra over a field of characteristic p. Then primitives are indecomposable in B, that is,*

$$P(B) \to Q(B)$$

is a monomorphism if and only if B is associative, graded commutative, and all p-th powers are zero.

(In the case of characteristic zero, the above is true with no condition on powers.)

Corollary 9.7.3. *Let B be a connected differential Hopf algebra over a field. Then*

(a) *B has indecomposable primitives implies that the homology Hopf algebra HB has indecomposable primitives.*

(b) *B is primitively generated (that is, $P(B) \to Q(B)$ is an epimorphism) implies that the homology Hopf algebra HB is primitively generated.*

(c) *B is biprimitive implies that the homology Hopf algebra HB is biprimitive.*

Proof of the corollary: Part (a) is a clear consequence of 9.7.2.

Part (b) needs the following result which is left to the exercises.

Lemma 9.7.4. *Any connected Hopf algebra B over a field is a direct limit of finite type Hopf algebras.*

Since homology commutes with direct limits, we may assume that B is primitively generated of finite type in order to prove Part b),

But, in this case, the dual Hopf algebra B^* has indecomposable primitives and, by Part (a), so does the homology $H(B^*)$. Hence, the homology $HB = (HB^*)^*$ is primitively generated.

Part (c) is a consequence of Parts (a) and (b). □

Let L be a connected graded Lie algebra, that is, $L_0 = 0$, and let $A = UL$ be the universal enveloping algebra.

Recall that the Lie filtration on A is the increasing filtration with

$$F_0 A = K, \quad F_1 A = K + L$$

$$F_n A = F_{n-1} A + \mathrm{im}\, (PA \otimes F_{n-1} A \xrightarrow{\mathrm{mult}} A).$$

9.7 The homology of acyclic free differential graded Lie algebras

This filtration is multiplicative, that is,
$$F_i A \cdot F_j A \subseteq F_{i+j} A,$$
and comultiplicative
$$\Delta(F_k A) \subseteq \sum_{i+j=k} F_i A \otimes F_j A.$$

Thus, the associated graded object $E^0_* A = F_* A / F_{*+1} A$ has an induced Hopf algebra structure.

$E^0_* A$ is primitively generated by $E^0_1 A = L$.

Moreover, $E^0_* A = S(L) = E(L_{\text{odd}}) \otimes P(L_{\text{even}})$ is associative, graded commutative (which includes that the squares of odd dimensional classes are zero).

Hence, if L is a connected differential graded Lie algebra, then the Lie filtration on the differential Hopf algebra $A = UL$ defines a spectral sequence of Hopf algebras abutting to the homology HUL. All the spectral sequence terms $E^r_* A$ are primitively generated differential Hopf algebras,
$$P(E^r_* A) \to Q(E^r_* A)$$
is an epimorphism, and, of course, the differentials satisfy
$$d^r(P(E^r_* A)) \subseteq P(E^r_* A).$$

Armed with the above remarks, we proceed to:

Proof of Proposition 9.7.1: The computation of $HL(V)$ is a consequence of the following facts:

(1) $E^r_* = E^r_* T(V)$ is a spectral sequence of primitively generated differential graded Hopf algebras.

(2)
$$E^\infty_* = K, \quad E^0_1 = L(V), \quad \text{and } E^1_1 = HL(V).$$

The idea is that, by the time we reach E^∞_*, the generators of $E^1_1 = HL(V)$ must be eliminated by differentials in the spectral sequence. Understanding the way in which this elimation can occur will compute the homology $HL(V)$.

Then
$$E^0_* = S(L(V)) = S(HL(V) \oplus K) = S(HL(V)) \otimes S(K),$$
$$\text{bidegree}(L(V))) = (1, *),$$
and the differential d^r maps bidegree $(t, *)$ to bidegree $(t - r, *)$.

Here is a detailed description of the effect of the differentials $dx_\alpha = y_\alpha$ and $dz_\beta = w_\beta$:

(1)
$$d^0 x_\alpha = y_\alpha$$

eliminates the generators x_α and y_α and, in the process, it creates as nontrivial homology classes the primitive generators of E_*^1, x_α^p and $x_\alpha^{p-1} y_\alpha = \eta(x_\alpha)$, both of bidegree $(p, *)$.

(2) In the universal example, $A = T(x, dx)$, d^0 does not eliminate $\sigma(x)$ since it is a nontrivial homology class in $HL(V)$. But it must be eliminated by an element with $p - 1$ occurences of x and 1 occurrence of dx. There are only two possibilities, $\tau(x)$ which cannot do it since $d\tau(x) = 0$, and $\eta(x) = x^{p-1}(dx)$ which therefore must do it, at least up to unit which can safely be ignored. Since $\eta(x)$ has bidegree $(p, *)$, this implies that, up to a unit,
$$d^{p-1} \eta(x) = \sigma(x).$$

If we interpret what this means in a spectral sequence, we can assert the more precise statement: There is a noncommutative polynomial
$$\kappa(x) \; \varepsilon \; T(x, dx)$$
which has Lie filtration $< p$ such that
$$d(\eta(x) + \kappa(x)) = \sigma(x).$$

(3) Therefore, for all x_α,
$$d(\eta(x_\alpha) + \kappa(x_\alpha)) = \sigma(x_\alpha), \text{ and } d^{p-1} \eta(x_\alpha) = \sigma(x_\alpha).$$
Since $d^{p-1}(\eta(x_\alpha) \sigma(x_\alpha)^j) = \sigma(x_\alpha)^{j+1}$, this differential eliminates the two generators $\eta(x_\alpha)$ and $\sigma(x)$ and it creates no new generators.

(4) Since $d(x_\alpha^p) = \tau(x_\alpha)$, we have $d^{p-1}(x_\alpha^p) = \tau(x_\alpha)$. This differential eliminates the generators x_α^p and $\tau(x_\alpha)$ and creates the two primitive generators $x_\alpha^{p^2}$ and $\eta(x_\alpha^p)$.

(5) Since
$$d(x_\alpha^{p^k}) = \tau(x_\alpha^{p^{k-1}})$$
and
$$d(\eta(x_\alpha^{p^{k-1}}) + \kappa(x_\alpha^{p^{k-1}})) = \sigma(x_\alpha^{p^{k-1}}),$$

9.7 The homology of acyclic free differential graded Lie algebras

with $\kappa(x_\alpha^{p^{k-1}})$ of Lie filtration $< p$, we have the differentials

$$d^{p^k-1}(x_\alpha^{p^k}) = \tau(x_\alpha^{p^{k-1}})$$

$$d^{p^k-1}(\eta(x_\alpha^{p^{k-1}})) = \sigma(x_\alpha^{p^{k-1}}).$$

The continuation of this pattern explains the existence of the basis elements $\tau(x_\alpha^{p^{k-1}}) = \tau_k(x_\alpha)$ and $\sigma(x_\alpha^{p^{k-1}})$ in the homology $HL(V)$.

(6)
$$d^0 z_\beta = w_\beta$$

eliminates the two generators z_β and w_β and, since $d^0(z_\beta w_\beta^j) = w_\beta^{j+1}$, it creates no new generators.

Since all generators must be eliminated, it follows that

$$\{\tau_k(x_\alpha) = \tau(x_\alpha^{p^{k-1}}), \sigma(x_\alpha^{p^{k-1}})\}_{\alpha, k \geq 1}$$

is a complete list of the basis of $HL(V)$.

The proof is completed by noting that, in the universal example, $L(x, dx)$, the nonzero homology classes $[\sigma(x^{p^{k-1}})]$ and $[\sigma_k(x)]$ both have weight $p^k - 2$ in x and weight 2 in dx. Therefore, they must be equal up to a unit.

Hence,

$$\{\tau_k(x_\alpha), \sigma_k(x_\alpha)\}_{\alpha, k \geq 1}$$

is a basis for $HL(V)$. □

Consider the short exact sequence

$$0 \to L(V) \to P(V) \to P(V)/L(V) \to 0.$$

Since $P(V)/L(V)$ has a basis of p–th powers, its basis is

$$\{x_\alpha^{p^j}, w_\beta^{p^j}, \sigma_k(x_\alpha)^{p^j}\}_{\alpha, j \geq 1}.$$

Furthermore, $H(P(V)/L(V)) = P(V)/L(V)$ and the long exact homology sequence shows that

Proposition 9.7.5.

(a) *The homology $HP(V)$ has a basis*

$$\{w_\beta^{p^k}, \sigma_k(x_\alpha)^{p^j}\}_\beta, \alpha, k \geq 1, j \geq 0\}.$$

(b) *The image of $HL(V) \to HP(V)$ has a basis $\{\sigma_k(x_\alpha)\}_{\alpha, k \geq 1}$.*

312 Applications of graded Lie algebras

Exercises

(1) Let x and y be elements in an associative algebra A and suppose that x has even degree.

(a) Show that
$$\mathrm{ad}(x^n)(y) = \sum_{j=0}^{n-1} \mathrm{ad}^{n-j}(x)(y)x^j.$$
(Hint: Use induction and the formula
$$\mathrm{ad}(x^n)(y)x + \mathrm{ad}(x)\mathrm{ad}(x^{n-1})(y) + \mathrm{ad}(x)(y)x^{n-1}.)$$

(b) If A is an associative algebra over a field of characteristic p, show that
$$\mathrm{ad}(x^p)(y) = \mathrm{ad}^p(x)(y).$$

(c) If A is an associative differential algebra over a field of characteristic p, show that
$$L(x^p, d(x^p)) \subseteq L(x, dx).$$

(2) Show that, for even degree elements x, the fact that $\kappa(x)$ has Lie filtration $< p$ in $UL(x, dx)$ implies that the element $\kappa(x^{p^k})$ has Lie filtration $< p$ in $UL(x, dx)$ for all k.

(3) Prove Lemma 9.7.4.

10 Differential homological algebra

Differential homological algebra is used in algebraic topology to study the homology of fibrations. The Serre spectral sequence [116] determines the homology of the total space given the homologies of the base and fibre. In general, given the homologies of two out of the three, base, fibre, and total space, various techniques have been developed to determine the homology of the third.

Two cases of historical interest are the following: (1) Given the homology of a space, determine the homology of the loop space. (2) Given the homology of a topological group, determine the homology of the classifying space. In a real sense, these problems are dual to one another and have specific solutions.

The problem of determining the homology of the loop space is solved with the help of the cobar construction introduced by Frank Adams [1]. This differential algebra, when applied to the chains on a space, is homologically equivalent to the chains on the loop space together with the multiplication induced by loop multiplication.

The problem of determining the homology of the classifying space is solved with the help of the bar construction introduced by Eilenberg and MacLane [41] and followed up in a geometric form by John Milnor [86]. This differential coalgebra, when applied to the chains on a topological group, is homologically equivalent to the chains on the classifying space together with the coalgebra structure given by the diagonal.

In order to study the cobar and bar constructions, we begin with some preliminary algebraic results on algebras and coalgebras. We describe derivations on algebras and coderivations on coalgebras. We pay special attention to the universal associative algebras and the universal associative coalgebras, the so-called tensor algebras and tensor coalgebras. This is used to describe the differentials in the cobar and bar constructions.

The chains on a principal bundle are modeled by twisted tensor products of differential algebras and differential coalgebras [20]. The chains on the path space are modeled by a universal twisted tensor product involving the cobar construction.

Dually, the chains on the universal bundle are modeled by a universal twisted tensor product involving the bar construction.

The limitations of specific models for fibrations are overcome by the introduction of relative homological algebra. To this end, we study modules over algebras and determine the projective objects. We study tensor products and their derived functors which will be used in the study of homogeneous spaces. We encounter certain problems with the homological algebra when applied to differential modules. These problems were solved by the introduction by Eilenberg and Moore of projective classes [43]. In a projective class, the definitions of projective objects, epimorphisms, and exact sequences are changed in a consistent manner which make them more suitable to application to topology.

Of more interest to us here in this chapter is the study of comodules over coalgebras. We introduce the notion of cotensor product of comodules and proceed to study the derived functors with the help of relative injective classes wherein the injective objects, the monomorphisms, and the exact sequences are modifed in a consistent manner.

In the language of Bill Singer, the main object of this chapter is the second quadrant Eilenberg–Moore spectral sequence. A special case of this spectral sequence computes the homology of a loop space and is used by us in the next chapter to study the homology of the fibre of the pinch map and thence to prove the fundamental splitting theorems into a product. These lead to the exponent theorems for spheres and Moore spaces at odd primes.

Much of the material in this chapter is inspired by the seminal Cartan seminar of John Moore [94] on the homology of classifying spaces. The first application of relative homological algebra to the study of classifying spaces occured there. And it introduced the use of the differential Tor of several variables to study coproducts in the homology of classifying spaces. But, since the chains on a topological group are a differential Hopf algebra, the diagonal, being a map of differential algebras, can induce the coproduct and thus make the use of the differential Tor of several variables optional.

We introduce here the use of the differential Cotor of several variables in order to define and to study products in the homology of the loop space. In this case, there is no pairing of the chains on a topological space which can induce the algebra structure in differential Cotor. The use of differential Cotor of several variables is essential to define an algebra structure in differential Cotor which corresponds to the multiplication in the loop space.

In the case of the second quadrant Eilenberg–Moore models, the dominant algebraic structure is the structure of an associative algebra in differential Cotor. Coalgebra structures may occur but they are not always present. When present for a natural reason, they are consistent with the dominant structure. For example,

10.1 Augmented algebras and supplemented coalgebras

if we start with a differential coalgebra which happens to be a differential Hopf algebra, then we get a coalgebra structure in differential Cotor and this combines with the algebra structure to give a Hopf algebra structure in differential Cotor.

The paper of Eilenberg–Moore [42] introduced the spectral sequence of the same name but also the more fundamental Cotor approximation to the chains on a homotopy pullback. Our treatment has also been influenced by Larry Smith's thesis [122] on the Eilenberg–Moore spectral sequence and by the paper of Husemoller, Moore, and Stasheff [64]. The latter is especially insightful in its treatment of twisted tensor products. And twisted tensor products have a nice summary in a paper of Hess and Levi [52].

In the last chapter of this book, we complete our presentation of differential homological algebra by presenting the so-called first quadrant Eilenberg–Moore spectral sequence which computes the homology of classifying spaces.

10.1 Augmented algebras and supplemented coalgebras

Throughout this chapter, we shall use the following sign convention for tensor products of maps $f : A \to C$ and $g : B \to D$ of possibly nonzero degree. A sign is introduced whenever elements of nonzero degree pass by one another, that is, the map $f \otimes g : A \otimes B \to C \otimes D$ is given on tensor elements by $(f \otimes g)(x \otimes y) = (-1)^{\deg(x)\deg(g)} f(x) \otimes g(y)$.

We begin by recalling some well known notions concerning augmented algebras and indecomposables. Although these notions are extremely well known, it is worthwhile reviewing them in order to make the dual notions transparent. Then we shall continue by considering the dual notions of supplemented coalgebras and primitives.

An algebra A is a positively graded module over a commutative ground ring R which has an associative multiplication $\mu : A \otimes A \to A$ and a unit $\eta : R \to A$.

In detail, a multiplication is a degree 0 linear map $\mu : A \otimes A \to A$, written $\mu(a \otimes b) = a \cdot b$, which is associative so that the diagram

$$\begin{array}{ccc} A \otimes A \otimes A & \xrightarrow{\mu \otimes 1} & A \otimes A \\ \downarrow 1 \otimes \mu & & \downarrow \mu \\ A \otimes A & \xrightarrow{\mu} & A \end{array}$$

commutes and η is a two-sided unit so that the diagram

$$\begin{array}{ccccc} R \otimes A & \xrightarrow{\eta \otimes 1} & A \otimes A & \xleftarrow{1 \otimes \eta} & A \otimes R \\ & {}_{\cong}\searrow & \downarrow \mu & \swarrow_{\cong} & \\ & & A & & \end{array}$$

commutes. The slant maps above are the canonical isomorphisms, that is, $1 \cdot a = a \cdot 1 = a$ for all $a \in A$.

The algebra is called (graded) commutative if $a \cdot b = (-1)^{\deg(a)\deg(b)} b \cdot a$ for all $a, b \in A$, that is,

$$\begin{array}{ccc} A \otimes A & \xrightarrow{T} & A \otimes A \\ {}_{\mu}\searrow & & \swarrow_{\mu} \\ & A & \end{array}$$

commutes where $T: A \otimes A \to A$ is the twist map, $T(a \otimes b) = (-1)^{\deg(a)\deg(b)} b \otimes a$.

Recall that the ground ring R is an algebra concentrated in degree 0 with multiplication $\mu: R \otimes R \cong R$, $\mu(1 \otimes 1) = 1$ and unit $\eta: R \cong R$, $\eta(1) = 1$.

An augmented algebra is an algebra A over R together with a map of algebras $\epsilon: A \to R$. We denote the augmentation ideal by $I(A) = \overline{A} = \text{kernel}(\epsilon)$. It can also be called the reduced algebra. Note that we have a direct sum decomposition $A = \overline{A} \oplus \eta(R)$. We identify $R \simeq \eta(R)$; it is the summand generated by 1.

Remarks. A differential algebra is just an algebra A with the extra structure of a degree -1 differential $d: A \to A$ for which the structure maps of multiplication and unit are maps of differential objects. The ground ring R is a differential algebra with the zero differential. And an augmented differential algebra is just a differential algebra with the augmentation $\epsilon: A \to R$ being a map of differential objects. In particular, $\overline{A}$ is closed under the differential.

Definition 10.1.1. If A is an augmented algebra, the module of indecomposables QA is the cokernel of the reduced multiplication, that is, $QA = \overline{A}/\overline{A} \cdot \overline{A} = \text{coker } \mu: \overline{A} \otimes \overline{A} \to \overline{A}$.

The product ideal $\overline{A} \cdot \overline{A} \subseteq \overline{A}$ is called the module of decomposables and, if A is connected, that is, if $\epsilon: A_0 \cong R$, then QA is related to choosing a set of algebra generators as follows. We shall say that we have chosen a generating module for A if we have chosen a linear map $\iota: QA \to \overline{A}$ which is a section of the natural projection $\pi: \overline{A} \to QA$, that is, $\pi \cdot \iota = 1_{QA}$. (Choosing a set of generators is equivalent to choosing ι and a set of module generators of QA.)

Lemma 10.1.2. *If A is a connected augmented algebra with a choice of generating module ι, then the iterated multiplication maps*

$$\mu_n: QA \otimes \ldots (n - times) \cdots \otimes QA$$
$$\xrightarrow{\iota \otimes \cdots \otimes \iota} \overline{A} \otimes \ldots (n - times) \cdots \otimes \overline{A} \xrightarrow{mult} \overline{A}$$

10.1 Augmented algebras and supplemented coalgebras

combine to give a surjective map

$$\bigoplus_{n \geq 0} \mu_n : \bigoplus QA \otimes \cdots \otimes QA \to \overline{A},$$

that is, every element of $\overline{A}$ can be written as a sum of elements of the form $\iota(a_1) \cdot \ldots \cdot \iota(a_n)$ with $a_i \in QA$.

Proof: Use induction on the degree and the split exact sequence

$$\overline{A} \otimes \overline{A} \to \overline{A} \to QA \to 0$$

to get $\overline{A} = \iota(QA) \oplus (\overline{A} \cdot \overline{A})$. □

Corollary 10.1.3. *Let $f, g : A \to B$ be two maps (homomorphisms) of augmented algebras and let $\iota : QA \to \overline{A}$ be a generating module. If $f \cdot \iota = g \cdot \iota$, then $f = g$.*

Definition 10.1.4. *Let $f : A \to B$ be a map of augmented algebras and let $g : A \to B$ be a linear map of possibly nonzero positive or negative degree. Then g is called a derivation with respect to f if $\epsilon \cdot g = 0$ and the following diagram commutes*

$$\begin{array}{ccc} A \otimes A & \xrightarrow{g \otimes f + f \otimes g} & B \otimes B \\ \downarrow \mu & & \downarrow \mu \\ A & \xrightarrow{g} & B. \end{array}$$

The next lemma shows that derivations are like homomorphisms in that they are determined by their values on generators.

Lemma 10.1.5. *Let $g, h : A \to B$ be two derivations with respect to a homomorphism f of augmented algebras and let $\iota : QA \to \overline{A}$ be a choice of a generating module. If $g \cdot \iota = h \cdot \iota$, then $g = h$.*

Proof: Merely note that

$$g(a_1 \cdot \ldots \cdot a_i \cdot \ldots \cdot a_n)$$
$$= \Sigma_{i=1}^n (-1)^{\deg(g)(\deg(a_1) + \cdots + \deg(a_{i-1}))} fa_1 \cdot \ldots \cdot ga_i \cdot \ldots \cdot fa_n$$

and similarly for h. □

Corollary 10.1.6. *Let $f : A \to B$ be a map of augmented algebras and suppose that A and B are in fact differential algebras. If $\iota : QA \to A$ is a choice of a generating module and $d \cdot f \cdot \iota = f \cdot d \cdot \iota$, then $f \cdot d = d \cdot f$, that is, f is a map of augmented differential algebras.*

Proof: Note that $d \cdot f$ and $f \cdot d$ are both derivations with respect to f and are therefore determined by their effect on generators. □

We now turn to supplemented coalgebras.

A coalgebra C is a positively graded module over a commutative ground ring R which has an associative diagonal $\Delta : C \to C \otimes C$ and a counit $\epsilon : C \to R$. We denote the so called reduced coalgebra by $\overline{C} = \text{kernel}(\epsilon)$.

In detail, a diagonal is a degree 0 linear map $\Delta : C \to C \otimes C$ which is associative in the sense that the diagram

$$\begin{array}{ccc} C & \xrightarrow{\Delta} & C \otimes C \\ \downarrow \Delta & & \downarrow \Delta \otimes 1 \\ C \otimes C & \xrightarrow{1 \otimes \Delta} & C \otimes C \otimes C \end{array}$$

commutes.

A counit is a degree 0 linear map $\epsilon : C \to R$ such that

$$\begin{array}{ccccc} & & C & & \\ & \cong \swarrow & \downarrow \Delta & \searrow \cong & \\ R \otimes C & \xleftarrow{\epsilon \otimes 1} & C \otimes C & \xrightarrow{1 \otimes \epsilon} & C \otimes R \end{array}$$

commutes where the slant maps are the canonical isomorphisms.

The coalgebra is called (graded) commutative if

$$\begin{array}{ccccc} & & C & & \\ & \Delta \swarrow & & \searrow \Delta & \\ C \otimes C & & \xrightarrow{T} & & C \otimes C \end{array}$$

commutes.

Recall that the ground ring R is a coalgebra concentrated in degree 0 with diagonal $\Delta : R \cong R \otimes R, \Delta(1) = 1 \otimes 1$ and counit $\epsilon : R \cong R$, $\epsilon(1) = 1$.

A supplemented coalgebra is a coalgebra C over R together with a map of coalgebras $\eta : R \to C$.

We have a direct sum decomposition $C = \overline{C} \oplus \eta(R)$. We identify $R \simeq \eta(R)$ and we note that the counit properties show that diagonal can be described via:

$$\Delta(1) = 1 \otimes 1,$$

and, if $c \epsilon \overline{C}$, then

$$\Delta(c) = c \otimes 1 + 1 \otimes c + \Sigma c' \otimes c''$$

where $c', c'' \epsilon \overline{C}$.

The reduced diagonal $\overline{\Delta} : \overline{C} \to \overline{C} \otimes \overline{C}$ is given by $\overline{\Delta}(c) = \Sigma c' \otimes c''$.

Remarks. A differential coalgebra C is just a coalgebra with the extra structure of a degree -1 differential $d : C \to C$ for which the structure maps of diagonal

10.1 Augmented algebras and supplemented coalgebras

and counit are maps of differential objects. The ground ring R is a differential coalgebra with the zero differential. And a supplemented differential coalgebra is just a differential coalgebra with the supplement $\eta : R \to C$ being a map of differential objects.

The following is the coalgebra notion which is dual to the notion of indecomposables for an augmented algebra.

Definition 10.1.7. If C is a supplemented coalgebra, the module of primitives PC is the kernel of the reduced diagonal, that is, $PC = \ker \overline{\Delta} : \overline{C} \to \overline{C} \otimes \overline{C}$. Thus, $c \in PC$ if and only if $\Delta(c) = c \otimes 1 + 1 \otimes c$.

The primitives form the first stage of an increasing filtration of the reduced coalgebra $\overline{C}$.

Let C be a supplemented coalgebra. For $n \geq 1$, the coalgebra C has a unique iterated diagonal given by

$$\Delta^{(n)} : C \to C \otimes \ldots (n-times) \cdots \otimes C$$

where

$$\Delta^{(1)} = 1_{\overline{C}}, \ \Delta^{(2)} = \Delta$$

$$\Delta^{(n)} = (1 \otimes \cdots \otimes 1 \otimes \Delta) \cdot \Delta^{(n-1)} = \cdots = \Delta \otimes 1 \otimes \cdots \otimes 1) \cdot \Delta^{(n-1)}.$$

This is well defined by Exercise 8.

Similarly, for $n \geq 2$, the reduced coalgebra $\overline{C}$ has a unique iterated diagonal given by

$$\overline{\Delta}^{(n)} : \overline{C} \to \overline{C} \otimes \ldots (n-times) \cdots \otimes \overline{C}$$

where

$$\overline{\Delta}^{(1)} = 1_{\overline{C}}, \ \overline{\Delta}^{(2)} = \Delta$$

$$\overline{\Delta}^{(n)} = (1 \otimes \cdots \otimes 1 \otimes \overline{\Delta}) \cdot \overline{\Delta}^{(n-1)} = \cdots = (\overline{\Delta} \otimes 1 \otimes \cdots \otimes 1) \cdot \overline{\Delta}^{(n-1)}.$$

Definition 10.1.8. If C is a supplemented coalgebra, the reduced coalgebra filtration is the increasing filtration of $\overline{C}$ given by $F_n C = \ker \overline{\Delta}^{(n)} : \overline{C} \to \overline{C} \otimes \cdots \otimes \overline{C}$.

Thus, $F_2 C = PC$ and if C is connected, that is, if $C_0 = R$, then $F_{n+1} C = \overline{C}$ in degree n.

Lemma 10.1.9. *Let C be a supplemented coalgebra and assume that C is flat as a module over R. Then, for $n \geq 2$, $F_n C =$ the inverse image of $PC \otimes \cdots \otimes PC \subseteq \overline{C} \otimes \cdots \otimes \overline{C}$ with respect to the map $\overline{\Delta}^{(n-1)}$.*

Proof: It is clear that the inverse image via the map $\overline{\Delta}^{n-1}$ of any

$$\overline{C} \otimes \cdots \otimes \overline{C} \otimes PC \otimes \overline{C} \otimes \cdots \otimes \overline{C}$$

is contained in $F_n C$. Thus, the inverse image of $PC \otimes \cdots \otimes PC$ is contained in $F_n C$. We need to show that $\overline{\Delta}^{n-1}$ factors through $PC \otimes \cdots \otimes PC$.

Assume inductively that we can factor

$$\overline{\Delta}^{(n-1)} : F_n C \to PC \otimes \cdots \otimes PC \otimes \overline{C} \otimes \cdots \otimes \overline{C}$$

with $k \leq n - 2$ factors of PC. Thus, $F_n C$ is contained in the inverse image of

$$PC \otimes \cdots \otimes PC \otimes \overline{C} \otimes \cdots \otimes \overline{C}$$

with respect to the map $\overline{\Delta}^{(n-1)}$ and $F_n C$ is precisely the kernel of the composition

$$\overline{C} \xrightarrow{\overline{\Delta}^{(n-1)}} PC \otimes \cdots \otimes PC \otimes \overline{C} \otimes \cdots \otimes \overline{C}$$

$$\xrightarrow{1 \otimes \ldots 1 \otimes \overline{\Delta} \otimes 1 \cdots \otimes 1} PC \otimes \cdots \otimes PC \otimes \overline{C} \otimes \overline{C} \otimes \cdots \otimes \overline{C}.$$

Since $0 \to PC \to \overline{C} \xrightarrow{\overline{\Delta}} \overline{C} \otimes \overline{C}$ is exact and $\overline{C}$ is flat, it follows that $F_n C$ factors through $PC \otimes \cdots \otimes PC \otimes \overline{C} \otimes \cdots \otimes \overline{C}$ with $k + 1 \leq n - 1$ factors of PC.

The case $k = n - 1$ shows that $F_n C$ is precisely the inverse image of $PC \otimes \cdots \otimes PC$ with respect to the map $\overline{\Delta}^{(n-1)}$. □

The dual notion to the choice of a generating module of an algebra is the choice of a retraction onto the module of primitives of a coalgebra. We have:

Corollary 10.1.10. *Let C be a supplemented coalgebra. Assume that C is flat as a module over R, that C is connected, and let $r : \overline{C} \to PC$ be a retraction onto the module of primitives. Then two elements $x, y \in \overline{C}$ are equal if and only if*

$$(r \otimes \cdots \otimes r) \cdot \overline{\Delta}^{(n)}(x) = (r \otimes \cdots \otimes r) \cdot \overline{\Delta}^{(n)}(y)$$

for all $n \geq 1$.

Proof: Consider the element $z = x - y$. If $z \neq 0$, then, since C is connected, there exists $n \geq 2$ such that $z \in F_n C - F_{n-1} C$. Since $\overline{\Delta}^{(n-1)}(z) \in PC \otimes \cdots \otimes PC$, $0 = (r \otimes \cdots \otimes r) \cdot \overline{\Delta}^{(n-1)}(z) = \overline{\Delta}^{(n-1)}(z)$. Thus, $z \in F_{n-1} C$, which is a contradiction. □

The next lemma is the coalgebra analogue to the statement that a map of algebras is determined by its values on a set of generators.

Lemma 10.1.11. *Let $f, g : C \to D$ be two maps of supplemented coalgebras where D is flat over R and connected. Let $r : \overline{D} \to PD$ be a retraction onto the module of primitives and assume that $r \cdot f = r \cdot g$. Then $f = g$.*

10.1 Augmented algebras and supplemented coalgebras

Proof: By corollary 10.1.10, it suffices to note that

$$(r \otimes \cdots \otimes r) \cdot \overline{\Delta}^{(n)} \cdot f = (r \cdot f \otimes \cdots \otimes r \cdot f) \cdot \overline{\Delta}^{(n)}$$
$$= (r \cdot g \otimes \cdots \otimes r \cdot g) \cdot \overline{\Delta}^{(n)} = (r \otimes \cdots \otimes r) \cdot \overline{\Delta}^{(n)} \cdot g$$

for all $n \geq 1$. $\square$

Definition 10.1.12. Let $f : C \to D$ be a map of supplemented coalgebras. Let $g : C \to D$ be a linear map of possibly nonzero degree. Then g is a coderivation with respect to f if $g(1) = 0$ and the following diagram commutes:

$$\begin{array}{ccc} C & \xrightarrow{g} & D \\ \downarrow \Delta & & \downarrow \Delta \\ C \otimes C & \xrightarrow{f \otimes g + g \otimes f} & D \otimes D \end{array}$$

The next lemma shows that coderivations are determined by the projection onto the module of primitives.

Lemma 10.1.13. *Let $g, h : C \to D$ be two coderivations of supplemented coalgebras with respect to a coalgebra map $f : C \to D$. Suppose that D is flat over R and connected. If $r : \overline{D} \to PD$ is a retraction onto the module of primitives and $r \cdot g = r \cdot h$, then $g = h$.*

Proof: The proof is similar to that of Lemma 10.1.5.

$$(r \otimes \cdots \otimes r) \cdot \overline{\Delta}^{(n)} \cdot g = (r \otimes \cdots \otimes r) \cdot (\Sigma\, f \otimes \cdots \otimes g \otimes \cdots \otimes f) \cdot \overline{\Delta}^{(n)}$$
$$= (r \otimes \ldots \otimes r) \cdot (\Sigma\, f \otimes \cdots \otimes h \otimes \cdots \otimes f) \cdot \overline{\Delta}^{(n)} = (r \otimes \cdots \otimes r) \cdot \overline{\Delta}^{(n)} \cdot h$$

for all $n \geq 1$. Both of the above sums contain n summands, one for each possible placement of g or h, respectively. $\square$

The next corollary is the coalgebra analogue of Corollary 10.1.6 and its proof is just the dual of that one.

Corollary 10.1.14. *Let $f : C \to D$ be a map of supplemented coalgebras with D connected and flat over the ground ring. Let $r : \overline{D} \to PD$ be a retraction onto the primitives and suppose that C and D are in fact supplemented differential coalgebras. Then, $r \cdot f \cdot d = r \cdot d \cdot f$ implies that $f \cdot d = d \cdot f$, that is, f is a map of differential supplemented coalgebras.*

Exercises

(1) Show that the sign convention for tensor product of maps satisfies the composition law
$$(f \otimes g) \cdot (h \otimes k) = (-1)^{\deg(g)\deg(h)}(f \cdot h) \otimes (g \cdot k).$$

(2) Let $d : A \to A$ be a derivation with respect to the identity map $1 : A \to A$. Show that $d(1) = 0$.

(3) Let A be an augmented algebra and let $\mathcal{D}(A)$ be the graded set of all derivations of A with respect to the identity map $1 : A \to A$. Show that $\mathcal{D}(A)$ is a graded Lie algebra. That is, show:

   (a) In each fixed degree $\mathcal{D}(A)$ is closed under linear combinations.

   (b) If $d, e \in \mathcal{D}(A)$, then the graded bracket $[d.e] = de - (-1)^{\deg(d)\deg(e)}ed$ is in $\mathcal{D}(A)$.

   (c) If $d \in \mathcal{D}(A)$ and $\deg(d)$ is odd, then $d^2 = dd$ is in $\mathcal{D}(A)$.

(4) Verify that in a supplemented coalgebra C the diagonal can be described by
$$\Delta(1) = 1 \otimes 1,$$
and, if $c \in \overline{C}$, then
$$\Delta(C) = c \otimes 1 + 1 \otimes c + \Sigma c' \otimes c''$$
where $c', c'' \in \overline{C}$.

(5) Show that the diagonal $\Delta : C \to C \otimes C$ is associative, that is, $(\Delta \otimes 1)\Delta = (1 \otimes \Delta)\Delta$ if and only if the reduced diagonal $\overline{\Delta} : \overline{C} \to \overline{C} \otimes \overline{C}$ is associative, that is, $(\overline{\Delta} \otimes 1)\overline{\Delta} = (1 \otimes \overline{\Delta})\overline{\Delta}$.

(6) Let $f : C \to D$ be a map of supplemented coalgebras and let $g : C \to D$ be a coderivation with respect to f.

   (a) Show that $f(1) = 1$ and f preserves reduced coalgebras and the module of primitives, $f(\overline{C}) \subseteq \overline{D}$, $f(PC) \subseteq PD$.

   (b) Show that g preserves the module of primitives, $g(PC) \subseteq PD$.

   (c) Show that the composition $\epsilon \cdot g : C \to D \to R$ is 0 and g preserves reduced coalgebras, $g(\overline{C}) \subseteq \overline{D}$.

(7) Let C be a supplemented coalgebra and let $cD(C)$ be the graded set of coderivations of C with respect to the identity map $1_C : C \to C$. Show that $cD(C)$ is a graded Lie algebra, that is, show that $cD(C)$ and the graded bracket satisfy the same laws as in exercise 3a,b,c.

(8) In an (associative) algebra A, give an inductive proof that any n-fold multiplication $\mu_n : A \otimes \cdots (n - times) \cdots \otimes A \to A$ of n elements is equal to the composition

$$\mu \cdot (\mu \otimes 1) \cdot \cdots \cdot (\mu \otimes 1 \otimes \ldots (n - 2 - times) \cdots \otimes 1)$$
$$(\mu \otimes 1 \otimes \ldots (n - 1 - times) \cdots \otimes 1).$$

Similarly, in an (associative) coalgebra C, give an inductive proof that any n-fold diagonal $\Delta^{(n)} : C \to C \otimes \ldots (n - times) \cdots \otimes C$ is equal to the composition

$$(\Delta \otimes 1 \otimes \ldots (n - 1 - times) \cdots \otimes 1)$$
$$\times (\Delta \otimes 1 \otimes \ldots (n - 2 - times) \cdots \otimes 1) \cdot (\Delta \otimes 1) \cdot \Delta.$$

10.2 Universal algebras and coalgebras

Let V be a positively graded module over a commutative ring R. The universal associative algebra generated by V is just the tensor algebra $T(V)$, that is:

Definition 10.2.1. The tensor algebra $T(V)$ is the augmented algebra

$$T(V) = R \oplus V \oplus (V \otimes V) \oplus (V \otimes V \otimes V) \oplus \cdots$$

with augmentation $\epsilon : T(V) \to R$ characterized by $\epsilon(V) = 0$ and multiplication given by justaposition of tensors

$$\mu[(a_1 \otimes \cdots \otimes a_n) \otimes (b_1 \otimes \cdots \otimes b_m)] = a_1 \otimes \cdots \otimes a_n \otimes b_1 \otimes \cdots \otimes b_m.$$

Up to isomorphism, $T(V)$ is uniquely characterized by the following universal property.

Lemma 10.2.2. *There is a linear map $\iota : V \to T(V)$ such that for all linear maps $f : V \to A$ into an associative algebra A, there is a unique map of algebras $\overline{f} : T(V) \to A$ such that $\overline{f} \cdot \iota = f$.*

Clearly, $\overline{f}$ is given by

$$\overline{f}(a_1 \otimes \cdots \otimes a_n) = (fa_1) \cdot \cdots \cdot (fa_n).$$

In fact, $T(V)$ is also universal for derivations:

Lemma 10.2.3. *Let $f : T(V) \to A$ be a map of augmented algebras and let $g : V \to \overline{A}$ be a linear map of possibly nonzero degree. Then there exists a unique derivation $\overline{g} : T(V) \to A$ with respect to f such that $\overline{g} \cdot \iota = g$.*

Clearly, $\bar{g}$ is given by

$$\bar{g}(a_1 \otimes \cdots \otimes a_n)$$
$$= \Sigma_{i=1}^n (-1)^{\deg(g)[\deg(a_1)+\cdots+\deg(a_{i-1})]} (fa_1) \cdot \cdots \cdot (ga_i) \cdot \cdots \cdot (fa_n).$$

We now present the notions which are dual to the above and which apply to coalgebras. Suppose V is a connected positively graded module, that is, $V_0 = 0$. Then the universal associative coalgebra cogenerated by V is the tensor coalgebra $T'(V)$ which is described as follows.

Definition 10.2.4.

(a) As a graded R module, the tensor coalgebra $T'(V)$ is the same as the tensor algebra, that is,

$$T'(V) = R \oplus V \oplus (V \otimes V) \oplus (V \otimes V \otimes V) \oplus \ldots.$$

We remark here that the fact that V is connected implies that this infinite direct sum is also the infinite product.

(b) The diagonal is given by

$$\Delta(a_1 \otimes \cdots \otimes a_n) = \Sigma_{i=0}^n (a_1 \otimes \cdots \otimes a_i) \otimes (a_{i+1} \otimes \cdots \otimes a_n).$$

We remark here that this diagonal does not agree with the multiplicative diagonal which makes the tensor algebra into a Hopf algebra.

(c) The counit $\epsilon : T'(V) \to R$ is given by

$$\varepsilon(1) = 1, \epsilon(a_1 \otimes \cdots \otimes a_n) = 0.$$

(d) The supplement $\eta : R \to T'(V)$ is given by $\eta(1) = 1$.

Up to isomorphism, $T'(V))$ is uniquely characterized by the universal property:

Proposition 10.2.5. *The standard linear map $\pi : T'(V) \to V$ such that $\pi \cdot \eta = 0$ has the following property. For all supplemented (associative) coalgebras C and all linear maps $f : C \to V$ with $f \cdot \eta = 0$, there is a unique map of supplemented coalgebras $\bar{f} : C \to T'(V)$ such that $\pi \cdot \bar{f} = f$. That is, $\bar{f}$ lifts f.*

Proof: Since the module of primitives $PT'(V) = V$, results of Section 10.1 show that $\bar{f}$ is unique if it exists. So it suffices to define a suitable $\bar{f}$ and check that it is a map of supplemented coalgebras.

We introduce the notation $f^{(n)} = f \otimes \ldots (n-times) \cdots \otimes f$.

Define $\bar{f}(1) = 1$ and, on the reduced coalgebra $\bar{C}$,

$$\bar{f} = f + f^{(2)}\bar{\Delta}^{(2)} + f^{(3)}\bar{\Delta}^{(3)} + \cdots.$$

To show that $\overline{f}$ is a map of supplemented coalgebras, we need to show that

$$\overline{\Delta}^{(2)} \cdot [\Sigma_{n \geq 1} f^{(n)} \overline{\Delta}^{(n)}]$$
$$= [\Sigma_{p \geq 1} f^{(p)} \overline{\Delta}^{(p)} \otimes \Sigma_{q \geq 1} f^{(q)} \overline{\Delta}^{(q)}] \cdot \overline{\Delta}^{(2)}$$
$$= [\Sigma_{p,q \geq 1} f^{(p)} \overline{\Delta}^{(p)} \otimes f^{(q)} \overline{\Delta}^{(q)}] \cdot \overline{\Delta}^{(2)}.$$

This follows from the fact that $\overline{\Delta}^{(n)} = (\overline{\Delta}^{(p)} \otimes \overline{\Delta}^{(q)}) \cdot \overline{\Delta}^{(2)}$ whenever $p + q = n$. $\square$

Tensor coalgebras are also universal for coderivations:

Proposition 10.2.6. *Let $f, g : C \to V$ be linear maps such that $f(1) = g(1) = 0$ and suppose that f has degree 0. Let $\overline{f} : C \to T'(V)$ be the map of supplemented coalgebras which lifts f. Then there exists a unique coderivation $\overline{g} : C \to T'(V)$ with respect to $\overline{f}$ such that $\pi \cdot \overline{g} = g$. That is, $\overline{g}$ lifts g.*

Proof: As in the proof of Proposition 10.2.5, the lift $\overline{g}$ is unique if it exists. We set

$$g^{[n]} = \Sigma_{i=1}^{n} f \otimes \cdots \otimes g \otimes \cdots \otimes f$$

with g in the i-th place. Now define $\overline{g}(1) = 0$ and, on $\overline{C}$,

$$\overline{g} = g + g^{[2]} \overline{\Delta}^{(2)} + g^{[3]} \overline{\Delta}^{(3)} + \cdots.$$

We need to check that $\overline{g}$ is a coderivation with respect to $\overline{f}$, that is, we need

$$\overline{\Delta}^{(2)} \cdot \overline{g} = (\overline{g} \otimes \overline{f} + \overline{f} \otimes \overline{g}) \cdot \overline{\Delta}^{(2)}.$$

This follows from

$$\overline{\Delta}^{(2)} \overline{g}^{[n]} \overline{\Delta}^{(n)}$$
$$= \Sigma_{p+q=n} (\overline{g}^{[p]} \otimes \overline{f}^{(q)} + \overline{f}^{(p)} \otimes \overline{g}^{[q]}) \cdot (\overline{\Delta}^{(p)} \otimes \overline{\Delta}^{(q)}) \cdot \overline{\Delta}^{(2)}.$$

Exercises

(1) Let $w_1 \otimes \cdots \otimes w_n \in T'(W)$. Show that $\overline{\Delta}^{(k)}(w_1 \otimes \cdots \otimes w_n) = 0$ if $k > n$ and, if $k \leq n$, it

$$= \Sigma (w_1 \otimes \cdots \otimes w_{i_1}) \otimes (w_{i_1+1} \otimes \cdots \otimes w_{i_2}) \otimes \cdots \otimes (w_{i_{k-1}+1} \otimes \cdots \otimes w_n)$$

the sum being over all partitions of length n tensors into k parts.

(2) Suppose that $f : T'(W) \to V$ is a degree 0 linear map which is nonzero only on length m tensors in $T'(W)$. If $\overline{f} : T'(W) \to T'(V)$ is the unique lift of f to a map of supplemented coalgebras, show that $\overline{f}(w_1 \otimes \cdots \otimes w_n) = 0$ if m

does not divide n and, if $n = mq$, then it

$$= f(w_1 \otimes \cdots \otimes w_m) \otimes f(w_{m+1} \otimes \cdots \otimes w_{2m})$$
$$\otimes \cdots \otimes f(w_{(m-1)q+1} \otimes \cdots \otimes w_n).$$

(3) Suppose that $g : T'(W) \to W$ is a degree 0 linear map which is nonzero only on length m tensors in $T'(W)$. If $\bar{g} : T'(W) \to T'(W)$ is the unique lift of g to a coderivation with respect to the identity map of $T'(W)$, show that $\bar{g}(w_1 \otimes \cdots \otimes w_n) = 0$ if $n < m$ and, if $n = mq$, then it

$$= \Sigma(-1)^{\deg(g)[\deg(w_1)+\cdots+\deg(w_i)]} w_1$$
$$\otimes \cdots \otimes w_i \otimes g(w_{i+1} \otimes \cdots \otimes w_{i+m}) \otimes w_{i+m+1} \otimes \cdots \otimes w_n,$$

the sum being over all subsets of m successive tensors.

10.3 Bar constructions and cobar constructions

The cobar construction is a functor from the category of connected differential coalgebras to the category of differential algebras which corresponds to the loop space functor on the category of pointed topological spaces. Dually, the bar construction is a functor from the category of differential algebras to the category of differential coalgebras which corresponds to the classifying space functor on the category of topological groups. Since the description of algebras is more intuitive, we begin with the cobar construction.

Let C be a connected differential coalgebra over a commutative ring R with counit $\epsilon : C \to R$ and let $\overline{C} = \text{kernel}(\epsilon)$ be the reduced coalgebra. Let $s^{-1}\overline{C}$ denote the desuspension, that is, in general,

$$(s^{-1}V)_n = V_{n+1}.$$

The cobar or loop construction on C is the following differential algebra $\Omega(C)$:

Definition 10.3.1.

(a) As an algebra, $\Omega(C) = T(s^{-1}\overline{C}) =$ the tensor algebra on the desuspension of the reduced coalgebra.

(b) $\Omega(C)$ has a so-called internal differential d_I which annihilates the unit, is a derivation of degree -1, and is given on generators by

$$d_I(s^{-1}c) = -s^{-1}(dc)$$

where d is the differential of C.

(c) $\Omega(C)$ has a so-called external differential d_E which annihilates the unit, is a derivation of degree -1, and is given on generators by
$$d_E(s^{-1}c) = \Sigma(-1)^{\deg(c_i')}s^{-1}c_i' \otimes s^{-1}c_i''$$
where the diagonal is
$$\Delta(c) = c \otimes 1 + \Sigma c_i' \otimes c_i'' + 1 \otimes c.$$

(d) $\Omega(C)$ has a so-called total differential given by $d = d_T = d_I + d_E$. The augmented algebra $\Omega(C)$ with total differential is called the cobar or loop construction on the coalgebra C.

Remarks. It is clear that $(d_I)^2 = 0$. An exercise shows that the associativity of the diagonal gives that $(d_E)^2 = 0$. Hence, both the internal and external differentials give $\Omega(C)$ the structure of a differential algebra. Since the internal and external differentials are derivations, the total differential is also a derivation. An exercise shows that $d_I d_E + d_E d_I = [d_I, d_E] = 0$ and hence the total differential d_T also gives $\Omega(C)$ the structure of a differential algebra.

Alternate notation

It is customary to denote the element $s^{-1}c_1 \otimes \cdots \otimes s^{-1}c_n$ by $[c_1|\ldots|c_n]$ or $[c_1]\ldots[c_n]$. We also introduce the notation $\bar{c} = (-1)^{\deg(c)-1}c$. Observe that with this notation, we have the formulas:
$$d_I[c_1|\ldots|c_i|\ldots|c_n] = \Sigma(-1)^{\deg(c_1)+\cdots+\deg(c_{i-1})-i}[c_1|\ldots|dc_i|\ldots|c_n]$$
$$= -\Sigma[\overline{c_1}|\ldots|\overline{c_{i-1}}|dc_i|\ldots|c_n]$$
and
$$d_E[c_1,\ldots,c_i,\ldots,c_n]$$
$$= \Sigma(-1)^{\deg(c_1)+\cdots+\deg(c_{i-1})+\deg(c_i')-i}[c_1|\ldots|c_i'|c_i''|\ldots|c_n]$$
$$= \Sigma[\overline{c_1}|\ldots|\overline{c_i'}|c_i''|\ldots|c_n]$$
where the reduced diagonal is:
$$\overline{\Delta}(c_i) = \Sigma c_i' \otimes c_i''.$$

We now turn to the functor which converts algebras into coalgebras, the so-called bar or classifying construction.

Let A be an augmented differential algebra over a commutative ring R with augmentation $\epsilon : A \to R$ and let $\overline{A}$ be the augmentation ideal. Let $s\overline{A}$ denote the suspension, that is, in general,
$$(sV)_n = V_{n-1}.$$

The bar or classifying construction on A is the following differential coalgebra $B(A)$ where the internal, external, and total differentials are all coderivations with respect to the identity:

Definition 10.3.2.

(a) As an algebra, $B(A) = T'(s\overline{A}) =$ the tensor coalgebra on the suspension of the augmentation ideal.

(b) $B(A)$ has a so-called internal differential d_I which annihilates the unit, is a coderivation of degree -1, and whose projection on the length 1 tensors is nonzero only on length 1 tensors and that projection is given by

$$d_I(sa) = -s(da)$$

where d is the differential of A.

(c) $B(A)$ has a so-called external differential d_E which annihilates the unit, is a coderivation of degree -1, and and whose projection on the length 1 tensors is nonzero only on length 2 tensors and that projection is given by

$$d_E((sa)(sb)) = (-1)^{\deg(a)} s(ab).$$

(d) $B(A)$ has a so-called total differential given by $d = d_T = d_I + d_E$. The coalgebra $B(A)$ with total differential is called the bar or classifying construction on the algebra A.

Remarks. It is clear that $(d_I)^2 = 0$. Since this is a coderivation, it is sufficient to check it on length 1 tensors. An exercise shows that the associativity of the multiplication gives that $(d_E)^2 = 0$ on length 3 tensors. Since this is a coderivation, this is sufficient to check it in general. Hence, both the internal and external differentials give $B(A)$ the structure of a differential coalgebra. Since the internal and external differentials are coderivations, the total differential is also a coderivation. An exercise shows that the coderivation $d_I d_E + d_E d_I = [d_I, d_E] = 0$ and hence the total differential d_T also gives $B(A)$ the structure of a supplemented differential coalgebra.

Alternate notation

It is customary to denote the element $sa_1 \otimes \cdots \otimes sa_n$ by $[a_1|\ldots|a_n]$. Observe that with this notation and with $\overline{a} = (-1)^{\deg(a)+1} a$, we have the formulas:

$$d_I[a_1|\ldots|a_i|\ldots|a_n] = \Sigma(-1)^{\deg(a_1)+\cdots+\deg(a_{i-1})+i}[a_1|\ldots|da_i|\ldots|a_n]$$
$$= -\Sigma([\overline{a_1}|\ldots|\overline{a_{i-1}}|da_i|\ldots|a_n]$$

and
$$d_E[a_1|\ldots|a_i|a_{i+1}|\ldots|a_n]$$
$$= \Sigma(-1)^{\deg(a_1)+\cdots+\deg(a_i)+i-1}[a_1,\ldots,a_i,a_{i+1},\ldots,a_n]$$
$$= -\Sigma[\overline{a_1},\ldots,\overline{a_i},a_{i+1},\ldots,a_n].$$

Exercises

(1) Show that the internal, external, and total differentials all make $\Omega(C)$ into an augmented differential algebra.

(2) Show that the internal, external, and total differentials all make $B(A)$ into a supplemented differential coalgebra.

10.4 Twisted tensor products

In this section we introduce twisted tensor products which are the algebraic analogue of principal bundles.

Recall the sign convention for maps
$$f: X \to Y, \quad g: \mathbb{Z} \to W$$
of possibly nonzero degree. On elements, we have
$$(f \otimes g)(x \otimes y) = (-1)^{\deg(g)\deg(x)}(fx \otimes gy)$$
and, for maps, we have
$$(f \otimes g)(h \otimes k) = (-1)^{\deg(g)\deg(h)}(fh \otimes gk).$$

In general, a sign is always introduced when two entities of nonzero degree are commuted past one another.

For example, if X and Y are chain complexes, then the tensor differential $d_\otimes$ on $X \otimes Y$ is given by this sign convention and the formula
$$d_\otimes = d \otimes 1 + 1 \otimes d.$$
The tensor differential on $X_1 \otimes \cdots \otimes X_i \otimes \cdots \otimes X_n$ is given by the formula
$$d_\otimes = \Sigma_{i=1,\ldots,n} 1 \otimes \cdots \otimes d_i \otimes \cdots \otimes 1.$$

Let A = an augmented differential algebra and let C = a supplemented differential coalgebra over a commutative ring R. Thus, the augmentation is a map of differential algebras $\epsilon: A \to R$ and the supplement is a map of differential coalgebras $\eta: R \to C$. Let $\mu: A \otimes A \to A$ be the multiplication and let $\Delta: C \to C \otimes C$ be the diagonal or comultiplication.

Definition 10.4.1. A twisting morphism $\tau : C \to A$ is a map of degree -1 such that

$$\epsilon \tau = 0$$

and

$$d\tau + \tau d = \mu(\tau \otimes \tau)\Delta.$$

Twisting morphisms allow us to define twisted tensor products $A \otimes_\tau C$ or $C \otimes_\tau A$. As a graded module, these are just $A \otimes C$ and $C \otimes A$, respectively, but the differential is altered by adding to the tensor differential another term $\pm \Gamma_\tau$.

Definition 10.4.2. If $\tau : C \to A$ is a twisting morphism, then the twisted tensor product $A \otimes_\tau C$ is $A \otimes C$ with the differential $d_\tau = d_\otimes + \Gamma_\tau$ where

$$\Gamma_\tau = (\mu \otimes 1)(1 \otimes \tau \otimes 1)(1 \otimes \Delta).$$

Similarly, the twisted tensor product $C \otimes_\tau A$ is $C \otimes A$ with the differential $d_\tau = d_\otimes - \Gamma_\tau$ where

$$\Gamma_\tau = (1 \otimes \mu)(1 \otimes \tau \otimes 1)(\Delta \otimes 1).$$

The fact that these definitions of d_τ give differentials are consequences of Lemma 10.4.4 below.

Three definitions which are useful in the statement of 10.4.4 are:

Definition 10.4.3.

(a) If $f : X \to Y$ is a map of degree n betweem graded objects, then $Df : X \to Y$ is the map of degree $n - 1$ given by

$$D(f) = df - (-1)^n fd.$$

(b) If $f, g : C \to A$ are maps of degrees n and m, respectively, then $f \cup g : C \to A$ is the map of degree $n + m$ given by

$$C \xrightarrow{\Delta} C \otimes C \xrightarrow{f \otimes g} A \otimes A \xrightarrow{\mu} A.$$

(c) If $f : C \to A$ is a map, then $\Gamma_f : A \otimes C \to A \otimes C$ is the composition

$$A \otimes C \xrightarrow{1 \otimes \Delta} A \otimes C \otimes C \xrightarrow{1 \otimes f \otimes 1} A \otimes A \otimes C \xrightarrow{\mu \otimes 1} A \otimes C.$$

The map $\Gamma_f : C \otimes A \to C \otimes A$ is defined in a symmetric fashion.

Thus, the fact that $\tau : C \to A$ is a twisting morphism asserts by definition that

$$D(\tau) = \tau \cup \tau.$$

10.4 Twisted tensor products 331

Lemma 10.4.4.

(a) *If $f, g : C \to A$ are maps of the same degree and a, b are scalars, then $\Gamma_{af+bg} = a\Gamma_f + b\Gamma_g$.*

(b) *If $f : C \to A$, then $D(\Gamma_f) = \Gamma_{D(f)}$ in both cases $A \otimes C$ and $C \otimes A$.*

(c) *In the case of $A \otimes C$, $\Gamma_f^2 = (-1)^{\deg(f)} \Gamma_{f \cup f}$. In the case of $C \otimes A$, $\Gamma_f^2 = \Gamma_{f \cup f}$.*

(d) *If $\tau : C \to A$ is a twisting morphism, $d_\tau^2 = 0$. That is, in the first case, $(d_\otimes + \Gamma_\tau)^2 = 0$, in the second case, $(d_\otimes - \Gamma_\tau)^2 = 0$.*

Twisted tensor products admit a Serre filtration and a Serre spectral sequence which are analogous to the concepts which go by the same name for fibrations. Let $\tau : C \to A$ be a twisting morphism and let $A \otimes_\tau C$ be the resulting twisted tensor product. We shall call A the fibre, C the base, and $A \otimes_\tau C$ the total space of the twisted tensor product.

Definition 10.4.5. The Serre filtration is the increasing filtration of $A \otimes_\tau C$ by differential submodules defined by $F_n(A \otimes_\tau C) = A \otimes C_{\leq n}$.

The resulting spectral sequence is the Serre spectral sequence and, if C is flat over the ground ring, it has the following properties:

(a)
$$E^0_{p,q} = A_q \otimes C_p, \quad d^0 = d_A \otimes 1.$$

(b)
$$E^1_{p,q} = H_q(A) \otimes C_p, \quad d^1 = 1 \otimes d_C.$$

(c)
$$E^2_{p,q} = H_p(H_q(A) \otimes C_*, d^1)$$

and, if either $H_q(A)$ is flat over R or if R is a principal ideal domain and C and $H_*(C)$ are both projective over R, then

$$E^2_{p,q} = H_q(A) \otimes H_p(C).$$

(d) The spectral sequence is a homology spectral sequence confined to the first quadrant and abuts to $H_*(A \otimes_\tau C)$.

Since the Serre filtration is a filtration by A-submodules, it follows that, from the E^1 term onwards, the spectral sequence is a spectral sequence of modules over $H_*(A)$. In particular, if $\alpha \in H_n(A)$ and $\beta \in E^r_{p,q}$ for $r \geq 1$, then the differentials commute up to sign with the action of $H_*(A)$, that is,

$$d^r(\alpha \beta) = (-1)^n \alpha \, d^r(\beta).$$

The Zeeman comparison theorem applies to this Serre spectral sequence and yields the following:

Let (A, C, τ) and (A', C', τ') be two triples consisting of augmented algebras, supplemented coalgebras, and twisting morphisms. A morphism $F : (A, C, \tau) \to (A', C', \tau')$ consists of a map of augmented algebras $f : A \to A'$ (a map of the fibres) and a map of supplemented coalgebras $g : C \to C'$ (a map of the bases) such that $\tau' \cdot g = f \cdot \tau$. It induces a map $f \otimes g : A \otimes_\tau C \to A' \otimes_{\tau'} C'$ of twisted tensor products (a map of the total spaces).

Proposition 10.4.6. *If C and C' are flat modules over the ground ring and $f \otimes g : A \otimes_\tau C \to A' \otimes_{\tau'} C'$ is a map of twisted tensor products, then, if two out of three of f, g, $f \otimes g$ are homology equivalences, so is the third.*

Exercises

(1) Prove lemma 10.4.4.

(2) (a) Show that the twisted tensor product $A \otimes_\tau C$ is a left differential A module, that is, $d_\tau((a \cdot b) \otimes c) = (da \cdot b) \otimes c + (-1)^{\deg(a)} a \cdot d_\tau(b \otimes c)$, $a, b \in A$, $c \in C$.

(b) Show that the twisted tensor product $A \otimes_\tau C$ is a right differential C comodule, that is, show that

$$\begin{array}{ccc} A \otimes_\tau C & \xrightarrow{1 \otimes \Delta} & (A \otimes_\tau C) \otimes C \\ \downarrow d_\tau & & \downarrow \\ A \otimes_\tau C & \xrightarrow{1 \otimes \Delta} & (A \otimes_\tau C) \otimes C \end{array} \qquad d_\tau \otimes 1 + (1_{A \otimes_\tau C}) \otimes d$$

commutes.

(c) Verify the analogues of (a) and (b) for the twisted tensor product $C \otimes_\tau A$.

10.5 Universal twisting morphisms

There are two universal twisting morphisms, one for coalgebras and one for algebras. Let A be an augmented differential algebra and let C be a supplemented differential coalgebra.

The two universal twisting morphisms are:

(1) The universal twisting morphism out of a supplemented coalgebra:

$$\tau_C : C \to \Omega(C), \quad \tau_C(1) = 0, \quad \tau_C(c) = s^{-1}\overline{c} = [c]$$

for $c \in \overline{C}$.

(2) The universal twisting morphism into an augmented algebra:
$$\tau_A : B(A) \to A, \quad \tau_A(1) = 0, \quad \tau_A[a] = a, \quad \tau_A[a_1|\ldots|a_n] = 0 \text{ if } n \geq 1.$$

It is easy to check that these are twisting morphisms, that is, they both satisfy
$$d\tau + \tau d = \tau \cup \tau, \quad \epsilon\tau = 0.$$

Let $\tau : C \to A$ be any twisting morphism and note that we can regard this as a map $\tau : \overline{C} \to \overline{A}$. Since this is a linear map, it has a unique extension to a map out of the the tensor algebra
$$\overline{\tau_C} : \Omega(C) \to A$$
and a unique lift to a map into the cotensor algebra
$$\overline{\tau_A} : C \to B(A).$$

That is, we have a commutative diagram

$$\begin{array}{ccc}
 & C & \xrightarrow{\tau_C} \Omega(C) \\
\overline{\tau_A} \swarrow & \downarrow \tau & \swarrow \overline{\tau_C} \\
B(A) & \xrightarrow{\tau_A} A &
\end{array}$$

The universal properties are:

(1) $\tau : C \to A$ is a twisting morphism if and only if the extension $\overline{\tau_C} : \Omega(C) \to A$ is a map of augmented differential algebras.

(2) $\tau : C \to A$ is a twisting morphism if and only if the lift $\overline{\tau_A} : C \to B(A)$ is a map of supplemented differential coalgebras.

Thus, the one-to-one correspondences
$$\overline{\tau_A} \leftrightarrows \tau \leftrightarrows \overline{\tau_C}$$
give natural bijections between the three sets:

Supplemented Diff Coalgebra maps$(C, B(A))$, Twisting morphisms(C, A), and Augmented Diff Algebra maps$(\Omega(C), A)$.

In particular, the natural bijections
$$\text{map}(C, B(A)) \xrightarrow{\alpha} \text{map}(\Omega(C), A)$$
show that the loop functor Ω and the classifying functor B are a pair of adjoint functors.

The explicit adjoint correspondence between maps is given by:

Lemma 10.5.1. *Given a map of supplemented differential coalgebras $f : C \to B(A)$, the map of augmented differential algebras $\alpha(f) : \Omega(C) \to A$ is given* by $\alpha(f)([c_1]\ldots[c_n]) = (fc_1)\ldots(fc_n)$. Given a map of augmented differential algebras $g: \Omega(C) \to A$, the map of supplemented differential coalgebras $\alpha^{-1}(g) = \beta(g): C \to B(A)$ is given by

$$\beta(g)(c) = [gc] + \Sigma_i[gc^i_{2,1}|gc^i_{2,2}] + \Sigma_i[gc^i_{3,1}|gc^i_{3,2}|gc^i_{3,3}]| + \cdots$$

where the reduced iterated diagonal is $\overline{\Delta}^{(n)}(c) = \Sigma_i c^i_{n,1} \otimes \cdots \otimes c^i_{n,n}$ for $c \in \overline{C}$.

As usual, the adjoint of the identity map $1_{B(A)}: B(A) \to B(A)$ is the adjunction map of differential algebras $\alpha: \Omega B(A) \to A$ and the adjoint of the identity map $1_{\Omega(C)}: \Omega(C) \to \Omega(C)$ is the adjunction map of differential coalgebras $\beta: C \to B\Omega(C)$. These are given explicitly by:

Lemma 10.5.2.

(a) *On generators of $\Omega B(A)$, $\alpha([a_1|\ldots|a_n]) = 0$ if $n > 1$ and $= a_1$ if $n = 1$.*

(b) *The projection of β onto primitives of $B\Omega(C)$ is given by $\mathrm{proj} \cdot \beta(c) = [[c]]$ and thus*

$$\beta(c) = [[c]] + \Sigma_i[[c^i_{2,1}]|[c^i_{2,2}]] + \Sigma_i[[c^i_{3,1}]|[c^i_{3,2}]|[c^i_{3,3}]] + \cdots$$

with the reduced iterated diagonal notation as in Lemma 10.5.1

Remarks. From the universal properties of the twisting morphisms we get the universal properties of the twisted tensor products $\Omega(C) \otimes_{\tau_C} C$ and $A \otimes_{\tau_A} B(A)$. The first is an initial object for twisted tensor products with fixed base C and the second is a terminal object for twisted tensor products with fixed fibre A. These are evident from the following commutative diagram:

$$\begin{array}{ccccc}
\Omega(C) & \xrightarrow{\tau_C} & A & \xrightarrow{=} & A \\
\downarrow & & \downarrow & & \downarrow \\
\Omega(C) \otimes_{\tau_C} C & \to & A \otimes_\tau C & \to & A \otimes_{\tau_A} B(A) \\
\downarrow & & \downarrow & & \downarrow \\
C & \xrightarrow{=} & C & \xrightarrow{\tau_A} & B(A).
\end{array}$$

In the next section we shall prove:

Proposition 10.5.3. *The twisted tensor products $A \otimes_{\tau_A} B(A)$ and $\Omega(C) \otimes_{\tau_C} C$ are acyclic, that is, their homologies are the ground field concentrated in degree 0.*

This has the following corollaries.

Corollary 10.5.4. *If the twisted tensor product $A \otimes_\tau C$ is acyclic, then there is a map of augmented differential algebras $\Omega(C) \to A$ and a map of supplemented differential coalgebras $C \to B(A)$ both of which are homology isomorphisms.*

10.6 Acyclic twisted tensor products

Proof: The twisting morphism $\tau : C \to A$ extends to a map of augmented differential algebras $\Omega(C) \to A$ and yields a map of acyclic twisted tensor products $\Omega(C) \otimes_{\tau_C} C \to A \otimes_\tau C$. Since this is a homology isomorphism on the base and total space, this is a homology isomorphism on the fibre.

Similarly, the twisting morphism $\tau : C \to A$ lifts to a map of supplemented differential algebras $C \to B(A)$ and yields a map of acyclic twisted tensor products $A \otimes_\tau C \to A \otimes_{\tau_A} B(A)$. Since this is a homology isomorphism on the fibre and total space, this is a homology isomorphism on the base. □

A special case of the above is:

Corollary 10.5.4. *The adjunction maps $\alpha : \Omega B(A) \to A$ and $\beta : C \to B\Omega(C)$ are homology equivalences.*

Proof: Since the twisting morphism $\tau_A : B(A) \to A$ is the composition $B(A) \xrightarrow{\tau_{BA}} \Omega B(A) \xrightarrow{\alpha} A$, there is a map of twisted tensor products

$$\Omega B(A) \otimes_{\tau_{BA}} B(A) \to A \otimes_\tau B(A).$$

Since this is a homology isomorphism on the total space and base, it is a homology isomorphism on the fibre.

Similarly, the twisting morphism τ_C is the composition $C \xrightarrow{\beta} B\Omega(C) \xrightarrow{\tau_{\Omega C}} \Omega(C)$, we have a map $\Omega(C) \otimes_{\tau_C} C \to \Omega(C) \otimes_{\tau_{\Omega C}} B\Omega(C)$ and that $C \to B\Omega(C)$ is a homology isomorphism follows much as before. □

Exercises

(1) Show that $\tau_C : C \to \Omega(C)$ and $\tau_A : B(A) \to A$ are twisting morphisms.

(2) Show that $\tau : C \to A$ is a twisting morphism if and only if the extension $\overline{\tau_C} : \Omega(C) \to A$ is a map of augmented differential algebras.

(3) Show that $\tau : C \to A$ is a twisting morphism if and only if the lift $\overline{\tau_A} : C \to B(A)$ is a map of supplemented differential coalgebras.

(4) Prove Lemma 10.5.1.

(5) Prove Lemma 10.5.2.

10.6 Acyclic twisted tensor products

In this section, we prove that the twisted tensor products $A \otimes_{\tau_A} B(A)$ and $\Omega(C) \otimes_{\tau_C} C$ are acyclic. We begin with:

Proposition 10.6.1. $A \otimes_{\tau_A} B(A)$ *is acyclic.*

Proof: Note that the twisted tensor product differential is a sum $d_\tau = d_E + d_I$ of so called external and internal differentials where

$$d_I(a[a_1|\ldots|a_n]) = da[a_1|\ldots|a_n] - \Sigma \bar{a}[\overline{a_1}|\ldots|\overline{a_{i-1}}|da_i|a_{i+1}|\ldots|a_n]$$

$$d_E(a[a_1|\ldots|a_n]) = aa_1[a_2|\ldots|a_n] + \Sigma \bar{a}[\overline{a_1}|\ldots|\overline{a_{i-1}}|\overline{a_i}a_{i+1}|\ldots|a_n].$$

Let $B_n(A) = s\bar{A} \otimes \ldots (n-times) \cdots \otimes s\bar{A} =$ the span of all $[a_1|\ldots|a_n]$. For $n \geq 0$, define a contracting homotopy $h : A \otimes_\tau B_n(A) \to A \otimes_\tau B_{n+1}(A)$ by

$$h(a[a_1|\ldots|a_n]) = \begin{cases} 0 & \text{if } a = 1 \\ [a|a_1|\ldots|a_n] & \text{if } a \in \bar{A} \end{cases}$$

We can easily check.

Lemma 10.6.2.

$$hd_E + d_E h = 1 - \eta \cdot \epsilon, \quad hd_I = -d_I h, \quad hd_\tau + d_\tau h = 1 - \eta \cdot \epsilon,$$

where $R \xrightarrow{\eta} A \otimes_\tau B(A) \xrightarrow{\epsilon} R$ are the obvious maps in degree 0 with $\eta(1) = 1[\,]$, $\epsilon(1[\,]) = 1$. Thus, $A \otimes_\tau B(A)$ is acyclic.

We now prove:

Proposition 10.6.3. $\Omega(C) \otimes_{\tau_C} C$ is acyclic.

Proof: The proof is similar to a dual of Proposition 10.6.1 but some care must be taken with the signs.

As before, note that the twisted tensor product differential is a sum $d_\tau = d_E + d_I$ of so called external and internal differentials where

$$d_I([c_1|\ldots|c_n]c) = [\overline{c_1}|\ldots|\overline{c_n}]dc - \Sigma[\overline{c_1}|\ldots|\overline{c_{i-1}}|dc_i|c_{i+1}|\ldots|c_n]c$$

$$d_E([c_1|\ldots|c_n]c) = [\overline{c_1}|\ldots|\overline{c_n}|c - \eta\epsilon(c)]1 + \Sigma_j[\overline{c_1}|\ldots|\overline{c_n}|c'_j]c''_j$$
$$- \Sigma_{i,j}[\overline{c_1}|\ldots|\overline{c_{i-1}}|\overline{c'_{i,j}}|c''_{i,j}|c_{i+1}|\ldots|c_n]c$$

where $\bar{\Delta}(c) = \Sigma c'_j \otimes c''_j$, $\bar{\Delta}(c_i) = \Sigma c'_{i,j} \otimes c''_{i,j}$.

Let $\Omega_n(C) = s^{-1}\bar{C} \otimes \ldots (n-times) \cdots \otimes s^{-1}\bar{C} =$ the span of all $[c_1|\ldots|c_n]$. For $n \geq 0$, define a contracting homotopy $h : \Omega_n(C) \otimes_\tau C \to \Omega_{n-1}(C) \otimes_\tau C$ by setting $h = 0$ if $n = 0$ and for $n > 1$, set

$$h([c_1|\ldots|c_n]c) = \begin{cases} [\overline{c_1}|\ldots|\overline{c_{n-1}}|c_n & \text{if } c = 1 \\ 0 & \text{if } c \in \bar{C} \end{cases}$$

We can easily check:

Lemma 10.6.4.
$$hd_E + dh_E = 1 - \eta \cdot \epsilon, \quad d_I h = -hd_I, \quad hd_\tau + d_\tau h = 1 - \eta \cdot \epsilon$$
where $R \xrightarrow{\eta} \Omega(C) \otimes_\tau C \xrightarrow{\epsilon} R$ are the obvious maps in degree 0 with $\eta(1) = [\,]1$, $\epsilon([\,]1) = 1$. Thus, $\Omega(C) \otimes_\tau C$ is acyclic.

Exercises

(1) Check Lemma 10.6.2.

(2) Check Lemma 10.6.4.

(3) (a) In the twisted tensor product $A \otimes_\tau B(A)$, check that the internal and external differentials d_I and d_E satisfy the formulas
$$d_E^2 = 0, \quad d_I^2 = 0, \quad d_E \cdot d_I = -d_I \cdot d_E.$$
(b) Let $\mathbb{Z}_I$ is the subset of $A \otimes_\tau B(A)$ consisting of all cycles with respect to the internal differential. Show that $d_E(\mathbb{Z}_I) \subseteq \mathbb{Z}_I$ so that $\mathbb{Z}_I$ is a chain complex with the differential d_E.

(4) Do the analogue of exercise 3 in the twisted tensor product $\Omega(C) \otimes_\tau C$.

(5) In both the twisted tensor products $A \otimes_\tau B(A)$ and $\Omega(C) \otimes_\tau C$, let $\mathbb{Z}_I$ denote the cycles with respect to the internal differential. Show that the external differential makes $\mathbb{Z}_I$ into an acyclic chain complex.

10.7 Modules over augmented algebras

Let A be an algebra with unit $\eta : R \to A$. A (left) module over A is a graded module M together with a degree 0 linear map $\mu : A \otimes M \to M$ called an action which is associative and has a unit with respect to the multiplication, that is, if we write $\mu(a \otimes m) = a \cdot m$, then we have $a \cdot (b \cdot m) = (a \cdot b) \cdot m$, $1 \cdot m = m$ for all $a, b \in A$, $m \in M$. In other words, the following diagrams are commutative:

$$\begin{array}{ccc} A \otimes A \otimes M & \xrightarrow{\mu \otimes 1} & A \otimes M \\ \downarrow 1 \otimes \mu & & \downarrow \mu \\ A \otimes M & \xrightarrow{\mu} & M \end{array}$$

$$\begin{array}{ccc} R \otimes M & \xrightarrow{\eta \otimes 1} & A \otimes M \\ & \cong \searrow & \downarrow \mu \\ & & M \end{array}$$

Remarks. As expected, a differential module over a differential algebra is a module with a differential in which the structure map is a differential map, that is, $d(a \cdot m) = da \cdot m + (-1)^{\deg(a)} a \cdot dm$ for all $a \in A$, $m \in M$.

Definition 10.7.1. If M is a module over an augmented algebra A with augmentation ideal $I(A) = \overline{A}$, then the module of indecomposables $Q_A(M) = M/\overline{A} \cdot M$.

A useful property of the functor $Q_A(M)$ is that it is additive and right exact, that is:

Lemma 10.7.2.

(a) $Q_A(M \oplus N) \cong Q_A(M) \oplus Q_A(N)$

(b) $M \to N \to P \to 0$ exact implies that $Q_A(M) \to Q_A(N) \to Q_A(P) \to 0$ is exact.

The graded version of Nakayama's lemma is the following.

Lemma 10.7.3. *If A is a connected augmented algebra and M is a module over A which is bounded below, then $Q_A(M) = 0$ implies that $M = 0$.*

Proof: It follows easily by induction on the degree and the fact that

$$\overline{A} \otimes M \to M \to Q_A(M) \to 0$$

is exact. □

Corollary 10.7.4. *If A is a connected augmented algebra and $f : M \to N$ is a map of modules over A with N bounded below, then f is an epimorphism if and only if the induced map of indecomposables $Q_A(f) : Q_A(M) \to Q_A(N)$ is an epimorphism.*

Definition 10.7.5. A choice of generators is a section $\iota : Q_A(M) \to M$ of the canonical projection $\pi : M \to Q_A(M)$, that is, $\pi \cdot \iota = 1_{Q_A(M)}$.

Lemma 10.7.6. *Let $f, g : M \to N$ be two maps of modules over a connected augmented algebra A and let $\iota : Q_A(M) \to M$ be a choice of generators with M bounded below. If $f \cdot \iota = g \cdot \iota$, then $f = g$.*

Proof: One uses induction on the degree and the split exact sequence $\overline{A} \otimes M \to M \to Q_A(M) \to 0$. □

For the remainder of this section, we will assume that the algebra A and the A modules M have no differential structure. In the presence of differential structures, the notions of projective and of free modules must be modified. One such modification is illustrated in the exercises to this section. But the modification we are really interested in is deferred to a later section.

A free module M over A is one which has a graded basis $\{x_\alpha\}$ so that

$$M = \bigoplus_\alpha A \cdot x_\alpha.$$

This is clearly equivalent to the existence of a free graded R module V such that $M = A \otimes V$.

It is evident that there are enough free modules in the sense that, for every module N, there is an epimorphism $P \to M$ with P free.

Furthermore, free modules P are projective in the sense that, given any epimorphism $f: M \to N$ and any map $g: P \to N$, there exists a map $h: P \to M$ such that $f \cdot h = g$. It follows easily that a module P is projective if and only if there exists a module Q such that $P \oplus Q$ is a free module.

Proposition 10.7.7. *Suppose A is a connected augmented algebra and M is a projective A module which is bounded below. Then $M \cong A \otimes Q_A(M)$ where $Q_A(M)$ is projective over R. In particular, if R is a principal ideal domain, then M is a free module.*

Proof: Let M be a summand of the free module $F = A \otimes V$. Then $Q_A(M)$ is a summand of $Q_A(F) \cong V$. Since V is a free R module, $Q_A(M)$ is a projective R module. Hence the standard projection $\pi: M \to Q_A(M)$ has a section $\iota: Q_A(M) \to M$ and we can consider the composition $f: A \otimes Q_A(M) \xrightarrow{1 \otimes \iota} A \otimes M \xrightarrow{\mu} M$. Since $Q_A(f): Q_A(A \otimes Q_A(M)) \to Q_A(M)$ is an isomorphism, it is also an epimorphism. Therefore $f: A \otimes Q_A(M) \to M$ is an epimorphism. Since M is projective, $A \otimes Q_A(M) = K \oplus M$ where K is the kernel of f. Since, $Q_A(A \otimes Q_A(M)) = Q_A(M) = Q_A(K) \oplus Q_A(M)$, it follows that $Q_A(K)$ and hence K itself are 0. Hence, $A \otimes Q_A(M) = M$. $\square$

Given an A module M, the usual definitions and facts concerning projective resolutions hold true and the proofs are standard:

10.7.8. There exist projective resolutions

$$\cdots \xrightarrow{d_{n+1}} P_n \xrightarrow{d_n} P_{n-1} \xrightarrow{d_{n-1}} \cdots \xrightarrow{d_1} P_0 \xrightarrow{\epsilon} M \to 0,$$

that is, complexes $P_* \xrightarrow{\epsilon} M \to 0$ which are resolutions in the sense that they are exact sequences (of R modules) and such that all the P_* are projective A modules.

10.7.9. Suppose we are given a map $f: M \to N$ of A modules. For any projective complex $P_* \xrightarrow{\epsilon} M \to 0$ and any exact complex $Q_* \xrightarrow{\epsilon} N \to 0$, there is a map of complexes $f_*: P_* \to Q_*$ covering f, that is, a commutative diagram

$$\begin{array}{ccccccccccc} \xrightarrow{d_{n+1}} & P_n & \xrightarrow{d_n} & P_{n-1} & \xrightarrow{d_{n-1}} & \cdots & \xrightarrow{d_1} & P_0 & \xrightarrow{\epsilon} & M \\ & \downarrow f_n & & \downarrow f_{n-1} & & & & \downarrow f_0 & & \downarrow f \\ \xrightarrow{d_{n+1}} & Q_n & \xrightarrow{d_n} & Q_{n-1} & \xrightarrow{d_{n-1}} & \cdots & \xrightarrow{d_1} & Q_0 & \xrightarrow{\epsilon} & N \end{array}$$

10.7.10. Given a map $f: M \to N$ of A modules and two choices for maps $f_*, g_*: P_* \to Q_*$ covering the same f with P_* a projective complex and Q_* an exact complex, the maps f_* and g_* are chain homotopic, that is, there exists a map $H_* : P_* \to Q_*$ of degree $+1$ such that $d_* H_* + H_* d_* = f_* - g_*$. More precisely, for $n \geq 0$, there are A-linear $H_n : P_n \to Q_{n+1}$ such that

$$d_1 \cdot H_0 = f_0 - g_0, \quad H_{n-1} \cdot d_n + d_{n+1} \cdot H_n = f_n - g_n, n \geq 1.$$

Exercises

(1) Prove Lemma 10.7.2.

(2) Prove Lemma 10.7.3, Corollary 10.7.4, and Lemma 10.7.6.

(3) Consider the category of positively (nonnegatively) graded differential modules over a commutative ring R, that is, the category of chain complexes over R.

   (a) Show that the following objects are projective in this category:

   (i) Any module M which is concentrated in degree 0 and has M_0 projective over R.

   (ii) Any acyclic module M which is concentrated in two nonnegative degrees n and $n - 1$ with M_n projective over R.

   (iii) Any arbitrary direct sum of objects of the forms (i) and (ii).

   (b) Show that any module M is a surjective image of a projective as in (3).

(4) Suppose A is an augmented differential algebra and consider the category of differential modules over A.

   (a) If V is a projective chain complex over R, then $A \otimes V$ is a projective differential module.

   (b) If M is a projective differential module over A, then M is a retract of some $A \otimes V$ where V is a projective chain complex over R.

   (c) If M is a projective differential module over A, show that HM is a retract of $HA \otimes W$ where W is a projective R module concentrated in degree 0. (This is a strong indication that this notion of projective differential A module is too restrictive!)

(5) Prove 10.7.8, 10.7.9, and 10.7.10.

10.8 Tensor products and derived functors

Let A be a (graded) algebra over a commutative ring R and let M and N be right and left modules, respectively, over A. The definition of the tensor product

10.8 Tensor products and derived functors

of graded modules over a graded algebra $M \otimes_A N$ has a definition which, by duality, will provide a template for the definition of the cotensor product of graded comodules over a graded coalgebra.

This definition of the tensor product $M \otimes_A N$ is based on the usual tensor product of modules over R and its extension to graded modules over R. We shall assume that this is known and that it satisfies the standard properties of associativity, distributivity over direct sums, and that the ground ring is a two sided unit.

To be specific, the tensor product $M \otimes_A N$ is the following quotient of the tensor product over R, $M \otimes N = M \otimes_R N$ where

$$(M \otimes N)_n = \bigoplus_{i+j=n} M_i \otimes N_j :$$

Definition 10.8.1. $M \otimes_A N$ is the coequalizer of the two action maps:

$$\mu \otimes 1, 1 \otimes \mu : M \otimes A \otimes N \to M \otimes N.$$

In other words, the sequence

$$M \otimes A \otimes N \xrightarrow{\delta} M \otimes N \to M \otimes_A N \to 0$$

is exact where $\delta = \mu \otimes 1 - 1 \otimes \mu$.

If $m \in M$ and $n \in N$, we denote the image of $m \otimes n$ in $M \otimes_A N$ by $m \otimes_A n$ and observe that these elements generate $M \otimes_A N$.

A fundamental property of the tensor product is that it is right exact.

Proposition 10.8.2. *If $M_1 \to M_2 \to M_3 \to 0$ is an exact sequence of right A modules and N is a left A module, then*

$$M_1 \otimes_A N \to M_2 \otimes_A N \to M_3 \otimes_A N \to 0$$

is exact. The corresponding statement is true when the roles of left and right modules are reversed.

Proof: Consider the commutative diagram

$$\begin{array}{ccccccc}
M_1 \otimes A \otimes N & \to & M_2 \otimes A \otimes N & \to & M_3 \otimes A \otimes N & \to & 0 \\
\downarrow \delta & & \downarrow \delta & & \downarrow \delta & & \\
M_1 \otimes N & \to & M_2 \otimes N & \to & M_3 \otimes N & \to & 0 \\
\downarrow & & \downarrow & & \downarrow & & \\
M_1 \otimes_A N & \to & M_2 \otimes_A N & \to & M_3 \otimes_A N & \to & 0 \\
\downarrow & & \downarrow & & \downarrow & & \\
0 & & 0 & & 0 & &
\end{array}$$

where the exactness of the vertical columns defines the tensor product. The right exactness of the usual tensor product of modules over R says that the top two rows are exact. Now the exactness of the bottom row is a simple exercise. □

The tensor product satisfies the following three formal algebraic properties in Proposition 10.8.3 below: it is associative, distributive over direct sums, and the algebra is a two-sided unit.

Let A and B be algebras. Let M, M_1, M_2 be right A modules, let K, K_1, K_2 be left A modules, let N be a left B module, and let P be an $A - B$ bimodule, that is, the left and right actions on P commute as follows:

$$\begin{array}{ccc} A \otimes P \otimes B & \xrightarrow{1 \otimes \mu} & A \otimes P \\ \downarrow \mu \otimes 1 & & \downarrow \mu \\ P \otimes B & \xrightarrow{\mu} & B \end{array}$$

Then:

Proposition 10.8.3.

(a) *There is a natural isomorphism*

$$M \otimes_A (P \otimes_B N) \cong (M \otimes_A P) \otimes_B N.$$

(b) *There are natural isomorphisms*

$$(M_1 \oplus M_2) \otimes_A K \cong (M_1 \otimes_A K) \oplus (M_2 \otimes_A K),$$
$$M \otimes_A (K_1 \oplus K_2) \cong (M \otimes_A K_1) \oplus (M \otimes_A K_2).$$

(c) *The module actions induce natural isomorphisms*

$$A \otimes_A N \cong N, \quad M \otimes_A A \cong M.$$

Proof:

(a) Let X be the cokernel of the natural map

$$\psi = \delta_1 \oplus \delta_2 : (M \otimes A \otimes P \otimes N) \oplus (M \otimes P \otimes B \otimes N) \to M \otimes P \otimes N.$$

Since tensor over R is right exact, the cokernel of δ_1 is

$$(M \otimes_A P) \otimes N.$$

Thus X is the cokernel of the composition

$$M \otimes P \otimes B \otimes N \to (M \otimes_A P) \otimes B \otimes N \to (M \otimes_A P) \otimes N$$

and this is clearly $(M \otimes_A P) \otimes_B N$.
Since the argument is symmetric, X is also isomorphic to $M \otimes_A (P \otimes_B N)$.

(b) This is an immediate consequence of the fact that tensor over R naturally distributes over direct sums.

(c) A natural isomorphism $M \otimes_A A \to M$ is given by $m \otimes_A a \mapsto ma$ with inverse given by $m \mapsto m \otimes_A 1$. □

Suppose A is an augmented algebra with augmentation ideal $\overline{A}$. The ground ring is a two-sided module over A and we have the following alternate description of the indecomposables of a module:

Proposition 10.8.4. *Let M and N be right and left modules over A, respectively. Then there are isomorphisms*

$$M \otimes_A R \cong M/M \cdot \overline{A} = Q_A(M), \quad R \otimes_A N \cong N/\overline{A} \cdot N = Q_A(N).$$

Proof: Since tensor is right exact, we have an exact sequence

$$M \otimes_A \overline{A} \to M \otimes_A A \to M \otimes_A R \to 0.$$

The fact that $M \otimes_A A \cong M$ completes the proof that $M \otimes_A R \cong M/M \cdot \overline{A} = Q_A(M)$. □

We now introduce the derived functors $\mathrm{Tor}_n^A(M, N)$ of tensor product and prove the important result that these are balanced.

Let $P_* \xrightarrow{\epsilon} M \to 0$ be a projective resolution by right A modules and let $Q_* \xrightarrow{\epsilon} N \to 0$ be a projective resolution by left A modules. If we grade P_* and Q_* by the so-called resolution degree, then $P_* \otimes Q_*$ can be graded by the addition of these degrees,

$$(P \otimes_A Q)_n = \bigoplus_{i+j=n} P_i \otimes_A Q_j$$

and we define

Definition 10.8.5. $\mathrm{Tor}_n^A(M, N) = H_n(P_* \otimes_A Q_*, d_\otimes)$ where $d_\otimes = d_P \otimes 1 + 1 \otimes d_Q$.

Up to natural isomorphism, $\mathrm{Tor}_n^A(M, N)$ is a well-defined functor. We shall give two proofs, the first depending on the uniqueness of lifts up to chain homotopy and the second depending on the so-called balanced property.

Proof 1: Let $f : M \to M_1$ and $g : N \to N_1$ be maps of A modules and choose lifts to maps of resolutions $\overline{f} : P_* \to P_{1*}$ and $\overline{g} : Q_* \to Q_{1*}$ which cover f and g, respectively. Then we have a chain map

$$\overline{f} \otimes_A \overline{g} : P_* \otimes Q_* \to P_{1*} \otimes_A Q_{1*}$$

which induces the homology maps
$$H_n(\overline{f} \otimes_A \overline{g}) = \operatorname{Tor}_n^A(f,g) : \operatorname{Tor}_n^A(M,N) \to \operatorname{Tor}_n^A(M_1, N_1).$$
$\operatorname{Tor}_n^A(f,g)$ is independent of the choice of lifts since:

If $\overline{f}, \overline{f}' : P_* \to P_{1*}$ and $\overline{g}, \overline{g}' : Q_* \to Q_{1*}$ are choices of lifts, there are chain homotopies H and K with $H : \overline{f} \simeq \overline{f}'$, $K : \overline{g} \simeq \overline{g}'$, that is,
$$dH + Hd = \overline{f} - \overline{f}', \quad dK + Kd = \overline{g} - \overline{g}'.$$
Then
$$H \otimes_A \overline{g} : \overline{f} \otimes_A \overline{g} \simeq \overline{f}' \otimes_A \overline{g}$$
and
$$\overline{f}' \otimes_A K : \overline{f}' \otimes_A \overline{g} \simeq \overline{f}' \otimes_A \overline{g}'$$
imply that the induced homology maps satisfy
$$H_n(\overline{f} \otimes_A \overline{g}) = H_n(\overline{f} \otimes_A \overline{g}) = H_n(\overline{f} \otimes_A \overline{g})$$
and hence $\operatorname{Tor}_n^A(f,g)$ is independent of the choice of lifts.

Now it is immediate that $\operatorname{Tor}_n^A(M,N)$ is a functor for all $n \geq 0$, uniquely defined up to natural isomorphism.

Proof 2: $\operatorname{Tor}_n^A(M,N)$ is a balanced functor in the sense that one of the two resolutions may be omitted, more precisely,

Lemma 10.8.6. *The following maps induce isomorphisms of homology*
$$P_* \otimes_A N \xleftarrow{1 \otimes_A \epsilon} P_* \otimes_A Q_* \xrightarrow{\epsilon \otimes_A 1} M \otimes_A Q_*$$

Since the P_i are projective, they are retracts of free A modules $V_i \otimes_R A$ with V_i projective over R and hence P_i is flat over A, that is, $P_i \otimes_A (\ )$ is an exact functor for all $i \geq 0$.

Filter the complex P_* by the increasing filtration $F_n(P_*) = P_{\leq n}$. This induces the filtrations $F_n(P_* \otimes_A N) = F_n(P_*) \otimes_A N$ and $F_n(P_* \otimes_A Q_*) = F_n(P_*) \otimes_A Q_*$. In the resulting spectral sequences we have:
$$E^0(P_* \otimes_A N) = P_* \otimes_A N, \ d^0 = 0,$$
$$E^1(P_* \otimes_A N) = P_* \otimes_A N, \ d^1 = d_P \otimes_A 1,$$
$$E^0(P_* \otimes_A Q_*) = P_* \otimes_A Q_*, \ d^0 = 1 \otimes_A d_Q,$$
$$E^1(P_* \otimes_A Q_*) = P_* \otimes_A N, \ d^1 = d_P \otimes_A 1$$
since P_* is flat over A.

Thus $P_* \otimes_A N \xleftarrow{1 \otimes_A \epsilon} P_* \otimes_A Q_*$ induces an isomorphism on E^1 and is an isomorphism of homology.

That $P_* \otimes_A Q_* \xrightarrow{\epsilon \otimes_A 1} M \otimes_A Q_*$ is an isomorphism of homology is similar.

Having proved the balanced lemma that any one of the two resolutions can be omitted, it is easy to see that maps $\operatorname{Tor}_n^A(f, g)$ are independent of the choice of lifts and the fact that the $\operatorname{Tor}_n^A(M, N)$ are functors well defined up to natural isomorphism follows as before. $\square$

Finally, we observe that:

Lemma 10.8.7. *There is a natural isomorphism*

$$\operatorname{Tor}_0^A(M, N) \cong M \otimes_A N.$$

Proof:

$$\operatorname{Tor}^A(M, N) \cong H_0(P_* \otimes_A N) \cong M \otimes_A N$$

since

$$P_1 \otimes_A N \xrightarrow{d_1 \otimes_A 1} P_0 \otimes_A N \xrightarrow{\epsilon \otimes_A 1} M \otimes_A N \to 0$$

is exact. $\square$

Remarks. If M and N are differential algebras over a differential algebra A, then the tensor product differential on $M \otimes_R N$ defines a differential on the quotient $M \otimes_A N$ and also on the derived functors $\operatorname{Tor}_n^A(M, N)$. The isomorphisms in this section are all isomorphisms of differential objects.

10.9 Comodules over supplemented coalgebras

Let C be a coalgebra with counit $\epsilon : C \to R$. A (left) comodule over C is a graded module M together with a degree 0 linear map $\Delta : M \to C \otimes M$ called a coaction. The coaction is associative and has a counit with respect to the diagonal of the coalgebra, that is, the following diagrams are commutative:

$$\begin{array}{ccc} M & \xrightarrow{\Delta} & C \otimes M \\ \downarrow \Delta & & \downarrow \Delta \otimes 1 \\ C \otimes M & \xrightarrow{1 \otimes \Delta} & C \otimes C \otimes M \end{array}$$

$$\begin{array}{ccc} & & M \\ & \cong \swarrow & \downarrow \Delta \\ R \otimes M & \xleftarrow{\eta \otimes 1} & C \otimes M \end{array}$$

Remarks. As expected, a differential comodule over a differential coalgebra is a comodule with a differential in which the structure map is a differential map, that is, $\Delta \cdot d_M = (d_C \otimes 1 + 1 \otimes d_M) \cdot \Delta : M \to C \otimes M$.

The primitives of a comodule is the dual of the indecomposables of a module:

Definition 10.9.1. If M is a comodule over a supplemented coalgebra C with reduced coalgebra $\overline{C} = \text{kernel}\,(\epsilon) : C \to R$, then the module of primitives $P_C(M) =$ the kernel of the reduced coaction $\overline{\Delta} : M \to \overline{C} \otimes M$. That is, $m \in M$ is in $P_C(M)$ if and only if the coaction is given by $\Delta(m) = 1 \otimes m$.

The functor $P_C(M)$ is additive and left exact under mild restrictions, that is:

Lemma 10.9.2.

(a) $P_C(M \oplus N) \cong P_C(M) \oplus P_C(N)$

(b) $0 \to M \to N \to P$ exact implies that $0 \to P_C(M) \to P_C(N) \to P_C(P)$ is exact, provided either that C is flat as an R module or that $0 \to M \to N$ is a split monomorphism of R modules.

Proof: Part (a) is clear. Since $C = R \oplus \overline{C}$, it follows that $\overline{C}$ is flat if C is flat. Hence part (b) follows from the diagram below with exact bottom two rows and exact columns:

$$\begin{array}{ccccccc}
& & 0 & & 0 & & 0 \\
& & \downarrow & & \downarrow & & \downarrow \\
0 & \to & P_C(M) & \to & P_C(N) & \to & P_C(P) \\
& & \downarrow & & \downarrow & & \downarrow \\
0 & \to & M & \to & N & \to & P \\
& & \downarrow \overline{\Delta} & & \downarrow \overline{\Delta} & & \downarrow \overline{\Delta} \\
0 & \to & \overline{C} \otimes M & \to & \overline{C} \otimes N & \to & \overline{C} \otimes P
\end{array}$$

$\square$

The graded version of the dual of Nakayama's lemma is the following.

Lemma 10.9.3. *If C is a connected supplemented coalgebra and M is a comodule over C which is bounded below, then $P_C(M) = 0$ implies that $M = 0$.*

Proof: If there is a nonzero element m in M, then there is such an element of least degree and $\Delta(m) = 1 \otimes m$. $\square$

Corollary 10.9.4. *Let C be a connected supplemented coalgebra which is flat over the ground ring R and let $f : M \to N$ be a map of comodules over C with M bounded below. Then f is a monomorphism if and only if the induced map of primitives $P_C(f) : P_C(M) \to P_C(N)$ is a monomorphism.*

Definition 10.9.5. A choice of cogenerators is a projection $\pi : M \to P_C(M)$, that is, if $\iota : P_C(M) \to M$ is the inclusion, then $\pi \cdot \iota = 1_{P_C(M)}$.

10.9 Comodules over supplemented coalgebras 347

Lemma 10.9.6. *Let $f, g : M \to N$ be two maps of comodules over a connected supplemented coalgebra C and let $\pi : N \to P_C(N)$ be a choice of cogenerators with N bounded below. If $\pi \cdot f = \pi \cdot g$, then $f = g$.*

Proof: One uses induction on the degree and the split exact sequence $0 \to P_C(N) \to N \to \overline{C} \otimes N$. $\square$

For the remainder of this section, we will assume that the coalgebra C and the C comodules M have no differential structure. Just as with algebras and modules, the presence of differential structures requires modification of the notions of injective and extended comodules. One such modification is illustrated in the exercises to this section. But the modification we are really interested in is deferred to the relative homological algebra of section 10.10.

Definition 10.9.7. An extended comodule M over C is one of the form $M = C \otimes V$ with coaction $\Delta = \Delta \otimes 1 : C \otimes V \to C \otimes C \otimes V$.

The basic facts about extended comodules are:

Lemma 10.9.8.

(a) $P_C(C \otimes V) = V$.

(b) *If N is a comodule, there is a one to one correspondence between comodule maps $g : N \to C \otimes V$ and module maps $\overline{g} : N \to V$ given by:*

$$\overline{g} = (\epsilon \otimes 1) \cdot g : N \to C \otimes V \to R \otimes V = V.$$
$$g = (1 \otimes \overline{g}) \cdot \Delta : N \to C \otimes N \to C \otimes V.$$

There are enough extended comodules in the sense that any comodule M can be embedded in an extended comodule. This can be done via the R-split embedding $\Delta : M \to C \otimes M$.

It follows from Lemma 10.9.8 that extended comodules $Q = C \otimes V$ are injective if and only if V is an injective R module. That is, given any monomorphism $f : M \to N$ and any map $g : M \to C \otimes V$ with V injective, there exists a map $h : N \to Q$ such that $h \cdot f = g$.

If C is flat over R, it follows easily that a comodule M can be embedded in an injective extended comodule $C \otimes V$ where V is an injective envelope of the underlying R module of M. Thus, a comodule M is injective if and only there is a comodule N such that $M \oplus N$ is an injective extended comodule.

Proposition 10.9.9. *Suppose that C is a connected supplemented coalgebra which is flat over R and that M is an injective C comodule which is bounded below. Then $M \cong C \otimes P_C(M)$ where $P_C(M)$ is injective over R, that is, M is an extended comodule.*

Proof: Let M be a summand of the injective extended comodule $P = C \otimes V$. Then $P_C(M)$ is a summand of $P_C(P) \cong V$. Since V is an injective R module, $P_C(M)$ is a injective R module. Hence the injection $\iota : P_C(M) \to M$ has a projection $\pi : M \to P_C(M)$ and we can consider the composition $f = (1 \otimes \pi) \cdot \Delta : M \to C \otimes M \to C \otimes P_C(M)$.

Since $P_C(f) : P_C(M) \to P_C(C \otimes P_C(M))$ is an isomorphism, it is also an monomorphism. Therefore $f : M \to C \otimes P_C(M)$ is an monomorphism. Since M is injective, $C \otimes P_C(M) = K \oplus M$ where K is the cokernel of f. Since, $P_C(C \otimes P_C(M)) = P_C(M) = P_C(K) \oplus P_C(M)$, it follows $P_C(K)$ and hence K itself are 0. Hence, $C \otimes P(M) = M$. □

Suppose C is flat over R. For all C comodules M, the usual definitions and facts concerning injective resolutions are valid and the proofs are mild variations of the standard proofs.

10.9.10. There exist injective resolutions of comodules

$$0 \to M \xrightarrow{\eta} Q_0 \xrightarrow{d_0} Q_1 \xrightarrow{d_1} \cdots \xrightarrow{d_{n-1}} Q_n \xrightarrow{d_n} \cdots,$$

that is, complexes of comodules $0 \to M \xrightarrow{\eta} Q_*$ which are resolutions in the sense that they are exact sequences and such that all the Q_* are injective comodules.

Given a comodule M, here is a procedure for making an injective resolution: Let $Q(M)$ be an injective envelope of M as an R module. Set $Q_0 = C \otimes Q(M)$ and embed M as a comodule into an injective comodule via

$$\eta : M \xrightarrow{\Delta} C \otimes M \to C \otimes Q(M).$$

Let $M_0 =$ the cokernel of η and embedd $M_0 \to C \otimes Q(M_0) = Q_1$ by the same procedure as was used to embedd M. Successive interation gives an exact sequence $0 \to M \to Q_0 \to Q_1 \to \ldots$ which is the injective resolution of M as a comodule.

10.9.11. Suppose we are given a map $f : M \to N$ of comodules. For any exact complex of comodules $0 \to M \xrightarrow{\eta} P_*$ and any complex of injective comodules $0 \to N \xrightarrow{\eta} Q_*$, there is a map of complexes $f_* : P_* \to Q_*$ lifting f, that is, there is a commutative diagram

$$\begin{array}{ccccccccc} M & \xrightarrow{\eta} & P_0 & \xrightarrow{d_0} & P_1 & \xrightarrow{d_1} & \cdots & \xrightarrow{d_{n-1}} & P_n & \xrightarrow{d_n} \\ \downarrow f & & \downarrow f_0 & & \downarrow f_1 & & & & \downarrow f_n & \\ N & \xrightarrow{\eta} & Q_0 & \xrightarrow{d_0} & Q_1 & \xrightarrow{d_1} & \cdots & \xrightarrow{d_{n-1}} & Q_n & \xrightarrow{d_n} \end{array}$$

10.9.12. Given a map $f : M \to N$ and two choices for maps $f_*, g_* : P_* \to Q_*$ lifting the same f with P_* an acylic complex and with Q_* an injective complex, the maps f_* and g_* are chain homotopic, that is, there exists a map $H_* : P_* \to Q_*$ of degree -1 such that $d_* H_* + H_* d_* = f_* - g_*$. More precisely, for $n \geq 0$, there

are A–linear $H_{n+1} : P_{n+1} \to Q_n$ such that
$$H_1 \cdot d_0 = f_0 - g_0, \quad H_{n+1} \cdot d_n + d_{n-1} \cdot H_n = f_n - g_n, n \geq 1.$$

Exercises

(1) Verify Lemmas 10.9.4 and 10.9.5.

(2) Verify Lemma 10.9.8.

(3) Consider the category of positively (nonnegatively) graded differential modules over a commutative ring R, that is, the category of chain complexes over R.

   (a) Show that the following objects are injective in this category:

   (i) Any acyclic module M which is concentrated in two nonnegative degrees n and $n - 1$ with M_n injective over R.

   (ii) Any arbitrary product of objects of the form (1).

   (b) Show that any R differential R module can be embedded in an injective differential R module.

   (c) Show that any injective differential module M is acyclic, that is, $HM = 0$.

(4) Suppose C is a supplemented differential coalgebra and consider the category of differential comodules over C.

   (a) If V is an injective chain complex over R, then the extended comodule $C \otimes V$ is an injective differential C comodule.

   (b) If M is an injective differential comodule over C, then M is a retract of some $C \otimes V$ where V is an injective chain complex over R.

   (c) If M is an injective differential comodule, show that $HM = 0$. (Hence, this notion of injective differential comodule is very restrictive. In the next section, we produce a more useful notion of injective projective comodule.)

(5) Prove 10.9.11 and 10.9.12.

10.10 Injective classes

Let $\mathcal{A}$ be an additive category which has cokernels. Usually, $\mathcal{A}$ is an abelian category. But the example of comodules over a coalgebra is important to us and this is not an abelian category unless the coalgebra is flat over the ground ring.

We follow Eilenberg and Moore [43] in defining what is called relative homological algebra in $\mathcal{A}$. This involves defining what are called relative monomorphisms and what are called relative injective objects. The essential point is that maps into a relative injective object can be extended over relative monomorphisms. In this manner, the relative monomorphisms determine the relative injective objects and vice versa.

Let $\mathcal{I}$ be a class of objects in $\mathcal{A}$ and let $\mathcal{M}$ be a class of morphisms in $\mathcal{A}$.

Let $\mathcal{I}^* =$ the set of morphisms $f : A \to B$ in $\mathcal{A}$ such that $f^* : \mathrm{map}(A, Q) \leftarrow \mathrm{map}(B, Q)$ is an epimorphism for all Q in $\mathcal{I}$.

Let $\mathcal{M}^* =$ the set of objects Q in $\mathcal{A}$ such that $f^* : \mathrm{map}(A, Q) \leftarrow \mathrm{map}(B, Q)$ is an epimorphism for all $f : A \to B$ in $\mathcal{M}$.

We note the following properties.

Lemma 10.10.1.

(a) $\mathcal{I} \subseteq \mathcal{I}'$ implies that $\mathcal{I}'^* \subseteq \mathcal{I}^*$.

(b) $\mathcal{I} \subseteq \mathcal{I}^{**}$.

(c) $\mathcal{I}^* = \mathcal{I}^{***}$.

(d) $\mathcal{M} \subseteq \mathcal{M}'$ implies that $\mathcal{M}'^* \subseteq \mathcal{M}^*$.

(e) $\mathcal{M} \subseteq \mathcal{M}^{**}$.

(f) $\mathcal{M}^* = \mathcal{M}^{***}$.

Lemma 10.10.2.

(a) *The class of objects $\mathcal{M}^*$ is closed under retracts, that is, if there exists an identity map $1 : P \to Q \to P$ and Q is in $\mathcal{M}^*$, then so is P.*

(b) *The class of morphisms $\mathcal{I}^*$ is closed under left factorization, that is, if a composition $A \xrightarrow{f} B \to C$ is in $\mathcal{I}^*$, then so is f.*

Definition 10.10.3. The pair $(\mathcal{I}, \mathcal{M})$ is called an injective class if the following three properties are satisfied:

(a) $\mathcal{M}^* = \mathcal{I}$.

(b) $\mathcal{I}^* = \mathcal{M}$.

(c) For all objects A in $\mathcal{A}$, there exists an object Q in $\mathcal{I}$ and a morphism $f : A \to Q$ in $\mathcal{M}$.

The objects in $\mathcal{I}$ are called relative injectives and the morphisms in $\mathcal{M}$ are called relative monomorphisms. Condition (c) above says that there are enough relative injective objects.

10.10 Injective classes 351

Remarks 10.10.4. Often an injective class is given in the following way. Specify the class of relative monomorphisms $\mathcal{M}$. Check that this is closed under left factorization. Define the relative injectives to be $\mathcal{I} = \mathcal{M}^*$. Check that there are enough relative injectives, that is, for all objects A in $\mathcal{A}$, there is an object Q in $\mathcal{I}$ and a morphism $F : A \to Q$ in $\mathcal{M}$. Then the pair $(\mathcal{I}, \mathcal{M})$ is an injective class. One needs to check that $\mathcal{M} = \mathcal{I}^*$.

The split injective class 10.10.5

In any abelian category, an injective class can be given by choosing the relative monomorphisms to be all split monomorphisms, in which case the relative injective objects are all objects in the category.

Reflection via adjoint functors 10.10.6

New injective classes arise from known injective classes by a process called reflection via adjoint functors.

We describe this process now. Suppose we are given two abelian categories $\mathcal{A}$ and $\mathcal{B}$ (more generally, it is sufficient that these categories be additive categories with cokernels), an injective class $(\mathcal{M}, \mathcal{I})$ in $\mathcal{B}$, and a pair of adjoint functors $S : \mathcal{A} \to \mathcal{B}$ and $T : \mathcal{B} \to \mathcal{A}$ with equivalences of morphism sets

$$\mathcal{A}(A, TB) \cong \mathcal{B}(SA, B)$$

and adjunction maps $STB \to B$ and $A \to TSA$. We know from Exercise 3 that the left adjoint S preserves the usual kernels and the right adjoint preserves the usual cokernels.

Define a new injective class $(\mathcal{M}', \mathcal{I}')$ in $\mathcal{A}$ as follows:

(a) The relative monomorphisms are $\mathcal{M}' = S^{-1}(\mathcal{M}) = \{A_1 \xrightarrow{f} A_2 \mid Sf \in \mathcal{M}\}$.

(b) The relative injective objects are $\mathcal{I}' = \overline{T(\mathcal{I})} = $ the set of all retracts of objects $T(Q)$ with Q in $\mathcal{I}$.

We need to verify that the pair $(\mathcal{M}', \mathcal{I}')$ is an injective class.

(c) $\mathcal{M}' = S^{-1}(\mathcal{M})$ is closed under left factorization, that is, $A_1 \to A_2 \to A_3$ in $S^{-1}(\mathcal{M})$ implies that $A_1 \to A_2$ is also in $S^{-1}(\mathcal{M})$. This follows from the fact that $\mathcal{M}$ is closed under left factorization.

(d) We note that, for all objects A in $\mathcal{A}$, there is a map $SA \to Q$ which is a relative monomorphism in $\mathcal{M}$ with Q a relative injective in $\mathcal{I}$. We claim that the adjoint $A \to TQ$ is a relative monomorphism in $\mathcal{M}'$ (and clearly TQ is in $\mathcal{I}'$). But $SA \to STQ \to Q$ is in $\mathcal{M}$ implies that $SA \to STQ$ is in $\mathcal{M}$ since $\mathcal{M}$ is closed under left factorization.

(e) Finally, we claim that $\overline{T(\mathcal{I}')} = \mathcal{M}'^*$.

Since $\mathcal{M}'^*$ is closed under retracts, to show that $\overline{T(\mathcal{I}')} \subseteq \mathcal{M}'^*$, it suffices to show that $T(\mathcal{I}') \subseteq \mathcal{M}'^*$. But, if $A_1 \to A_2$ is in $\mathcal{M}'$ and Q is in $\mathcal{I}$, then the map

$$\mathcal{A}(A_1, TQ) \leftarrow \mathcal{A}(A_2, TQ)$$

is equivalent to the epimorphism

$$\mathcal{B}(SA_1, Q) \leftarrow \mathcal{B}(SA_2, Q).$$

Thus, TQ is in $\mathcal{M}'^*$.

To show that $\mathcal{M}'^* \subseteq \overline{T(\mathcal{I})}$, let M be in $\mathcal{M}'^*$ and consider the map $M \to TQ$ in $\mathcal{M}'$ guaranteed by (d). Since M is in $\mathcal{M}'^*$, this map is split and M is a retract of TQ.

Thus Remarks 10.10.4 shows that the pair $(\mathcal{M}' = S^{-1}(\mathcal{M}), \mathcal{I}' = \overline{T(\mathcal{I})})$ is an injective class in $\mathcal{A}$.

The proper injective class for nondifferential comodules 10.10.7

Let $\mathcal{B} = \mathcal{M}_R$ be the category of all graded R modules and endow it with the injective class where the relative monomorphisms are all split monomorphisms and all objects are relative injective objects. Let C be a graded coalgebra over R and let $\mathcal{A} = \mathcal{M}_C$ be the category of all graded C comodules.

We have a pair of adjoint functors $S: \mathcal{M}_C \to \mathcal{M}_R$ and $T: \mathcal{M}_R \to \mathcal{M}_C$ given by

$$S(M) = M \text{ regarded just as an } R\text{-module}, M \in \mathcal{M}_C$$

$$T(N) = C \otimes N, \quad N \in \mathcal{M}_R,$$

with the coalgebra structure given by $\Delta \otimes 1 : C \otimes N \to C \otimes C \otimes N$.

The specific one-to-one correspondence of maps is as follows:

(a) Given a comodule M, an R module N, and an R module map $f: M \to N$, we have the comodule map

$$\overline{f}: M \xrightarrow{\Delta} C \otimes M \xrightarrow{1 \otimes f} C \otimes N.$$

(b) Given a comodule map $\overline{f}: M \to C \otimes N$, we have the R module map

$$f: M \xrightarrow{\overline{f}} C \otimes N \to R \otimes N = N.$$

Thus, the process of reflection via adjoint functors gives an injective class in $\mathcal{M}_C$, the so-called proper injective class, with:

(A) A map $f: M \to N$ of comodules is a relative monomorphism if it is a split monomorphism of R modules. This is also called a proper monomorphism. Note that proper monomorphisms are also monomorphisms in the usual sense.

10.10 Injective classes 353

(B) A relative injective comodule is any retract of an extended comodule $C \otimes N$ where N is any R module. This is also called a proper injective. Note that proper injective comodules are injective in the usual sense only if N is an injective R module.

(C) For any comodule M, the coaction map $M \xrightarrow{\Delta} C \otimes M$ is a relative monomorphism into a relative injective.

Our next and most important example is the differential version of the above, the injective class in which the relative monomorphisms and the relative injective differential comodules are also called proper. This will lead to no confusion with the above since forgetting the differential transforms these new relative monomorphisms into R split monomorphisms and transforms these new relative injective comodules into retracts of extended comodules.

The proper injective class for differential comodules 10.10.8

Let $\mathcal{B} = \mathcal{DM}_R$ be the category of all differential graded R modules and endow it with the injective class where the relative monomorphisms are all split monomorphisms and all objects are relative injective objects. Let C be a differential graded coalgebra over R and let $\mathcal{A} = \mathcal{DM}_C$ be the category of all graded C comodules. We have a pair of adjoint functors $S : \mathcal{DM}_C \to \mathcal{DM}_R$ and $T : \mathcal{DM}_R \to \mathcal{DM}_C$ given by

$$S(M) = M \text{ regarded just as a differential graded } R\text{-module}, \ M \in \mathcal{DM}_C$$
$$T(N) = C \otimes N, \ N \in \mathcal{DM}_R$$

with the tensor product differential and with the coalgebra structure given by $\Delta \otimes 1 : C \otimes N \to C \otimes C \otimes N$.

The specific one-to-one correspondence of maps is as before:

(a) Given a differential comodule M, a differential R module N, and a map of differential R modules $f : M \to N$, we have the differential comodule map

$$\overline{f} : M \xrightarrow{\Delta} C \otimes M \xrightarrow{1 \otimes f} C \otimes N.$$

(b) Given a differential comodule map $\overline{f} : M \to C \otimes N$, we have the map of differential R modules

$$f : M \xrightarrow{\overline{f}} C \otimes N \to R \otimes N = N.$$

Thus, the process of reflection via adjoint functors gives an injective class in $\mathcal{DM}_C$, the so-called proper injective class, with:

(A) A map $f : M \to N$ of differential comodules is a relative monomorphism or a proper monomorphism if it is a split monomorphism of differential R modules.

(B) A relative injective comodule or a proper injective comodule is any retract of $C \otimes N$ where N is any differential R module.

(C) For any differential comodule M, the coaction map $M \xrightarrow{\Delta} C \otimes M$ is a proper monomorphism into a proper injective.

Whenever we have an injective class we have a good theory of injective resolutions as follows and the proofs are identical to the standard proofs. More precisely, we define:

Definition 10.10.9. A sequence $0 \to M \to N \to P \to 0$ is called a relative short exact sequence if P is the cokernel of $M \to N$ and $M \to N$ is a relative monomorphism.

A complex $0 \to M \to Q_*$ is a relative acyclic complex or a relative exact sequence if it can be factored into relative short exact sequences. We can do homological algebra using injective classes since we have relative injective resolutions which are functorial up to chain homotopy:

10.10.10. There exist relative injective resolutions

$$0 \to M \xrightarrow{\eta} Q_0 \xrightarrow{d_0} Q_{-1} \xrightarrow{d_{-1}} \cdots \xrightarrow{d_{-n+1}} Q_{-n} \xrightarrow{d_{-n}} \cdots,$$

that is, complexes $0 \to M \xrightarrow{\eta} Q_*$ which are relative resolutions in the sense that they are relative exact sequences and such that all the Q_* are relative injective objects.

10.10.11. The fact that $\Delta : M \to C \otimes M$ is always a proper monomorphism allows us to construct functorial proper injective resolutions of differential comodules:

$$0 \to M \xrightarrow{\Delta} C \otimes M \to \operatorname{coker}(\Delta) \to 0$$

$$0 \to \operatorname{coker}(\Delta) \xrightarrow{\Delta = d_{-1}} C \otimes \operatorname{coker}(\Delta) \to \operatorname{coker}(d_{-1}) \to 0$$

$$0 \to \operatorname{coker}(d_{-1}) \xrightarrow{\Delta = d_{-2}} C \otimes \operatorname{coker}(d_{-1}) \to \operatorname{coker}(d_{-2}) \to 0$$

$\ldots$

$$Q_0 = C \otimes M$$
$$Q_{-1} = C \otimes \operatorname{coker}(\Delta)$$
$$Q_{-2} = C \otimes \operatorname{coker}(d_{-1}).$$

$\ldots$

In general, $Q_{-n-1} = C \otimes \mathrm{coker}(d_{-n})$. This functorial resolution is called the categorical cobar resolution and we denote it by $0 \to M \to Q_*(M)$.

10.10.12. Suppose we are given a map $f: M \to N$ of objects. For any relative exact complex $0 \to M \xrightarrow{\eta} P_*$ and any relative injective complex $0 \to N \xrightarrow{\eta} Q_*$, there is a map of complexes $f_*: P_* \to Q_*$ extending f, that is, there is a commutative diagram

$$\begin{array}{ccccccccc} M & \xrightarrow{\eta} & P_0 & \xrightarrow{d_0} & P_{-1} & \xrightarrow{d_{-1}} & \cdots & \xrightarrow{d_{-n+1}} & P_{-n} & \xrightarrow{d_{-n}} \\ \downarrow f & & \downarrow f_0 & & \downarrow f_{-1} & & & & \downarrow f_{-n} & \\ N & \xrightarrow{\eta} & Q_0 & \xrightarrow{d_0} & Q_{-1} & \xrightarrow{d_{-1}} & \cdots & \xrightarrow{d_{-n+1}} & Q_{-n} & \xrightarrow{d_{-n}} \end{array}$$

10.10.13. Given a map $f: M \to N$ and two choices for maps $f_*, g_*: P_* \to Q_*$ extending the same f with P_* a relative acyclic complex and with Q_* a relative injective complex, the maps f_* and g_* are chain homotopic, that is, there exists a map $H_*: P_* \to Q_*$ of degree $+1$ such that $d_* H_* + H_* d_* = f_* - g_*$. More precisely, for $n \geq 0$, there are A−linear $H_{-n-1}: P_{-n-1} \to Q_{-n}$ such that

$$H_{-1} \cdot d_0 = f_0 - g_0, \quad H_{-n-1} \cdot d_{-n} + d_{-n+1} \cdot H_{-n} = f_{-n} - g_{-n}, n \geq 1.$$

Exercises

(1) Prove Lemma 10.10.1.

(2) Prove Lemma 10.10.2.

(3) Suppose we are given two abelian categories $\mathcal{A}$ and $\mathcal{B}$, and a pair of adjoint functors $S: \mathcal{A} \to \mathcal{B}$ and $T: \mathcal{B} \to \mathcal{A}$ with equivalences of morphism sets

$$\mathcal{A}(A, TB) \cong \mathcal{B}(SA, B).$$

Show that:

(a) The right adjoint T preserves kernels, that is, $0 \to B_1 \to B_2 \to B_3$ exact in $\mathcal{B}$ implies that $0 \to TB_1 \to TB_2 \to TB_3$ is exact in $\mathcal{A}$, that is, T is left exact in the usual sense.

(b) The left adjoint S preserves cokernels, that is, $A_1 \to A_2 \to A_3 \to 0$ exact in $\mathcal{A}$ implies that $SA_1 \to SA_2 \to SA_3 \to 0$ is exact in $\mathcal{B}$, that is, S is right exact in the usual sense.

(c) Suppose that the injective class $(\mathcal{M}', \mathcal{I}')$ in $\mathcal{A}$ arises by reflection via the adjoint functors from the injective class $(\mathcal{M}, \mathcal{I})$ in $\mathcal{B}$. Then a sequence $0 \to A_1 \to A_2 \to A_3 \to 0$ is a relative exact sequence in $\mathcal{A}$ if and only if $0 \to SA_1 \to SA_2 \to SA_3 \to 0$ is a relative exact sequence in $\mathcal{B}$.

(4) Show that the procedure in Remarks 10.10.4 always gives an injective class.

(5) Verify that the functors in Example 10.10.7 are an adjoint pair and that the relative injective embedding in property (C) is valid.

(6) Verify 10.10.9. 10.10.10, and 10.10.12

(7) Let A be an abelian category with the injective class of split monomorphisms. Show that a sequence

$$0 \to M \xrightarrow{\eta} Q_0 \xrightarrow{d_0} Q_{-1} \xrightarrow{d_{-1}} \cdots\cdots$$

is a split exact sequence if and only if there exists a contracting homotopy $h_0 : Q_0 \to M$, $h_{-i} : Q_{-i} \to Q_{-i+1}$, $i > 0$ such that

$$h_0 \cdot \eta = 1_M, \quad d_{-i+1} \cdot h_{-i} + h_{-i-1} \cdot d_{-i} = 1_{Q_{-i}}, \quad i \geq 0.$$

(8) **Nice proper injective resolutions of differential comodules:** Let M be a (right) differential comodule over a differential coalgebra C over a commutative ring R. Show that:

(a) If M and C are flat modules over R, then there exists a proper injective resolution $0 \to M \to Q_*$ which is extended and R projective in the sense that all the $Q_{-i} = \overline{Q_{-i}} \otimes C$ with $\overline{Q_{-i}}$ and hence all the Q_{-i} are projective R modules.

(b) Do part (a) with R projective replacing R flat.

(c) If C is k-connected, that is, $\overline{C}_i = 0$ for all $i \leq k$, then Q_* can be chosen to be tapered in the sense that each Q_{-i} is a $(k+1)i - 1$ connected R module. And this can be done while doing parts (a) or (b).

(9) If C is a coalgebra which is flat over the ground ring show that the category of comodules over C is an abelian category. If the coalgebra is not flat over the ground ring, then show that the category is an additive category which has cokernels.

10.11 Cotensor products and derived functors

Let C be a (graded) coalgebra over a commutative ring R and let M and N be right and left comodules, respectively, over C. The definition of the cotensor product of graded comodules over a graded coalgebra $M \otimes_C N$ is the dual of the definition of the tensor product of graded modules over a graded algebra.

The cotensor product $M \square_C N$ is the following submodule of the tensor product over R, $M \otimes N = M \otimes_R N$ where

$$(M \otimes N)_n = \bigoplus_{i+j=n} M_i \otimes N_j :$$

10.11 Cotensor products and derived functors

Definition 10.11.1. $M\square_C N$ is the equalizer of the two coaction maps:
$$\Delta \otimes 1, 1 \otimes \Delta : M \otimes N \to M \otimes C \otimes N.$$

In other words, the sequence
$$0 \to M\square_C N \to M \otimes N \xrightarrow{\delta} M \otimes C \otimes N$$
is exact where $\delta = \Delta \otimes 1 - 1 \otimes \Delta$.

A fundamental property of the cotensor product is that it is left exact with respect to the injective class where relative monomorphisms are the proper monomorphisms, that is, the R split monomorphisms:

Proposition 10.11.2. *Suppose that $0 \to M_1 \to M_2 \to M_3$ is a proper exact sequence of right C comodules and that N is a left C module. Then*
$$0 \to M_1\square_C N \to M_2\square_C N \to M_3\square_C N$$
is exact. The corresponding statement is true when the roles of left and right comodules are reversed.

Proof: Consider the commutative diagram

$$
\begin{array}{ccccccc}
 & & 0 & & 0 & & 0 \\
 & & \downarrow & & \downarrow & & \downarrow \\
0 & \to & M_1\square_C N & \to & M_2\square_C N & \to & M_3\square_C N \\
 & & \downarrow & & \downarrow & & \downarrow \\
0 & \to & M_1 \otimes N & \to & M_2 \otimes N & \to & M_3 \otimes N \\
 & & \downarrow & & \downarrow & & \downarrow \\
0 & \to & M_1 \otimes C \otimes N & \to & M_2 \otimes C \otimes N & \to & M_3 \otimes C \otimes N
\end{array}
$$

where the exactness of the vertical columns defines the cotensor product. The proper exactness guarantees the left exactness of the usual tensor product of modules over R and hence the bottom two rows are exact. Now the exactness of the top row is a simple exercise. $\square$

The cotensor product satisfies the following three formal algebraic properties in Proposition 10.11.3 below: it is associative, distributive over direct sums, and the coalgebra is a two-sided unit.

Let C and D be coalgebras. Let M, M_1, M_2 be right C comodules, let K, K_1, K_2 be left C modules, let N be a left D comodule, and let P be a $C - D$ bicomodule, that is the coactions on P commute as follows:

$$
\begin{array}{ccc}
P & \xrightarrow{\Delta} & P \otimes D \\
\downarrow \Delta & & \downarrow \Delta \otimes 1 \\
C \otimes P & \xrightarrow{1 \otimes \Delta} & C \otimes P \otimes D
\end{array}
$$

Then:

Proposition 10.11.3.

(a) *If M, N, and C are flat over R, there is a natural isomorphism*

$$M \square_C (P \square_D N) \cong (M \square_C P) \square_D N.$$

(b) *There are natural isomorphisms*

$$(M_1 \oplus M_2) \square_C K \cong (M_1 \square_C K) \oplus (M_2 \square_C K),$$
$$M \square_C (K_1 \oplus K_2) \cong (M \square_C K_1) \oplus (M \square_C K_2).$$

(c) *The comodule actions induce natural isomorphisms*

$$C \square_C N \cong N, \quad M \square_C C \cong M,$$

that is, there are isomorphisms

$$\Delta : N \to C \square_C N, \quad \Delta : M \to M \square_C C.$$

Proof:

(a) Let X be the kernel of the natural map

$$\psi = \delta_1 \oplus \delta_2 : M \otimes P \otimes N \to (M \otimes C \otimes P \otimes N) \oplus (M \otimes P \otimes D \otimes N)$$

where $\delta_1 = \delta \otimes 1_N$ and $\delta_2 = 1_M \otimes \delta$. Since N is flat over R, the kernel of δ_1 is $(M \square_C P) \otimes N$. Since $(M \square_C P) \otimes D \otimes N \to (M \otimes P) \otimes D \otimes N$ is a monomorphism, the intersection of the kernels of δ_1 and δ_2 is $(M \square_C P) \square_D N$.

But, since the argument is symmetric, X is also $M \square_C (P \square_D N)$.

(b) This is an immediate consequence of the fact that tensor over R naturally distributes over direct sums.

(c) The associativity of the coaction shows that the image of $\Delta : M \to M \otimes C$ is contained in $M \square_C C$. And clearly Δ is a monomorphism.

Let $\alpha = \Sigma m \otimes c \in M \square_C C \subseteq M \otimes C$. We need to show that $\alpha \in \text{image}(\Delta)$. Since $(1 \otimes \Delta)\alpha = (\Delta \otimes 1)\alpha$, we have $\Sigma m \otimes \Delta(c) = \Sigma \Delta(m) \otimes c$ and hence $(1 \otimes 1 \otimes \epsilon)(1 \otimes \Delta)\alpha = (1 \otimes 1 \otimes \epsilon)(\Delta \otimes 1)\alpha$ implies that $\Sigma m \otimes c \otimes 1 = \Sigma \Delta(m) \otimes \epsilon(c)$. Therefore $\alpha = \Sigma m \otimes c = \Sigma \Delta(m \cdot \epsilon(c))$.

Thus, $\Delta : M \to M \square_C C$ is an isomorphism. □

Suppose C is an supplemented coalgebra with reduced coalgebra $\overline{C}$. The ground ring is a two-sided comodule over C and we have the following alternate description of the primitives of a comodule:

10.11 Cotensor products and derived functors

Proposition 10.11.4. *Let M and N be right and left comodules over C, respectively. Then there are isomorphisms*

$$M\square_C R \cong P_C(M) = \text{kernel}(M \to M \otimes \overline{C}),$$
$$R\square_C N \cong P_C(N) = \text{kernel}(N \to \overline{C} \otimes N).$$

Proof: Since cotensor is left exact (applied to sequences which are R split), we have an exact sequence

$$0 \to M\square_C R \to M\square_C C \to M\square_C \overline{C}.$$

Since $\Delta : M \to M\square_C C$ is an isomorphism, $M\square_C R$ is isomorphic to the kernel $P_C(M)$ of $\overline{\Delta} : M \to M\square_C \overline{C} \subseteq M \otimes \overline{C}$. $\square$

We now introduce the derived functors $\text{Cotor}^C_{-n}(M, N)$ of cotensor product and prove the important result that these are balanced.

Let $0 \to M \xrightarrow{\eta} P_*$ be a proper injective resolution by right C comodules and let $0 \to N \xrightarrow{\eta} Q_*$ be a proper injective resolution by left C comodules. If we grade P_* and Q_* by the so-called resolution degree, then $P_* \otimes Q_*$ can be graded by the addition of these degrees,

$$(P \otimes_A Q)_{-n} = \bigoplus_{i+j=-n} P_i \otimes_A Q_j$$

and we define

Definition 10.11.5. $\text{Cotor}^C_{-n}(M, N) = H_{-n}(P_* \otimes_A Q_*, d_\otimes)$ where $d_\otimes = d_P \otimes 1 + 1 \otimes d_Q$.

Up to natural isomorphism, $\text{Cotor}^C_{-n}(M, N)$ is a well-defined functor. We shall give two proofs, the first depending on the uniqueness of extensions up to chain homotopy and the the second depending on the so-called balanced property.

Proof 1: Let $f : M \to M_1$ and $g : N \to N_1$ be maps of A modules and choose extensions to maps of resolutions $\overline{f} : P_* \to P_{1*}$ and $\overline{g} : Q_* \to Q_{1*}$ which cover f and g, respectively. Then we have a chain map $\overline{f}\square_C\overline{g} : P_*\square_C Q_* \to P_{1*}\square_C Q_{1*}$ which induce the homology maps $H_{-n}(\overline{f}\square_C\overline{g}) = \text{Cotor}^C_{-n}(f,g) : \text{Cotor}^C_{-n}(M,N) \to \text{Cotor}^C_{-n}(M_1,N_1)$.

$\text{Cotor}^C_{-n}(f,g)$ is independent of the choice of extensions since:

If $\overline{f}, \overline{f}' : P_* \to P_{1*}$ and $\overline{g}, \overline{g}' : Q_* \to Q_{1*}$ are choices of extensions, there are chain homotopies H and K with $H : \overline{f} \simeq \overline{f}'$, $K : \overline{g} \simeq \overline{g}'$, that is,

$$dH + Hd = \overline{f} - \overline{f}', \quad dK + Kd = \overline{g} - \overline{g}'.$$

Then

$$H\square_C\overline{g} : \overline{f}\square_C\overline{g} \simeq \overline{f}'\square_C\overline{g}$$

and
$$\overline{f'} \square_C K : \overline{f'} \square_C \overline{g} \simeq \overline{f'} \square_C \overline{g'}$$
imply that the induced homology maps satisfy
$$H_{-n}(\overline{f} \otimes_A \overline{g}) = H_{-n}(\overline{f} \square_C \overline{g}) = H_{-n}(\overline{f} \square_C \overline{g})$$
and hence $\text{Cotor}_{-n}^C(f,g)$ is independent of the choice of extensions.

Now it is immediate that $\text{Cotor}_{-n}^C(M,N)$ is a functor for all $n \geq 0$, uniquely defined up to natural isomorphism.

Proof 2: $\text{Cotor}_{-n}^C(M,N)$ is a balanced functor in the sense that one of the two resolutions may be omitted, more precisely,

Lemma 10.11.6. *The following maps induce isomorphisms of homology*
$$P_* \square_C N \xrightarrow{1 \square_C \eta} P_* \square_C Q_* \xleftarrow{\eta \square_C 1} M \square_C Q_*$$

Since the P_i are relative injectives, they are retracts of extended C comodules $V_i \otimes_R C$ and hence P_i is coflat over A, that is, $P_i \square_C (\ ) = V_i \otimes_R (\ )$ preserves R split exact sequences for all $i \leq 0$.

Filter the complex P_* by the increasing filtration $F_{-n}(P_*) = P_{\leq -n}$. Note that increasing filtration means that $F_{-n-1} \subseteq F_{-n}$.

This induces the increasing filtrations $F_{-n}(P_* \square_C N) = F_{-n}(P_*) \square_C N$ and $F_{-n}(P_* \square_C Q_*) = F_{-n}(P_*) \square_C Q_*$. In the resulting spectral sequences we have:
$$E^0(P_* \square_C N) = P_* \square_C N, \ d^0 = 0,$$
$$E^1(P_* \square_C N) = P_* \square_C N, \ d^1 = d_P \square_C 1,$$
$$E^0(P_* \square_C Q_*) = P_* \square_C Q_*, \ d^0 = 1 \square_C d_Q,$$
$$E^1(P_* \square_C Q_*) = P_* \square_C N, \ d^1 = d_P \square_C 1$$
since P_* is coflat over C and $0 \to N \xrightarrow{\eta} Q_*$ is split exact over R.

Thus $P_* \square_C N \xrightarrow{1 \square_C \eta} P_* \square_C Q_*$ induces an isomorphism on E^1 and is an isomorphism of homology.

That $P_* \otimes_A Q_* \xleftarrow{\eta \square_C 1} M \square_C Q_*$ is an isomorphism of homology is similar.

Having proved the balanced lemma that any one of the two resolutions can be omitted, it is easy to see that maps $\text{Cotor}_{-n}^C(f,g)$ are independent of the choice of extensions and the fact that the $\text{Cotor}_{-n}^C(M,N)$ are functors well defined up to natural isomorphism follows as before. $\square$

Observe that:

Lemma 10.11.7. *There is a natural isomorphism*
$$\mathrm{Cotor}_0^C(M, N) \cong M \square_C N.$$

Proof:
$$\mathrm{Cotor}_0^C(M, N) \cong H_0(P_* \square_C N) \cong M \square_C N$$

since
$$0 \to M \square_C N \xrightarrow{\eta \square_C 1} P_0 \square_C N \xrightarrow{d_0 \square_C 1} P_{-1} \square_C N$$

is exact.

Lemma 10.11.8. *If $M = K \otimes C$ is an extended right comodule, then $\mathrm{Cotor}_{-n}^C(M, N) = 0$ for all $n > 0$ and all left comodules N. And a similar result holds with the roles of left and right comodules reversed.*

Proof: Since M is a proper injective, this is clear.

A partial converse to this lemma is given in Exercise 3 below.

Remarks. If M and N are differential comodules over a differential coalgebra C, then the tensor product differential on $M \otimes_R N$ defines a differential on the subobject $M \square_C N$ and also on the derived functors $\mathrm{Cotor}_{-n}^C(M, N)$. The isomorphisms in this section are all isomorphisms of differential objects.

Remarks on duality. Algebras and coalgebras are dual in the well known way. This duality extends to a duality between modules and comodules and also to tensor products and cotensor products and to their derived functors Tor and Cotor.

In detail, let C be a coalgebra, M a right comodule over C and N a left comodule over C. Assume that in each fixed degree all are projective and finitely generated R modules. Then the dual $C^* = \hom(C, R)$ is an algebra via the multiplication
$$\mu : C^* \otimes C^* = (C \otimes C)^* \xrightarrow{\Delta^*} C^*$$

and the unit
$$\eta : R = R^* \xrightarrow{\epsilon^*} C^*.$$

Similarly, the dual M^* us a right module over C^* via
$$M^* \otimes C^* = (M \otimes C)^* \to M^*.$$

In the same way, the dual N^* is a left module.

Furthermore, the dual of cotensor is the tensor:
$$M^* \otimes_{C^*} N^* = (M \square_C N)^*.$$

The dual of Cotor is Tor:

$$\operatorname{Tor}_n^{C^*}(M^*, N^*) = (\operatorname{Cotor}_{-n}^C(M, N))^*.$$

Of course, the duality goes the other way too, where we start with algebras and modules which are finitely generated and projective in each degree and and we end with coalgebras and comodules.

For conceptual reasons, we try to avoid overusing this duality but at times it is very convenient to use it for computational convenience.

Exercises

(1) Suppose that M, M_1 and N, N_1 are right and left comodules, respectively, over a coalgebra C. Maps $H : M \to M_1$ and $K : N \to N_1$ of possibly nonzero degree are maps of comodules if $\Delta \cdot H = (H \otimes 1) \cdot \Delta$ and $\Delta \cdot K = (1 \otimes K) \cdot \Delta$ with the usual sign convention.

Show:

(a) If $f : M \to M_1$ and $g : N \to N_1$ are maps of comodules (of possibly nonzero degree), then $f \otimes g$ restricts to a map of R modules $f \square_C g : M \square_C N \to M_1 \square_C N_1$.

(b) If M and N are differential comodules over a differential coalgebra C, then $d_\otimes = d_M \otimes 1 + 1 \otimes d_N$ restricts to a differential $d_\square$ on $M \square_C N$.

(c) If $f : M \to M_1$ and $g : N \to N_1$ are maps of differential comodules, then $f \square_C g : M \square_C N \to M_1 \square_C N_1$ is a map of differential modules.

(d) If $f, f_1 : M \to M_1$ and $g, g_1 : N \to N_1$ are chain homotopic maps of differential comodules with chain homotopies $H : M \to M_1$ and $K : N \to N_1$ which are maps of comodules, that is, if

$$H : f \simeq f_1, \; Hd + dH = f - f_1, \; K : g \simeq g_1, \; Kd + dK = g - g_1,$$

then

$$H \square_C g + f_1 \square_C K : f \square_C g \simeq f_1 \square_C g_1.$$

(2) Suppose that $0 \to M \to N \to P \to 0$ is a proper exact sequence of right C comodules and that $0 \to P \to Q_*$ is a proper injective resolution.

(a) Show that there are proper injective comodules $\overline{Q_{-n}}$ such that there is a proper injective resolution $0 \to N \to \overline{Q_*} \oplus Q_*$ with the standard projection $\overline{Q_*} \oplus Q_* \to Q_*$ being a map of resolutions which extends $N \to P$.

(b) If $0 \to M \to N \to P \to 0$ is a proper exact sequence, show that
$$0 \to \overline{Q_*} \to \overline{Q_* \oplus Q_*} \to \overline{Q_*} \to 0$$
is a proper short exact sequence of proper injective resolutions.

(c) If L is a left C comodule, show that there is a long exact sequence
$$0 \to M \square_C L \to N \square_C L \to P \square_C L \to$$
$$\mathrm{Cotor}^C_{-1}(M, L) \to \mathrm{Cotor}^C_{-1}(N, L) \to \mathrm{Cotor}^C_{-1}(P, L) \to$$
$$\mathrm{Cotor}^C_{-2}(M, L) \to \mathrm{Cotor}^C_{-2}(N, L) \to \mathrm{Cotor}^C_{-2}(P, L) \to \ldots .$$

(3) Suppose C is a coalgebra over a Dedekind ring (that is, a ring in which submodules of projectives are projective.) Suppose M is a right comodule over C and that M and C are projective over R.

(a) Show that the module of primitives PM is a retract of M.

(b) If M is bounded below and $\mathrm{Cotor}^C_{-1}(M, R) = 0$, then show that there is an isomorphism
$$M \to PM \otimes C.$$

10.12 Injective resolutions, total complexes, and differential Cotor

Let $Q_* \xleftarrow{\eta} M \to 0$ be an augmented chain complex of differential comodules over a differential coalgebra C, that is, we have a chain complex of comodules and comodule maps
$$\ldots \xleftarrow{d_{-i-1}} Q_{-i-1} \xleftarrow{d_{-i}} Q_{-i} \xleftarrow{d_{-i+1}} Q_{-i+1} \xleftarrow{d_{-i+2}} \ldots \xleftarrow{d_0} Q_0 \xleftarrow{\eta} M \to 0.$$

This is a double complex with an internal differential d_I in each Q_{-i} and an external differential d_E given by the resolution differential $d_{-i} : Q_{-i} \to Q_{-i-1}$. We have
$$d_I \cdot d_I = 0, \; d_E \cdot d_E = 0, \; d_E \cdot d_I = d_I \cdot d_E,$$
and d_I is part of a differential comodule structure for which $d_E : Q_* \to Q_{*-1}$ is a map of differential comodules.

We assume that C is k-connected with $k \geq 0$ so that the resolution can be assumed to be tapered in the sense that each Q_{-p} is at least $(k+1)p - 1$ connected. We apply the standard process of assembling this double complex into a single chain complex called the augmented total complex $T(Q_*) \xleftarrow{\eta} M \to 0$.

As an R module,
$$T(Q_*) = \bigoplus_{p \geq 0} s^{-p} Q_p$$

where $(s^{-1}Y)_n = s^{-1}Y_{n+1} = \{s^{-1}y \mid y \in Y_{n+1}\}$ is the desuspension. Since we are assuming that the resolution is suitably tapered, the total complex is concentrated in positive degrees.

Differentials in the total complex

The differential of $T(Q_*)$ is the sum $d_T = d_I + d_E$ where:

$$d_I : s^{-p}Q_{-p} \to s^{-p}Q_{-p} \text{ and } d_E : s^{-p}Q_{-p} \to s^{-p-1}Q_{-p-1}$$

are defined by the commutation formulas

$$d_I \cdot s^{-p} = (-1)^p s^{-p} \cdot d_I, \quad d_E \cdot s^{-p} = s^{-p-1} \cdot d_E.$$

One easily checks that:

Lemma 10.12.1. *In the complex $T(Q_*)$,*

$$d_I \cdot d_I = 0, \quad d_E \cdot d_E = 0, \quad d_E \cdot d_I = -d_I \cdot d_E$$

and

$$d_T \cdot d_T = 0.$$

We note that $T(Q_*)$ is supplemented by the chain map $\eta : M \to Q_0 \subseteq T(Q_*)$.

Differential comodule structures in the total complex

We make $T(Q_*)$ into a differential comodule via:

If right comodules,

$$\Delta : s^{-p}Q_{-p} \to s^{-p}Q_{-p} \otimes C, \text{ is given by } \Delta \cdot s^{-p} = (s^{-p} \otimes 1) \cdot \Delta.$$

If left comodules,

$$\Delta : s^{-p}Q_{-p} \to C \otimes s^{-p}Q_{-p}, \text{ is given by } \Delta \cdot s^{-p} = (1 \otimes s^{-p}) \cdot \Delta.$$

One easily checks:

Lemma 10.12.2. *In the case of right comodules, the above maps satisfy:*

(a)

$$(d_E \otimes 1) \cdot \Delta \cdot s^{-p} = \Delta \cdot d_E s^{-p},$$
$$(d_I \otimes 1 + 1 \otimes d) \cdot \Delta \cdot s^{-p} = \Delta \cdot d_I \cdot s^{-p}$$
$$(d_T \otimes 1 + 1 \otimes d) \cdot \Delta \cdot s^{-p} = \Delta \cdot d_T \cdot s^{-p}$$

(b)

$$(\Delta \otimes 1) \cdot \Delta \cdot s^{-p} = (1 \otimes \Delta) \cdot \Delta \cdot s^{-p}.$$

10.12 Injective resolutions, total complexes, and differential Cotor 365

(c)
$$(1 \otimes \epsilon) \cdot \Delta \cong s^{-p}.$$

and similar formulas hold in the case of left comodules.

Thus, $T(Q_*)$, d_T is a differential comodule over the differential coalgebra C and the supplement $\eta : M \to T(Q_*)$ is clearly a map of differential comodules.

Chain maps and chain homotopies in the total complex

Suppose we have a map of supplemented chain complexes of differential comodules

$$\begin{array}{ccc} M & \xrightarrow{f} & N \\ \downarrow \eta & & \downarrow \eta. \\ Q_* & \xrightarrow{f} & P_* \end{array}$$

Then we get a map of supplemented total complexes

$$\begin{array}{ccc} M & \xrightarrow{f} & N \\ \downarrow \eta & & \downarrow \eta \\ T(Q_*) & \xrightarrow{f} & T(P_*) \end{array}$$

defined by the commutation formulas $f \cdot s^{-p} = s^{-p} \cdot f$. One checks:

Lemma 10.12.3.

$$f \cdot d_I \cdot s^{-p} = d_I \cdot f \cdot s^{-p},$$
$$f \cdot d_E \cdot s^{-p} = d_E \cdot f \cdot s^{-p},$$
$$f \cdot d_T \cdot s^{-p} = d_T \cdot f \cdot s^{-p},$$

and, in the case of right comodules,

$$(f \otimes 1) \cdot \Delta \cdot s^{-p} = \Delta \cdot f \cdot s^{-p}$$

with a similar formula in the case of left comodules.

Thus, maps of supplemented chain complexes of differential comodules induce differential comodule maps of supplemented total complexes.

Suppose we have maps $f, g : Q_* \to P_*$ of chain complexes of differential comodules and a chain homotopy $H : Q_* \to P_{*+1}$ such that each $H_{-p} : Q_{-p} \to Q_{-p+1}$ is a map of differential comodules and $H \cdot d_E + d_E \cdot H = f - g$.

We define a degree $+1$ chain homotopy of total complexes $H : T(Q_*) \to T(P_*)$, $H : s^{-p}Q_{-p} \to s^{-p+1}P_{-p+1}$ by $H \cdot s^{-p} = s^{-p+1} \cdot H$. One checks:

Lemma 10.12.4.
$$H \cdot d_I \cdot s^{-p} = -d_I \cdot H \cdot s^{-p},$$
$$(H \cdot d_E + d_E \cdot H) \cdot s^{-p} = f - g,$$
$$(H \cdot d_T + d_T \cdot H) \cdot s^{-p} = f - g,$$

and, in the case of right comodules,
$$(H \otimes 1) \cdot \Delta \cdot s^{-p} = \Delta \cdot H \cdot s^{-p}$$
with a similar formula in the case of left comodules.

Thus, the chain maps on the total complex are chain homotopic via a degree $+1$ chain homotopy of comodules.

Contracting homotopies in the total complex

If $0 \to M \xrightarrow{\eta} Q_*$ is proper exact, then it has an R linear contracting homotopy
$$h_0 : Q_0 \to M, \; h_{-p} : Q_{-p} \to Q_{-p+1}, \; p \geq 0, \; h : Q_* \to Q_{*+1}, \; * < 0$$
such that
$$d_M \cdot h_0 = h_0 \cdot d_I, \; d_I \cdot h_{-p} = h_{-p} \cdot d_I, \; p > 0$$
and
$$h_0 \cdot \eta = 1_M, \; d_E \cdot h_{-p} + h_{-p-1} \cdot d_E = 1$$

If we set
$$h_0 = h_0, \; h_{-p} \cdot s^{-p} = s^{-p+1} \cdot h_{-p}$$
we get a contracting homotopy for the augmented total complex, $h \cdot d_T + d_T \cdot h = 1 - \eta \cdot \epsilon$ where ϵ is the composition $\epsilon : T(Q_*) \to Q_0 \xrightarrow{h_0} M$. One checks:

Lemma 10.12.5.
$$h \cdot d_I \cdot s^{-p} = -d_I \cdot h \cdot s^{-p},$$
$$(h \cdot d_E + d_E \cdot h) \cdot s^{-p} = 1 - \eta \cdot \epsilon,$$
$$(h \cdot d_T + d_T \cdot h) \cdot s^{-p} = 1 - \eta \cdot \epsilon.$$

This implies that $\eta : M \to T(Q_*)$ is a homology equivalence, in fact, it has a (strong) deformation retraction $\epsilon : T(Q_*) \to M$ such that $\eta \cdot \epsilon \simeq 1$.

The total complex of a relative injective complex

Suppose finally that $0 \to M \xrightarrow{\eta} Q_*$ is a proper injective resolution of differential comodules. We will assume that have right comodules and leave modifications in the case of left comodules to the reader.

10.12 Injective resolutions, total complexes, and differential Cotor

We can assume that each $Q_{-p} = \overline{Q_{-p}} \otimes C$ is an extended differential comodule with differential $d_\otimes = d \otimes 1 + 1 \otimes d$ as a differential module. In other words, $d_I = d \otimes 1 + 1 \otimes d$.

The differential comodule structure on the total complex is given by:

$$s^{-p}Q_{-p} = s^{-p}\overline{Q_{-p}} \otimes C,$$

$$d_I \cdot (s^{-p} \otimes 1) = (d \otimes 1 + 1 \otimes d) \cdot (s^{-p} \otimes 1),$$

$$\Delta \cdot (s^{-p} \otimes 1) = (1 \otimes \Delta) \cdot (s^{-p} \otimes 1)$$

In particular, each $s^{-p}Q_{-p} = s^{-p}(\overline{Q_{-p}}) \otimes C$ is an extended comodule and a differential comodule with respect to the internal differential d_I.

The case of the external differential is a little more complicated:

Recall that $d_E : Q_* \to Q_{*-1}$ is a map of differential comodules and let τ be the composition:

$$\overline{Q}_* \otimes C \xrightarrow{d_E} \overline{Q}_{*-1} \otimes C \xrightarrow{1 \otimes \epsilon} \overline{Q}_{*-1} \otimes R = \overline{Q}_{*-1}.$$

We get

Lemma 10.12.6. *The external differential d_E is the composition*

$$\overline{Q}_* \otimes C \xrightarrow{1 \otimes \Delta} \overline{Q}_* \otimes C \otimes C \xrightarrow{\tau \otimes 1} \overline{Q}_{*-1} \otimes C.$$

That is, if

$$\Delta(c) = c \otimes 1 + 1 \otimes c + \Sigma c' \otimes c''$$

then

$$d_E(x \otimes c) = \tau(x \otimes 1) \otimes c + \tau(x \otimes c) \otimes 1 + \Sigma \tau(x \otimes c') \otimes c''.$$

In particular,

$$d_E(x \otimes 1) = \tau(x \otimes 1) \otimes 1$$

and

$$d_I(x \otimes 1) = d_I(x) \otimes 1.$$

If we define $\overline{d}_E : \overline{Q}_* \to \overline{Q}_{*-1}$ by

$$d_E(x) = \tau(x \otimes 1)$$

and $\overline{d}_I : \overline{Q}_* \to \overline{Q}_*$ by

$$d_I(x \otimes 1) = \overline{d}_I(x) \otimes 1,$$

then

Lemma 10.12.7. $(\overline{Q_*}, \overline{d_I}, \overline{d_E})$ *is a subcomplex of the double complex* (Q_*, d_I, d_E) *via the embedding*

$$\overline{Q_*} = \overline{Q_*} \otimes R \subseteq \overline{Q_*} \otimes C.$$

Lemma 10.12.8. *If C is simply connected (connected and $C_1 = 0$), then in the total complex $T(Q_*) = \overline{Q_*} \otimes C$, the total differential $d_T = d_I + d_E$ and:*

(a) *the internal differential d_I is the tensor product differential*

$$d_\otimes = \overline{d_I} \otimes 1 + 1 \otimes d_C.$$

(b) *the external differential d_E is given by*

$$d_E(x \otimes c) = \overline{d_E} \otimes c + \tau(x \otimes c) \otimes 1 + \Sigma\tau(x \otimes c') \otimes c''$$

where the diagonal is

$$\Delta(c) = c \otimes 1 + 1 \otimes c + \Sigma c' \otimes c''.$$

Finally we come to the definition of differential Cotor. Let $0 \to M \to Q_*$ and $0 \to N \to P_*$ be proper injective resolutions of right and left C comodules M and N, respectively. Then:

Definition of differential Cotor 10.12.8. *If C is connected, then*

$$\mathrm{Cotor}_n^C(M, N) = H_n(T(Q_*) \square_C T(P_*))$$

As usual we need to know that this is a well defined functor, unique up to canonical isomorphism and, for this, the following suffices:

If $f: M \to M_1$ and $g: N \to N_1$ have lifts to $\overline{f}: Q_* \to P_*$ and $\overline{g}: Q_{1,*} \to P_{1,*}$, we set

$$\mathrm{Cotor}_*^C(f, g) = H_*(\overline{f} \square_C \overline{g}).$$

This is well defined since

$$H \square_C \overline{g} + \overline{f_1} \square_C K : \overline{f} \square_C \overline{g} \simeq \overline{f_1} \square_C \overline{g_1}$$

when we are given comodule homotopies of maps of resolutions $H: \overline{f} \simeq \overline{f_1}$, $K: \overline{g} \simeq \overline{g_1}$.

Remarks. If M, N, and C have zero differentials, then the relationship between the nondifferential $\mathrm{Cotor}_{-p}^C(M, N)$ and the differential $\mathrm{Cotor}_*^C(M, N)$ is simply given by the internal differential $d_I = 0$ and thus

$$\mathrm{Cotor}_*^C(M, N) = \bigoplus_{p \geq 0} s^{-p} \mathrm{Cotor}_{-p}^C(M, N).$$

Remarks. If M, N, and C are projective R modules, then we know that the resolutions Q_* and P_* can be chosen to be extended and R projective in the sense of Exercise 8 in Section 10.10. In fact, it could be reasonably argued that we should restrict the definition of differential Cotor to the category of R projective differential comodules over an R projective differential coalgebra. There would then be no loss in assuming that the definition of a proper injective resolution included the requirement that it be extended and R projective. In fact, it would be useful to assume that R is a principal ideal domain or, at least, a Dedekind domain in which submodules of projective modules are projective modules.

10.13 Cartan's constructions

We relate total complexes to Cartan's constructions. Cartan constructions are a generalization of twisted tensor products and of total complexes which can substitute for the total complex of a resolution in the definition of differential Cotor.

Let C be a simply connected differential coalgebra which is flat as an R module. Consider twisted tensor products $A \otimes_\tau C$ with A a differential algebra and total complexes $T(Q_*)$ with Q_* a proper injective resolution. These are both examples of (right) constructions in the sense of Cartan:

A differential module E is a construction in the sense of Cartan if:

(a) E is a differential comodule over the differential coalgebra C. E is called the total space and C is called the base of the construction.

(b) There is a chain complex F such that $E = F \otimes C$ as R modules.

(c) F is a differential subcomplex of E. F is called the fibre of the construction.

(d) E is a differential comodule over C via $\Delta = 1 \otimes \Delta : F \otimes C \to F \otimes C \otimes C$.

(e) There is a degree -1 linear map $\tau : F \otimes C \to F$ with $\tau(f \otimes 1) = 0$ related to the differential on E as follows: $d(f \otimes 1) = df \otimes 1$ and, if $\Delta(c) = c \otimes 1 + 1 \otimes c + \Sigma c' \otimes c''$, then

$$d(f \otimes c) = df \otimes c + (-1)^{\deg(f)} f \otimes dc + \tau(f \otimes c) \otimes 1 + \Sigma \tau(f \otimes c') \otimes c''.$$

(f) The following identity holds

$$d\tau(f \otimes c) + \tau(df \otimes c) + (-1)^{\deg(f)} \tau(f \otimes dc) + \Sigma \tau[\tau(f \otimes c') \otimes c''] = 0.$$

Hence, with respect to the Serre filtration

$$F_n(E) = F \otimes C_{\leq n},$$

we have

$$E^0(E) = F \otimes C, \, d^0 = d_F \otimes 1,$$
$$E^1(E) = H(F) \otimes C, d^1 = 1 \otimes d_C.$$

Thus, Cartan's constructions have properties reminiscent of Serre's computation of the E^1 term of his spectral sequence for an orientable fibration.

We leave the following Proposition to the exercises:

Proposition 10.13.1. *Suppose C is a simply connected differential coalgebra which is free over the coefficient ring R. Suppose that F is a chain complex over R. Let $E = F \otimes C$ as a C comodule and suppose that E is a differential comodule over C and $F \equiv F \otimes R \subseteq E$ is a subcomplex. Then E is a construction, that is, there exists a degree -1 linear map*

$$\tau : F \otimes \overline{C} \to F$$

satisfying the identities in (e) and (f) above. In fact, the identities in (e) are equivalent to the statements that F is a subcomplex and that E is a differential comodule and the identities in (f) are equivalent to the statement that $d^2 = 0$ in E.

The following lemma is important for considering constructions for bicomodules.

Lemma 10.13.2. *Let C and D be simply connected differential coalgebras and suppose that E is a (right) construction over $C \otimes D$ with fibre F. Then E is also a construction over D with fibre $F \otimes C$.*

Proof: That E is a construction over $C \otimes D$ gives us a degree -1 linear map $\tau : F \otimes C \otimes D \to F$ with the property that $\tau(f \otimes 1 \otimes 1) = 0$. The differential on E is then defined by the formulas above.

To show that E is a construction over D, we need to define a differential on $G = F \otimes C$ and to give a degree -1 linear map $\sigma : G \otimes D \to G$ with the property that $\sigma(g \otimes 1) = 0$ and which will also define the differential on E by the formulas above.

First of all, we define $d(f \otimes 1) = df \otimes 1$. We set $\Delta(c) = c \otimes 1 + 1 \otimes c + \Sigma c' \otimes c''$ and define

$$d(f \otimes c) = df \otimes c + (-1)^{\deg(f)} f \otimes dc$$
$$+ \tau(f \otimes c \otimes 1) \otimes 1 + \tau(f \otimes c' \otimes 1) \otimes c''.$$

This defines the differential on G which is the restriction of the differential on E via the embedding $G \equiv G \otimes R \subseteq E$.

10.13 Cartan's constructions   371

Now define a linear map $\sigma : G \otimes D \to G$ by
$$\sigma(f \otimes 1 \otimes d) = \tau(f \otimes 1 \otimes d) \otimes 1$$
and for $c \in \overline{C}$
$$\sigma(f \otimes c \otimes d) = \tau(f \otimes c \otimes d) \otimes 1 + \tau(f \otimes 1 \otimes d) \otimes c + \tau(f \otimes c' \otimes d) \otimes c''.$$

One now checks that this σ and the differential on G together define the differential on E. □

We recall that the opposite coalgebra of a differential coalgebra C is the differential coalgebra C^{opp} which is C as a graded differential R module and which has the same counit and the twisted diagonal
$$C \xrightarrow{\Delta} C \otimes C \xrightarrow{T} C \otimes C.$$

A left differential comodule M over C is then the same thing as a right differential comodule M over C^{opp}. The correspondence is given by twisting the coaction
$$M \xrightarrow{\Delta} M \otimes C^{\mathrm{opp}} \xrightarrow{T} C \otimes M.$$

In the same way, a differential bimodule M over C on the right and over D on the left is just a right differential comodule over $C \otimes D^{\mathrm{opp}}$. Thus, we immediately have the existence of proper injective bimodules and of proper exact sequences of bimodules, including the existence of proper injective resolutions of bimodules. We can form the associated total complexes of these resolutions of bimodules and get biconstructions. These are just right constructions over $C \otimes D^{\mathrm{opp}}$.

In particular, Lemma 10.13.2 says that a biconstruction is simultaneously a right and left construction. In addition, a biconstruction is an example of the following.

Definition 10.13.3. If E is a $D - C$ bimodule which is a right C construction with fibre F, then E is called a right C construction bimodule if F is a left differential comodule over D such that $E \cong F \otimes C$ as a left differential comodule over D.

Remarks. The most important observation related to 10.13.2 is that, if M is a $D - C$ bimodule and we form the functorial resolution as a right C comodule
$$0 \to M \xrightarrow{\Delta} M \otimes C \xrightarrow{d_0} \mathrm{coker}(\Delta) \otimes C \xrightarrow{d_{-1}} \mathrm{coker}(d_0) \otimes C \xrightarrow{d_{-2}} \ldots,$$
then it is a complex of $D - C$ bimodules and, since each of the cokernels is a left D comodule, the total complex is a right C construction bimodule.

Lemma 10.13.4. *Suppose that E_1 is a right construction with fibre F over a differential coalgebra C and that E_2 is a $C - D$ differential bimodule which is a right D construction bimodule with fibre G. Then $E_1 \square_C E_2$ is a right construction with fibre $F \otimes G$ over D.*

372 Differential homological algebra

Proof: Write $E_1 = F \otimes C$ and $E_2 = G \otimes D$ where G is a left differential C comodule. Then $F \otimes G \otimes D \xrightarrow{1 \otimes \Delta \otimes 1} F \otimes C \otimes G \otimes D$ is a monomorphism of differential objects onto $E_1 \square_C E_2$.

Let $\tau : F \otimes C \to F$ define the differential on the construction E_1. Define a differential on $F \otimes G$ by

$$\tau' : F \otimes G \xrightarrow{1 \otimes \Delta} F \otimes C \otimes G \xrightarrow{\tau \otimes 1} F \otimes G$$

that is, the differential is

$$d(f \otimes g) = df \otimes g + (-1)^{\deg(f)} f \otimes dg + \Sigma \tau(f \otimes c') \otimes g''$$

where

$$\Delta(g) = \Sigma c' \otimes g''.$$

Since $\Delta(dg) = d\Delta(g)$,

$(1 \otimes \Delta \otimes 1) \cdot d(f \otimes g \otimes 1)$

$= (1 \otimes \Delta \otimes 1) \cdot (df \otimes g \otimes 1 + (-1)^{\deg(f)}(f \otimes dg \otimes 1 + \Sigma \tau(f \otimes c') \otimes g''))$

$= df \otimes \Delta(g) \otimes 1 + (-1)^{\deg(f)}(f \otimes d\Delta(g) \otimes 1,$

hence the differential on $F \otimes G \subseteq E_1 \square_C E_2$ is the same as the differential on $F \otimes G$.

If $\sigma : G \otimes D \to G$ is gives E_2 the structure of a construction via the differential d_{E_2} on E_2 defined by the usual formula, then define the construction structure on $F \otimes G \otimes D$ by the composition

$$\sigma' : F \otimes G \otimes D \xrightarrow{1 \otimes \sigma} F \otimes G$$

and let the differential d on $F \otimes G \otimes D$ be given by the usual formula.

Of course, $\sigma'(f \otimes g \otimes 1) = 0$ and one checks that

$F \otimes G \otimes D$	$\xrightarrow{d}$	$F \otimes G \otimes D$
$\downarrow 1 \otimes \Delta \otimes 1$		$\downarrow 1 \otimes \Delta \otimes 1$
$F \otimes C \otimes G \otimes D$	$\xrightarrow{d_{E_1} \otimes 1 \otimes 1 + 1 \otimes 1 \otimes d_{E_2}}$	$F \otimes C \otimes G \otimes D$

commutes. Thus, σ gives the correct differential on $E_1 \square_C E_2$ and $E_1 \square_C E_2$ is a construction over D. $\square$

Remarks. If you have a Cartan construction, then you can use it to define differential Cotor, but Cartan constructions are hard to get without using proper injective resolutions. Of course, proper injective resolutions are convenient for functoriality and, in particular, they are a nice device for proving general homological invariance when you change the coalgebras and comodules by homology isomorphisms.

10.13 Cartan's constructions 373

Exercises

(1) Prove Proposition 10.13.1.

(2) Show that Proposition 10.13.1 implies that following result on twisted tensor products:

Proposition. *Suppose C is a simply connected differential coalgebra which is free over the coefficient ring R. Suppose that A is a differential algebra over R. Let $E = A \otimes C$ as a C comodule and as an A module. Suppose that E is a differential object which is simultaneously a differential comodule over C and a differential module over A. Suppose also that $1 \otimes 1$ is not a boundary. Then E is a twisted tensor product $A \otimes_\tau C$, that is, there exists a degree -1 linear map*

$$\tau : \overline{C} \to \overline{A}$$

which is a twisting morphism in the sense that it satisfies the two properties

$$d(a \otimes c) = da \otimes c + (-1)^{\deg(a)}\{a \otimes dc + a(\tau c) \otimes 1 + \Sigma a(\tau c') \otimes c''\}$$

where

$$\Delta(c) = c \otimes 1 + 1 \otimes c + \Sigma c' \otimes c''$$

and

$$d\tau + \tau d = (\mu \otimes 1)(1 \otimes \tau \otimes 1)(1 \otimes \Delta).$$

Hint: Define $\tau : C \to A$ by $\tau(c) = \tau(1 \otimes c)$ and show that $\tau(a \otimes c) = (-1)^{\deg(a)} \tau(c)$. Use the fact that $d^2 = 0$.

(3) Let $A \to B \to C$ be a sequence of maps of differential Hopf algebras with C simply connected and free over the coefficient ring. Suppose that

$$B = A \otimes C$$

as a left A module and right C comodule. Then there is a twisting morphism $\tau : C \to A$ such that B is a twisted tensor product,

$$B = A \otimes_\tau C.$$

(4) A differential comodule M over C is said to be augmented if there exists a differential map $M \to C$ which is a map of differential comodules.

   (a) If $E \to B$ is any map of simplicial sets, show that the normalized chain complex $C(E)$ is an augmented differential comodule over $C(B)$.

   (b) If M is an augmented differential comodule over a differential coalgebra C, show that there is a proper injective resolution $0 \to M \to Q_*$ such that the total complex $T(Q_*)$ is an augmented differential comodule over

C and that the map $M \to T(Q_*)$ is a map of augmented differential comodules.

10.14 Homological invariance of differential Cotor

The cotensor product is closely related to geometric pullbacks. We begin this section with some remarks on geometric pullbacks. Consider a pullback diagram of topological spaces

$$\begin{array}{ccc} X & \to & B \\ \downarrow & & \downarrow \\ A & \to & C \end{array}$$

We shall adopt the notation $X = A \times_C B$ for this pullback of A and B over C. Consider the functor $T_A(B) = A \times_C B$.

The functor T_A is said to be homotopically invariant if: Whenever $B_1 \to B_2$ is a map of spaces over C which is a homotopy equivalence then $A \times_C B_1 \to A \times_C B_2$ is a homotopy equivalence.

If $A \to C$ is not a fibration, then the functor T_A need not be homotopically invariant. Consider the example of $* \to PC$ over C and $T_C(*) = * \to T_C(PC) = \Omega(C)$.

But

Lemma 10.14.1. *If $A \to C$ is a fibration and $B_1 \to B_2$ is a map of spaces over C which is a homotopy equivalence, then $A \times_C B_1 \to A \times_C B_2$ is a homotopy equivalence.*

The quickest proof uses the long exact sequences of the fibrations

$$A \times_C B_1 \to B_1, \quad A \times_C B_2 \to B_2$$

and is left to the reader.

A homological variation of the above is also left as an exercise.

Lemma 10.14.2. *If $A \to C$ is an orientable fibration and $B_1 \to B_2$ is a map of spaces over C which is a homology equivalence, then $A \times_C B_1 \to A \times_C B_2$ is a homology equivalence.*

Now consider the following variation involving right differential comodules M and left differential comodules N over a differential comodule C. When is the functor $T_M(N) = M \square_C N$ homologically invariant? It is certainly not always the case. Consider the case $M = R$ and $R \to C \otimes_\tau \Omega(C)$ where the latter is the acyclic tensor product. Then $T_R(R) = R \to T_R(C \otimes_C \Omega(C)) = \Omega(C)$ is usually not a homology equivalence.

10.14 Homological invariance of differential Cotor

But Cartan's constructions are analogous to fibrations in the following sense:

Proposition 10.14.3. *Assume that R is a principal ideal domain. Let E be a right construction in the sense of Cartan over a differential coalgebra C with fibre F. Then T_E is a homologically invariant functor in the following sense: if $N_1 \to N_2$ is a map of left differential comodules over C which is a homology equivalence and if N_1 and N_2 are R projective, then $E \square_C N_1 \to E \square_C N_2$ is a homology equivalence.*

Proof: Assume that N is any R projective left differential comodule over C.

Filter E by the Serre filtration and N and C by the skeletal filtration, that is, filter by degree, $F_n(N) = N_{\leq n}$, $F_n(C) = C_{\leq n}$.

We have
$$E^0(N) = N, \ E^0(d_N) = 0,$$
$$E^1(N) = N, \ E^1(d_N) = d_N$$

and
$$E^2(N) = H(N).$$

And also $E = F \otimes C$ as R modules,
$$E^0(E) = F \otimes C, \ E^0(d_E) = d_F \otimes 1,$$
$$E^1(E) = HF \otimes C, \ E^1(d_E) = 1 \otimes d_C$$

and
$$E^2(E) = H(HF \otimes C).$$

Filter $E \square_C N$ by the product filtration:
$$F_n(E \otimes N) = \bigoplus_{i+j=n} F_i(E) \otimes F_j(N)$$

and let
$$F_n(E \square_C N) = E \square_C N \cap F_n(E \otimes N).$$

As R modules, the short exact sequence
$$0 \to E \square_C N \to E \otimes N \xrightarrow{\delta} E \otimes C \otimes N$$

is
$$0 \to F \otimes N \xrightarrow{1 \otimes \Delta} F \otimes C \otimes N \to F \otimes C \otimes C \otimes N$$

and $d_{E \otimes N} = d_E \otimes 1 + 1 \otimes d_N$. Hence,
$$E^0(d_{E \otimes N}) = E^0(d_E) \otimes 1 + 1 \otimes E^0(d_N) = d_F \otimes 1 \otimes 1.$$

Therefore,
$$E^0(d_{E\square_C N}) = d_F \otimes 1$$
and
$$E^1(E\square_C N) = HF \otimes N.$$
That is,
$$E^1(E\square_C N) \to E^1(E \otimes N) \xrightarrow{E^1(\delta)} E^1(E \otimes C \otimes N)$$
is the short exact sequence
$$0 \to HF \otimes N \xrightarrow{1\otimes\Delta} HF \otimes C \otimes N \xrightarrow{\delta} HF \otimes C \otimes C \otimes N.$$
Since
$$E^1(d_{E\otimes N}) = 1 \otimes d_C \otimes 1 + 1 \otimes 1 \otimes d_N$$
and, since $\Delta: N \to C \otimes N$ is a map of differential comodules, it follows that
$$E^1(d_{E\square_C N}) = 1 \otimes d_N.$$
Thus, $E^2(E\square_C N) = H(HF \otimes N)$ and, since R is a principal ideal domain, the universal coefficient theorem gives a functorial short exact sequence
$$0 \to H(F) \otimes H(N) \to E^2(E\square_C N) \to \mathrm{Tor}^{\mathbb{Z}}(H(F), s^{-1}H(N)) \to 0.$$
The homological invariance of $E\square_C N$ follows from this identification of E^2. □

Remarks. If we accept the identification of differentials, then a quick summary of the above proof is: With the Serre filtration on E and the skeletal filtration on N,
$$E^0(E\square_C N) = F \otimes N, \quad d^0 = d_F \otimes 1,$$
$$E^1(E\square_C N) = HF \otimes N, \quad d^1 = 1 \otimes d.$$
And
$$0 \to H(F) \otimes H(N) \to E^2(E\square_C N) \to \mathrm{Tor}^{\mathbb{Z}}(H(F), s^{-1}H(N)) \to 0$$
is short exact.

When applied to a construction, the cotensor product is homologically equivalent to differential Cotor. More precisely,

Corollary 10.14.4. *If R is a principal ideal domain, C is an R projective simply connected coalgebra, E is an R projective right construction over C, and N is an R projective left differential comodule over C, then there is an isomorphism*
$$H(E\square_C N) \to \mathrm{Cotor}^C(E, N).$$

10.14 Homological invariance of differential Cotor

Proof: Let $0 \to E \to Q_*$ and $0 \to N \to P_*$ be proper injective resolutions which we can assume are R projective and tapered so that the respective total complexes are constructions concentrated in positive degrees. In particular, both $E \to T(Q_*)$ and $N \to T(P_*)$ are homology isomorphisms and $T(P_*)$ is a construction. Hence Proposition 10.14.3 asserts that there are isomorphisms

$$H(E \square_C N) \to H(E \square_C T(P_*)) \to H(T(Q_*) \square_C T(P_*)) = \mathrm{Cotor}^C(E, N).$$

$\square$

Remark. For example, in Corollary 10.14.4, the construction E could be a twisted tensor product $A \otimes_\tau C$. We would then get

$$HA = H(A \otimes_\tau C \square_C R) = \mathrm{Cotor}^C(A \otimes_\tau C, R).$$

The fact that resolutions are constructions shows that differential Cotor is balanced, that is:

Corollary 10.14.5. *If M, N, and C are R projective with C simply connected and $0 \to M \to Q_*$ and $0 \to N \to P_*$ are R projective tapered proper injective resolutions, then there are homology equivalences*

$$T(Q_*) \square_C N \to T(Q_*) \square_C T(P_*) \leftarrow M \square_C T(P_*).$$

Let C be an R-projective simply connected coalgebra. The cobar construction $\Omega C \otimes_\tau C$ is the total complex of an R projective proper injective resolution of R and hence:

Corollary 10.14.6.

$$\mathrm{Cotor}^C(R, N) = H(\Omega C \otimes_\tau C \square_C N) = H(\Omega C \otimes N)$$

or, when $N = R$,

$$\mathrm{Cotor}^C(R, R) = H(\Omega C).$$

Remark. If C is a coalgebra with zero differential, there is no internal differential in ΩC and hence

$$\mathrm{Cotor}^C_{-1}(R, R) = \ker \overline{C} \to \overline{C} \otimes \overline{C}$$

$= PC =$ the module of primitives of the coalgebra.

Exercise

(1) Show that the fact that differential $\mathrm{Cotor}^C(M, N)$ is balanced implies that it is well defined functor when M, N, and C are all R projective and C is simply connected.

10.15 Alexander–Whitney and Eilenberg–Zilber maps

Let Δ be the category whose objects are the finite sets $[n] = \{0, 1, \ldots, n\}$ and whose morphisms are (weakly) monotone maps $\alpha : [n] \to [m]$. Let $\mathcal{S}$ be the category of sets and functions.

Recall that a simplicial set is a contravariant functor $X : \Delta \to \mathcal{S}$.

It is customary to write $X_n = X([n])$ for the functor on objects and $\alpha^* = X(\alpha) : X_m \to X_n$ for the functor on morphisms. The elements of X_n are called n-simplices.

Among the monotone maps there are two special sets of maps, the coface maps and the codegeneracy maps:

(1) for all $0 \leq i \leq n$, the coface maps are the injections $\epsilon^i : [n-1] \to [n]$ defined by
$$\epsilon^i(k) = \begin{cases} k & \text{if } 0 \leq k < i \\ k+1 & \text{if } i \leq k \leq n-1 \end{cases}.$$

(2) for all $0 \leq j \leq n$, the codegeneracy maps are the surjections $\eta^j : [n+1] \to [n]$ are defined by
$$\eta^j(k) = \begin{cases} k & \text{if } 0 \leq j \leq k \\ k-1 & \text{if } j < k \leq n+1 \end{cases}.$$

Given a simplicial set X, we write
$$(\epsilon^i)^* = d_i : X_n \to X_{n-1}, \ 0 \leq i \leq n$$
and
$$(\eta^j)^* = s_j : X_{n+1} \to X_n, \ 0 \leq j \leq n$$
and call these maps face and degeneracy operators, respectively.

Note that every monotone map $\alpha : [n] \to [m]$ has a unique factorization into a composition of monotone maps $\alpha = \beta \cdot \gamma : [n] \to [k] \to [m]$ where γ is a surjection and β is an injection.

We note that every monotone surjection can be written uniquely as a composition
$$\gamma = \eta^{\iota_1} \cdot \eta^{\iota_2} \cdots \cdots \eta^{\iota_r}$$
where $\iota_1 < \iota_2 < \cdots < \iota_r$. We also note that every monotone injection can be written uniquely as a composition
$$\beta = \epsilon^{\delta_s} \cdots \cdots \epsilon^{\delta_2} \cdot \epsilon^{\delta_1}$$
where $\delta_s > \delta_{s-1} > \cdots > \delta_1$.

10.15 Alexander–Whitney and Eilenberg–Zilber maps

Hence, every monotone map α has a unique expression

$$\alpha = \beta \cdot \gamma = \epsilon^{\delta_s} \cdot \ldots \cdot \epsilon^{\delta_2} \cdot \epsilon^{\delta_1} \cdot \eta^{\iota_1} \cdot \eta^{\iota_2} \cdot \ldots \cdot \eta^{\iota_r}.$$

It follows that the maps in a simplicial set $\alpha^* : X_m \to X_n$ have canonical expressions as

$$\alpha^* = s_{\iota_r} \cdot \ldots \cdot s_{\iota_2} \cdot s_{\iota_1} \cdot d_{\delta_1} \cdot d_{\delta_2} \cdot \ldots \cdot d_s$$

where $\iota_1 < \iota_2 < \cdots < \iota_r$ and $\delta_s > \delta_{s-1} > \cdots > \delta_1$.

One can check the standard simplicial identities:

Lemma 10.15.1. *The face and degeneracy operators satisfy the identities*

$$d_i d_j = d_{j-1} d_i, \quad i < j$$

$$s_i s_j = s_{j+1} s_i, \quad i \leq j$$

$$d_i s_j = \begin{cases} s_{j-1} d_i, & i < j \\ 1, & i = j, \; i = j+1 \\ s_j d_{i-1}, & i > j+1. \end{cases}$$

The chains on a simplicial set X is the free graded R module $C(X)$ with graded basis X and differential $d : C_n(X) \to C_{n-1}(X)$ defined on generators by

$$d(\sigma) = \Sigma_{i=0}^n (-1)^i d_i(\sigma), \quad \sigma \in X_n.$$

Definition 10.15.2. An n-simplex $\sigma \in X_n$ is called degenerate if it is of the form $s_j \tau$ for some $\tau \in X_{n+1}$.

The identities in Lemma 10.15.1 show that the degenerate simplices span a subcomplex $D(X)$ which is closed under the differential. In MacLane's **Homology**, it is shown that $D(X)$ is acyclic, $HD(X) = 0$.

Definition 10.15.3. If X is a simplicial set, the complex of normalized chains is the quotient complex $C^N(X) = C(X)/D(X)$.

Of course, $C(X) \to C^N(X)$ is a homology isomorphism and $C^N(X)$ is the free module generated by nondegenerate simplices. This makes the normalized chains more convenient in the sense that, if X is a k-reduced simplicial set, that is, if $X_i = *$ for $i \leq k$, then the normalized chain complex $C^N(X)$ is k connected, $\overline{C}_i^N(X) = 0$ for $i \leq k$.

Universal models for simplicial sets are Δ^n, $n \geq 0$:

$$(\Delta^n)_k = \mathrm{map}([k],[n]) = \text{monotone maps}$$

We denote a monotone map $\alpha : [k] \to [n]$ by the ordered $n+1$ tuple $\alpha = (\alpha(0), \ldots, \alpha(n))$. We note that the face operators are

$$d_i(\alpha(0), \ldots, \alpha(n)) = (\alpha(0), \ldots, \widehat{\alpha(i)}, \ldots, \alpha(n))$$

and the degeneracy operators are

$$s_j(0, \ldots, n) = (\alpha(0), \ldots, \alpha(j), \alpha(j), \ldots, \alpha(n)).$$

Hence α is nondegenerate if and only if the entries are all distinct.

If $x \in X_n$ is any n-simplex, there is a unique simplicial map $\bar{x} : \Delta^n \to X$ such that $\bar{x}(0, \ldots, n) = x$.

We now turn to the Alexander–Whitney maps $\Delta : C(X \times Y) \to C(X) \otimes C(Y)$ and the Eilenberg–Zilber maps $\nabla : C(X) \otimes C(Y) \to C(X \times Y)$. Once again, the main reference is MacLane's **Homology**.

If $\sigma \in X_n$ is an n-simplex, the front i-face is $_i\sigma = d_{i+1} \ldots d_{n-1} d_n \sigma = \bar{\sigma}(0, \ldots, i)$ and the back j-face is $\sigma_j = (d_0)^{n-j} \sigma = \bar{\sigma}(j, \ldots, n)$. Thus, on universal examples,

$$_i(0, \ldots, n) = (0, \ldots, i), \quad (0, \ldots, n)_j = (j, \ldots, n)$$

The Alexander–Whitney maps are based on this choice of front and back faces.

Definition 10.15.4. If $(\sigma, \tau) \in X_n \times Y_n$ is an n-simplex, the Alexander–Whitney map is

$$\Delta(\sigma, \tau) = \Sigma_{i=0}^n {}_i\sigma \otimes \tau_{n-i}.$$

The universal examples for the Alexander–Whitney maps are $\Delta : C(\Delta^n \times \Delta^n) \to C(\Delta^n) \otimes C(\Delta^n)$ with

$$\Delta((0, \ldots, n), (0, \ldots, n)) = \Sigma_{i=0}^n (0, \ldots, i) \otimes (i, \ldots, n).$$

We state without proof the key properties of the Alexander–Whitney maps:

Proposition 10.15.5. *The Alexander–Whitney maps $\Delta : C(X \times Y) \to C(X) \otimes C(Y)$ satisfy:*

(1) *they are natural chain equivalences*

(2) *they are associative, that is,*

$$\begin{array}{ccc} C(X \times Y \times Z) & \xrightarrow{\Delta} & C(X) \otimes C(Y \times Z) \\ \downarrow \Delta & & \downarrow 1 \otimes \Delta \\ C(X \times Y) \otimes C(Z) & \xrightarrow{\Delta \otimes 1} & C(X) \otimes C(Y) \otimes C(Z) \end{array}$$

(3) *they induce a map of normalized chains*

$$\Delta : C^N(X \times Y) \to C^N(X) \otimes C^N(Y)$$

that is, if (σ, τ) is a degenerate n-simplex and $i + j = n$, then so is at least one of the front face $_i\sigma$ or the back face τ_j.

The Alexander–Whitney maps define the differential coalgebra structures on the chains $C(X)$ and the normalized chains $C^N(X)$ via the Alexander–Whitney diagonal formed by composition with the induced map of the diagonal $X \to X \times X$, that is,

$$\Delta = \Delta \cdot C(\Delta) : C(X) \to C(X \times X) \to C(X) \otimes C(X)$$

$$\Delta = \Delta \cdot C^N(\Delta) : C^N(X) \to C^N(X \times X) \to C^N(X) \otimes C^N(X)$$

The Eilenberg–Zilber maps $\nabla : C(X) \otimes C(Y) \to C(X \times Y)$ are based on the classical triangulation of a product of two simplices.

Consider the product $[n] \times [m]$.

A monotone path p from $(0,0)$ to (n,m) is a piecewise linear path from lattice point to lattice point and such that the horizontal moves are to the right and vertical moves are upward. Thus, p is a function from $[n+m]$ to the lattice points of $[n] \times [m]$ which is monotone in each coordinate $\pi_1 \cdot p$ and $\pi_2 \cdot p$, $p(0) = (0,0)$, $p(n+m) = (n,m)$ and such that each step has length 1.

We have the standard monotone path p_0 which goes via:

$$(0,0), \ldots, (n,0), (n,1), \ldots, (n,m)$$

and we associate to any monotone path p the integer k which is the area enclosed by p and p_0. Define the sign of p by $\text{sgn}(p) = (-1)^k$.

Now the Eilenberg–Zilber map is the following:

$\nabla(x \otimes y)$
$= \Sigma_p \text{sgn}(p)(\overline{x}(\pi_1 p(0), \ldots, \pi_1 p(n+m)), \overline{y}(\pi_2 p(0), \ldots, \pi_2 p(n+m)))$,

the summation being taken over all monotone maps p.

The universal example for the Eilenberg–Zilber maps are

$$\nabla : C(\Delta^n) \otimes C(\Delta^m) \to C(\Delta^n \times \Delta^m)$$

and

$$\nabla((0, \ldots, n) \otimes (0, \ldots, m)) = \Sigma_p \text{sgn}(p) s^*(0, \ldots, n) \times t^*(0, \ldots, m)$$

where s^* is an interated degeneracy inserting repetitions whenever the first coordinate of the path is constant, that is, whenever $\pi_1 p(i) = \pi_1 p(i+1)$, and t^* is an interated degeneracy inserting repetitions whenever the second coordinate of the path is constant, that is, whenever $\pi_2 p(i) = \pi_2 p(i+1)$.

We state without proof the key properties of the Eilenberg–Zilber maps.

Proposition 10.15.6. *The Eilenberg–Zilber maps* $\nabla : C(X) \otimes C(Y) \to C(X \times Y)$ *satisfy:*

(1) *they are natural chain equivalences*

(2) *they are associative, that is,*

$$\begin{array}{ccc} C(X) \otimes C(Y) \otimes C(\mathbb{Z}) & \xrightarrow{\nabla \otimes 1} & C(X \times Y) \otimes \mathbb{Z} \\ \downarrow 1 \otimes \nabla & & \downarrow \nabla \\ C(X) \otimes C(Y \times \mathbb{Z}) & \xrightarrow{\nabla} & C(X \times Y \times \mathbb{Z}) \end{array}$$

(3) *they are maps of differential coalgebras*

(4) *they induce a map of normalized chains*

$$\nabla : C^N(X) \otimes C^N(Y) \to C^N(X \times Y)$$

that is, if σ in X or τ in Y are degenerate simplices, then $\nabla(\sigma \otimes \tau)$ is a sum of $\pm$ degenerate simplices.

(5) *the Eilenberg–Zilber maps and the Alexander–Whitney maps are inverses up to natural chain homolopies, that is, $\Delta \cdot \nabla \simeq 1$, $\nabla \cdot \Delta \simeq 1$. In the case of normalized chains, $\Delta \cdot \nabla = 1$, that is, the Alexander–Whitney map is a left inverse to the Eilenberg–Zilber map.*

Remarks. All of the above properties of Eilenberg–Zilber maps except for (3) may be found in MacLane's book [77]. Property (3) may be found in the paper by Eilenberg and Moore on Homology and Fibrations I [42].

If C and D are differential coalgebras, then $C \otimes D$ is a differential coalgebra with the tensor product differential and the diagonal

$$C \otimes D \xrightarrow{\Delta \otimes \Delta} C \otimes C \otimes D \otimes D \xrightarrow{1 \otimes T \otimes 1} C \otimes D \otimes C \otimes D$$

where T is the twist map $T(c \otimes d) = (-1)^{\deg(c)\deg(d)} d \otimes c$.

We now define the skeletal filtration.

Definition 10.15.7. An n-simplex σ has skeletal filtration $\mathrm{sk}(\sigma) \leq k$ if there is a monotone map $\alpha : [n] \to [k]$ such that σ is in the image

$$\alpha^* : X_k \to X_n.$$

Let us denote the skeletal filtration on a simplicial set X by

$$F_n X = \{\sigma \in X | \mathrm{sk}(\sigma) \leq n\}.$$

Then

$$F_n X_k = \begin{cases} X_k & \text{if } k \leq n \\ \bigcup s^* X_n & \text{if } k > n \text{ where } s^* \text{ runs over } k - n \text{ fold iterated degeneracies.} \end{cases}$$

One checks:

Lemma 10.15.8. *The skeletal fitration $F_n X$ is a subsimplicial set of X and hence generates a subcomplex $F_n C(X)$ of the chains $C(X)$ and a subcomplex $F_n C^N(X)$ of the normalized chains $C^N(X)$.*

Note that $F_n C^N(X)$ equals 0 in dimensions greater than n.

One can check:

Lemma 10.15.9. *The Alexander–Whitney map preserves the skeletal filtration in the sense that, if (σ, τ) is a simplex of dimension n in $X \times Y$, $n = i+j$, and $\mathrm{sk}(\sigma, \tau) \leq k$, then $\mathrm{sk}(_i\sigma) + \mathrm{sk}(\tau_j) \leq k$, that is, the Alexander–Whitney map satisfies*

$$\Delta F_n C(X \times Y) \subseteq F_n(C(X) \otimes C(Y)) = \Sigma_{i+j=n} F_i C(X) \otimes C_j(Y).$$

Suppose that $\pi : E \to B$ is a fibration of simplicial sets. The Serre filtration on E is the inverse image of the skeletal filtration on B, that is, $F_n E = \pi^{-1} F_n B$.

Exercise

(1) Verify Lemma 10.15.1.

10.16 Eilenberg–Moore models

We shall denote pullback diagrams of simplicial sets or of topological spaces by

$$\begin{array}{ccc} Y \times_X \mathbb{Z} & \overset{k}{\to} & \mathbb{Z} \\ \downarrow h & & \downarrow g \\ Y & \overset{f}{\to} & X \end{array}.$$

Assume that g is a fibration with fibre F and hence that h is also a fibration with fibre F.

We filter X and Y by the skeletal filtrations and we filter $Y \times_X \mathbb{Z}$ and $\mathbb{Z}$ by the Serre filtrations,

The diagonal maps give maps

$$\Delta : Y \to Y \times Y \to Y \times X,$$

$$\Delta : \mathbb{Z} \to \mathbb{Z} \times \mathbb{Z} \to \mathbb{Z} \times X.$$

Composition with the Alexander–Whitney map

$$\Delta : C(Y \times \mathbb{Z}) \to C(Y) \otimes C(\mathbb{Z})$$

defines a map $\overline{\Delta} : C(Y \times_X \mathbb{Z}) \to C(Y) \otimes C(\mathbb{Z})$ given on n-simplices by

$$\overline{\Delta}(\sigma) = \Sigma_{i+j=n} h(_i\sigma) \otimes k(\sigma_j).$$

Lemma 10.16.1. *The map $\overline{\Delta}$ is filtration preserving.*

Proof: If σ is in the n-th Serre filration of $Y \times_X \mathbb{Z}$, that is, $serre(\sigma) \leq n$, then $h\sigma$ is in the n-th skeletal filtration of Y, that is, $\text{sk}(h\sigma) \leq n$. Since the Alexander–Whitney map preserves the skeletal filtration, $\text{sk}(_ih\sigma) + \text{sk}(h\sigma_j) \leq n$ for $i + j = n$. Thus,

$$\text{sk}(_ih\sigma) + \text{sk}(fh\sigma_j) \leq n$$
$$\text{sk}(_ih\sigma) + \text{sk}(gk\sigma_j) \leq n$$
$$\text{sk}(_ih\sigma) + serre(k\sigma_j) \leq n.$$

$\square$

The chain complexes $C(Y)$ and $C(\mathbb{Z})$ are differential comodules over the differential coalgebra $C(X)$ via the maps

$$\Delta : C(Y) \to C(Y \times Y) \to C(Y) \otimes C(Y) \to C(Y) \otimes C(X)$$
$$\Delta : C(\mathbb{Z}) \to C(\mathbb{Z} \times \mathbb{Z}) \to C(\mathbb{Z}) \otimes C(\mathbb{Z}) \to C(X) \otimes C(\mathbb{Z}).$$

The associativity and naturality of the Alexander–Whitney maps show that there is a map

$$\Delta : C(Y \times X \times \mathbb{Z}) \to C(Y) \otimes C(X) \otimes C(\mathbb{Z})$$

such that both diagrams below commute (of course, $\Delta = (1_Y \otimes \Delta_{X,\mathbb{Z}})$ $\Delta_{Y,X \times \mathbb{Z}} = (\Delta_{Y,X} \otimes 1_\mathbb{Z})\Delta_{Y \times X, \mathbb{Z}}).$

$$\begin{array}{ccc} C(Y \times \mathbb{Z}) & \xrightarrow{\Delta \times 1} & C(Y \times X \times \mathbb{Z}) \\ \downarrow \Delta & & \downarrow \Delta \\ C(Y) \otimes C(\mathbb{Z}) & \xrightarrow{\Delta \otimes 1} & C(Y) \otimes C(X) \otimes C(\mathbb{Z}) \end{array}$$

and

$$\begin{array}{ccc} C(Y \times \mathbb{Z}) & \xrightarrow{1 \times \Delta} & C(Y \times X \times \mathbb{Z}) \\ \downarrow \Delta & & \downarrow \Delta \\ C(Y) \otimes C(\mathbb{Z}) & \xrightarrow{1 \otimes \Delta} & C(Y) \otimes C(X) \otimes C(\mathbb{Z}). \end{array}$$

Hence, we have a unique map $C(Y \times_X \mathbb{Z}) \to C(Y) \square_{C(X)} C(\mathbb{Z})$ such that

$$\begin{array}{ccccc} C(Y \times_X \mathbb{Z}) & \xrightarrow{(h,k)} & C(Y \times \mathbb{Z}) & \xrightarrow{\Delta \times 1 - 1 \times \Delta} & C(Y \times X \times \mathbb{Z}) \\ \downarrow & & \downarrow \Delta & & \downarrow \Delta \\ C(Y) \square_{C(X)} C(\mathbb{Z}) & \to & C(Y) \otimes C(\mathbb{Z}) & \xrightarrow{\Delta \otimes 1 - 1 \otimes \Delta} & C(Y) \otimes C(X) \otimes C(\mathbb{Z}) \end{array}$$

commutes.

We give $C(Y) \square_{C(X)} C(\mathbb{Z})$ the subspace filtration of the tensor product, that is,

$$F_n(C(Y) \square_{C(X)} C(\mathbb{Z})) = F_n(C(Y) \square_{C(X)} C(\mathbb{Z})) \bigcap (C(Y) \square_{C(X)} C(\mathbb{Z})).$$

Lemma 10.16.1 says that the map

$$\overline{\Delta}: C(Y \times_X \mathbb{Z}) \to C(Y) \square_{C(X)} C(\mathbb{Z})$$

is filtration preserving where the domain has the Serre filtration and the range has the filtration coming from the skeletal fitration on Y and the Serre filtration on $\mathbb{Z}$.

Let X be a one-reduced simplicial set so that the normalized chains $C^N(X)$ are a simply connected differential coalgebra. We shall change notation so that $C(X)$ denotes the normalized chains.

With this understanding, let

$$0 \to C(Y) \to Q_*$$

be an R projective proper injective resolution of the right differential module $C(Y)$ over the differential coalgebra $C = C(X)$. We recall that $Q_* = \overline{Q_*} \otimes C$ as comodules.

Eilenberg–Moore geometric approximation theorem 10.16.2. *The composite map*

$$C(Y \times_X \mathbb{Z}) \to C(Y) \square_{C(X)} C(\mathbb{Z}) \to T(Q_*) \square_{C(X)} C(\mathbb{Z})$$

is a homology isomorphism, that is, there is a natural isomorphism

$$H(Y \times_X \mathbb{Z}) \to \mathrm{Cotor}^{C(X)}(C(Y), C(\mathbb{Z})).$$

Proof: Serre's computation for a fibration asserts that, in the Serre spectral sequence of an orientable fibration sequence $F \to E \to B$, one has

$$E^1 C(E) = HF \otimes C(B), \quad d^1 = 1 \otimes d_{C(B)}$$

and hence there is a short exact sequence

$$0 \to HF \otimes H_* B \to E^2 C(E) \to \mathrm{Tor}^{\mathbb{Z}}(HF, H_{*-1}(B)) \to 0.$$

In particular, if F is the fibre of $\mathbb{Z} \to X$, we have a short exact sequence

$$0 \to H_* Y \otimes HF \to E^2 C(Y \times_X \mathbb{Z}) \to \mathrm{Tor}^{\mathbb{Z}}(H_{*-1}(Y), HF) \to 0.$$

We filter $T(Q_*)$ by the skeletal or degree filtration and $C(Y)$ by the skeletal filtration. The augmentation map $C(Y) \to T(Q_*)$ is filtration preserving.

Hence, if we filter $T(Q_*) \square_C C(\mathbb{Z})$ by the filtration induced by the skeletal fitration on $T(Q_*)$ and by the Serre filtrations on $C(\mathbb{Z})$ and $C(Y \times_X \mathbb{Z})$, we get that the composition map

$$\overline{\Delta}: C(Y \times_X \mathbb{Z}) \to C(Y) \square_{C(X)} C(\mathbb{Z}) \to T(Q_*) \square_{C(X)} C(\mathbb{Z})$$

is filtration preserving.

The theorem will follow when we show that, in the spectral sequence associated with the filtration on $T(Q_*)\square_{C(X)}C(\mathbb{Z})$, we have a short exact sequence

$$0 \to H_*Y \otimes HF \to E^2(C(T(Q_*)\square_C C(\mathbb{Z})) \to \text{Tor}^{\mathbb{Z}}(H_{*-1}(Y), HF) \to 0,$$

similar to the above sequence for $E^2 C(Y \times_X \mathbb{Z})$. Since we have an isomorphism at E^2, we have a homology isomorphism.

In brief summary, the proof goes as follows:

$$E^0(T(Q_*)\square_C C(\mathbb{Z})) = T(\overline{Q}_*) \otimes C(\mathbb{Z}), \ E^0(d) = 1 \otimes d^0.$$
$$E^1(T(Q_*)\square_C C(\mathbb{Z})) = T(\overline{Q}_*) \otimes C \otimes HF = T(Q_*) \otimes HF, \ E^1(d) = d_T \otimes 1.$$

Hence, we get the above short exact sequence and this would complete the proof. We now go into more detail on the identification of the differentials.

First of all, we record the following observation: We shall denote by $d_\square$ the restriction of the tensor product differential $d_\otimes = d \otimes 1 + 1 \otimes d$ on $M \otimes N$ to the differential on the cotensor product $M\square_C N$. Then, under the isomorphism $\Delta: M \to M\square_C C$, the differential $d_\square$ corresponds to the differential d_M on M.

With the skeletal filtration, we have

$$E^0(T(Q_*)) = T(Q_*), \ E^0(d_T) = 0,$$
$$E^1(T(Q_*)) = T(Q_*), \ E^1(d_T) = d_T,$$
$$E^2(T(Q_*)) = HT(Q_*) \cong H(Y).$$

Also with the skeletal filtration, we have

$$E^0(C) = C, \ E^0(d_C) = d_C^0 = 0,$$
$$E^1(C)) = C, \ E^1(d_C) = d_C,$$
$$E^2(C) = H(C).$$

With the Serre filtration of $C(\mathbb{Z})$, we have

$$E^1(C(\mathbb{Z})) = C \otimes HF, \ d^1 = d_C \otimes 1.$$

Hence, with the filtration of $T(Q_*) \otimes C(\mathbb{Z})$, we have

$$E^0(T(Q_*)) \otimes C(\mathbb{Z})) = T(Q_*) \otimes C(\mathbb{Z}),$$
$$E^0(d) = E^0(d_T) \otimes 1 + 1 \otimes d_{C(\mathbb{Z})}^0 \equiv 1 \otimes d_{C(\mathbb{Z})}^0,$$
$$E^1(T(Q_*)) \otimes C(\mathbb{Z})) = T(Q_*) \otimes C \otimes HF,$$
$$E^1(d) = d_T \otimes 1 \otimes 1 + 1 \otimes d_C \otimes 1.$$

With the filtration of $T(Q_*))\square_C C(\mathbb{Z})$, we have

$$E^0(T(Q_*)\square_C C(\mathbb{Z})) = T(\overline{Q}_*) \otimes C\square_C C(\mathbb{Z}) \cong T(\overline{Q}_*) \otimes C(\mathbb{Z}),$$

via the differential isomorphism $\Delta : C(\mathbb{Z}) \to C \square_C C(\mathbb{Z})$. Hence
$$E^0(d) = 1 \otimes 1 \otimes d^0_{C(\mathbb{Z})} \equiv 1 \otimes 1 \otimes d^0_{C(\mathbb{Z})} + 1 \otimes d^0_C \otimes 1 \cong 1 \otimes d^0_{C(\mathbb{Z})}.$$
Thus,
$$E^1(T(Q_*)) \square_C C(\mathbb{Z})) = T(\overline{Q}_*) \otimes H(C(\mathbb{Z}), d^0_{C(\mathbb{Z})})$$
$$= T(\overline{Q}_*) \otimes C \otimes HF \cong T(Q_*) \otimes HF,$$
via the isomorphism $T(Q_*) \square_C C \otimes HF \cong T(Q_*) \otimes HF$. It follows that
$$E^1(d) \cong d_T \otimes 1.$$
Hence, $E^2(T(Q_*)) \square_C C(\mathbb{Z})) = H(T(Q_*) \otimes HF, d_T \otimes 1)$ and there is a short exact sequence
$$0 \to H_*(T(Q_*)) \otimes HF \to E^2(T(Q_*) \otimes HF)$$
$$\to \operatorname{Tor}^{\mathbb{Z}}(H_{*-1}(T(Q_*)), HF) \to 0.$$
Since $HT(Q_*) = H(Y)$, the above short exact sequence is isomorphic to what we want and the proof is complete. $\square$

Remarks. There are two ways to replace simplicial sets by topological spaces in the statement of the geometric approximation Theorem 10.16.2.

First, one can assume that Y and X are both simply connected spaces. Then replace the singular complexes $S(Y)$ and $S(X)$ by the first Eilenberg subcomplexes which are the subsimplicial sets where the 1-skeleton is restricted to a point. Then restrict the pullback to these complexes in the obvious way. Since the homotopy types of all the simplicial sets in the pullback does not change, this translates into a true topological statement.

Alternatively, one can assume that X is a simply connected space and that both $Y \to X$ and $\mathbb{Z} \to X$ are fibrations. Now, replace the simplicial set $S(X)$ by the first Eilenberg subcomplex and restrict all the singular commmplexes to being over $S(X)$. Since fibrations restrict to fibrations, no homotopy types change and, once again, this translates into a true topological statement.

10.17 The Eilenberg–Moore spectral sequence

Let M and N be differential comodules over a simply connected differential coalgebra C. The Eilenberg-Moore spectral sequence is a purely algebraic object which relates the nondifferential Cotor on the homologies to the differential Cotor.

Proposition 10.17.1. *If C is a simply connected differential coalgebra which is projective over the ground ring and which has flat homology HC, then there* *is a functorial convergent second quadrant spectral sequence with abutment* $\mathrm{Cotor}^C(M, N)$ *and with*

$$E^2_{-p,q} = (\mathrm{Cotor}^{HC}_{-p}(HM, HN))_q.$$

Proof: Let

$$0 \to M \to Q_*$$

be a proper injective resolution of the right comodule M over the differential coalgebra C. With no loss of generality, we can assume that the resolution is tapered in the sense that each Q_{-p} is at least $2p - 1$ connected. We can also assume that we can write each $Q_{-p} = \overline{Q}_{-p} \otimes C$ as extended differential comodules with respect to the internal differential d_I and that each $\overline{Q}_{-p}$ is a flat R module.

Since a proper injective resolution is split as a resolution of differential modules, it follows that the internal homology of the resolution

$$0 \to HM \to H_I Q_*$$

is also split exact. By the Kunneth theorem $H_I Q_* = H_I \overline{Q}_* \otimes HC$ and the resolution

$$0 \to HM \to HQ_*$$

is in fact a proper projective resolution of the comodule HM over the coalgebra HC.

We filter $T(Q_*) \square_C N$ by the resolution degree, that is,

$$F_n(T(Q_*) \square_C N) = T(Q_{-p \leq n}) \square_C N = T(\overline{Q}_{-p \leq n}) \otimes N$$

and note that the tapering of the resolution guarantees that this filtration is finite in each fixed degree. Hence the spectral sequence based on this filtration will converge.

Note that

$$E^0_{-p} = s^{-p}\overline{Q}_{-p} \otimes C \otimes N, \quad d^0 = d_I \otimes 1 + 1 \otimes d_N$$

in the spectral sequence of $T(Q_*) \otimes N$. And

$$E^1_{-p} = Hs^{-p}\overline{Q}_{-p} \otimes HC \otimes HN = Hs^{-p}Q_{-p} \otimes HN, \quad d^1 = d_E \otimes 1$$

in this spectral sequence.

Hence,

$$E^0 = s^{-p}\overline{Q}_{-p} \otimes N, \quad d^0 = d_I \otimes 1 + 1 \otimes d_N$$

in the spectral sequence of $T(Q_*) \square_C N$. And

$$E^1 = Hs^{-p}\overline{Q}_{-p} \otimes HN = Hs^{-p}Q_{-p} \square_{HC} HN, \quad d^1 = \overline{d}_E \square_C 1$$

in this spectral sequence.

Thus,
$$E^2 = \operatorname{Cotor}^{HC}(HM, HN)$$
in the spectral sequence which abuts to $\operatorname{Cotor}^C(M, N)$. We are done. □

Remarks. From the construction of the Eilenberg–Moore spectral sequence, there is an increasing filtration of $\operatorname{Cotor}^C(M, N)$ indexed by the nonpositive integers
$$\cdots \subseteq F_{-n-1} \subseteq F_{-n} \subseteq \cdots \subseteq F_{-1} \subseteq F_0 = \operatorname{Cotor}^C(M, N).$$

Since C is simply connected, this filtration is strongly convergent in the sense that it is finite in each fixed degree.

Note the important edge homomorphism
$$\operatorname{Cotor}^C(M, N) = F_0 \to F_0/F_{-1} = E_{0,*}^\infty \subseteq E_{0,*}^2 = HM \square_{HC} HN.$$

Remarks. Consider the algebraic structures of associative algebras, associative coalgebras, and (bi)associative Hopf algebras. In this book, it is invariably the case that, if a particular differential Cotor has this structure, then the whole Eilenberg–Moore spectral sequence, including the edge homomorphisms, respects this structure. The reason for this is that the Eilenberg–Moore spectral sequence is defined by the filtration by resolution degree and the various algebra structures in differential Cotor are defined by the Künneth theorem and maps of total complexes of resolutions, both of which are compatible with the filtration by resolution degree. In particular, these algebraic structures induce the same structures on the E^2 term, that is, on the nondifferential Cotor defined by the homologies. Important examples are:

(1) If
$$\begin{array}{ccc} E & \to & \mathbb{Z} \\ \downarrow & & \downarrow \\ Y & \to & X \end{array}$$

is a homotopy pullback of simplicial sets with X being 1-reduced and, if all the spectral sequence terms are R projective, then the Eilenberg–Moore spectral sequence is a spectral sequence of differential coalgebras and the edge homomorphism $HE \to HY \square_{HX} HZ$ is a morphism of coalgebras.

(2) If, in addition, the above example is a homotopy pullback of simplicial monoids with X being 1-reduced and, if all the spectral sequence terms are R projective, then the Eilenberg–Moore spectral sequence is a spectral sequence of differential Hopf algebras and the edge homomorphism $HE \to HY \square_{HX} HZ$ is a morphism of Hopf algebras.

(3) If C is a simply connected supplemented differential coalgebra, then the Eilenberg–Moore spectral sequence abutting to $\mathrm{Cotor}^C(R,R)$ is a spectral sequence of differential algebras and the edge homomorphism $\mathrm{Cotor}^C(R,R) \to R \square_{HC} R = R$ is an augmentation morphism of algebras.

(4) If $C = C(X)$ is the normalized chains on a 1-reduced simplicial set and if all the spectral sequence terms are R projective, then the Eilenberg–Moore spectral sequence abutting to $\mathrm{Cotor}^{C(X)}(R,R) = H(\Omega X)$ is a spectral sequence of differential Hopf algebras and the edge homomorphism $H(\Omega X) \to R \square_{HCX} R = R$ is an augmentation morphism of Hopf algebras.

The Eilenberg–Moore spectral sequence is often used for computation but its functoriality has another use in proving a strong form of homological invariance. Let M and N be differential comodules over a simply connected differential coalgebra C such that C is R projective and HC is R flat. Let M_1 and N_1 be differential comodules over a differential coalgebra C_1 satisfying the same hypothesis.

Corollary 10.17.2. *Let $C \to C_1$ be a map of differential coalgebras as above and let $M \to M_1$ and $N \to N_1$ be maps of differential comodules with respect to the coalgebra map. If these maps are all homology isomorphisms, then the induced map*

$$\mathrm{Cotor}^C(M,N) \to \mathrm{Cotor}^{C_1}(M_1,N_1)$$

is an isomorphism.

10.18 The Eilenberg–Zilber theorem and the Künneth formula

The Alexander–Whitney map $\Delta : C(X \times Y) \to C(X) \otimes C(Y)$ and the Eilenberg–Zilber map $\nabla : C(X) \otimes C(Y) \to C(X \times Y)$ are two natural chain equivalences which have different uses and virtues. The use of the Alexander–Whitney map is that, in composition with the induced map of the diagonal,

$$C(X) \xrightarrow{C(\Delta)} C(X \times X) \xrightarrow{\Delta} C(X) \otimes C(X),$$

it endows the chains with the structure of a differential coalgebra. The virtue of the Eilenberg–Zilber map

$$\nabla : C(X) \otimes C(Y) \to C(X \times Y)$$

is that it is a map of differential coalgebras and a chain equivalence (homology isomorphism).

10.18 The Eilenberg–Zilber theorem and the Künneth formula

Let

$$\begin{array}{ccc} E & \xrightarrow{k} & \mathbb{Z} \\ \downarrow h & & \downarrow g \\ Y & \xrightarrow{f} & X \end{array} \quad \text{and} \quad \begin{array}{ccc} E_1 & \xrightarrow{k_1} & \mathbb{Z}_1 \\ \downarrow h_1 & & \downarrow g_1 \\ Y_1 & \xrightarrow{f_1} & X_1 \end{array}$$

be two pullback squares where g and g_1 are fibrations. Let

$$\begin{array}{ccc} E \times E_1 & \xrightarrow{k \times k_1} & \mathbb{Z} \times \mathbb{Z}_1 \\ \downarrow f \times f_1 & & \downarrow g \times g_1 \\ Y \times Y_1 & \xrightarrow{f \times f_1} & X \times X_1 \end{array}$$

be the product pullback square.

Homological invariance and the fact the Eilenberg–Zilber maps are maps of differential coalgebras immediately implies that:

Proposition 10.18.1. *The Eilenberg–Zilber map defines an isomorphism*

$$\nabla = \mathrm{Cotor}^\nabla(\nabla, \nabla) : \mathrm{Cotor}^{C(X) \otimes C(X_1)}(C(Y) \otimes C(Y_1), C(\mathbb{Z}) \otimes C(\mathbb{Z}_1))$$
$$\to \mathrm{Cotor}^{C(X \times X_1)}(C(Y \times Y_1), C(\mathbb{Z} \times \mathbb{Z}_1)).$$

This is compatible with the Eilenberg–Zilber map

$$\nabla : C(E) \otimes C(E_1) \to C(E \times E_1).$$

Before discussing this further we shall compress the notation and denote by X the normalized chains $C(X)$.

Using this notation, let $0 \to Y \xrightarrow{\eta} Q_*$ and $0 \to Y_1 \xrightarrow{\eta} Q_{1*}$ be nice (= tapered and R-projective) proper injective resolutions of right differential comodules over the respective differential coalgebras X and X_1. And let

$$\begin{array}{ccccc} 0 & \to & Y \times Y_1 & \xrightarrow{\eta \otimes \eta} & Q_* \otimes Q_{1*} \\ & & \downarrow & & \downarrow \overline{\nabla} \\ 0 & \to & Y \times Y_1 & \xrightarrow{\eta} & P_* \end{array}$$

be a map from one proper injective resolution to a proper injective resolution. (See Exercise 1 below.)

Recall that the Eilenberg–Moore geometric approximation theorem says that the composite map

$$E \xrightarrow{\Delta} Y \square_X \mathbb{Z} \xrightarrow{\eta \square 1} T(Q_*) \square_X \mathbb{Z}$$

is an homology isomorphism where $\Delta: E \to Y\square_X Z$ is the restriction to E of the Alexander–Whitney map $\Delta: Y \times Z \to Y \otimes Z$ and $T(Q_*)$ is the total complex of the resolution.

Consider the diagram

$$\begin{array}{ccccccc}
E \otimes E_1 & \xrightarrow{\Delta \otimes \Delta} & (Y\square_X Z) \otimes (Y_1 \square_{X_1} Z_1) & \xrightarrow{\eta \square \eta \otimes \eta \square \eta} & T(Q_*)\square_X Z \otimes T(Q_{1*})\square_{X_1} Z_1 \\
\downarrow & & \downarrow 1 \otimes T \otimes 1 \cong & B & \downarrow 1 \otimes T \otimes 1 \cong \\
|\nabla & C & (Y \otimes Y_1)\square_{X \otimes X_1}(Z \otimes Z_1) & \xrightarrow{\eta \otimes \eta \otimes 1 \otimes 1} & T(Q_*) \otimes T(Q_{1*})\square_{X \otimes X_1}(Z \otimes Z_1) \\
\downarrow & & \downarrow \nabla \square \nabla & A & \downarrow \overline{\nabla} \square \nabla \\
E \times E_1 & \xrightarrow{\Delta} & (Y \times Y_1)\square_{X \times X_1}(Z \times Z_1) & \xrightarrow{\eta \square 1} & T(P_*)\square_{X \times X_1}(Z \times Z_1)
\end{array}$$

The top and bottom horizontal compositions are homology isomorphisms by the geometric approximation theorem. The right and left vertical maps are homology isomorphisms by the usual Eilenberg–Zilber theorem. (See Exercise 2.)

The square A commutes since we have a map of resolutions extending $\nabla \square \nabla$. The square B commutes since we are using the same twist maps on the left and right. The square C commutes since the Eilenberg–Zilber map is a map of differential coalgebras and this diagram is covered by the commutative diagram

$$\begin{array}{ccc}
E \otimes E_1 & \xrightarrow{\Delta \otimes \Delta} & E \otimes E \otimes E_1 \otimes E_1 \\
\downarrow & & \downarrow 1 \otimes T \otimes 1 \\
|\nabla & & E \otimes E_1 \otimes E \otimes E_1 \\
\downarrow & & \downarrow \nabla \otimes \nabla \\
E \times E_1 & \xrightarrow{\Delta} & (E \times E_1) \otimes (E \times E_1).
\end{array}$$

If we take the homology of the above commutative diagram, we get:

Proposition 10.18.2. *We have a commutative diagram*

$$\begin{array}{ccc}
HE \otimes HE_1 & \xrightarrow{\eta \otimes \eta} & \mathrm{Cotor}^X(Y, Z) \otimes \mathrm{Cotor}^{X_1}(Y_1, Z_1) \\
\downarrow H\nabla & & \downarrow \nabla \\
H(E \times E_1) & \xrightarrow{\eta} & \mathrm{Cotor}^{X \times X_1}(Y \times Y_1, Z \times Z_1)
\end{array}$$

in which the horizontal maps are isomorphisms and, if the homologies HE and HE_1 are flat over R, the vertical maps are isomorphisms.

We shall call the left-hand vertical map the Eilenberg–Zilber map for differential Cotor.

For the sake of completeness, we record from the above proof the purely algebraic Künneth formula.

Proposition 10.18.3. *The twisting map gives a morphism*

$$\mathrm{Cotor}^C(M, N) \otimes \mathrm{Cotor}^{C_1}(M_1, N_1) \to \mathrm{Cotor}^{C \otimes C_1}(M \otimes M_1, N \otimes N_1)$$

which is an isomorphism if $C, M, N, \operatorname{Cotor}^C(M, N)$ and C_1, M_1, N_1, $\operatorname{Cotor}^{C_1}(M_1, N_1)$ are all R projective.

Exercises

(1) If $0 \to M \to Q_*$ and $0 \to M_1 \to Q_{1*}$ are proper injective resolutions of right differential comodules over the differential coalgebras C and C_1, respectively, then
$$0 \to M \otimes M_1 \to Q_* \otimes Q_{1*}$$
is a proper injective resolution of right differential comodules over the differential coalgebra $C \otimes C_1$.

(2) Prove the map $1 \otimes T \otimes 1$ restricts to an isomorphism
$$1 \otimes T \otimes 1 : (M \square_C N) \otimes (M_1 \square_{C_1} N_1) \to (M \otimes M_1) \square_{C \otimes C_1} (N \otimes N_1).$$

10.19 Coalgebra structures on differential Cotor

Consider the diagonal map of pullback squares

$$\begin{array}{ccc} E & \xrightarrow{k} & \mathbb{Z} \\ \downarrow h & & \downarrow g \\ Y & \xrightarrow{f} & X \end{array} \xrightarrow{\Delta} \begin{array}{ccc} E \times E & \xrightarrow{k \times k} & \mathbb{Z} \times \mathbb{Z} \\ \downarrow h \times h & & \downarrow g \times g \\ Y \times Y & \xrightarrow{f \times f} & X \times X, \end{array}$$

where the map g is a fibration. When HE is R flat, this induces a coalgebra structure which is compatible with the Eilenberg–Moore model in the following manner.

Whenever the homology HE of a space is flat over the ground ring it has a coalgebra structure defined by the composition of the Alexander–Whitney map with the induced map of the diagonal

$$\Delta : HE \xrightarrow{H\Delta} H(E \times E) \xrightarrow{H\Delta = (H\nabla)^{-1}} H(E \otimes E) \cong HE \otimes HE.$$

Since we always have a commutative diagram

$$\begin{array}{ccccc} HE & \xrightarrow{H\Delta} & H(E \times E) & \xleftarrow{H\nabla} & HE \otimes HE \\ \downarrow \eta & & \downarrow \eta & & \downarrow \eta \otimes \eta \\ \operatorname{Cotor}^X(Y, \mathbb{Z}) & \xrightarrow{\Delta} & \operatorname{Cotor}^{X \times X}(Y \times Y, \mathbb{Z} \times \mathbb{Z}) & \xleftarrow{\nabla} & \operatorname{Cotor}^X(Y, \mathbb{Z}) \otimes \operatorname{Cotor}^X(Y, \mathbb{Z}) \end{array}$$

in which the vertical maps are isomorphisms, it follows that, if $HE \cong \operatorname{Cotor}^X(Y, \mathbb{Z})$ is R flat, then we have coalgebra structures on HE and $\operatorname{Cotor}^X(Y, \mathbb{Z})$ which are isomorphic.

Remarks. We can strenghten the Künneth Theorem 10.18.2 as follows:

If the homologies HE and HE_1 are flat over R, then we have a commutative diagram

$$\begin{array}{ccc} HE \otimes HE_1 & \xrightarrow{\eta \otimes \eta} & \mathrm{Cotor}^X(Y, \mathbb{Z}) \otimes \mathrm{Cotor}^{X_1}(Y_1, \mathbb{Z}_1) \\ \downarrow H\nabla & & \downarrow \nabla \\ H(E \times E_1) & \xrightarrow{\eta} & \mathrm{Cotor}^{X \times X_1}(Y \times Y_1, \mathbb{Z} \times \mathbb{Z}_1) \end{array}$$

in which the horizontal maps and vertical maps are isomorphisms of coalgebras. All that is new is that the right hand map is a map of coalgebras. The proof of this fact is that we already know that the horizontal maps and the left-hand map are isomorphisms of coalgebras and hence so is the right-hand map.

Remarks. It is important to observe that there is no intrinsically defined coalgebra structure in a differential Cotor, $\mathrm{Cotor}^C(M, N)$, defined for differential comodules and differential coalgebras. The coalgebra structure in this section depends on the geometry of the Eilenberg–Zilber map.

For example, if C is a commutative differential coalgebra then the comultiplication $C \to C \otimes C$ is a map of coalgebras and hence, if $\mathrm{Cotor}^C(R.R)$ is R flat, then the induced map $\mathrm{Cotor}^C(R.R) \to \mathrm{Cotor}^{C \otimes C}(R \otimes R, R \otimes R) \equiv \mathrm{Cotor}^C(R.R) \otimes \mathrm{Cotor}^C(R.R)$ gives a coalgebra structure. But there is no guarantee that this structure has any relation to the coalgebra structure discussed in this section. Suppose that there is a map of differential coalgebras $C \to C(X)$ which is a homology isomorphism. The isomorphism of differential Cotors

$$\mathrm{Cotor}^C(R.R) \to \mathrm{Cotor}^{C(X)}(R.R)$$

need not be a morphism of coalgebras.

Remarks. Suppose we have a pullback square

$$\begin{array}{ccc} E & \to & \mathbb{Z} \\ \downarrow & & \downarrow \\ Y & \to & X \end{array}$$

with the vertical maps fibrations and HE a flat R module. Since the Eilenberg–Moore spectral sequence is defined by the filtration by resolution degree and since the maps which define the coalgebra structure on the differential Cotor, that is, the diagonal map on a space and the Eilenberg–Zilber map, are both compatible with this filtration, it follows that,this Eilenberg–Moore spectral sequence is a spectral sequence of coalgebras

$$E^2 = \mathrm{Cotor}^{HX}(HY, H\mathbb{Z}) \Rightarrow \mathrm{Cotor}^{CX}(CY, C\mathbb{Z}).$$

Exercises

(1) Show that the map

$$\mathrm{Cotor}^X(Y, \mathbb{Z}) \to \mathrm{Cotor}^R(R, R)$$

is a counit for the coalgebra structure in this section.

10.20 Homotopy pullbacks and differential Cotor of several variables

Pullbacks of several variables are the geometric analoques of iterated cotensor products and homotopy pullbacks of several variables are the geometric analogues of the several variable derived functors. We introduce these ideas in this section.

For us here a pullback of several variables is just the pullback of a zigzag diagram of maps as follows

$$X_1 \to A_1 \leftarrow X_2 \to A_2 \leftarrow X_3 \to \ldots A_{n-1} \leftarrow X_{n-1} \to A_{n-1} \leftarrow X_n.$$

We write the pullback as

$$X_1 \times_{A_1} X_2 \times_{A_2} X_3 \times_{A_3} \ldots X_{n-1} \times_{A_{n-1}} X_n.$$

These pullbacks of several variables can be regarded as iterated pullbacks:

$$X_1 \times_{A_1} X_2 \times_{A_2} X_3 \times_{A_3} \ldots X_{n-1} \times_{A_{n-1}} X_n$$
$$\equiv X_1 \times_{A_1} (X_2 \times_{A_2} (X_3 \times_{A_3} \ldots (X_{n-1} \times_{A_{n-1}} X_n)\ldots)).$$

Consider maps $X \to A \leftarrow Y \to B \leftarrow Z$. We note the following identities.

Lemma 10.20.1.

(a) *Pullbacks of several variables are associative:*

$$X \times_A (Y \times_B Z) \equiv (X \times_A Y) \times_B Z.$$

(b) *There are left and right units:*

$$A \times_A X \equiv X, \quad X \times_A A \equiv X.$$

(c) *If $*$ is a point, then*

$$X \times_A * \equiv F, \quad * \times_A X \equiv F$$

where F is the fibre of $X \to A$.

(d) *Points induce splittings into a product:*

$$X \times_A * \times_B Z \equiv (X \times_A *) \times (* \times_B Z).$$

We define the homotopy pullback of several variables by replacing the maps out of the X_i by fibrations with total space E_i in the standard way:

$$X_1 \simeq E_1 \to A_1, \; X_2 \simeq E_2 \to A_1 \times A_2, \; X_3 \simeq E_3 \to A_2 \times A_3, \ldots X_{n-1}$$
$$\simeq E_{n-1} \to A_{n-2} \times A_{n-1}, \; X_n \simeq E_n \to A_{n-1}.$$

Then the homotopy pullback is

$$E_1 \times_{A_1} E_2 \times_{A_2} E_3 \times_{A_3} \ldots E_{n-1} \times_{A_{n-1}} E_n.$$

We remark that this is balanced in the sense that we get something homotopy equivalent to the homotopy pullback if we replace all the maps $X_i \to A_{i-1} \times A_i$ by fibrations except possibly for one (which could be one of the ends), that is,

Lemma 10.20.2.

$$X_1 \times_{A_1} X_2 \times_{A_2} X_3 \times_{A_3} \cdots \times_{A_{n-1}} X_n$$
$$\to E_1 \times_{A_1} E_2 \times_{A_2} E_3 \times_{A_3} \cdots \times_{A_{n-1}} E_n$$

is a homotopy equivalence if all of the maps $X_i \to A_{i-1} \times A_i$ except possibly one are fibrations.

There are several other ways in which we can get something equivalent to the homotopy pullback.

Lemma 10.20.3. *The above map is a homotopy equivalence if for all $2 \le i \le n$ the maps $X_i \to A_{i-1}$ are fibrations (or if for all $1 \le i \le n-1$ the maps $X_i \to A_i$ are fibrations).*

Proof: Lemmas 10.20.2 and 10.20.3 are easy consequences of the fact that, if $X \to A$ is a fibration, then so is $X \times_A Y \to Y$ and that the inclusion of $X \times_A Y$ into the homotopy pullback is a homotopy equivalence. See the diagram below in the proof of Theorem 10.21.1 and the first part of that proof for guidance. The point is that all the squares in that diagram are forced to be homotopy equivalent to homotopy pullbacks by either of the hypotheses of Lemmas 10.20.2 or 10.20.3. □

We note the following properties of the pathspace fibration $PA \to A$ in the homotopy pullback:

Lemma 10.20.4.

(a) *The homotopy pullback $X \times_A PA$ is just the homotopy theoretic fibre of $X \to A$.*

(b) *Contractible spaces induce splittings of the homotopy pullback*

$$X \times_A P(A \times B) \times_B \mathbb{Z} \equiv (X \times_A PA) \times (PB \times_B \mathbb{Z}).$$

We recall the striking similarity between pullbacks and cotensor products. Let M, N, and P be differential comodules over differential coalgebras C and D. We have associativity, left and right units, primitives, and splittings induced by the ground ring R:

10.20 Homotopy pullbacks and differential Cotor of several variables

Lemma 10.20.5.

$$(M\square_C N)\square_D P \cong M\square_C (N\square_D P)$$
$$M\square_C C \cong M, \ C\square_C N \cong N, M\square_C C\square_C N \cong M\square_C N$$
$$M\square_C R \cong PM, \ R\square_C N \cong PN$$
$$M\square_C R\square_D N \cong (M\square_C R) \otimes (R\square_D N)$$

Remarks. It is worth noting in the above Lemma 10.20.5 that the two isomorphisms

$$\Delta\square 1, 1\square\Delta : M\square_C N \to M\square_C C\square_C N$$

are in fact the same.

Now let $C_1, C_2, \ldots, C_{n-1}$ be differential coalgebras over a commutative ring R. Let M_1 be a right differential comodule over C_1, for $2 \leq i \leq n-1$ let M_i be a $C_{i-1} - C_i$ differential bimodule, and let M_n be a left differential comodule over C_{n-1}. Let $0 \to M_i \to Q_{i*}$ be proper injective resolutions and let $0 \to M_i \to T(Q_{i*})$ be the associated total complexes. We define the several variable differential Cotor as follows:

Definition 10.20.6. The several variable differential Cotor is

$$\text{Cotor}^{C_1,\ldots,C_{n-1}}(M_1, \ldots, M_n)$$
$$= H(T(Q_{1*})\square_{C_1} T(Q_{2*})\square_{C_2} \ldots \square_{C_{n-1}} T(Q_{n*})).$$

This is balanced in the sense that the iterated cotensor product is homology equivalent to the several variable differential Cotor if all the differential comodules are constructions except possibly for one (which could be one of the ends), that is:

Lemma 10.20.7.

$$M_1 \square_{C_1} M_2 \square_{C_2} M_3 \square_{C_3} \ldots \square_{C_{n-1}} M_n$$
$$\to T(Q_{1*})\square_{C_1} T(Q_{2*})\square_{C_2} T(Q_{3*})\square_{C_3} \ldots \square_{C_{n-1}} T(Q_{n*})$$

is a homology equivalence if all of the differential comodules M_i except possibly one are (bi)constructions.

There are several other ways in which we can get something equivalent to the differential Cotor.

Lemma 10.20.8. *The above map is a homology equivalence if for all $2 \leq i \leq n$ the M_i are right bicomodule constructions over C_i (or if for all $1 \leq i \leq n-1$ the M_i are left bicomodule constructions over C_{i-1}).*

Proof: Lemmas 10.20.7 and 10.20.8 are easy consequences of the fact that, if M is a right construction over C and N is a $C - D$ differential bicomodule which is a right bicomodule construction over D, then $M \square_C N$ is a right construction over D and hence $P \mapsto M \square_C N \square_D P$ preserves homology isomorphisms.

For example, suppose that $n = 3$. Abbreviate the total complexes by T_1, T_2, T_3.

Assume that M_2 is not necessarily a construction but that M_1 and M_3 are. Then $M_2 \to T_2$ and $M_1 \to T_1$ are equivalences and M_1, T_2 are constructions. Hence

$$M_1 \square_{C_1} M_2 \to M_1 \square_{C_1} T_2 \to T_1 \square_{C_1} T_2$$

are equivalences.

Likewise, since M_3 and $T_1 \square_{C_1} T_2$ are constructions,

$$M_1 \square_{C_1} M_2 \square_{C_2} M_3 \to T_1 \square_{C_1} T_2 \square_{C_2} M_3 \to T_1 \square_{C_1} T_2 \square_{C_2} T_3$$

are equivalences.

Or assume that M_2, M_3 are right bicomodule constructions. Then

$$M_1 \square_{C_1} M_2 \to T_1 \square_{C_1} M_2 \to T_1 \square_{C_1} T_2$$

are equivalences, as are

$$M_1 \square_{C_1} M_2 \square_{C_2} M_3 \to T_1 \square_{C_1} M_2 \square_{C_2} M_3 \to T_1 \square_{C_1} T_2 \square_{C_2} T_3$$

$\square$

Just as for homotopy pullbacks we have splittings and collapse theorems for several variable differential Cotor. If $C_1 = \cdots = C_{n-1}$, we shall write

$$\mathrm{Cotor}^{C_1,\ldots,C_{n-1}}(M_1,\ldots,M_n) = \mathrm{Cotor}^C(M_1,\ldots,M_n).$$

Lemma 10.20.9.

(a) *If $M_i = C$, then differential Cotor reduces to differential Cotor of one less variable, that is,*

$$\mathrm{Cotor}^C(M_1,\ldots,M_n) \cong \mathrm{Cotor}^C(M_1,\ldots,\hat{M}_i,\ldots,M_n)$$

(b) *If $M_i = R$, if C is R projective, and if all of the differential Cotors are R flat, then*

$$\mathrm{Cotor}^C(M_1,\ldots,M_n) \cong \mathrm{Cotor}^C(M_1,\ldots,M_{i-1},R)$$
$$\otimes \mathrm{Cotor}^C(R,M_{i+1},\ldots,M_n).$$

Proof: Both parts of the lemma are consequences of the balanced property of differential Cotor, namely, we get the same result if we neglect to resolve one of the variables. Hence, both parts reduce to the corresponding properties of the cotensor product. $\square$

10.20 Homotopy pullbacks and differential Cotor of several variables

Remarks. In terms of constructions, Lemma 10.20.9 can be phrased in the somewhat stronger form as follows:

Let T_i be a construction. for $1 \leq i \leq 3$.

If $C \to T_2$ is a homology equivalence of bicomodules, then

$$T_1 \square_C T_2 \square_C T_3 \leftarrow T_1 \square_C C \square_C T_3 \leftarrow T_1 \square_C T_3$$

are homology equivalences. Furthermore, if T_1 is a biconstruction, then the above are all equivalences of right constructions. And, if T_3 is a biconstruction then the above are all left constructions. The obvious modifications hold true if $C = M_1$ or if $C = M_3$. In particular, all the complexes are biconstructions if T_1, T_2 and T_3 are.

If $R \to T_2$ homology equivalence of bicomodules, then T_2 can be chosen to be $T \otimes S$ where T is the total complex of a left resolution of R and S is the total complex of a right resolution of R. Then

$$T_1 \square_C T_2 \square_C T_3 \leftarrow T_1 \square_C (T \otimes S) \square_C T_3 \leftarrow$$
$$(T_1 \square_C T) \otimes (S \square_C T_3) \leftarrow (T_1 \square_C R) \otimes (R \square_C T_3)$$

are homology equivalences. And, if T_1 and T_3 are biconstructions, then

$$(T_1 \square_C R) \otimes (R \square_C T_3)$$

is a biconstruction.

These collapsing and splitting results are compatible with the Eilenberg–Moore models as follows:

Let $X_1, \ldots, X_n$ and A be simplicial sets with A being 1-reduced. We also use these abbreviations to denote the corresponding normalized chains.

Lemma 10.20.10. *Suppose all the differential Cotors are R projective, then there are isomorphisms of coalgebras:*

(a)

$$\begin{array}{ccc} H(X_1 \times_A \cdots * \cdots \times_A X_n) & \xrightarrow{\cong} & H(X_1 \times_A \ldots X_i \times_A *) \otimes H(* \times_A X_{i+1} \times_A \cdots \times_A X_n) \\ \downarrow \cong & & \downarrow \cong \\ \operatorname{Cotor}^A(X_1, \ldots, R, \ldots X_n) & \xrightarrow{\cong} & \operatorname{Cotor}^A(X_1, \ldots, X_i, R) \otimes \operatorname{Cotor}^A(R, X_{i+1}, \ldots, X_n) \end{array}$$

(b)

$$\begin{array}{ccc} H(X_1 \times_A \ldots X_i \times_A A \times_A X_{i+1} \times_A \cdots \times_A X_n) & \xleftarrow{\cong} & H(X_1 \times_A \ldots X_i \times_A X_{i+1} \times_A \cdots \times_A X_n) \\ \downarrow \cong & & \downarrow \cong \\ \operatorname{Cotor}^A(X_1, \ldots, X_i, A, X_{i+1}, \ldots, X_n) & \xleftarrow{\cong} & \operatorname{Cotor}^A(X_1, \ldots, X_i, X_{i+1}, \ldots, X_n) \end{array}$$

Proof: The vertical maps are isomorphisms of coalgebras and so are the top maps. The two bottom maps are induced by the isomorphisms of complexes

$$T_1 \square_C \ldots \square_C R \square_C \ldots \square_C T_n \to T_1 \square_C \ldots \square_C R \otimes R \square_C T_{i+1} \square_C \ldots \square_C T_n$$

and
$$T_1 \square_C \ldots \square_C C \square_C \ldots \square_C T_n \xleftarrow{1\square\ldots\square\Delta\square\ldots1} T_1 \square_C \ldots \square_C T_n$$
and hence are part of commutative diagrams. Therefore, the bottom maps are both coalgebra isomorphisms. □

Just as with the two variable differential Cotor, there are two possible proofs of the fact that the several variable version of differential Cotor is a well defined functor of several variables, independent of the choice of resolutions and independent of extensions of maps of comodules to maps of resolutions. We can use the balanced fact that we can omit one of the resolutions and thus one of the choices of extensions. Or we can use the fact that extensions of comodule maps are unique up to chain homotopy.

We note that filtering total complexes of resolutions by the resolution degree gives an Eilenberg–Moore spectral sequence:

Let $C_1, \ldots, C_{n-1}$ be differential coalgebras and let $M_1, \ldots, M_n$ be differential comodules with M_1 a right C_1 comodule, M_2 a $C_2 - C_3$ bicomodule, ..., and M_n a left C_{n-1} comodule.

Then the same proof as in the two variable case gives the following proposition.

Proposition 10.20.11. *If the C_i are simply connected differential coalgebras which are projective over the ground ring and which have flat homologies HC_i, then there is a functorial convergent second quadrant spectral sequence with abutment* $\mathrm{Cotor}^{C_1,\ldots,C_{n-1}}(M_1, \ldots, M_n)$ *and with*
$$E^2_{-p,q} = (\mathrm{Cotor}^{HC_1,\ldots,HC_{n-1}}_{-p}(HM_1, \ldots, HM_n))_q.$$

This Eilenberg–Moore spectral sequence is used to prove the homological invariance.

Corollary 10.20.12. *With hypotheses as in Lemma 10.20.10, let $C_i \to D_i$ be a maps of differential coalgebras and let $M_i \to N_i$ be maps of differential comodules with respect to the coalgebra maps. If these maps are all homology isomorphisms, then the induced map*
$$\mathrm{Cotor}^{C_1,\ldots,C_{n-1}}(M_1, \ldots, M_n) \to \mathrm{Cotor}^{D_1,\ldots,D_{n-1}}(N_1, \ldots, N_n)$$
is an isomorphism.

10.21 Eilenberg–Moore models of several variables

Consider a pullback of several variables
$$E = X_1 \times_{A_1} X_2 \times_{A_2} \ldots X_{n-1} \times_{A_{n-1}} X_n$$

10.21 Eilenberg–Moore models of several variables

where the X_i and A_i are simplicial sets and the A_i are 1-reduced in the sense that their 1-skeletons consist of a single point. (Of course, we can translate the results below to simply connected spaces by considering Eilenberg subcomplexes.)

Assume that this is a homotopy pullback in the sense that the maps $X_i \to A_i$ are fibrations for all $1 \le i \le n-1$. We will compress notation by writing X for the normalized chains $C(X)$.

The Eilenberg–Moore geometric approximation theorem for several variable homotopy pullbacks 10.21.1. *There is an isomorphism of R modules*

$$H(E) \cong \operatorname{Cotor}^{A_1, A_2, \ldots, A_{n-1}}(X_1, X_2, \ldots, X_n).$$

Proof: For $1 \le i \le j \le n$, write

$$E_{i,j} = X_i \times_{A_i} X_{i+1} \times_{A_{i+1}} \cdots \times_{A_{j-1}} X_j.$$

For example, $X_i = E_{i,i}$ and $E_{1,n} = E$. In general, these fit into a honeycomb diagram in which the squares are all homotopy pullbacks. For example, all the squares below are pullback diagrams:

$$\begin{array}{ccccccccc}
X_1 & \leftarrow & E_{1,2} & \leftarrow & E_{1,3} & \leftarrow & E_{1,4} & \leftarrow & \cdots \\
\downarrow & & \downarrow & & \downarrow & & \downarrow & & \\
A_1 & \leftarrow & X_2 & \leftarrow & E_{2,3} & \leftarrow & E_{2,4} & \leftarrow & \cdots \\
& & \downarrow & & \downarrow & & \downarrow & & \\
& & A_2 & \leftarrow & X_3 & \leftarrow & E_{3,4} & \leftarrow & \cdots \\
& & & & \downarrow & & \downarrow & & \\
& & & & A_3 & \leftarrow & X_4 & \leftarrow & \cdots \\
& & & & & & \downarrow & & \\
& & & & & & A_4 & \leftarrow & \cdots
\end{array}$$

It is evident that, if a vertical map is a fibration, then all the squares to its left are homotopy pullbacks and all the vertical maps to its left are fibrations. (Of course, the same remarks apply to horizontal maps being fibrations and to homotopy pullback squares.)

In our case, every square is a homotopy pullback square and all the vertical maps are fibrations.

Let $0 \to X_i \to T_i$ be the supplemented total complexes of proper injective resolutions.

The proof of geometric approximation theorem for several variables now follows by successive application of the known two variable theorem. In detail:

$$E_{1,2} \xrightarrow{\simeq} X_1 \square_{A_1} T_2$$

is a homology equivalence by the two variable theorem.

Likewise,
$$E_{1,3} \xrightarrow{\simeq} E_{1,2}\square_{A_2}T_3$$
is a homology equivalence by the two variable theorem. Since T_3 is a construction, the composite
$$E_{1,3} \xrightarrow{\simeq} E_{1,2}\square_{A_2}T_3 \simeq X_1\square_{A_1}T_2\square_{A_3}T_3$$
is a homology equivalence.

If we continue we see that the composite
$$E_{1,n} \xrightarrow{\simeq} E_{1,n-1}\square_{A_{n-1}}T_n \simeq X_1\square_{A_1}T_2\square_{A_3}T_3\square_{A_4}\ldots\square_{A_{n-1}}T_n$$
is a homology equivalence. $\square$

For the differential Cotors where the coalgebras are the normalized chains on a simplicial set, we have coalgebra structures coming from the fact that the Eilenberg–Zilber maps are maps of differential coalgebras. This is just as in the two variable case and corresponds to the geometry via the Eilenberg–Moore geometric approximation. In detail, the coalgebra structure is defined by the following three ingredients:

(1) There is the diagonal map of simplicial sets $\Delta : X \to X \times X$ and hence the induced maps of differential Cotor
$$\Delta = \text{Cotor}^{\Delta,\ldots,\Delta}(\Delta,\ldots,\Delta):$$
$$\Delta : \text{Cotor}^{A_1,\ldots,A_{n-1}}(X_1,\ldots,X_n)$$
$$\to \text{Cotor}^{A_1\times A_1,\ldots,A_{n-1}\times A_{n-1}}(X_1\times X_1,\ldots,X_n\times X_n).$$

(2) There is the Eilenberg–Zilber map $\nabla : Y \otimes Z \to Y \times Z$ which is both a chain equivalence and a map of differential coalgebras. It induces an isomorphism
$$\nabla = \text{Cotor}^{\nabla,\ldots,\nabla}(\nabla,\ldots,\nabla):$$
$$\text{Cotor}^{A_1\otimes B_1,\ldots,A_{n-1}\otimes B_{n-1}}(X_1\otimes Y_1,\ldots,X_n\otimes Y_n)$$
$$\to \text{Cotor}^{A_1\times B_1,\ldots,A_{n-1}\times B_{n-1}}(X_1\times Y_1,\ldots,X_n\times Y_n).$$

(3) There is the Künneth theorem which is valid when the differential Cotors are R flat (given R projective coalgebras):
$$\text{Cotor}^{C_1\otimes D_1,\ldots,C_{n-1}\otimes D_{n-1}}(M_1\otimes N_1,\ldots,M_n\otimes N_n)$$
$$\cong \text{Cotor}^{C_1,\ldots,C_{n-1}}(M_1,\ldots,M_n)\otimes \text{Cotor}^{D_1,\ldots,D_{n-1}}(N_1,\ldots,N_n),$$

then we get just as in the two variable case:

Let
$$E = X_1 \times_{A_1} X_2 \times_{A_2} \ldots X_{n-1} \times_{A_{n-1}} X_n$$

be a homotopy pullback over 1-reduced simplicial sets $A_1, \ldots, A_{n-1}$. The maps

$$E \xrightarrow{\Delta} E \times E \xleftarrow{\nabla} E \otimes E$$

commute with their extensions to maps of resolutions, hence:

Proposition 10.21.2. *If the homology HE is R flat, then the isomorphism in 10.21.1,*

$$H(E) \cong \mathrm{Cotor}^{A_1, A_2, \ldots, A_{n-1}}(X_1, X_2, \ldots, X_n)$$

is an isomorphism of coalgebras.

Remarks. The Eilenberg–Moore spectral sequence converging to the above coalgebra is a spectral sequence of coalgebras.

10.22 Algebra structures and loop multiplication

We begin this section by interpreting loop multiplication in terms of homotopy pullbacks. Afterwards, we will discuss how this leads to a natural associative multiplication in differential Cotor.

But first a question: Where does the loop multiplication come from in the Eilenberg–Moore models? For most spaces X, there is no multiplication

$$C(X) \otimes C(X) \to C(X)$$

but there is one in the Eilenberg–Moore model for the loop multiplication

$$\Omega X \times \Omega X \to \Omega X.$$

Where did it come from? The answer is that it came from these several variable differential Cotors.

Consider the right and left pathspace fibrations:

$$P_R X = \{\omega : I \to X \mid \omega(0) = *\}, \quad \pi : P_R X \to X, \ \pi(\omega) = \omega(1),$$

$$P_L X = \{\gamma : I \to X \mid \omega(1) = *\}, \quad \pi : P_L X \to X, \ \pi(\gamma) = \gamma(0).$$

Let $\tilde{\Omega} X = \{(\omega, \gamma) \mid \omega(0) = \gamma(1) = *, \ \omega(1) = \gamma(0)\}$ and note that the map $\tilde{\Omega} X \to \Omega X$, $(\omega, \gamma) \mapsto \omega * \gamma$ is a homotopy equivalence.

We have a map of the homotopy pullback diagram

$$\begin{array}{ccccc} P_R X & \leftarrow & \Omega X & \leftarrow & \Omega X \times \Omega X \\ \downarrow & & \downarrow & & \downarrow \\ X & \leftarrow & * & \leftarrow & \Omega X \\ & & \downarrow & & \downarrow \\ & & X & \leftarrow & P_L X \end{array}$$

to the homotopy pullback diagram

$$\begin{array}{ccccc} P_R X & \leftarrow & P_R X & \leftarrow & \tilde{\Omega} X \\ \downarrow & & \downarrow & & \downarrow \\ X & \leftarrow & X & \leftarrow & P_L X \\ & & \downarrow & & \downarrow \\ & & X & \leftarrow & P_L X \end{array}$$

where the composition

$$\Omega X \times \Omega X \to \tilde{\Omega} X \to \Omega X$$

is the standard loop multiplication $(\omega, \gamma) \mapsto \omega * \gamma$.

Let C be a simply connected supplemented differential coalgebra and suppose that $\operatorname{Cotor}^C(R, R)$ is R projective. The above description of loop multiplication suggests that we use the splitting and collapsing of differential Cotor to define a multiplication in $\operatorname{Cotor}^C(R, R)$ which is natural on the category of supplemented differential coalgebras. In this respect, it is very different from the coalgebra structure on $\operatorname{Cotor}^{C(X)}(R, R)$ which is defined and natural only on the category of simplicial sets.

Definition 10.22.1. The multiplication in differential $\operatorname{Cotor}^C(R, R)$ is:

$$\mu : \operatorname{Cotor}^C(R, R) \otimes \operatorname{Cotor}^C(R, R) \cong \operatorname{Cotor}^C(R, R, R)$$
$$\to \operatorname{Cotor}^C(R, C, R) \cong \operatorname{Cotor}^C(R, R).$$

In this definition we use the fact that $R \to C$ is a map of differential bicomodules over C. We now describe this multiplication explicitly in terms of resolutions.

Let T_i be the total complex of a proper injective resolution of R and choose $T_2 = S \otimes T$ where S and T are the total complexes of left and right resolutions of R. Let $\tilde{T}_i$ be the total complex of a resolution of C. Then the multiplication is induced by the maps of complexes

$$(T_1 \square_C S) \otimes (T \square_C T_3) = T_1 \square_C T_2 \square_C T_3 \to T_1 \square_C \tilde{T}_2 \square_C T_3$$
$$\leftarrow T_1 \square_C C \square_C T_3 \leftarrow T_1 \square_C T_3.$$

Associativity amounts to the easily verified statement that the two maps

$$\operatorname{Cotor}^C(R, R, R, R) \to \operatorname{Cotor}^C(R, C, C, R)$$

are induced by equal maps on the level of complexes.

10.22 Algebra structures and loop multiplication

The unit is given by the map $R = R\square_C R \to \text{Cotor}^C(R,R)$. In terms of the maps of resolutions, the fact that this is a left unit is proved by the observation that, for $x \in T_3$, $1\square 1 \otimes 1\square x \mapsto 1\square 1\square x$ represents the identity map in homology $R \otimes R\square_C T_3 \to T_1 \square_C C \square_C T_3$.

This algebra structure is compatible with the Künneth theorem.

Proposition 10.22.2. *If C and D are simply connected supplemented differential coalgebras, then*

$$\nu : \text{Cotor}^C(R,R) \otimes \text{Cotor}^D(R,R) \to \text{Cotor}^{C \otimes D}(R,R)$$

is an morphism of algebras. That is, if we write $\nu(\alpha \otimes \beta) = \alpha \otimes \beta$ for the morphism and $\mu(\alpha \otimes \gamma) = \alpha * \gamma$ for the multiplication, then

$$(\alpha * \gamma) \otimes (\beta * \epsilon) = (-1)^{\deg(\beta)\deg(\gamma)}(\alpha \otimes \beta) * (\gamma * \epsilon)$$

for α, γ in $\text{Cotor}^C(R,R)$ and β, ϵ in $\text{Cotor}^D(R,R)$. (If the two Cotors are R projective, the map ν is an isomorphism of algebras.)

Proof: The sign in the formula comes from the fact that isomorphisms of chain complexes $X \otimes Y \to Y \otimes X$ use the twist morphism $T(x \otimes y) = (-1)^{\deg(x)\deg(y)}(y \otimes x)$. In terms of resolutions, the relevant multiplication is given by the commutative diagram

$$\begin{array}{ccc}
T_1^C \square_C R \square_C T_3^C \otimes T_1^D \square_D R \square_D T_3^D & \to & T_1^C \otimes T_1^D \square_{C \otimes D} R \otimes R \square_{C \otimes D} T_3^C \otimes T_3^D \\
\downarrow & & \downarrow \\
T_1^C \square_C \tilde{T}^C \square_C T_3^C \otimes T_1^D \square_D \tilde{T}^D \square_D T_3^D & \to & T_1^C \otimes T_1^D \square_{C \otimes D} \tilde{T}^C \otimes \tilde{T}^D \square_{C \otimes D} T_3^C \otimes T_3^D \\
\uparrow & & \uparrow \\
T_1^C \square_C C \square_C T_3^C \otimes T_1^D \square_D D \square_D T_3^D & \to & T_1^C \otimes T_1^D \square_{C \otimes D} C \otimes D \square_{C \otimes D} T_3^C \otimes T_3^D
\end{array}$$

The sign comes from the top square. $\square$

The above multiplication on differential Cotor is chosen so that it is automatically compatible with the Eilenberg-Moore model for the homotopy pullbacks:

Proposition 10.22.3. *Let X be a 1-reduced simplicial set. Then the diagram below commutes*

$$\begin{array}{ccccc}
H(\Omega X) \otimes H(\Omega X) & \to & H(\Omega X \times \Omega X) & \xrightarrow{\mu_*} & H(\Omega X) \\
\downarrow \cong & & \downarrow \cong & & \downarrow \cong \\
\text{Cotor}^X(R,R) \otimes \text{Cotor}^X(R,R) & \to & \text{Cotor}^X(R,R,R) & \xrightarrow{\mu} & \text{Cotor}^X(R,R)
\end{array}$$

Furthermore, if $H(\Omega X)$ is R projective. then the horizontal maps are maps of coalgebras so that

$$H(\Omega X) \to \text{Cotor}^X(R,R)$$

is an isomorphism of Hopf algebras.

Finally we relate this multiplication to the multiplication in the cobar or loop construction ΩC.

Proposition 10.22.4. *If C is a simply connected supplemented differential coalgebra, then there is an isomorphism of algebras*

$$\mathrm{Cotor}^C(R,R) \to H\Omega C.$$

Proof: Since the twisted tensor product $\Omega C \otimes_\tau C$ is an acyclic construction, it follows that $\mathrm{Cotor}^C(R,R) \cong H(\Omega C \otimes_\tau C \square_C R) \cong H(\Omega C)$ as R modules. Hence, we need to show that it is an isomorphism of algebras.

Since the ayclic twisted tensor products $\Omega C \otimes_\tau C$ and $C \otimes_\tau \Omega C$ are the total complexes of resolutions of R, the multiplication is given by the map of complexes

$$\Omega C \otimes_\tau \square_C R \square_C C \otimes_\tau \Omega C \xrightarrow{\mu} \Omega C \otimes_\tau \square_C C \square_C C \otimes_\tau \Omega C$$

and this reduces by isomorphisms to

$$\Omega C \otimes R \otimes \Omega C \xrightarrow{\mu} \Omega C \otimes_\tau C \otimes_\tau \Omega C$$

where the differential in the domain is the tensor product differential and the differential in the range is a special case of the following lemma whose proof is left as an exercise.

Lemma 10.22.5. *If A and B are augmented differential algebras and C is a supplemented differential coalgebra with twisting cochains $\tau : C \to A$ and $\sigma : C \to B$, then the differential in*

$$A \otimes_\tau C \otimes_\sigma B \cong A \otimes_\tau C \square_C C \otimes_\sigma B$$

is given on the left by

$$d = d_\tau \otimes 1 + 1 \otimes d_\sigma - (1 \otimes d_C \otimes 1).$$

(Of course the differential on the right is the restriction of the tensor product differential $d_\tau \otimes 1 \otimes 1 + 1 \otimes 1 \otimes d_\sigma$.

There are three relevant maps of differential objects.

First, there are the obvious two inclusions which play the role of left and right units

$$R \otimes R \otimes \Omega C \xrightarrow{\alpha} \Omega C \otimes R \otimes \Omega C \xleftarrow{\beta} \Omega C \otimes R \otimes R$$

and, second, the map below which uses the multiplication in ΩC

$$\Omega C \otimes_\tau C \otimes_\tau \Omega C \xrightarrow{\gamma} \Omega C$$

where

$$\gamma(x \otimes c \otimes y) = \begin{cases} x \cdot y & \text{if } c = 1 \\ 0 & \text{if } c \in \overline{C}. \end{cases}$$

It is easy to check that γ is a chain map and one sees that the difference in the signs of the left and right twisted tensor products has an essential role.

It is obvious that the compositions $\gamma \cdot \mu \cdot \alpha$ and $\gamma \cdot \mu \cdot \beta$ are the identity.

We claim that $\mu \cdot \alpha$ and $\mu \cdot \beta$ are homology isomorphisms, but, for example, $\mu \cdot \alpha$ is the composition

$$R \otimes R \square_C C \otimes_\tau \Omega C \to \Omega C \otimes_\tau C \square_C C \otimes_\tau \Omega C$$

where $R \otimes R \to \Omega C \otimes_\tau C$ is a homology isomorphism and $C \otimes_\tau \Omega C$ is a construction. Hence, $\mu \cdot \alpha$ and likewise $\mu \cdot \beta$ are homology isomorphisms.

It follows that $\text{Cotor}^C(R, R)$ and $H\Omega C$ are isomorphic as algebras. □

It follows from Corollary 10.5.4 that:

Corollary 10.22.6. *If A is a augmented differential algebra and $\tau : C \to A$ is a twisting morphism such that $A \otimes_\tau C$ is acylic, then there is an isomorphism of algebras*

$$\text{Cotor}^C(R, R) \to HA.$$

Exercise

(1) Prove Lemma 10.22.5.

10.23 Commutative multiplications and coalgebra structures

In special circumstances, there are ways of getting algebra and coalgebra structures on differential Cotor. These methods are often accessible to computation.

Let C be a simply connected differential Hopf algebra. Then the multiplication $\nu : C \otimes C \to C$ is a map of differential coalgebras and induces a multiplication

$$\nu : \text{Cotor}^C(R, R) \otimes \text{Cotor}^C(R, R)$$
$$\to \text{Cotor}^{C \otimes C}(R, R) \xrightarrow{\text{Cotor}^\nu(R,R)} \text{Cotor}^C(R, R).$$

This multiplication has the usual unit

$$R \cong R \otimes R = \text{Cotor}^R(R, R) \to \text{Cotor}^C(R, R)$$

as a two-sided unit. Write $\alpha \diamond \beta = \nu(\alpha \otimes \beta)$ for this multiplication and write $\alpha * \beta = \mu(\alpha \otimes \beta)$ for the natural multiplication introduced in the previous section.

Since ν is defined by the Künneth theorem and maps of coalgebras, ν is a morphism of the algebra structures defined by μ. This implies that:

Proposition 10.23.1. *The multiplications $*$ and $\diamondsuit$ are the same and both are graded commutative.*

Proof: Since ν is a morphism of the natural algebra structure $*$, we have the formula
$$\nu((\alpha \otimes \beta) * (\gamma \otimes \epsilon)) = \nu(\alpha \otimes \beta) * \nu(\gamma \otimes \epsilon)$$
which can also be written as
$$(-1)^{\deg(\gamma)\deg(\beta)}(\alpha * \gamma)\diamondsuit(\beta * \epsilon) = (\alpha\diamondsuit\beta) * (\gamma\diamondsuit\epsilon).$$
Use the fact that the two multiplications have a common two-sided unit. Hence
$$(\alpha * 1)\diamondsuit(1 * \epsilon) = (\alpha\diamondsuit 1) * (1\diamondsuit\epsilon)$$
$$\alpha\diamondsuit\epsilon = \alpha * \epsilon.$$
And hence
$$(-1)^{\deg(\gamma)\deg(\beta)}(1 * \gamma)\diamondsuit(\beta * 1) = (1\diamondsuit\beta) * (\gamma\diamondsuit 1)$$
$$(-1)^{\deg(\gamma)\deg(\beta)}\gamma\diamondsuit\beta = \beta * \gamma.$$
$\square$

Remarks. The above proof is essentially the same proof as one of the usual proofs used to show that the fundamental group of an H-space has two equal multiplications both of which are commutative.

The above has a slight extension to a geometric version. Suppose that we have a homotopy pullback diagram of simplicial monoids with strictly multiplicative maps:
$$\begin{array}{ccc} E & \to & \mathbb{Z} \\ \downarrow & & \downarrow \\ Y & \to & X \end{array}.$$
For example, this could be a homotopy pullback diagram of simplicial loop spaces. If we make the standard assumption that X is 1-reduced, then
$$\mathrm{Cotor}^X(Y, \mathbb{Z}) \otimes \mathrm{Cotor}^X(Y, \mathbb{Z})$$
$$\to \mathrm{Cotor}^{X \times X}(Y \times Y, \mathbb{Z} \times \mathbb{Z}) \to \mathrm{Cotor}^X(Y, \mathbb{Z})$$
defines an associative multiplication with unit
$$R = \mathrm{Cotor}^*(*, *) \to \mathrm{Cotor}^X(R, R).$$
The proof of the following is left as an exercise.

Proposition 10.23.2. *The map $HE \to \operatorname{Cotor}^X(Y, \mathbb{Z})$ is an isomorphism of algebras and, if HE is R projective, it is an isomorphism of Hopf algebras.*

Remarks. We can apply the above proposition to homotopy pullback diagrams of topological loop spaces as follows. Suppose the homotopy pullback E over the simply connected space X is the loop of the homotopy pullback $\overline{E}$ over the two-connected space $\overline{X}$. Apply the singular complex functor to $\overline{E}$ and consider the diagram which is the restriction to the second Eilenberg subcomplex of $\overline{X}$, that is, the maximal subcomplex whose two-skeleton is a point. Loop this via the Kan loop group construction and take the normalized chains.

The following lemma is also left as an exercise.

Lemma 10.23.3. *If C and D are (graded) commutative differential coalgebras, then the diagonal $\Delta : C \to C \otimes C$ and the twist map $T : C \otimes D \to D \otimes C$ are maps of differential coalgebras.*

Since maps of differential coalgebras induce maps of differential Cotor compatible with the natural multiplication, this lemma immediately gives

Proposition 10.23.4. *If C is a simply connected commutative differential coalgebra with $\operatorname{Cotor}^C(R, R)$ projective over R, then*

$$\operatorname{Cotor}^C(R, R) \xrightarrow{\operatorname{Cotor}^\Delta(R,R)} \operatorname{Cotor}^{C \otimes C}(R, R)$$
$$\cong \operatorname{Cotor}^C(R, R) \otimes \operatorname{Cotor}^C(R, R)$$

and

$$\operatorname{Cotor}^C(R, R) \to \operatorname{Cotor}^R(R, R) = R$$

give $\operatorname{Cotor}^C(R, R)$ the structure of a Hopf algebra with commutative diagonal and the natural multiplication.

Exercise

(1) Prove Proposition 10.23.2.

(2) Prove Lemma 10.23.3.

10.24 Fibrations which are totally nonhomologous to zero

Let $F \to E \to B$ be a fibration sequence and assume that F and B are connected and that the homologies HB and HF are R flat. Then the homology HE is a comodule over HB via the coaction

$$\Delta : HE \to H(E \times E) \to H(E \times B) \cong HE \otimes HB.$$

The following lemma is immediate.

Lemma 10.24.1. *The image of $HF \to HE$ is contained in the module of primitives of the comodule $P_{HB}(HE) = $ kernel $HE \to HE \otimes \overline{H}B$.*

Definition 10.24.2. A fibration sequence is totally nonhomologous to zero if $HF \to HE$ is an R split monomorphism.

Proposition 10.24.3. *If $F \to E \to B$ is totally nonhomologous to zero, then $HF \to P_{HB}(HE)$ is an isomorphism and there is an isomorphism of right HB comodules*

$$HE \to HF \otimes HB.$$

Proof: The following lemma is an exercise.

Lemma 10.24.4. *If C is a connected differential coalgebra and x is an element of least degree such that $dx \neq 0$, then dx is primitive.*

We claim that the homology Serre spectral sequence collapses. Since $E^2 = HF \otimes HB$ as differential coalgebras, the coalgebra module of primitives $PE^2 = PHF \oplus PHB$. Since elements in HB cannot be in the image of a differential, the only possibility for targets of nonzero differentials is in HF. But this is ruled out by the hypothesis of totally nonhomologous to zero. Hence, $E^2 = E^3$ and likewise

$$E^2 = E^3 = \cdots = E^\infty = HF \otimes HB$$

as coalgebras. Hence, the edge homomorphism $HF \to P_{HB}(E^\infty)$ is an isomorphism. Thus, the inclusion $HF \to P_{HB}(HE)$ is also an isomorphism.

The splitting $HE \to HF$ defines a unique map of HB comodules

$$HE \to HE \otimes HB \to HF \otimes HB$$

such that $P_{HB}(HE) \to P_{HB}(HF \otimes HB) = HF$ is an isomorphism. Since HB is R projective, it follows from Corollary 10.9.4 that $HE \to HF \otimes HB$ is a monomorphism of HB modules. But this is not enough. Filter HE by the filtration coming from the Serre filtration on the chains of E. Then the map of associated graded objects is the isomorphism

$$E^0 HE = E^\infty \to E^0(HF \otimes HB) = HF \otimes HB.$$

Hence, the map $HE \to HF \otimes HB$ is an isomorphism of comodules. $\square$

Definition 10.24.5. Let $f : B \to C$ be a morphism of connected coalgebras and assume that B is R flat. Then $A = B \square_C R = $ kernel $B \to B \otimes B \to B \otimes C \to B \otimes \overline{C}$ is called the coalgebra kernel of f.

We need a lemma to make some sense of this.

Lemma 10.24.6.

(a) *If B has a commutative diagonal, then A is a subcoalgebra of B.*

(b) *The composition $A \to B \to C$ factors through the ground ring R.*

(c) *If $g : D \to B$ is a morphism of coalgebras such that the composition factors as $f \cdot g : D \to R \to C$, then g factors through A.*

Proof:

(a) If the diagonal is commutative, then
$$A = B \square_C R = R \square_C B.$$
We need to show that $\Delta(A) \subseteq A \otimes A$.
Since
$$A \otimes A = B \otimes A \bigcap B \otimes A,$$
symmetry implies that it is sufficient to show that $\Delta(A) \subseteq B \otimes A$.
Let $x \in A$ and write
$$\Delta(x) = x \otimes 1 + 1 \otimes x + \Sigma x' \otimes x''$$
$$\Delta(x') = x' \otimes 1 + 1 \otimes x' + \Sigma y' \otimes y''.$$
Note that
$$B \otimes A \xrightarrow{\Delta} B \otimes B \xrightarrow{1 \otimes \overline{\Delta}} B \otimes \otimes \overline{C} \otimes B$$
is exact. Associativity gives the equality
$$f = (1 \otimes \overline{\Delta}) \cdot \Delta = (\overline{\Delta} \otimes 1) \cdot \Delta.$$
Since $\Sigma y' \otimes y''$ is a linear function of x', observe that
$$f(x) = x \otimes 1 \otimes 1 + 1 \otimes x \otimes 1 + 1 \otimes 1 \otimes x + \Sigma 1 \otimes x' \otimes x''$$
$$+ \Sigma(x' \otimes 1 \otimes x'' + 1 \otimes x' \otimes x'' + \Sigma y' \otimes y'' \otimes x'') = 0.$$
Hence, $\Delta(A) \subseteq A \otimes A$.

(b) It is clear that $\overline{A} \subseteq \text{kernel}(f)$.

(c) Since the composition $\overline{f} \cdot g : D \to B \to \overline{C}$ is 0 and since g is a map of coalgebras, it follows that $g(D) \subseteq A$. □

Hence we should use the notion of coalgebra kernel only if the diagonal is commutative. Let $f : B \to C$ be a morphism of connected Hopf algebras with commutative diagonals. Then the above coalgebra kernel A is actually a sub-Hopf algebra of B and is called the Hopf algebra kernel. We need to check:

Lemma 10.24.7. *A is a subalgebra of B.*

Proof: Let $x, y \in \overline{B}$ and write

$$\Delta(x) = x \otimes 1 + 1 \otimes x + \Sigma x' \otimes x''$$
$$\Delta(y) = y \otimes 1 + 1 \otimes y + \Sigma y' \otimes y''.$$

If $x \in A$, then $B \otimes B \to B \otimes C$ sends both $1 \otimes x$ and $\Sigma x' \otimes x''$ to 0. If $y \in A$ a similar statement holds.

We compute $\Delta(xy) = \Delta(x) \cdot \Delta(y)$ and see that $xy \in A$. Hence A is a subalgebra. $\square$

Proposition 10.24.8. *Let $F \to E \to B$ be a fibration sequence of connected simplicial monoids with B 1-reduced. Assume that there is a short exact sequence of graded Lie algebras*

$$0 \to L' \to L \to L'' \to 0$$

which are R free and such that the map $HE \to HB$ is isomorphic to the map of universal enveloping algebras $UL \to UL''$. Then there is an isomorphism of Hopf algebras

$$HF \to UL'.$$

Proof: Since there is an isomorhism of coalgebras

$$UL \cong UL' \otimes UL''$$

and of UL' modules, it follows that $UL \cong UL' \otimes UL''$ as UL'' comodules and that

$$\text{Cotor}_{-n}^{HB}(HE, R) = \text{Cotor}_{-n}^{UL''}(UL, R)$$

$$= \begin{cases} 0 & \text{if } n \geq 0 \\ HE \square_{HB} R = UL \square_{UL''} R = UL' & \text{if } n = 0 \end{cases}$$

Hence, the edge homomorphism of the Eilenberg–Moore spectral sequence

$$HF = \text{Cotor}^{CB}(CE, R) \to \text{Cotor}_0^{HB}(HE, R) = HE \square_{HB} R = UL'$$

is an isomorphism of Hopf algebras.

Note that one can be certain that $HF = UL'$ as Hopf algebras since it is embedded as a subHopf algebra of $HE = UL$. $\square$

Remark. Proposition 10.24.8 has the following algebraic version. Suppose that

$$A \to B \to C$$

is a sequence of maps of simply connected commutative differential coalgebras and that, for all $D = A, B, C$, we have that $\text{Cotor}^D(R, R)$ is R-projective. Then all $\text{Cotor}^D(R, R)$ are Hopf algebras with commutative diagonal. Suppose in addition that the above sequence is a sequence of differential Hopf algebras and that $B = A \otimes C$ as an A module and C comodule. Then Exercise 3 in Section 10.13 says that there is a twisting morphism $\tau : C \to A$ such that $B = A \otimes_\tau C$. Hence, the homology of the fibre is

$$HA = H(B \square_C R) = \text{Cotor}^C(B, R).$$

If, in addition, all HD are projective over R and there is a Hopf algebra X such that

$$HB = X \otimes HC$$

as Hopf algebras, then the edge homomorphism in the Eilenberg–Moore spectral sequence is an isomorphism

$$\text{Cotor}^C(B, R) \to \text{Cotor}_0^{HC}(HB, R) = HB \square_{HC} R = X$$

and hence

$$HA = \text{Cotor}^C(B, R) = X.$$

Since $HA = X$ embedds in the Hopf algebra HB, we know that $HA = X$ as Hopf algebras.

Exercises

(1) Prove Lemma 10.24.1.

(2) Prove Lemma 10.24.4.

10.25 Suspension in the Eilenberg–Moore models

Let $\Omega X \xrightarrow{\iota} PX \xrightarrow{\pi} X$ be the path space fibration sequence. Then the homology suspension is the map

$$\sigma : \overline{H}_{n-1}(\Omega X) \xrightarrow{\partial^{-1}} H_n(PX, \Omega X) \xrightarrow{\pi} \overline{H}_{n-1}(X).$$

The homology suspension provides an important connection between the Eilenberg–Moore models and geometry. Under some circumstances it can be used to compute the coalgebra structure in the homology of a loop space. In what follows, we shall relate this homology suspension to the Eilenberg–Moore spectral sequence and prove the result originally due to George Whitehead [134].

Proposition 10.25.1. *The homology suspension factors through the indecomposables of the loop space and the primitives of the base, that is, we have a factorization*

$$\sigma : \overline{H}(\Omega X) \to QH(\Omega X) \xrightarrow{\overline{\sigma}} PHX \subseteq \overline{H}(X).$$

The homology suspension is inverse to the transgression τ in the homology Serre spectral sequence in the sense that the following diagram commutes:

$$\begin{array}{ccc} H_{n-1}\Omega X & \xrightarrow{\sigma} & H_n X \\ \downarrow & & \cup | \\ E^n_{0,n-1} & \xleftarrow{\tau=d^n} & E^n_{n,0} \end{array}$$

Anticipating Proposition 10.25.1 for the moment, we shall define:

Definition 10.25.2. If the map $\overline{\sigma} : QH(\Omega X) \to PHX$ is an isomorphism, we shall say that the loop space is transgressive and, in this case, any lift of the transgression to a map $\tau : PHX \to \overline{H}(\Omega X)$ generates the homology of the loop space as an algebra. In general, elements in the image of the homology suspension are called transgressive elements.

Recall also that the homology suspension is related to the ordinary suspension ΣX of a space in the following manner.

The suspension of a space is defined by the cofibration sequence $X \to CX \to \Sigma X = CX/X$ where $CX = X \times I / * \times I \cup X \times 0$ is the cone and $X \equiv X \times 1$. This sequence maps to the path space fibration sequence $\Omega \Sigma X \to P\Sigma X \to \Sigma X$ via the three maps:

$$1 : \Sigma X \to \Sigma X,$$

$$CX \to P\Sigma X, \ \langle x,t \rangle \mapsto \omega_{\langle x,t \rangle},$$

where $\omega_{\langle x,t \rangle}(s) = \langle x,s \rangle$, $0 \le s \le t$ and $\omega_{\langle x,t \rangle}(s) = \langle x,t \rangle$, $t \le s \le 1$ is the path which stops at $\langle x,t \rangle$ and

$$\Sigma : X \to \Omega \Sigma X$$

is the map which is the suspension, $x \mapsto \omega_{\langle x,1 \rangle} = \langle x, \ \rangle$.

In particular, given a map $f : \Sigma X \to Y$, one gets a commutative diagram

$$\begin{array}{ccccc} X & \to & CX & \to & \Sigma X \\ \downarrow \Sigma & & \downarrow & & \downarrow = \\ \Omega \Sigma X & \to & P\Sigma X & \xrightarrow{\pi} & \Sigma X \\ \downarrow \Omega f & & \downarrow & & \downarrow f \\ \Omega Y & \to & PY & \xrightarrow{\pi} & Y \end{array}$$

Hence, if $g : X \to \Omega Y$ is the adjoint of f, then the composition $\sigma \cdot g : \overline{H}X \to \overline{H}\Omega Y \to \overline{H}Y$ is the same as the composition of Σf with the usual suspension isomorphism $\Sigma f \cdot s : \overline{H}X \to \overline{H}\Sigma X \to \overline{H}Y$. In particular, the Bott–Samelson theorem gives:

10.25 Suspension in the Eilenberg–Moore models 415

Proposition 10.25.3. *If HX is a free R module, then the loop space $\Omega\Sigma X$ is transgressive and the suspension $\Sigma : \overline{H}X \to H(\Omega\Sigma X)$ provides a lift of the transgression.*

It is clear from the above that a homology suspension arises whenever we have a sequence of augmented complexes $A \xrightarrow{\iota} B \xrightarrow{\pi} C$ (that is, the composition $\pi \cdot \iota = 0$ on the augmentation ideals) where A is a subcomplex of an acyclic complex B. The homology suspension given by the composition

$$\overline{H}A \xleftarrow{\partial \simeq} H(B, A) \xrightarrow{\pi} \overline{H}C$$

and is clearly a natural transformation on the category of such sequences.

In particular, the sequence of normalized chain complexes $C(\Omega X) \to C(PX) \to C(X) = C$ is such a sequence. If we embed the C comodule $C(PX)$ into an augmented acyclic construction $E = F \otimes C$ with fibre F and base C, we get another such sequence $F \to E \to C$ and a map of the first such sequence to this sequence. The map $C(\Omega X) \to F$ is a homology isomorphism.

Hence, the geometric version of the homology suspension defined above is isomorphic to this algebraic homology suspension $\sigma : \overline{H}F \to H(E, F) \to \overline{H}C$ defined for an acyclic construction $E = F \otimes C$.

In more detail, this clearly agrees with the following definition of the homology suspension:

Given $x \in F$ with $dx = 0$, the fact that E is acyclic means that we can write $x \otimes 1 = dz$ where $z = 1 \otimes c + w \otimes 1 + \Sigma y \otimes e$ and where $y, w \in \overline{F}$ and $e, c \in \overline{C}$.

If $\pi : E \to C$ is the natural map, we define the homology suspension of x to be the homology class of $\pi(z) = c$, that is, $\sigma[x] = [c] \in \overline{H}C$.

Finally, let C be an R free differential coalgebra and consider the Eilenberg–Moore spectral sequence associated to the differential Cotor, $\mathrm{Cotor}^C(R, R)$. If $\mathrm{Cotor}^C(R, R) = H(\Omega X)$, then we have the homology suspension originally considered. Or $\mathrm{Cotor}^C(R, R)$ could be HF for an acyclic construction $F \otimes C$ as above. In both cases, we want to relate these homology suspensions to the Eilenberg–Moore spectral sequence. We have:

Proposition 10.25.4. *The homology suspension is the composition*

$$\sigma : \mathrm{Cotor}^C(R, R) = F_{-1} \to F_{-1}/F_{-2} = E^\infty_{-1} \subseteq E^2_{-1}$$
$$= \mathrm{Cotor}^{HC}_{-1}(R, R) = PHC \subseteq \overline{H}C$$

where F_{-n} is the filtration induced by resolution degree.

To see this, we use the acyclic construction $\Omega C \otimes_\tau C$. Given a cycle x in ΩC, write as above

$$x \otimes 1 = dz$$

where $z = 1 \otimes c + w \otimes 1 + \Sigma y \otimes e$ and where $y, w \in \overline{\Omega C}$ and $e, c \in \overline{C}$. Reduce this equation mod F_{-2} and get

$$x \otimes 1 = dz \equiv s^{-1}c \otimes 1 + \Sigma(\pm s^{-1}c' \otimes c'') + 1 \otimes dc + dw \otimes 1$$

where $\overline{\Delta}(c) = \Sigma c' \otimes c''$ is the reduced diagonal. That is, $dc = 0$ and c is primitive in C. Thus

$$[x] \otimes 1 = [s^{-1}c] \otimes 1 = d(1 \otimes [c])$$

in $\Omega HC \otimes_\tau HC$. Thus, the homology suspension is as described in 10.25.4.

Remark. Note that $[s^{-1}c]$ is a lift of the transgression of the primitive cycle c.

We immediately get the factorization through the primitives in Whitehead's Proposition 10.25.1. On the other hand, the suspension factors through $QH(\mathrm{Cotor}^C(R, R))$ since the square of F_{-1} is contained in F_{-2}. This proves Proposition 10.25.1.

For future reference, we record the following property of the Eilenberg–Moore spectral sequence:

Proposition 10.25.5. *If $E^2 = \mathrm{Cotor}^{HC}(R, R)$ is generated as an algebra by homological degree -1, that is, by $E^2_{-1} = s^{-1}\mathrm{Cotor}^{HC}(R, R) = s^{-1}PHC$, and the homology suspension $\overline{\sigma}: Q\mathrm{Cotor}^C(R, R) \to PHC$ is surjective onto the primitives of the homology, then the Eilenberg–Moore spectral sequence collapses, $E^2 = E^\infty$, and $\mathrm{Cotor}^C(R, R)$ is transgressive, that is, $Q\mathrm{Cotor}^C(R, R) \to PHC$ is an isomorphism.*

Exercises

(1) Prove Proposition 10.25.5.

(2) Let Y be a simply connected space and suppose there is a map $f: \Sigma X \to Y$ which in homology is surjective onto the primitives. Show that, if E^2 is generated as an algebra by homological degree -1, then the Eilenberg–Moore spectral sequence which abuts to $H(\Omega Y)$ collapses, that is, $E^2 = E^\infty$.

10.26 The Bott–Samelson theorem and double loops of spheres

We are going to use the Eilenberg–Moore models to compute the homology of various loop spaces. Throughout this section, we are going to do computations of loop space homology, not with spectral sequences, but by replacing the normalized

10.26 The Bott–Samelson theorem and double loops of spheres

chains on the base space with homology isomorphic differential coalgebras with zero differential.

As a first example, we are going to use the Eilenberg–Moore models to give another proof of the Bott–Samelson theorem. We begin by showing that the normalized chains on any connected suspension with R free homology is homology isomorphic to a differential coalgebra with zero differential.

Lemma 10.26.1. *If X is connected and HX is R free, then there is a homology isomorphism of differential coalgebras $H\Sigma X \to C_N(\Sigma X)$.*

Proof: We may assume that X is a simplicial set with a unique 0-vertex. Form the model for the suspension $\Sigma X = x * X/X$, which is the quotient of the cone $CX = x * X$. Simplices in the cone are of two kinds, n-simplices τ in X and cones $x * \tau$ which are $n+1$-simplices. The face operators are the usual on τ and, on $x * \tau$, they are $d_0(x * \tau) = \tau$ and, for $i > 0$, $d_i(x * \tau) = x * d_i\tau$. There are similar formulas for the degeneracies.

A back $n + 1 - i$-face on an $n + 1$-simplex $x * \tau$ always involves d_0 unless it is a back $n + 1$ face. Hence, in ΣX, it is a point. Hence, in ΣX it is degenerate unless it has dimension $n + 1$ or 0. Hence, in ΣX, the Alexander–Whitney map

$$\Delta(x * \tau) = \Sigma_{i=0}^{n+1}{}_i(x * \tau) \otimes (x * \tau)_{n-i}$$
$$= {}_{n+1}(x * \tau) \otimes (x * \tau)_0 + {}_0(x * \tau) \otimes (x * \tau)_{n+1}$$
$$= (x * \tau) \otimes 1 + 1 \otimes (x * \tau).$$

Hence, the differential coalgebra $C_N(\Sigma X)$ is primitive, that is, it consists entirely of the unit and primitive elements. Since HX is free, we can pick an embedding $HX \subseteq \mathbb{Z}C_N(\Sigma X) \subseteq C_N(\Sigma X)$ which is clearly both a homology isomorphism and an inclusion of coalgebras. $\square$

Bott–Samelson theorem 10.26.2. *If X is a connected space and HX is R free, then there is an isomorphism of algebras*

$$H(\Omega\Sigma X) = T(\overline{H}X).$$

Furthermore, the Hopf algebra structure is determined by the suspension map $\Sigma : X \to \Omega\Sigma X$, that is, the inclusion $HX \to T(\overline{H})$ is a map of coalgebras.

Proof: Since HX is R free, the homological invariance of differential Cotor and Lemma 10.26.1 imply that

$$H(\Omega\Sigma X) = \mathrm{Cotor}^{C(\Sigma X)}(R, R) = \mathrm{Cotor}^{HX}(R, R)$$

as algebras. Since HX has zero differential and HX is a primitive coalgebra, the internal and external differentials in $\Omega H\Sigma X$ are both zero. Hence,

$$H(\Omega\Sigma X) = \Omega H\Sigma X = T(s^{-1}\overline{H}X)$$

as algebras where $s^{-1}: \overline{H}X \to \overline{H}\Sigma X$ is any lift of the transgression. Since $\Sigma: \overline{H}X \to H(\Omega\Sigma X)$ is a canonical lift of the transgression,

$$H(\Omega\Sigma X) = T(\overline{H}X)$$

as algebras.

Since $\Sigma: X \to \Omega\Sigma X$ is a map of spaces, it must be a map of coalgebras and the Hopf algebra structure follows. $\square$

Remark. It should be noted that, in the above proof, the algebra structure of $H(\Omega\Sigma X)$ came directly from the algebra structure of differential Cotor. This depended entirely on the coalgebra structure of the base $H(\Sigma X)$, which in this case was always trivial. But the result implies that the coalgebra structure of $H(\Omega\Sigma X)$ depends on the coalgebra structure of HX. To determine this coalgebra structure, geometric properties of the homology suspension had to be invoked.

We turn to the homology of the double loop spaces of spheres. We assert:

Proposition 10.26.3. *With any coefficients R, there is a homology isomorphism of differential coalgebras*

$$H(\Omega S^{m+1}) \to C(\Omega S^{m+1})$$

and hence

$$H(\Omega^2 S^{m+1}) = \mathrm{Cotor}^{H(\Omega S^{m+1})}(R, R)$$

as algebras.

Recalling that $H(\Omega S^{m+1}) = T(\iota)$, this proposition has the following corollaries which determine the mod p homology of the double loops on any sphere. We begin with the computation of the odd dimensional case.

Corollary 10.26.4. *Let p be a prime.*

(a) *If p is odd, there is an isomorphism of primitively generated Hopf algebras*

$$H(\Omega^2 S^{2n+1}; \mathbb{Z}/p\mathbb{Z}) = E(\tau_0, \tau_1, \tau_2, \dots) \otimes P(\sigma_1, \sigma_2, \dots)$$

where $\deg(\tau_i) = 2p^i n - 1$ and $\deg(\sigma_i) = 2p^i n - 2$. The homology suspensions are $\sigma\tau_i = \iota^{p^i}$ and $\sigma\sigma_i = 0$.

The mod p homology Bockstein spectral sequence has first Bockstein differentials given by $\beta^1 \tau_0 = 0$ and $\beta^1 \tau_i = \sigma_i$ for $i \geq 1$.

(b) *If $p = 2$, there is an isomorphism of primitively generated Hopf algebras*

$$H(\Omega^2 S^{m+1}; \mathbb{Z}/2\mathbb{Z}) = P(\xi_0, \xi_2, \dots)$$

where $\deg(\xi_i) = 2^i m - 1$. The homology suspensions are $\sigma\xi_i = \iota^{2^i}$.

10.26 The Bott–Samelson theorem and double loops of spheres

If m is even, the mod 2 homology Bockstein spectral sequence has first Bockstein differentials given by $\beta^1 \xi_0 = 0$ and $\beta^1 \xi_i = (\xi_i)^2$ for $i \geq 1$.

The Bockstein spectral sequence enables us to determine the order of the torsion in the integral homology. This is a consequence of the fact that the first Bockstein differentials in Corollary 10.26.4 make the E^1 term of the homology Bockstein spectral sequence into an essential acyclic Hopf algebra. For p odd, it is a tensor product of $E(\tau_0)$ and the acyclic subHopf algebras $E(\tau_i) \otimes P(\sigma_i)$ for $i \geq 1$. For $p = 2$ and m odd, introduce the two stage filtration which starts with all even degrees and ends with everything. Then the associated graded object of E^1 is a tensor product of $E(\xi_0)$ and the acylic subHopf algebras $E(\xi_i) \otimes P(\xi_{i-1}^2)$ for $i \geq 1$.

Corollary 10.26.5. *If p is any prime and we write the homology localized at p as*

$$\overline{H}(\Omega^2 S^{2n+1}) = \mathbb{Z}_{(p)} \oplus V$$

then $pV = 0$, that is, the integral homology of this double loop space has all p torsion of order exactly p.

Finally, the double loop space on even dimensional spheres reduces to the odd dimensional case:

Corollary 10.26.6. *If p is any prime, then there is an isomorphism of primitively generated Hopf algebras over $\mathbb{Z}/p\mathbb{Z}$*

$$H(\Omega^2 S^{2n+2}) = H(\Omega S^{2n+1}) \otimes H(\Omega^2 S^{4n+3}).$$

Proof of 10.26.3: Let x be a primitive cycle in the normalized chains $C(\Omega S^{m+1})$ which represents the generator ι of the homology $H(\Omega S^{m+1}) = T(\iota)$. Such exists since it is contained in the $m-1$ connected Eilenberg subcomplex which is homotopy equivalent to the simplicial set of singular simplices of ΩS^{m+1}. Now define a homology isomorphism of differential coalgebras

$$\theta : T(\iota) \to C(\Omega S^{m+1})$$

by $\theta(\iota^k) = x^k$ using the fact that the chains are an associative differential Hopf algebra. This map is clearly a homology isomorphism of differential coalgebras.

Remark. Let G be the chains on an associative H-space and recall the fact that the Eilenberg–Zilber map $\nabla : G \otimes G \to G \times G$ is a map of differential coalgebras. Hence, the composition $G \otimes G \xrightarrow{\nabla} G \times G \xrightarrow{\mu} G$ gives G the structure of a differential Hopf algebra.

Since the homology $H(\Omega S^{m+1})$ is free over R, we have

$$H(\Omega^2 S^{m+1}) = \operatorname{Cotor}^{C(\Omega S^{m+1})}(R, R) = \operatorname{Cotor}^{H(\Omega S^{m+1})}(R, R)$$

as algebras. This proves 10.26.3.

Proof of 10.26.4: Given an element x and a positive integer k, let
$$T_k(x) = \langle 1, x, x^2, \ldots, x^{k-1} \rangle \subseteq T(x)$$
be the subcoalgebra of the tensor Hopf algebra spanned by the powers $1, x, x^2, \ldots, x^{k-1}$. Observe that:

Lemma 10.26.7. *Let the coefficients be $\mathbb{Z}/p\mathbb{Z}$ where p is a prime.*

(a) *If p is odd, then*
$$H(\Omega S^{2n+1}) = T(\iota) = \bigotimes_{k=0}^{\infty} T_p(\iota^{p^k})$$
as coalgebras.

(b) *If $p = 2$, then*
$$H(\Omega S^{m+1}) = T(\iota) = \bigotimes_{k=0}^{\infty} T_2(\iota^{2^k})$$
as coalgebras.

The proof of the above lemma is immediate. For example, when p is odd, observe that each $T_p(\iota^{p^k})$ is a subcoalgebra of $T(\iota)$ and note that writing $n = a_k p^k + a_{k-1} p^{k-1} + \cdots + a_0$ in terms of its dyadic expansion and then using this to express x^n as a product says that multiplication gives an isomorphism
$$\bigotimes_{k=0}^{m} T_p(\iota^{p^k}) \to T(\iota)$$
on each finite tensor product. The map is a map of coalgebras since $T(\iota)$ is a Hopf algebra. The isomorphism extends to the infinite tensor product since it is the union of the finite tensor products.

The Künneth theorem implies that with $\mathbb{Z}/p\mathbb{Z}$ coefficients:

Lemma 10.26.8.

(a) *If p is odd, then*
$$H(\Omega^2 S^{2n+1}) = \bigotimes_{k=0}^{\infty} \mathrm{Cotor}^{T_p(\iota^{p^k})}(\mathbb{Z}/p\mathbb{Z}, \mathbb{Z}/p\mathbb{Z})$$
as algebras.

(b) *If $p = 2$, then*
$$H(\Omega^2 S^{m+1}) = \bigotimes_{k=0}^{\infty} \mathrm{Cotor}^{T_2(\iota^{2^k})}(\mathbb{Z}/2\mathbb{Z}, \mathbb{Z}/2\mathbb{Z})$$
as algebras.

10.26 The Bott–Samelson theorem and double loops of spheres

Caution should be taken with the coalgebra structures above. It might be that these are not isomorphisms of Hopf algebras. It turns out that they in fact are isomorphisms of Hopf algebras but, until we know better, it could be otherwise.

The basic computation is:

Lemma 10.26.9. *As an algebra:*

(a) *If p is an odd prime and x has even degree $2n$, then*
$$\operatorname{Cotor}^{T_p(x)}(\mathbb{Z}/p\mathbb{Z}, \mathbb{Z}/p\mathbb{Z}) = E(s^{-1}x) \otimes P(z)$$
where $\deg(s^{-1}x) = 2n - 1, \deg(z) = 2pn - 2$ and the homology suspensions are $\sigma(s^{-1}x) = x$, $\sigma(z) = 0$.

(b) *If $p = 2$ and x has arbitrary degree m, then*
$$\operatorname{Cotor}^{T_2(x)}(\mathbb{Z}/2\mathbb{Z}, \mathbb{Z}/2\mathbb{Z}) = P(s^{-1}x)$$
where $\deg(s^{-1}x) = m - 1$, and the homology suspension is $\sigma(s^{-1}x) = x$.

This lemma proves the part of Corollary 10.26.4 that is concerned with the algebra structure. We proceed to prove this lemma using duality.

If C is a finite type commutative coalgebra which is free over R, then the dual $A = C^*$ is a commutative algebra and we have that $\operatorname{Tor}_A(R, R)$ is a Hopf algebra which is dual to the Hopf algebra $\operatorname{Cotor}^C(R, R)$. If we use $\mathbb{Z}_{(p)}$ coefficients and set $C = T_p(x)$ we get that $A = P_p(y) = P(y)/(y^p) = $ the truncated polynomial algebra with $\deg(y) = \deg(x)$.

With $\mathbb{Z}_{(p)}$ coefficients, consider the complex
$$\mathcal{R} = A \otimes E(s) \otimes \Gamma(t)$$
with $\deg(s) = \deg(x) - 1$, $\deg(t) = p \deg(x)$ and differential the derivation defined by
$$dy = 0, \ ds = y, \ dt = sy^{p-1}.$$

$\mathcal{R}$ is an acyclic complex and a differential algebra. It provides a resolution of $\mathbb{Z}_{(p)}$ over the algebra A and the multiplication is a map of resolutions
$$\mathcal{R} \otimes \mathcal{R} \to \mathcal{R}$$
and induces a multiplication on
$$\mathcal{R} \otimes_A \mathbb{Z}_{(p)} = E(s) \otimes \Gamma(t).$$

Since the latter has zero differential,
$$\operatorname{Tor}_A(\mathbb{Z}_{(p)}, \mathbb{Z}_{(p)}) = E(s) \otimes \Gamma(t)$$
as algebras.

If p is odd, both the elements s and t must be primitive for degree reasons. Hence, in this case, we have

$$\operatorname{Tor}_A(\mathbb{Z}_{(p)}, \mathbb{Z}_{(p)}) = E(s) \otimes \Gamma(t)$$

as Hopf algebras. It follows that

$$\operatorname{Cotor}^{T_p(x)}(\mathbb{Z}_{(p)}, \mathbb{Z}_{(p)}) = E(s^{-1}x) \otimes P(z)$$

as Hopf algebras and the homology suspensions are as indicated. Of course, we can reduce this result mod p to the desired result with $\mathbb{Z}/p\mathbb{Z}$ coefficients.

If $p = 2$, then we use $\mathbb{Z}/2\mathbb{Z}$ coefficients and make $\mathcal{R}$ into a differential Hopf algebra by making y and s primitive but setting

$$\Delta(t) = t \otimes 1 + s \otimes s + 1 \otimes t.$$

It follows that, mod 2,

$$\operatorname{Tor}_A(\mathbb{Z}/2\mathbb{Z}, \mathbb{Z}/2\mathbb{Z}) = E(s) \otimes \Gamma(t)$$

as Hopf algebras with s primitive but with $\Delta(t) = t \otimes 1 + s \otimes s + 1 \otimes t$. We can lift this result up to the Hopf algebra with coefficients $\mathbb{Z}_{(2)}$ and get that over this ring, we also have

$$\operatorname{Tor}_A(\mathbb{Z}_{(2)}, \mathbb{Z}_{(2)}) = E(s) \otimes \Gamma(t)$$

as Hopf algebras with s primitive but with $\Delta(t) = t \otimes 1 + s \otimes s + 1 \otimes t$. It follows that this Hopf algebra structure is completely determined and that

$$\operatorname{Cotor}^{T_p(x)}(\mathbb{Z}_{(2)}, \mathbb{Z}_{(2)}) = P(s^{-1}x).$$

As before, we complete the computation by reducing the result mod 2.

The fact that the Hopf algebras in Corollary 10.26.4 are primitively generated follows from a result of Milnor and Moore which we do not prove here.

Proposition 10.26.10. *Suppose that B is a connected Hopf algebra over $\mathbb{Z}/p\mathbb{Z}$ with commutative multiplication. Let $\xi : B \to B$ be the p-th power operation $\xi(x) = x^p$ and consider the subHopf algebra $\xi B \subseteq B$ of all p-th powers. There is an exact sequence*

$$0 \to P(\xi B) \to P(B) \to Q(B).$$

In particular, the kernel of the map $P(B) \to Q(B)$ is concentrated in degrees divisible by p.

The dual of this proposition implies that all the generators in Corollary 10.26.4 are primitive.

We complete the proof of Corollary 10.26.4 by determining the first Bockstein differentials β^1. We start with an explicit computation over $\mathbb{Z}_{(p)}$.

10.26 The Bott–Samelson theorem and double loops of spheres

Lemma 10.26.11. *For all primes p,*
$$H_{2pn-2}(\Omega^2 S^{2n+1}; \mathbb{Z}_{(p)}) = \mathbb{Z}/p\mathbb{Z}.$$

Proof: We use the cohomology version of the Eilenberg–Moore model, that is,
$$H^*(\Omega^2 S^{2n+1}; \mathbb{Z}_{(p)}) = \mathrm{Tor}_{H^*(\Omega S^{2n+1})}(\mathbb{Z}_{(p)}, \mathbb{Z}_{(p)})$$
where $H^*(\Omega S^{2n+1}) = \Gamma(y) = A$ with $\deg(y) = 2n$.

Let $\mathcal{R}$ be the free A resolution of $\mathbb{Z}_{(p)}$ which is the differential algebra
$$A \otimes E(s) \otimes E(t) \otimes \Gamma(w) \otimes \cdots$$
with differentials
$$dy = 0, \ d(\gamma_i(y)) = 0,$$
$$ds = y, \ d(s\gamma_i(y)) = \gamma_1(y)\gamma_i(y) = (i+1)\gamma_{i+1},$$
$$d(t) = \gamma_p(y), \ d(w) = pt - s\gamma_{p-1}, \cdots$$
and degrees
$$\deg(y) = 2n, \ \deg(s) = 2n-1, \ \deg(t) = 2pn-1, \ \deg(w) = 2pn-2, \ldots.$$
The generators y, t and w are all introduced to kill cycles which occur in the resolution.

Then $H^*(\Omega^2 S^{2n+1})$ with $\mathbb{Z}_{(p)}$ coefficients is the homology of
$$\mathbb{Z}_{(p)} \otimes_A \mathcal{R} = E(s) \otimes E(t) \otimes \Gamma(w) \otimes \cdots$$
with differentials
$$d(s) = 0, \ d(t) = 0, \ d(w) = pt, \ \cdots.$$
It follows that
$$H^{2pn-1}(\Omega^2 S^{2n+1}; \mathbb{Z}_{(p)}) = \mathbb{Z}/p\mathbb{Z}$$
and thus
$$H_{2pn-2}(\Omega^2 S^{2n+1}; \mathbb{Z}_{(p)}) = \mathbb{Z}/p\mathbb{Z}.$$
$\square$

Lemma 10.26.11 is just another way of saying that the first Bockstein differentials in the mod p homology of $\Omega^2 S^{2n+1}$ are
$$\beta^1 \tau_1 = \sigma_1 \quad \text{if } p \text{ is odd}$$
$$\beta^1 \xi_1 = \xi_0^2 \quad \text{if } p = 2 \text{ and } m \text{ is even.}$$

Now consider the Hopf invariant fibration sequence

$$J_{p-1}(S^{2n}) \to \Omega S^{2n+1} \xrightarrow{h_p} \Omega S^{2pn+1}$$

of spaces localized at p. Recall that

$$J_{p-1}(S^{2n}) = S^{2n} \cup e^{4n} \cup \cdots \cup e^{2(p-1)n}$$

(so that, if $p = 2$, $J_1(S^{2n}) = S^{2n}$) and that the mod p homology of the Hopf fibration sequence is

$$T_p(\iota) \to T(\iota) \to T(\iota^p).$$

Since $T(\iota) = T_p(\iota) \otimes T(\iota^p)$ as coalgebras, the fibration is totally nonhomologous to zero and

$$H(\Omega S^{2n+1}) = H(J_{p-1}(S^{2n})) \otimes H(\Omega S^{2pn+1})$$

with mod p coefficients. Since $J_{p-1}(S^{2n})$ is a skeleton of ΩS^{2n+1}, we observe that the mod p homology isomorphism of differential coalgebras

$$T(\iota) = H(\Omega S^{2n+1}) \to C\Omega S^{2n+1})$$

restricts to a mod p homology isomorphism of coalgebras

$$T_p(\iota) = H(J_{p-1}(S^{2n})) \to C(J_{p-1}(S^{2n})).$$

Hence, in the next lemma, all the homologies of the loop spaces are given by Cotor with respect to the homologies of the bases and we get

Lemma 10.26.12. *As algebra with mod p coefficients*

$$H(\Omega^2 S^{2n+1}) = H(|\Omega J_{p-1}(S^{2n})) \otimes H(\Omega^2 S^{2pn+1})$$

as algebras and the loops on the p-th Hopf invariant

$$\Omega h_p : \Omega^2 S^{2n+1} \to \Omega^2 S^{2np+1}$$

induces in homology the projection on the second tensor factor.

If p is odd, the above states that

$$H(\Omega^2 S^{2n+1}) = \{E(\tau_0) \otimes P(\sigma_1)\} \otimes \left\{ E(\tau_1) \otimes \bigotimes_{i=2}^{\infty} E(\tau_i) \otimes P(\sigma_i) \right\}.$$

where the projection on the tensor factors beginning with $E(\tau_1)$ is induced by the loops on the Hopf invariant. We know that $\beta^1 \tau_1 = \sigma_1$ in the case of $\Omega^2 S^{2n+1}$. Since we know this result also for $\Omega^2 S^{2pn+1}$, we know that $\beta^1 \tau_2$ must project to σ_2. But these are primitive classes and so the choice is unique, $\beta^1 \tau_2 = \sigma_2$ in the case of $\Omega^2 S^{2n+1}$ also. Induction shows that $\beta^1 \tau_i = \sigma_i$ for all $i \geq 1$.

If $p = 2$ and m is even, a similar proof shows that $\beta^1 \xi_i = \xi_{i-1}^2$ for all $i \geq 1$.

This completes the proof of Corollary 10.26.4 when m is even.

Finally, we prove the result in Corollary 10.26.6 on double loop spaces of even dimensional spheres. If $p = 2$, this is already done in Corollary 10.26.4. If p is odd, then 10.26.6 follows from the fact that

$$\Omega S^{2n+2} \simeq S^{2n+1} \times \Omega S^{4n+3}$$

localized at an odd prime p.

Exercise

(1) (a) If p is an odd prime, then show that, with respect to the first Bockstein differential β^1, the isomorphism of mod p homology

$$H(\Omega^2 S^{2n+2}) = H(\Omega S^{2n+1}) \otimes H(\Omega^2 S^{4n+3})$$

is an isomorphism of differential Hopf algebras.

(b) If $p = 2$, show that

$$H(\Omega^2 S^{2n+2}) = H(\Omega S^{2n+1}) \otimes H(\Omega^2 S^{4n+3})$$

is an isomorphism of Hopf algebras and that, in the notation of Corollary 10.26.4, the Bockstein differentials are given by

$$\beta^1 \xi_0 = \beta^1 \xi_1 = 0$$

and for $i \geq 2$, $\beta^1 \xi_i = \xi_{i-1}^2$ modulo the ideal generated by ξ_0. (Hint: Use the computation of rational homology and the Hopf fibration sequence

$$S^{2n+1} \to \Omega S^{2n+2} \to \Omega S^{4n+3}.)$$

10.27 Special unitary groups and their loop spaces

In this section we compute the homology of some of the special unitary groups and then use this to compute the homology of their loop spaces. It is very useful to find a generating complex for the Lie group if there is one. Loosely speaking, a generating complex is a subspace whose reduced homology generates the homology Pontrjagin ring of the Lie group. It is often the case that the homology of the Lie group is the exterior algebra generated by the homology of the generating complex. The generating complex has two uses. First, it controls the coalgebra structure of the homology of the Lie group and, second, via the coalgebra structure and the homology suspension, it enables the computation of the homology of the loop space.

First consider the case of the unitary group and the special unitary group. Let V be a finite dimensional vector space over the complex number field C and suppose we are given a positive definite Hermitian form on V, that is, for all x, $y \in V$, there is a complex number $\langle x, y \rangle$ which is biadditive,

$$\langle x_1 + x_2, y \rangle = \langle x_1, y \rangle + \langle x_2, y \rangle,$$

conjugate symmetric, for all x,

$$\langle y, x \rangle = \overline{\langle x, y \rangle},$$

bilinear, for all complex numbers α,

$$\langle \alpha x, y \rangle = \alpha \langle x, y \rangle$$

and positive definite

$$\langle x, x \rangle \geq 0$$

with

$$\langle x, x \rangle = 0$$

only if $x = 0$.

Definition 10.27.1. The unitary group $U(V)$ is the group of all invertible linear operators T on V which preserve the Hermitian form, that is, $\langle Tx, Ty \rangle = \langle x, y \rangle$ for all x, $y \in V$. The special unitary group $SU(V)$ is the subgroup of $U(V)$ consisting of all unitary transformations of determinant 1.

In the case $x = (x_1, \ldots, x_n)$, $y = (y_1, \ldots, y_n) \in V = C^n$, we set

$$\langle x, y \rangle = \Sigma_{i=1}^n x_i \overline{y_i}$$

and write

$$U(n) = U(C^n), \quad SU(n) = SU(C^n).$$

In particular, the standard basis $e_1, \ldots, e_n$ for C^n is orthonormal

$$\langle e_i, e_j \rangle = \delta_{ij}.$$

The generating complexes for the special unitary groups are given by the suspensions of the complex projective spaces.

Definition 10.27.2. The complex projective space $CP(V)$ is the space of all complex lines in V. If x is a nonzero vector in V, we shall use $\langle x \rangle = \{\lambda x \mid \lambda \in C\}$ to denote the complex line spanned by x. We write $CP^{n-1} = CP(C^n)$.

Given a unit complex number $u \in S^1 \subseteq C$, $|u| = 1$ and a complex line $\langle x \rangle$ represented by a unit vector x in V, we define a unitary operator $T(u, \langle x \rangle)$ in $U(V)$ by

$$T(u, \langle x \rangle)(y) = [u \langle y, x \rangle x] + [y - \langle y, x \rangle x],$$

10.27 Special unitary groups and their loop spaces

that is, $T(u, \langle x \rangle)$ multiplies the x component of y by the scalar u and is the identity on the component of y which is orthogonal to x. Clearly, the determinant of $T(u, \langle x \rangle)$ is u. In order to get a continous map

$$\Theta : S^1 \times CP^{n-1} \to SU(n),$$

we define

$$\Theta(u, \langle x \rangle) = T(u, \langle x \rangle) \cdot T(u^{-1}, \langle e_1 \rangle).$$

We note that we have a factorization

$$\Theta = \overline{\Theta} \cdot \pi : S^1 \times CP^{n-1} \to S^1 \wedge CP^{n-1} \to SU(n)$$

and we claim that $\overline{\Theta}$ is the embedding of a generating complex.

Consider the principal bundles

$$SU(n-1) \to SU(n) \xrightarrow{\pi} S^{2n-1}$$

where $n \geq 2$ and $\pi(T) = Te_n$: The following two propositions show that $S^1 \wedge CP^{n-1}$ is a generating complex for $SU(n)$ with any coefficients:

Proposition 10.27.3. *The map $\pi : SU(n) \to S^{2n-1}$ restricts to a factorization*

$$S^1 \wedge CP^{n-1} \to S^1 \wedge CP^{n-1}/S^1 \wedge CP^{n-2} \to S^{2n-1}$$

where the second map is a homeomorphism.

Proposition 10.27.4. *With any coefficients, the homology*

$$HSU(n) = E[x_3, \ldots, x_{2n-1}]$$

as an algebra where $\deg(x_{2i-1}) = 2i - 1$ and the map $\pi : SU(n) \to S^{2n-1}$ sends x_{2n-1} to a generator of the homology of the sphere.

The two propositions above can be combined so that the Hopf algebra structure is determined.

Corollary 10.27.5. *With any coefficients, there is an isomorphism of Hopf algebras*

$$E[\overline{H}(S^1 \wedge CP^{n-1})] \to HSU(n)$$

where $E[\overline{H}(S^1 \wedge CP^{n-1})]$ is the exterior algebra with the coalgebra structure being the unique multiplicative structure given by the diagonal on the generators in $H(S^1 \wedge CP^{n-1})$. But, since $S^1 \wedge CP^{n-1}$ is a suspension, this is a primitively generated Hopf algebra.

Proof of Proposition 10.27.3: First we note that $n \geq 2$ implies

$$\pi \cdot \Theta(u, \langle x \rangle) = [u \langle e_n, x \rangle x + [e_n - \langle e_n, x \rangle x]$$
$$= (u - 1) \langle e_n, x \rangle x + e_n,$$

from which it follows that the inverse image of e_n is precisely when $u = 1$ or $x \perp e_n$, that is, it consists of

$$1 \times CP^{n-1} \cup S^1 \times CP^{n-2}.$$

Since the domain is compact,

$$S^1 \wedge CP^{n-1}/S^1 \wedge CP^{n-2} \to S^{2n-1}$$

will be a homeomorphism if and only if it is one-to-one and onto away from the basepoint.

To see that it is one-to-one, let $u \neq 1$ and $\langle x, e_n \rangle \neq 0$, $|x| = 1$ and suppose that

$$e_n + (u - 1)\langle e_n, x \rangle x = e \neq e_n.$$

Then

$$(u - 1)\langle e_n, x \rangle x = e - e_n$$

and the line $\langle x \rangle = \langle e - e_n \rangle$ is determined by e.

That is,

$$x = \alpha(e - e_n)$$

from which we get

$$|\alpha|^2 = \frac{1}{|e - e_n|^2}.$$

Substituting these into the previous equation we get

$$u = 1 + \frac{|e - e_n|^2}{\langle e_n, e \rangle - 1}$$

and thus u is determined by e. Thus the map is one-to-one away from the basepoint.

To see that the map is onto away from the basepoint, let $e \neq e_n$ $|e| = 1$ and define u and x by the formulas

$$x = \alpha(e - e_n), \ |\alpha|^2 = \frac{1}{|e-e_n|^2}$$

$$u = 1 + \frac{|e-e_n|^2}{\langle e_n, e \rangle - 1}.$$

Clearly, $\langle e_n, x \rangle \neq 0$ and $|x| = 1$, $u \neq 1$. If it is in the domain, it is clear that $\pi \cdot \Theta(u, \langle x \rangle) = e$. So it suffices to show that $|u| = 1$. Write $u = 1 + z$ so that

$$|u|^2 = u\bar{u} = 1 + z + \bar{z} + z\bar{z}.$$

The verification that $z + \bar{z} + z\bar{z} = 0$ is a simple exercise and finishes the proof that the map is onto. $\square$

10.27 Special unitary groups and their loop spaces

Proof of Proposition 10.27.4 and Corollary 10.27.5: Let $x_1, \ldots, x_{2n-1}$ be a basis for $\overline{H}CP^{n-1}$ with x_{2i-1} in dimension $2i - 1$. The inductive statement is to show that

$$HSU(n) = E[x_{2n-1}] \otimes HSU(n - 1)$$

as Hopf algebras.

Since $SU(1) = \{e\}$, it is trivially true for $n = 1$. Assume that the result is true for $n - 1$ and that $n \geq 2$.

Since $SU(n - 1) \to SU(n) \xrightarrow{\pi} S^{2n-1}$ is a principal bundle, the homology Serre spectral sequence is a spectral sequence of $HSU(n - 1)$ modules and also of coalgebras. Hence, the existence of a generating complex immediately gives that the homology Serre spectral sequence collapses, that is, $E^2 = E^\infty$, and hence that

$$HSU(n) = E[x_{2n-1}] \otimes HSU(n - 1)$$

as $HSU(n - 1)$ modules. (This can also be seen by the fact that $d^{2n-1}x_{2n-1} \in H_{2n-2}SU(n-1))$ must be a primitive element and that there are none in that dimension.)

Hence, the cohomology Serre spectral sequence (which is the dual of the homolgy Serre spectral sequence) also collapses and

$$H^*SU(n) = E[z_{2n-1}] \otimes H^*SU(n - 1)$$

as $E[z_{2n-1}]$ modules. Because of the graded commutativity of the cohomology ring, this will be an isomorphism of algebras if we know that $(z_{2n-1})^2 = 0$. But graded commutativity implies that $2z^2 = 0$ for odd dimensional cohomology classes. Since there is no 2 torsion, $(z_{2n-1})^2 = 0$ and

$$H^*SU(n) = E[z_{2n-1}] \otimes H^*SU(n - 1)$$

as algebras.

Dually,

$$HSU(n) = E[x_{2n-1}] \otimes HSU(n - 1)$$

as coalgebras. In particular, they are all primitively generated. Now, in order to show that this is an isomorphism of Hopf algebras, it suffices to check that

$$[x_{2n-1}, x_{2j-1}] = x_{2n-1}x_{2j-1} + x_{2j-1}x_{2n-1} = 0 \quad \text{for } j \leq n$$

and

$$x_{2n-1}^2 = 0$$

Since these are even dimensional primitive elements, they must be zero.

This proves Proposition 10.27.4 and Corollary 10.27.5. □

430 Differential homological algebra

The fact that the special unitary group has a suspension as its generating complex makes the computation of the homology of its loop space an immediate consequence of the Eilenberg–Moore spectral sequence. Recall the commutative diagram related to the suspension

$$\begin{array}{ccccc} CP^{n-1} & \to & C(CP^{n-1}) & \to & \Sigma CP^{n-1} \\ \downarrow & & \downarrow & & \downarrow \\ \Omega SU(n) & \to & PSU(n) & \to & SU(n) \end{array}$$

where the left and right vertical maps are adjoints.

In the Eilenberg–Moore spectral sequence with any coefficients R, we have

$$E^2_{-p} = \mathrm{Cotor}^{HSU(n)}(R,R) = \mathrm{Cotor}^{E[x_3,\ldots,x_{2n-1}]}(R,R) = P[y_2,\ldots,y_{2n-2}]$$

as algebras where the y_{2i} have total degree $2i$, homological degree -1 and the homology suspensions are $\sigma(y_{2i}) = x_{2i+1}$. Since E^2 is generated as an algebra by homological degree -1 and since the homology suspension is surjective onto the primitives, it follows that $E^2 = E^\infty$ in the Eilenberg–Moore spectral sequence. Since $\Omega SU(n)$ is homotopy commutative, so is its Pontrjagin ring, there are no extension problems, and hence

Proposition 10.27.6. *With any coefficients,*

$$H\Omega SU(n) = P[y_2,\ldots,y_{2n-2}] = P[\overline{H}CP^{n-1}]$$

as (nonprimitively generated) Hopf algebras, that is, the diagonal is given on generators by

$$\Delta(y_{2k}) = \Sigma_{i+j=k}\, y_{2i} \otimes y_{2j}$$

where $y_0 = 1$.

Note that it is a generating complex which determines the coalgebra structure in the above.

10.28 Special orthogonal groups

The homology of the special orthogonal group looks very different depending on whether we are localized at 2 or away from 2.

In the first case, we choose mod 2 coefficients and, much like the previous case of the special unitary group, the special orthogonal group has a generating complex which is a projective space, not the suspension of one as in the previous case. This means that the mod 2 homology of the special orthogonal group is not primitively generated.

10.28 Special orthogonal groups

Let V be a finite dimensional real vector space with a positive definite symmetric bilinear form $\langle x, y \rangle$. Then the orthogonal group $O(V)$ is the group of all invertible linear operators T on V which preserve this form, that is, $\langle Tx, Ty \rangle = \langle x, y \rangle$ for all x, y in V, and the special orthogonal group $SO(V)$ is the subgroup of all such linear operators of determinant one. We write $O(n)$ and $SO(n)$ for the case when $V = R^n$ with the standard positive definite form

$$\langle x, y \rangle = \Sigma_{i=1}^{n} x_i y_i$$

where $x = (x_1, \ldots, x_n)$ and $y = (y_1, \ldots, y_n)$.

We define for $n \geq 1$ continuous maps $T : RP^{n-1} \to O(n)$ by

$$T(\langle x \rangle)(y) = -\langle y, x \rangle x + [y - \langle y, x \rangle x = -2\langle y, x \rangle x + y$$

where x is a unit vector which represents the real line $\langle x \rangle$ spanned by x and y is any vector in R^n. Thus, $T(\langle x \rangle)$ is multiplication by -1 on the line $\langle x \rangle$ and is the identity on the orthogonal complement.

If $n \geq 2$ and $e_1, \ldots, e_n$ is the standard orthonormal basis, we define $\Theta : RP^{n-1} \to SO(n)$ by

$$\Theta(\langle x \rangle) = T(\langle x \rangle) \cdot T(\langle e_1 \rangle).$$

For $n \geq 2$ let $\pi : SO(n) \to S^{n-1}$ be the map defined by $\pi(T) = T(e_n)$ and we have the principal bundle sequence $SO(n-1) \to SO(n) \xrightarrow{\pi} S^{n-1}$.

We have the following analogues of 10.27.3, 10.27.4, and 10.27.5.

Proposition 10.28.1. *The map* $\pi : SO(n) \to S^{n-1}$ *restricts to a factorization*

$$RP^{n-1} \to RP^{n-1}/RP^{n-2} \to S^{n-1}$$

where the second map is a homeomorphism.

Since this is a simpler variation of Proposition 10.27.3, we leave this proposition as an exercise.

Proposition 10.28.2. *With mod 2 coefficients, the homology*

$$HSO(n) = E[x_1, \ldots, x_{n-1}]$$

as an algebra where $\deg(x_i) = i$ *and the map* $\pi : SO(n) \to S^{n-1}$ *sends* x_{n-1} *to a generator of the homology of the sphere.*

We shall prove Proposition 10.28.2 but first we note that, since RP^{n-1} is a generating complex for the mod 2 homology of $SO(n)$, it follows that Propositions 10.28.1 and 10.28.2 have as an immediate consequence the Hopf algebra structure of the mod 2 homology of $SO(n)$.

Corollary 10.28.3. *With mod 2 coefficients, there is an isomorphism of Hopf algebras*

$$E[\overline{H}RP^{n-1}] \to HSO(n)$$

where $E[\overline{H}RP^{n-1}]$ is the exterior algebra with the coalgebra structure being the unique multiplicative structure given by the diagonal on the generators in HRP^{n-1}. Thus,

$$\Delta(x_k) = \Sigma_{i+j=k}\, x_i \otimes x_j$$

where $x_0 = 1$.

Proof of Proposition 10.28.2: We use induction on n, the case $n = 1$ being trivial.

The existence of a generating complex gives that

$$HSO(n) = E[x_{n-1}] \otimes HSO(n-1)$$

as $HSO(n-1)$ modules. In fact, in the homology Serre spectral sequence of the principal bundle $SO(n-1) \to SO(n) \to S^{n-1}$, we have that

$$E^\infty = E[x_{n-1}] \otimes HSO(n-1)$$

as coalgebras. Hence, the primitives satisfy $PE^\infty = \langle x_{n-1}\rangle \oplus PHSO(n-1)$ and, since primitives of $HSO(n)$ must be detected in the associated graded $E^0 HSO(n) = E^\infty$, we have

$$PHSO(n) \subseteq \langle x_{n-1}\rangle \oplus PHSO(n-1).$$

Since the first violation of the following equations must be primitive, it follows for dimensional reasons that they are valid:

$$[x_{n-1}, x_{j-1}] = 0,\ x_{n-1}^2 = 0,\ j \leq n.$$

Thus,

$$HSO(n) = E[x_{n-1}] \otimes HSO(n-1)$$

as algebras. $\square$

We now turn to the consideration of the homology of the special orthogonal groups with $\mathbb{Z}[\frac{1}{2}]$ coefficients.

Proposition 10.28.4. *With $\mathbb{Z}[\frac{1}{2}]$ coeficients,*

(a)

$$HSO(2n+1) = E[t_3, \ldots, t_{4n-1}]$$

as primitively generated Hopf algebras where the generators t_{4i-1} have degree $4i - 1$.

10.28 Special orthogonal groups

(b)
$$HSO(2n+2) = E[t_3, \ldots, t_{4n-1}, x_{2n+1}]$$
as primitively generated Hopf algebras where the generator x_{2n+1} has degree $2n+1$.

We remark that (a) implies (b) in a straightforward manner. Since the previous generating complex $RP^{2n+1} \subseteq SO(2n+2)$ is an odd dimensional projective space, we have that $H_{2n+1}(RP^{2n+1}; \mathbb{Z}) \simeq \mathbb{Z}$ and that the composite map $RP^{2n+1} \to SO(2n+2) \to S^{2n+1}$ maps the generator x_{2n+1} of this homology group to the generator of the homology group of the sphere. Hence,

$$HSO(2n+2) = E[x_{2n+1}] \otimes HSO(2n+1)$$

as $HSO(2n+1)$ modules. The same arguments as in the proof of Proposition 10.27.4 show that

$$H^*SO(2n+2) = E[z_{2n+1}] \otimes H^*SO(2n+1)$$

as algebras and that

$$HSO(2n+2) = E[x_{2n+1}] \otimes HSO(2n+1)$$

as algebras and finally as primitively generated Hopf algebras. We leave to the reader this routine verification of (b) and proceed to the proof of (a).

The proof of (a) requires that we do an induction using the principal bundles

$$SO(2n-1) \to SO(2n+1) \to SO(2n+1)/SO(2n-1)$$

and we start by computing the homology of the unit tangent bundles $T(S^{2n})$ of the even dimensional spheres,

$$T(S^{2n}) = SO(2n+1)/SO(2n-1).$$

Lemma 10.28.5. *With $\mathbb{Z}$ coefficients*

$$H_k T(S^{2n}) = \begin{cases} \mathbb{Z}, & k = 0 \\ \mathbb{Z}/2\mathbb{Z}, & k = 2n-1 \\ \mathbb{Z}, & k = 4n-1 \\ 0, & otherwise. \end{cases}$$

Proof: Consider the homology Serre spectral sequence of the unit tangent bundle sequence

$$S^{2n-1} \to T(S^{2n}) \to S^{2n}$$

and compute the transgression in this spectral sequence by the commutative diagram

$$\begin{array}{ccccccc}
S^{2n-1} & \xrightarrow{=} & S^{2n-1} & \to & D^{2n} & \to & S^{2n} \\
\downarrow = & & \downarrow 2 & & \downarrow & & \downarrow = \\
S^{2n-1} & \xrightarrow{2} & S^{2n-1} & \to & RP^{2n}/RP^{2n-2} & \to & S^{2n} \\
& & \downarrow = & & \downarrow & & \downarrow = \\
& & S^{2n-1} & \to & T(S^{2n}) & \to & S^{2n}
\end{array}$$

The top two rows are the standard cofibration sequences and the compatibility of cofibration sequences with the transgression gives that $d_{2n}x_{2n} = 2x_{2n-1}$ in the homology Serre spectral sequence with

$$E^2 = HS^{2n} \otimes HS^{2n-1} = E[x_{2n}] \otimes E[x_{2n-1}].$$

The result of the lemma follows. $\square$

We now finish the proof of (a). The case of $n = 1$ being trivially true, we can start by using the inductive assumption that

$$HSO(2n - 1) = E[t_3, \ldots, t_{4n-5}]$$

as Hopf algebras.

With $\mathbb{Z}[\frac{1}{2}]$ coefficients, the homology Serre spectral sequence of the principal bundle sequence

$$SO(2n-1) \to SO(2n+1) \to T(S^{2n})$$

shows that

$$HSO(2n+1) = E[t_{4n-1}] \otimes HSO(2n-1)$$

as $HSO(2n-1)$ modules. Once again, arguments identical to the ones in the proof of Proposition 10.27.4 give that this is an isomorphism of primitively generated Hopf algebras.

This completes the proof of Proposition 10.28.4. $\square$

Let us consider the loop spaces of the special orthogonal groups.

First of all, suppose that $G \to H$ is any covering of connected topological groups and let K be the kernel. If K is finite of order n we can consider the transfer $\tau : C(H) \to C(G)$ given on simplices σ of H by

$$\tau(\sigma) = \Sigma_{g \epsilon K}\ g\bar{\sigma}$$

where $\bar{\sigma}$ is any choice of a lift of σ to a simplex in G. Clearly, the transfer τ is a chain map and $\pi \cdot \tau = n \cdot 1_H$. Since G is connected we also have that

$$\tau \cdot \pi = \Sigma_{g \epsilon K}\ \ell_g \simeq n \cdot 1_G$$

where ℓ_g denotes left multiplication by g. Thus,

Lemma 10.28.5. *If $G \to H$ is a finite n-sheeted covering of connected topological groups, then $H_*G \to H_*H$ is an isomorphism of homologies with $\mathbb{Z}[\frac{1}{n}]$ coefficients.*

Hence, Proposition 10.28.4 also computes the homology with $\mathbb{Z}[\frac{1}{2}]$ coefficients of the double coverings $\mathrm{Spin}(n) \to SO(n)$ which are the universal covers of the special orthogonal groups. Since the groups $\mathrm{Spin}(n)$ are simply connected, the Eilenberg–Moore spectral sequence applies directly to these groups and $\Omega\mathrm{Spin}(n)$ is the component of the identity of $\Omega SO(n)$. We get:

Proposition 10.28.6. *With $\mathbb{Z}[\frac{1}{2}]$ coefficients,*

(a)
$$H\Omega\mathrm{Spin}(2n+1) = P[y_2, \ldots, y_{4n-2}]$$

(b)
$$H\Omega\mathrm{Spin}(2n+2) = P[y_2, \ldots, y_{4n-1}, z_{2n}]$$

as algebras where the generators have the indicated degrees.

Proof: Consider part (a). The Eilenberg–Moore spectral sequence has

$$E^2 = \mathrm{Cotor}^{H\mathrm{Spin}(n)}\left(\mathbb{Z}\left[\frac{1}{2}\right], \mathbb{Z}\left[\frac{1}{2}\right]\right)$$

$$= \mathrm{Cotor}^{E[x_3, \ldots, x_{4n-1}]}\left(\mathbb{Z}\left[\frac{1}{2}\right], \mathbb{Z}\left[\frac{1}{2}\right]\right) = P[y_2, \ldots, y_{4n-2}]$$

as algebras and all the generators are in filtration -1. Since this is a spectral sequence of Hopf algebras, these generators must be primitive and so must their differentials. But since the coefficient ring has no torsion, there are no nonzero primitives in filtrations less than -1. Hence, the spectral sequence collapses and $E^2 = E^\infty$. Since these loop spaces are homotopy commutative H-spaces and since E^∞ is a free commutative algebra, there are no extension problems and we have computed case (a).

The computation of case (b) is similar. □

Exercises

(1) (a) Show that there is a homeomorphism $RP^3 \to SO(3)$. (Hint: Show that every element of $SO(3)$ has at least one eigenvector of eigenvalue 1. Thus every element of $SO(3)$ can be represented as a rotation through some angle with this eigenvector as an axis, the rotation being

clockwise or counterclockwise depending on the choice of the pole of the axis.)

(b) Use the above to show that Corollary 10.28.3 implies that $H^*(RP^3; \mathbb{Z}/2\mathbb{Z})$ = a polynomial algebra on a 1 dimensional generator u truncated at height 4, $u^4 = 0$.

(2) Verify Proposition 10.28.1.

11 Odd primary exponent theorems

In order to prove the odd primary exponent theorems for spheres and Moore spaces, we have to use techniques from most of the chapters of this book.

We have to use localization. The theorems are not true without localization.

We have to use homotopy groups with coefficients and the structure of Samelson products on these groups in order to construct the maps which will give product decomposition theorems for certain loop spaces.

We have to use the fibre extensions of cubes and squares in order to tie together in fibration sequences different product decompositions.

We have to use Bockstein spectral sequences to analyse the torsion in the homology and homotopy of these loop spaces.

We have to use exact sequences of free differential graded Lie algebras in order to construct tensor product decompositions of universal enveloping algebras. In fact, it might be said that the main result of this chapter is the geometric realization via an infinite weak product of just such a tensor product decomposition of the universal enveloping algebra of a free differential graded Lie algebra.

We have to use differential homological algebra, not so much in the form of the Eilenberg–Moore spectral sequence, but in the chain equivalences with differential Cotor which underlie the Eilenberg–Moore theory.

Surprisingly, the only chapters in this book which we do not use much of here are the chapters on the Hopf invariants of Hilton, James, and Toda. These are exactly the things which were used in the proofs of the original exponents theorems of James and Toda and even of Selick.

We show here that, for a prime p greater than 3, the p-primary component of $\pi_*(S^{2n+1})$ has exponent p^n. This theorem is true for $p = 3$ but the failure of the Jacobi identity for Samelson products in mod 3 homotopy theory makes the proof much harder in this case. The details for the case $p = 3$ will be omitted but the odd primary exponent theorem and related product decompositions will be assumed to be proven for all odd primes. Since Gray has produced infinitely elements of order p^n in $\pi_*(S^{2n+1})$, the odd primary exponent theorem for spheres is the best possible [46].

The question of whether there is an exponent for all of the homotopy groups of a Moore space was first raised by Michael Barratt [9]. Minus the proof of an important lemma, we show that p^{r+1} annihilates the homotopy groups of an odd primary Moore space $P^m(p^r)$ with $m \geq 1$ [102]. Since Cohen, Moore, and Neisendorfer [27] have shown that $\pi_*(P^m(p^r))$ contains infinitely many elements of order p^{r+1}, the result on the exponent of the homotopy groups of odd primary Moore spaces is also the best possible. By the way, we do give complete details of the original upper bound that p^{2r+1} annihilates the homotopy groups of an odd primary Moore space $P^m(p^r)$ with $m \geq 1$ [28].

Finally, we show that double loops are necessary in order to have any H-space exponents for Moore spaces. On the single loop space of a Moore space, no power maps are null homotopic [28].

11.1 Homotopies, NDR pairs, and H-spaces

In this section, we review some fundamental notions connected with homotopies and use these to show that some homotopies can be simplified in the definitions related to H-spaces.

Recall the definition of a homotopy between two pointed maps. For the sake of simplicity, all maps and all homotopies will preserver basepoints.

Definition 11.1.1. A homotopy $H : X \times I \to Y$ from a continuous pointed map $f : X \to Y$ to a continuous pointed map $g : X \to Y$ is a continuous map $H : X \times I \to Y$ such that

(a)
$$H(x,0) = f(x), \quad H(x,1) = g(x). \quad \text{for all } x \in X$$

(b)
$$H(*, t) = * \quad \text{for all } t \in I.$$

If $A \subseteq X$ is a subspace the homotopy is said to be stationary on A if
$$H(a, t) = f(a) = g(a)$$
for all $(a, t) \in A \times I$.

Of course, a homotopy from one pointed map to another is just a path in the space of pointed maps $map_*(X, Y)$.

Definition 11.1.2. Given homotopies $H : X \times I \to Y$ from $f : X \to Y$ to $g : X \to Y$ and $K : X \times I \to Y$ from $g : X \to Y$ to $h : X \to Y$, the composition

11.1 Homotopies, NDR pairs, and H-spaces

or product homotopy $H * G : H \times I \to Y$ is the homotopy from f to h defined by

$$(H * G)(x,t) = \begin{cases} H(x, 2t) & \text{if } 0 \leq t \leq \frac{1}{2} \\ G(x, 2t-1) & \text{if } \frac{1}{2} \leq t \leq 1 \end{cases}$$

Definition 11.1.3. Given a homotopy $H : X \times I \to Y$ from f to g, the inverse or reverse homotopy is the homotopy $H^{-1} : X \times I \to Y$ from g to f defined by

$$H^{-1}(x,t) = H(x, 1-t).$$

Definition 11.1.4. Given homotopies $H : X \times I \to Y$ and $K : X \times I \to Y$, both from $f : X \to Y$ to $g : X \to Y$, a homotopy L from H to K is a continuous map

$$L : X \times I \times I \to Y$$

which satisfies

$$L(x, s, 0) = H(x, s), \quad L(x, s, 1) = K(x, s) \quad \text{for all } (x, s)$$

$$L(x, 0, t) = f(x), \quad L(x, 1, t) = g(x) \quad \text{for all } (x, t).$$

Thus, a homotopy of homotopies is a path in the space of homotopies from f to g,

$$\text{map}_{f,g}(X \times I, Y) = \{H : X \times I \to Y | H(x, 0) = f(x), H(x, 1) = g(x)\}.$$

The fundamental lemma is the generalization of the fact that the composition of a path with its inverse path is homotopic to the constant path:

Lemma 11.1.5. *If H is a homotopy from f to g, then the composite homotopy $H * H^{-1}$ is homotopic to the stationary or constant homotopy from f to f.*

Proof: Let $L : X \times I \times I \to Y$ be defined by

$$L(x, s, t) = \begin{cases} H(x, 2s), & 0 \leq s \leq \frac{1}{2}(1-t), \\ H(x, 2-2s), & \frac{1}{2}(1+t) \leq s \leq 1 \\ H(x, 1-t), & \frac{1}{2}(1-t) \leq s \leq \frac{1}{2}(1+t). \end{cases}$$

Then $(H * H^{-1})(x, s) = L(x, s, 0)$ is homotopic to the constant homotopy $L(x, s, 1) = H(x, 0) = f(x)$. $\square$

Definition 11.1.6. If $A \subseteq X$ is a subspace, then the pair (X, A) is called an NDR pair if the inclusion $\iota : A \to X$ is a cofibration, in other words, if the pair (X, A) has the homotopy extension property. In detail, given a map $f : X \times 0 \to Z$ and a compatible homotopy $K : A \times I \to Z$, there is a continuous extension of f and K to a homotopy $H : X \times I \to Z$.

Of course, (X, A) is an NDR pair if and only if the inclusion

$$A \times I \cup X \times 0 \subseteq X \times I$$

admits a retraction.

We refer to the basic paper of Steenrod [126] or to the book of George Whitehead [134] for the proof of the following basic proposition.

Proposition 11.1.7. *If (X, A) and (Y, B) are two NDR pairs, then the product*
$$(X, A) \times (Y, B) = (X \times Y, X \times B \cup A \times Y)$$
is an NDR pair.

An immediate corollary is:

Corollary 11.1.8. *If X is a space with a nondegenerate basepoint $*$, that is, if $(X, *)$ is an NDR pair, then the inclusion of the wedge*
$$(X \times X, X \vee X) = (X \times X, X \times * \cup * \times X)$$
and the inclusion of the fat wedge
$$(X \times X \times X, X \times X \times * \cup X \times * \times X \cup * \times X \times X)$$
are both NDR pairs.

Let X be an H-space with multiplication
$$\mu : X \times X \to X, \quad \mu(x, y) = xy$$
and suppose that the basepoint is nondegenerate. Thus, the fact that the basepoint is a homotopy unit says that the maps given by
$$x \mapsto x * \quad \text{and} \quad x \mapsto *x$$
are both homotopic to the identity map of X to itself. In other words, the composition
$$X \vee X \subseteq X \times X \xrightarrow{\mu} X$$
is homotopic to the fold map given by
$$(x, *) \mapsto x \quad \text{and} \quad (*, x) \mapsto x.$$

Since the inclusion $X \vee X \subseteq X \times X$ is an NDR pair, it follows that

Lemma 11.1.9. *In an H-space X with nondegenerate basepoint, the multiplication is homotopic to a multiplication $\mu : X \times X \to X$ for which the basepoint is a strict unit, that is,*
$$x* = x \quad \text{and} \quad *x = x.$$

Hence, it is no real loss of generality to assume that an H-space X has a nondegenerate basepoint which is a strict unit. In the remainder of this section, we will make this assumption. We will then show that, if the H-space is homotopy commutative, then the commuting homotopy can be assumed to be stationary on the wedge. Similarly, we will show that, if the H-space is homotopy associative,

11.1 Homotopies, NDR pairs, and H-spaces

then the associating homotopy can be assumed to be stationary on the fat wedge. These are both consequences of the next lemma.

Lemma 11.1.10. *Suppose* $(\mathbb{Z}, B)$ *is an NDR pair and* $H : \mathbb{Z} \times I \to W$ *is a homotopy from* f *to* g. *Suppose that the restriction* $K : B \times I \to W$ *is homotopic to a stationary homotopy. Of course, this requires that*

$$f(b) = g(b) = K(b,0) = K(b,1) \quad \forall b \in B.$$

Then H *is homotopic to a homotopy* L *which is stationary on* B.

Proof: H can be regarded as defining a map

$$H' : \mathbb{Z} \times I \times 0 \to W$$

such that

$$H'(z,0,0) = f(z), \quad H'(z,1,0) = g(z)$$

for all $z \epsilon \mathbb{Z}$. The fact that K is homotopic to a stationary homotopy can be regarded as defining a compatible map

$$K' : B \times I \times I \to W$$

such that

$$K'(b,s,0) = K(b,s), \quad K'(b,s,1) = f(b) = g(b).$$

Thus $K'(b,s,1)$ defines a homotopy which is stationary on B.

We let $A(z,1,t) = g(t)$ and note that this defines a compatible map

$$A : \mathbb{Z} \times 1 \times I \to W.$$

Similarly, we let $B(z,0,t) = f(z)$ and note that this defines a compatible map

$$B \times 0 \times I \to W.$$

We now have a map

$$C : \mathbb{Z} \times I \times 0 \cup \mathbb{Z} \times 0 \times I \cup \mathbb{Z} \times 1 \times I \cup B \times I \times I \to W.$$

From the standard fact that the pair $(I \times I, I \times 0 \cup 0 \times I \cup 1 \times I)$ is homeomorphic to the pair $(I \times I, I \times 0)$ and the fact that $(\mathbb{Z} \times I, B \times I)$ is an NDR pair, it follows that we have an extension to a map

$$D : \mathbb{Z} \times I \times I \to W.$$

The restriction $D(z,s,1) = L(z,t)$ is the required homotopy, □

Here are some facts which I learned from Frank Adams and Michael Barratt. They are consequences of the above Lemma.

Lemma 11.1.11. *Let* X *and* Y *be H-spaces with strict units and with respective multiplications* μ_X *and* μ_Y.

(a) If $f : X \to Y$ is a basepoint preserving H-map, then the homotopy $f\mu_x \simeq \mu_Y(f \times f)$ can be chosen to be stationary on the wedge $X \vee X$.

(b) If X is homotopy commutative, then the commuting homotopy $\mu_X T \simeq \mu_X$ can be chosen to be stationary on the wedge $X \vee X$.

(c) If X is homotopy associative, then the associating homotopy $\mu_X(\mu_X \times 1) \simeq \mu_X(1 \times \mu_X)$ can be chosen to be stationary on the fat wedge $* \times X \times X \cup X \times * \times X \cup X \times X \times *$.

Proof: Let F be a homotopy from $f\mu_X$ to $\mu_X(f \times f)$, that is, $F : f(ab) \simeq f(a)f(b)$ for $a, b \in X$. As usual we abreviate multiplication in an H-space by juxtaposition, $\mu(a, b) = ab$. Let G_1 be the homotopy from $\mu_Y(f \times f)$ to itself given by $G_1(a, b, t) = F(a, *, t)f(b)$. The composite homotopy $F * G_1^{-1}$ is homotopic to a homotopy F_1 from $f\mu_Y$ to $\mu_Y(f \times f)$ which is stationary on $X \times *$. Let G_2 be the homotopy from $\mu_Y(f \times f)$ to itself given by $G_2(a, b, t) = f(a)F_1(*, b, t))$. The composite homotopy $F_1 * G_1^{-1}$ is homotopic to a homotopy from $f\mu_X$ to $\mu_Y(f \times f))$ which is stationary on the wedge $X \vee X$. This is the required homotopy to prove a).

Part (b) follows from part (a) by considering $X = Y$ with $\mu_Y = \mu_X \circ T$ and $f = 1$.

To prove part c), let $F : a(bc) \simeq (ab)c$ be a homotopy. Let $G_1(a, b, c, t) = (F(a, *, *, t)b)c$. Then $G_1 : (ab)c \simeq (ab)c$ is a homotopy which agrees with F on $X \times * \times *$. Thus the composite homotopy $F * G_1^{-1} : a(bc) \simeq (ab)c$ is defined and is homotopic to a homotopy $F_1 : a(bc) \simeq (ab)c$ which is stationary on $X \times * \times *$.

Construct homotopies $F_1, F_2, \ldots, F_6 : a(bc) \simeq (ab)c$ which are respectively stationary on the increasing unions of

$$X \times * \times *, * \times X \times *, * \times * \times X,$$

$$* \times X \times X, X \times * \times X, X \times X \times *$$

such that the homotopy F_i is homotopic to the composite homotopy $F_{i-1} * G_i^{-1}$ for $i \leq 4$ or $i = 6$ and F_5 is homotopic to $G_5^{-1} * F_4$.

Then F_6 is the required homotopy which is stationary on the fat wedge. $\square$

Exercises

(1) Fill in the details in the proof of part c) of Lemma 11.1.11.

(2) Let X be a pointed space and let the space of Moore loops LX be the space of pairs (ω, a) where $a \geq 0$ is positive real number and $\omega : [0, \infty] \to X$ is a path with $\omega(0) = *$ and $\omega(t) = *$ for all $t \geq a$. Let LX have the product topology where the positive reals have the usual topology and the paths have the compact open topology. We say that the pair (ω, a) is a loop which ends at a. Let (ω, a) be a loop which ends at a and (γ, b) be a loop which ends at b. Define

11.1 Homotopies, NDR pairs, and H-spaces

the product loop $(\omega, a) * (\gamma, b) = (\omega * \gamma, a + b)$ which ends at $a + b$ by

$$(\omega * \gamma)(t) = \begin{cases} \omega(t), & 0 \le t \le a, \\ \gamma(t - a), & a \le t \le \infty \end{cases}$$

Show that

(a) The space of Moore loops LX is a strictly associative H-space with the basepoint being a strict unit.

(b) The space of Moore loops LX is homotopy equivalent to the loop space ΩX. Hint: Show that the sequence $LX \to MX \to X$ is a Serre fibration sequence with contractible total space.

(3) Suppose that $F \xrightarrow{\iota} E \xrightarrow{\pi} B$ is a fibration sequence of pointed spaces and that $f: X \to E$ and $g: X \to E$ are pointed maps.

(a) Show that

$$\mathrm{map}_{f,g}(X \times I, E) \to \mathrm{map}_{\pi f, \pi g}(X \times I, B)$$

is a fibration.

(b) If the compositions $\pi f = *$ and $\pi g = *$, then

$$\mathrm{map}_{f,g}(X \times I, F) \to \mathrm{map}_{f,g}(X \times I, E) \to \mathrm{map}_{*,*}(X \times I, B)$$

is a fibration sequence.

(4) Suppose that $F \to E \to B$ is an orientable fibration sequence. Two homotopies $X \times I \to E$ define a map

$$X \times I \times \{0, 1\} \to E.$$

Suppose that $A \subseteq X$ is a subspace and that

$$A \times I \times I \to E$$

is homotopy of the restrictions of the two homotopies to $A \times I$. Project the above homotopies into E to two homotopies into the base B and let

$$X \times I \times I \to B$$

be a homotopy of these two homotopies which is compatible with the projected homotopy of homotopies on A. This leads to the lifting problem

$$\begin{array}{ccc} X \times I \times \{0,1\} \cup A \times I \times I & \to & E \\ \downarrow & \nearrow & \downarrow \\ X \times I \times I & \to & B. \end{array}$$

The solution to this lifting problem is the existence of a lift of a homotopy of homotopies. Show that the obstructions to this lifting problem lie in the cohomology groups

$$H^n(X, A; \pi_n F).$$

11.2 Spheres, double suspensions, and power maps

In this section, we will introduce some specific H-space structures on localized spheres S^{2n-1} and on related spaces such as on localized homotopy theoretic fibres $C(n)$ of the double suspension $\Sigma^2 : S^{2n-1} \to \Omega^2 S^{2n+1}$ and on localized homotopy theoretic fibres $S^{2n-1}\{k\}$ of k-th power maps on spheres $k : S^{2n-1} \to S^{2n-1}$, $k(x) = x \ldots x = x^k$. (If the H-space structure is not homotopy associative, then x^k denotes some choice of a k-th power, for example, $x^3 = x(xx)$. And these k-th power maps on spheres are of course degree k maps.)

Throughout this section, we will assume that all spaces are localized at an odd prime p. For example, S^{2n-1} will denote the localized sphere. We have two goals. First, we wish to introduce H-space structures on the three examples above such that the obvious maps between them are H-maps. Second, we wish to check that these structures are homotopy commutative and homotopy associative and that the homotopy theoretic fibres $S^{2n-1}\{k\}$ have H-space exponent k. (That is, the k-th power maps are null homotopic on $S^{2n-1}\{k\}$.)

We begin by recalling the following well known general fact which will enable us to show that the homotopy theoretic fibres $C(n)$ and $S^{2n-1}\{k\}$ are H-spaces.

Lemma 11.2.1. *The homotopy theoretic fibre F of an H-map $f : X \to Y$ is an H-space.*

Proof: Up to homotopy equivalence, replace f by a fibration $\tilde{f} : \tilde{X} \to Y$. Then F is the actual fibre of $\tilde{f}$. Furthermore, $\tilde{X}$ is an H-space with a multiplication μ and $\tilde{f}$ is an H-map. The diagram below is homotopy commutative

$$\begin{array}{ccc} \tilde{X} \times \tilde{X} & \xrightarrow{\mu} & \tilde{X} \\ \downarrow \tilde{f} \times \tilde{f} & & \downarrow \tilde{f}. \\ Y \times Y & \xrightarrow{\text{mult}} & Y \end{array}$$

Since $\tilde{f}$ is a fibration, it follows that μ is homotopic to a multiplication ν which makes the diagram strictly commutative. Hence, with this multiplication, $\tilde{f}$ is a strict H-map and clearly ν restricts to a multiplication on the fibre F. $\square$

Loosely speaking, the multiplication on the double loop space $\Omega^2 S^{2n+1}$ restricts to a multiplication on the bottom cell S^{2n-1}. But, the fact that these are localized spaces and not cells makes it better to convert the double suspension $\Sigma^2 : S^{2n-1} \to \Omega^2 S^{2n+1}$ to a homotopy equivalent fibration $\tilde{\Sigma}^2 : \tilde{S}^{2n-1} \to \Omega^2 S^{2n+1}$ and to consider the lifting problem

$$\begin{array}{ccc} \tilde{S}^{2n-1} \times \tilde{S}^{2n-1} & \xrightarrow{\mu} & \tilde{S}^{2n-1} \\ \downarrow \tilde{\Sigma}^2 \times \tilde{\Sigma}^2 & & \tilde{\Sigma}^2 \\ \Omega^2 S^{2n+1} \times \Omega^2 S^{2n+1} & \xrightarrow{\text{mult}} & \Omega^2 S^{2n+1} \end{array}$$

11.2 Spheres, double suspensions, and power maps

Since we can assume that we have strict units, the obstructions to this lifting problem lie in

$$H^{m+1}(\tilde{S}^{2n-1} \times \tilde{S}^{2n-1}, \tilde{S}^{2n-1} \vee \tilde{S}^{2n-1}; \pi_m C(n)).$$

These groups can be nonzero only if $m+1 = 4n-2$ and $\pi_m C(n) = \pi_{4n-3} C(n) \neq 0$.

Since

$$H_*(\Omega^2 S^{2n+1}; \mathbb{Z}/p\mathbb{Z}) = E(\tau_0, \tau_1, \tau_2, \dots) \otimes P(\sigma_1, \sigma_2, \dots)$$

with degree $\tau_i = 2p^i n - 1$ and degree $\sigma_i = 2p^i n - 2$, the localized pair $(\Omega^2 S^{2n+1}, S^{2n-1})$ is $2pn - 3$ connected, that is, the fibre $C(n)$ is $2pn - 4$ connected. Hence, $\pi_{4n-3} C(n) = 0$ for $4n - 3 \leq 2pn - 4$, that is, for $1 \leq 2(p-2)n$ which is always the case if p is an odd prime and $n \geq 1$. Hence, the lift exists. Since obstructions to uniqueness of lifts up to fibre homotopy lie in

$$H^m(\tilde{S}^{2n-1} \times \tilde{S}^{2n-1}, \tilde{S}^{2n-1} \vee \tilde{S}^{2n-1}; \pi_m C(n)),$$

it follows that

Proposition 11.2.2. *If p is an odd prime and $n \geq 1$, then there is an H-space structure (unique up to homotopy if p is an odd prime and $n \geq 2$ or if $n \geq 1$ and $p \geq 5$) such that the double suspension $\Sigma^2 : S^{2n-1} \to \Omega^2 S^{2n+1}$ is an H-map.*

Now, since the homotopy theoretic fibre of an H-map is an H-space, Proposition 11.2.2 and Lemma 11.2.1 give that

Corollary 11.2.3. *If p is an odd prime and $n \geq 1$, then there is an H-space structure on $C(n)$ which makes $C(n) \to \tilde{S}^{2n-1} \to \Omega^2 S^{2n+1}$ into a fibration sequence of H-spaces and strict H-maps.*

Let k be a positive integer. Consider the power maps $k : S^{2n-1} \to S^{2n-1}$ and let $S^{2n-1}\{k\}$ be the homotopy theoretic fibre. Replacing k by a fibration, we have a fibration sequence $S^{2n-1}\{k\} \to \tilde{S}^{2n-1} \xrightarrow{k} S^{2n-1}$. Since the power map is an H-map when the H-space is homotopy commutative and homotopy associative, we get from Lemma 11.2.1 and Exercise 1 below

Proposition 11.2.4. *If $p \geq 5$ and $n \geq 1$, then there is an H-space structure on $S^{2n-1}\{k\}$ which makes the fibration sequence*

$$S^{2n-1}\{k\} \to \tilde{S}^{2n-1} \xrightarrow{k} S^{2n-1}$$

into a sequence of H-spaces and strict H-maps.

If k is a positive integer, then $k : \tilde{S}^{2n-1} \to S^{2n-1}$ being a fibration implies that we can assume that the homotopy commutative diagram

$$\begin{array}{ccc} \tilde{S}^{2n-1} & \xrightarrow{k} & \tilde{S}^{2n-1} \\ \downarrow k & & \downarrow k \\ S^{2n-1} & \xrightarrow{k} & S^{2n-1} \end{array}$$

is strictly commutative. Since we are localized at an odd prime, $\pi_{2n-1}S^{2n-1}\{k\} = 0$. The only possibly nonzero obstruction group $H^{2n-1}(\tilde{S}^{2n-1}; \pi_{2n-1}S^{2n-1}\{k\})$ to the fibre homotopy uniqueness of the lift $\tilde{S}^{2n-1} \xrightarrow{k} \tilde{S}^{2n-1}$ is in fact zero. Hence, the lift is unique up to fibre homotopy and above diagram must factor as

$$\begin{array}{ccccc} \tilde{S}^{2n-1} & \xrightarrow{k} & \tilde{S}^{2n-1} & \xrightarrow{1} & \tilde{S}^{2n-1} \\ \downarrow k & & \downarrow 1 & & \downarrow k \\ S^{2n-1} & \xrightarrow{1} & S^{2n-1} & \xrightarrow{k} & S^{2n-1} \end{array}$$

Hence:

Proposition 11.2.5. *Localized at an odd prime, the k-th power map on the fibre $k: S^{2n-1}\{k\} \to S^{2n-1}\{k\}$ is null homotopic. That is, the H-space $S^{2n-1}\{k\}$ has H-space exponent $\leq k$.*

Exercises

(1) (a) If p is an odd prime and $n \geq 2$ (or if $n \geq 1$ and $p \geq 5$), then the H-space structure on S^{2n-1} is homotopy commutative.

(b) If $p \geq 5$ and $n \geq 1$, then the H-space structure on S^{2n-1} is homotopy associative.

(c) If p is an odd prime and $n \geq 2$ (or if $n \geq 1$ and $p \geq 5$), then the H-space structure on $C(n)$ is homotopy commutative.

(d) If $p \geq 5$ and $n \geq 1$, then the H-space structure on $C(n)$ is homotopy associative.

(Hints: By Exercise 4 in Section 11.1, obstructions to lifting commuting homotopies lie in the groups

$$H^m(S^{2n-1} \times S^{2n-1}, S^{2n-1} \vee S^{2n-1}; \pi_m C(n))$$

and obstructions to lifting associating homotopies lie in the groups

$$H^m(S^{2n-1} \times S^{2n-1} \times S^{2n-1}, fat\ wedge; \pi_m C(n)).)$$

(2) If k is a positive integer and $k: S^{2n-1} \to S^{2n-1}$ is any k-th power map in any H-space structure on S^{2n-1}, show that this map is a map of degree k.

(3) Let X be an H-space, let k be an integer, and let $k: X \to X$ be any k-th power map. Show that

(a) The map k induces multiplication by k on the homotopy groups $\pi_*(X)$.

(b) The two multiplications on the loop space ΩX are equal to each other and both are homotopy associative and homotopy commutative.

(c) The loop on the k-th power map is a k-th power map, that is, $\Omega k = k$.

11.3 The fibre of the pinch map

Throughout this chapter, we assume that p is an odd prime and that all spaces are localized at p. Consider the pinch map $q : P^{2n+1}(p^r) \to S^{2n+1}$ which collapses the bottom cell of the Moore space $P^{2n+1}(p^r) = S^{2n} \cup_{p^r} e^{2n+1}$ to a point. Let $F^{2n+1}\{p^r\}$ be the homotopy theoretic fibre of q. In this section, we will determine the homology of this fibre and give a small model for its differential coalgebra of normalized chains.

Up to homotopy equivalence, there is a fibration sequence

$$\Omega S^{2n+1} \xrightarrow{\partial} F^{2n+1}\{p^r\} \xrightarrow{i} P^{2n+1}(p^r) \xrightarrow{\pi} S^{2n+1}$$

in which the first three spaces are the sequence of a principal bundle. Thus, there is an associative action

$$\mu : \Omega S^{2n+1} \times F^{2n+1}\{p^r\} \to F^{2n+1}\{p^r\}$$

which is compatible with the loop multiplication in the sense that the following diagram commutes:

$$\begin{array}{ccc} \Omega S^{2n+1} \times \Omega S^{2n+1} & \xrightarrow{\text{mult}} & \Omega S^{2n+1} \\ \downarrow 1 \times \partial & & \downarrow \partial \\ \Omega S^{2n+1} \times F^{2n+1}\{p^r\} & \xrightarrow{\mu} & F^{2n+1}\{p^r\} \end{array}$$

With any coefficients, the group of normalized chains $C(F^{2n+1}\{p^r\})$ is a module over the group of normalized chains $C(\Omega S^{2n+1})$ via the composition

$$\mu : C(\Omega S^{2n+1}) \otimes C(\Omega S^{2n+1}) \xrightarrow{\nabla} C(S^{2n+1} \times F^{2n+1}\{p^r\}) \xrightarrow{\mu} C(\Omega F^{2n+1}\{p^r\})$$

Since the diagram below commutes

$$\begin{array}{ccc} C(\Omega S^{2n+1}) \otimes C(\Omega S^{2n+1}) & \xrightarrow{\text{mult}} & C(\Omega S^{2n+1}) \\ \downarrow 1 \times \partial & & \downarrow \partial \\ C(\Omega S^{2n+1}) \times C(F^{2n+1}\{p^r\}) & \xrightarrow{\mu} & C(F^{2n+1}\{p^r\}) \end{array}$$

it follows that

$$x * 1 = \mu(x \otimes 1) = \partial_*(x)$$

for all x in $C(\Omega S^{2n+1})$.

Recall that, with any coefficients, $H(\Omega S^{2n+1}) = T(\iota) = $ a tensor Hopf algebra on a primitive generator ι of degree $2n$.

Proposition 11.3.1. *With coefficients $\mathbb{Z}_{(p)}$:*

(a) *the homology $H(F^{2n+1}\{p^r\})$ is torsion free with a basis $\{1, g_1, g_2, \ldots, g_n, \ldots\}$ such that degree g_k equal to $2kn$ for all $k \geq 1$.*

(b) *the coproduct is given by*

$$\Delta(g_k) = g_k \otimes 1 + 1 \otimes g_k + \Sigma_{i=1}^{k-1} p^r(i, k-i) g_i \otimes g_{k-i}.$$

(c) *the action of* $H(\Omega S^{2n+1}) = T(\iota)$ *on* $H(F^{2n+1}\{p^r\})$ *is given by*

$$\iota^j * g_k = g_{k+j}$$

for $k \geq 1$. *Thus, the reduced homology* $\overline{H}(F^{2n+1}\{p^r\})$ *is a free* $H(\Omega S^{2n+1})$ *module.*

(d) *the inclusion of the fibre* $\partial : \Omega S^{2n+1} \to F^{2n+1}\{p^r\}$ *induces* $\partial_*(\iota^k) = p^r g_k$ *in homology.*

Proof: Consider the mod p^r homology Serre spectral sequence of the principal bundle sequence

$$\Omega S^{2n+1} \xrightarrow{\partial} F^{2n+1}\{p^r\} \xrightarrow{i} P^{2n+1}(p^r).$$

It is a left module over $H(\Omega S^{2n+1})$ and has

$$E^2 = H(\Omega S^{2n+1}) \otimes H(P^{2n+1}(p^r)) = T(\iota) \otimes \langle 1, u, v \rangle$$

with degree $u = 2n$ and degree $v = 2n + 1$.

We claim that the first nonzero differentials are given by

$$d^{2n+1}(1 \otimes v) = \iota \otimes 1, \quad d^{2n+1}(1 \otimes u) = 0.$$

This is a consequence of $q_*(v) = e$ where e is a generator of $H_{2n+1}(S^{2n+1})$, of naturality, of the map of fibration sequences

$$\begin{array}{ccc} \Omega S^{2n+1} & \xrightarrow{=} & \Omega S^{2n+1} \\ \downarrow \partial & & \downarrow \\ F^{2n+1}\{p^r\} & \to & PS^{2n+1} \\ \downarrow i & & \downarrow \\ P^{2n+1}(p^r) & \to_q & S^{2n+1} \end{array}$$

into the path fibration, and of the transgression $d^{2n+1}(1 \otimes e) = \iota \otimes 1$.

Since the spectral sequence is a spectral sequence of $H(\Omega S^{2n+1})$ modules,

$$d^{2n+1}(\iota^k \otimes u) = 0, \quad d^{2n+1}(\iota^k \otimes v) = \iota^{k+1} \otimes 1$$

and E^{2n+1} has a $\mathbb{Z}/p^r\mathbb{Z}$ basis $\{1 \otimes 1, 1 \otimes u, \iota \otimes u, \iota^2 \otimes u, \dots\}$ concentrated in even degrees $2nk$.

It follows that $E^{2n+2} = E^\infty$ in the mod p^r spectral sequence and that the homology $H(F^{2n+1}\{p^r\})$ with localized coefficients $\mathbb{Z}_{(p)}$ is torsion free of rank 1 in each even degree $2nk$ with $k \geq 0$.

11.3 The fibre of the pinch map

With coefficients $\mathbb{Z}_{(p)}$, consider the homology Serre spectral sequence of the principal bundle sequence

$$\Omega S^{2n+1} \xrightarrow{\partial} F^{2n+1}\{p^r\} \xrightarrow{i} P^{2n+1}(p^r).$$

Since the reduced $\mathbb{Z}_{(p)}$ homology $\overline{H}(P^{2n+1}(p^r)) = \mathbb{Z}/p^r\mathbb{Z} \cdot u =$ a torsion module concentrated in degree $2n$, we have

$$E^2 = H(\Omega S^{2n+1}) \otimes H(P^{2n+1}) = E^2_{0,*} \oplus E^2_{2n,*} = T(\iota) \otimes 1 \oplus T(\iota) \otimes u.$$

Since the first summand is torsion free and the second summand is torsion, we have that $E^2 = E^\infty$.

Since we know that $H_{2nk}(F^{2n+1}\{p^r\}$ is torsion free of rank 1, we have short exact sequences

$$0 \to \overline{H}_{2nk}(\Omega S^{2n+1}) \xrightarrow{\partial} \overline{H}_{2nk}(F^{2n+1}) \to \mathbb{Z}/p^r\mathbb{Z} \to 0.$$

Hence, if g_k is a generator of $\overline{H}_{2nk}(F^{2n+1})$, we have $\partial_*(\iota^k) = p^r g_k$.

Since the coproduct

$$\Delta(\iota^k) = \iota^k \otimes 1 + 1 \otimes \iota^k + \Sigma_{i=1}^{k-1}(i, k-i)\iota^i \otimes \iota^{k-i},$$

naturality and the preceding formula yields

$$\Delta(g_k) = g_k \otimes 1 + 1 \otimes g_k + \Sigma_{i=1}^{k-1} p^r(i, k-i) g_i \otimes g_{k-i}.$$

Finally, $p^r g_k = \partial_*(\iota^k) = \iota^{k-1} * \partial_*(\iota) = \iota^{k-1} * p^r g_1 = p^r \iota^{k-1} * g_1$ implies that $g_k = \iota^{k-1} * g_1$ and that $\iota^j g_k = g_{k+j}$. □

Remark 11.3.2. It follows from 11.3.1 (b) and (d) that the embedding of reduced $\mathbb{Z}_{(p)}$ homologies $\overline{H}(\Omega S^{2n+1}) \to \overline{H}(F^{2n+1}\{p^r\})$ can be identified with the embedding $\overline{H}(\Omega S^{2n+1}) \subseteq \frac{1}{p^r}\overline{H}(\Omega S^{2n+1})$ inside the reduced coalgebra $\overline{H}(\Omega S^{2n+1}) \otimes \mathbb{Q}$. In particular, the reduced coproduct in $\overline{H}(F^{2n+1}\{p^r\})$ is compatible with the reduced coproduct in $\frac{1}{p^r}\overline{H}(\Omega S^{2n+1})$.

We now construct small coalgebras which are chain equivalent to the normalized chains $C(F^{2n+1}\{p^r\})$, that is, we construct small diffferential coalgebra models which are linked to the chains by homology equivalences.

We begin with some remarks on twisted tensor products.

Let A be a Hopf algebra and let C be a coalgebra. Recall that the tensor product $A \otimes C$ is a coalgebra via the usual coproduct (or diagonal):

$$\Delta : A \otimes C \xrightarrow{\Delta \otimes \Delta} A \otimes A \otimes C \otimes C \xrightarrow{1 \otimes T \otimes 1} A \otimes C \otimes A \otimes C.$$

Note also that this structure makes $A \otimes C$ into an A module coalgebra. That is, if A acts on $A \otimes C \otimes A \otimes C$ via the Hopf algebra diagonal of A and the twist as follows,

$$\mu : A \otimes (A \otimes C \otimes A \otimes C) \xrightarrow{\Delta \otimes 1 \otimes 1 \otimes 1 \otimes 1} A \otimes A \otimes (A \otimes C) \otimes A \otimes C \xrightarrow{1 \otimes T \otimes 1 \otimes 1}$$
$$A \otimes (A \otimes C) \otimes A \otimes A \otimes C \xrightarrow{\mu \otimes 1 \otimes \mu \otimes 1} A \otimes C \otimes A \otimes C,$$

then the coproduct $\Delta : A \otimes C \to A \otimes C \otimes A \otimes C$ is a map of A modules.

Suppose now that A is a differential Hopf algebra, C is a differential coalgebra, and $\tau : C \to A$ is a twisting morphism. Here is a special case when the twisted tensor product $A \otimes_\tau C$ is a differential A module coalgebra, that is, $A \otimes_\tau C$ is a differential coalgebra via the above coproduct structure on the tensor product and the action of A is given by a map of differential objects:

Lemma 11.3.3. *Suppose that A is a differential Hopf algebra with zero differential and that C is a differential coalgebra with trivial coproduct, that is, $\overline{C} = PC$. Assume that the image of the twisting morphism is primitive, $\tau : C \to PA \subseteq A$, then the twisted tensor product $A \otimes_\tau C$ is a differential A module coalgebra.*

We leave the proof of this as a simple exercise.

We now construct such a twisted tensor product to model the normalized chains $C(F^{2n+1}\{p^r\})$. We do this by modeling the fibration sequence

$$\Omega S^{2n+1} \xrightarrow{\partial} F^{2n+1}\{p^r\} \xrightarrow{i} P^{2n+1}(p^r).$$

Throughout the remainder of this section, the coefficient ring is $\mathbb{Z}_{(p)}$.

The chains on P^{2n+1} are modeled by the differential coalgebra $C(r)$ which is a free module with basis $\{1, u, v\}$. The respective degrees are 0, $2n$, $2n+1$ and the respective diagonals are

$$\Delta(1) = 1 \otimes 1, \quad \Delta(u) = u \otimes 1 + 1 \otimes u, \quad \Delta(v) = v \otimes 1 + 1 \otimes v.$$

The chains on ΩS^{2n+1} are modeled by the tensor Hopf algebra $T(\iota)$ with zero differential and ι primitive of degree $2n$.

Define a twisting morphism $\tau : C(r) \to T(\iota)$ by

$$\tau 1 = 0, \quad \tau u = 0, \quad \tau v = -\iota.$$

Thus the twisted tensor product $T(\iota) \otimes_\tau C(r)$ satisfies the hypotheses of Lemma 11.3.3 and is a $T(\iota)$ module coalgebra. Its differential is given by

$$d_\tau(\iota^k \otimes 1) = 0, \quad d_\tau(\iota^k \otimes u) = 0, \quad d_\tau(\iota^k \otimes v) = p^r \iota^k \otimes u - \iota^{k+1} \otimes 1.$$

The next result shows that this twisted tensor product and the homology are chain equivalent to the normalized chains of $C(F^{2n+1}\{p^r\})$.

11.3 The fibre of the pinch map   451

Proposition 11.3.4. *With coefficients $\mathbb{Z}_{(p)}$, there are homology isomorphisms of differential coalgebras*

$$H(F^{2n+1}\{p^r\}) \xleftarrow{\Psi} T(\iota) \otimes_\tau C(r) \xrightarrow{\Theta} C(F^{2n+1}\{p^r\}).$$

Proof: As in Chapter 10, we can pick a primitive cycle $x \in C_{2n}(\Omega S^{2n+1})$ which represents the homology class ι. It is primitive since we pick a cycle in the $2n-1$ connected Eilenberg subcomplex which is equivalent to the whole complex. Then the map

$$\theta' : T(\iota) \to C(\Omega S^{2n+1}), \quad \iota^k \mapsto x^k$$

is a homology isomorphism of differential Hopf algebras.

Similarly, there is a primitive cycle $z \in C_{2n}(F^{2n+1}\{p^r\})$ which represents the homology generator $g_1 \in H_{2n}(F^{2n+1}\{p^r\})$. And let $w \in C_{2n+1}(F^{2n+1}\{p^r\})$ be a primitive chain such that $dw = p^r z - \partial_*(x)$. Both z and w are primitive since they come from the Eilenberg subcomplex.

Now define $\theta'' : C(r) \to C(F^{2n+1}\{p^r\})$ by

$$\theta''1 = 0, \quad \theta''u = z, \quad \theta''v = w$$

and define $\Theta : T(\iota) \otimes_\tau C(r) \to C(F^{2n+1}\{p^r\})$ as the composition with the action

$$T(\iota) \otimes_\tau C(r) \xrightarrow{\theta' \otimes \theta''} C(\Omega S^{2n+1}) \otimes C(F^{2n+1}\{p^r\}) \xrightarrow{\mu} C(F^{2n+1}\{p^r\}).$$

Hence,

$$\Theta(\iota^k \otimes 1) = x^k * 1, \ \Theta(\iota^k \otimes u) = x^k * z, \ \Theta(\iota^k \otimes v) = x^k * w.$$

We claim that Θ is a map of differential coalgebras.

First, Θ is a map of coalgebras since θ', θ'', and the action μ are all maps of coalgebras.

Second, Θ is a map of $T(\iota)$ modules and sends the $T(\iota)$ basis elements $1 \otimes 1, 1 \otimes u, 1 \otimes v$ to $1 * 1 = 1, 1 * z = z, 1 * w = w$. We check that

$$d\Theta(1 \otimes 1) = d1 = 0 = \Theta d_\tau(1 \otimes 1)$$

$$d\Theta(1 \otimes u) = dz = 0 = \Theta d_\tau(1 \otimes u)$$

$$d\Theta(1 \otimes v) = dw = p^r z - \partial_*(x) = p^r z - x * 1 = \Theta(p^r 1 \otimes u - \iota \otimes 1)$$

$$= \Theta d_\tau(1 \otimes v)$$

Therefore, the Θ commutes with differentials and is a map of differential coalgebras.

We claim that the elements $1 \otimes 1$, $\iota^k \otimes u$ represent a basis for the homology of $T(\iota) \otimes_\tau C(r)$. We can see this directly from the formulas for the differential.

On the other hand, the elements 1, $x^k * z$ represent 1, g_{k+1}, which is a basis for the homology of $C(F^{2n+1}\{p^r\})$. Thus, Θ is a homology isomorphism.

Finally, let $\xi: C(r) \to H(F^{2n+1}\{p^r\})$, $1 \mapsto 1$, $u \mapsto z$, $v \mapsto 0$ and define the map

$$\Psi: T(\iota) \otimes C(r) \to H(F^{2n+1}\{p^r\})$$

as the composition

$$T(\iota) \otimes C(r) \to T(\iota) \otimes H(F^{2n+1}\{p^r\}) \xrightarrow{\mu} H(F^{2n+1}\{p^r\}).$$

$$\Psi(1 \otimes 1) = 1 * 1 = 1, \quad \Psi(\iota^k \otimes 1) = \iota^k * 1 = p^r g_k,$$

$$\Psi(\iota^k \otimes u) = \iota^k * g_1 = g_{k+1}, \quad \Psi(\iota^k \otimes v) = 0.$$

Since ξ and the action μ are maps of coalgebras, the map Ψ is a map of coalgebras. Since Ψ is a map of $T(\iota)$ modules and

$$d\Psi(1 \otimes 1) = d1 = 0 = \Psi d(1 \otimes 1), \quad d\Psi(1 \otimes u) = dg_1 = 0 = \Psi d(1 \otimes u),$$

$$d\Psi(1 \otimes v) = d0 = 0 = \Psi(p^r 1 \otimes u - \iota \otimes 1) = \Psi d(1 \otimes v),$$

it follows that Ψ is a chain map and a map of differential coalgebras.

By inspection, Ψ is a homology isomorphism. $\square$

Exercises

(1) Prove Lemma 11.3.3.

(2) Let p be an odd prime and let $S^{2n+1}\{p^r\}$ be the homotopy theoretic fibre of the degree p^r power map $p^r: S^{2n+1} \to S^{2n+1}$.

  (a) Show that the mod p homology

  $$H(S^{2n+1}\{p^r\}) = H(\Omega S^{2n+1}) \otimes H(S^{2n+1}) = T(\iota_{2n}) \otimes E(v_{2n+1})$$

  as a $H(\Omega S^{2n+1}) = T(\iota_{2n})$ module and the r-th the Bockstein differential is given by $\beta^r v_{2n+1} = \iota_{2n}$.

  (b) Show that the integral homology is

  $$\overline{H}_k(S^{2n+1}\{p^r\}) = \begin{cases} (\mathbb{Z}/p^r\mathbb{Z})u_k & \text{if } k = 2nj, j \geq 1 \\ 0 & \text{if } k \neq 2nj. \end{cases}$$

  (c) Show that the map $\Omega S^{2n+1} \to S^{2n+1}\{p^r\}$ induces an epimorphism in integral homology.

(d) Show that the isomorphism with coefficients $\mathbb{Z}/p\mathbb{Z}$ in part a) is an isomorphism of primitively generated differential Hopf algebras.

(e) Show that the coalgebra structure of the integral homology is given by
$$\Delta(u_k) = u_k \otimes 1 + 1 \otimes u_k + \Sigma_{i=1}^{k-1}(i, k-i)u_i \otimes u_{k-i}.$$

(f) Let $C = \langle 1, v \rangle$ be the coalgebra with zero differential and with v primitive and of degree $2n + 1$. Check that $\tau : C \to T(\iota_{2n})$, $\tau v = p^r \iota_{2n}$ is a twisting morphism.

(g) Show that, with any coefficients, there are homology isomorphisms of differential coalgebras
$$T(\iota_{2n}) \otimes_\tau C \to C(S^{2n+1}\{p^r\}).$$

(h) Show that, with coefficients $\mathbb{Z}/p\mathbb{Z}$, there are homology isomorphisms of differential coalgebras
$$T(\iota_{2n}) \otimes_\tau C \to H(S^{2n+1}\{p^r\}).$$

11.4 The homology exponent of the loop space

Throughout this section, suppose that p is an odd prime and that all spaces have been localized at p.

The fibration sequence
$$\Omega F^{2n+1}\{p^r\} \to \Omega P^{2n+1}(p^r) \to \Omega S^{2n+1}$$
shows that the rationalizations
$$\Omega F^{2n+1}\{p^r\} \otimes \mathbb{Q} \simeq \Omega^2 S^{2n+1} \otimes \mathbb{Q} \simeq S^{2n-1} \otimes \mathbb{Q}$$
have the same homotopy type. Hence, the integral homology satisfies
$$H_k(\Omega F^{2n+1}\{p^r\}) = \begin{cases} \mathbb{Z} & \text{if } k = 0, 2n \\ p\text{-torsion} & \text{if } k \neq 0, 2n. \end{cases}$$

We determine the exponent of this p-torsion in homology.

Proposition 11.4.1. *The p-torsion in $H(\Omega F^{2n+1}\{p^r\})$ has exponent p^{r+1}.*

Proof: Let $C' = H(F^{2n+1}\{p^r\})$ and let $C = H(\Omega S^{2n+1})$.

Since there are homology isomorphisms of differential coalgebras
$$C' \leftarrow T(\iota) \otimes_\tau C(r) \to C(F^{2n+1})$$

there are isomorphisms

$$\begin{aligned}H(\Omega F^{2n+1}\{p^r\}) &= \text{Cotor}^{C(F^{2n+1}\{p^r\})}(\mathbb{Z}_{(p)}, \mathbb{Z}_{(p)}) \\ &= \text{Cotor}^{T(\iota) \otimes_\tau C(r)}(\mathbb{Z}_{(p)}, \mathbb{Z}_{(p)}) \\ &= \text{Cotor}^{C'}(\mathbb{Z}_{(p)}, \mathbb{Z}_{(p)}) \\ &= H\Omega C'.\end{aligned}$$

Similarly, the homology isomorphism

$$C = T(\iota) \to C(\Omega S^{2n+1})$$

yields

$$\begin{aligned}H(\Omega^2 S^{2n+1}) &= \text{Cotor}^{C(\Omega S^{2n+1})}(\mathbb{Z}_{(p)}, \mathbb{Z}_{(p)}) \\ &= \text{Cotor}^{C}(\mathbb{Z}_{(p)}, \mathbb{Z}_{(p)}) \\ &= H\Omega C.\end{aligned}$$

Since C has zero differential, the cobar construction ΩC splits into a direct sum of tensors of length k,

$$\Omega C = \bigoplus_{k \geq 0} (\Omega C)_{-k}$$

where $(\Omega C)_{-k} = \overline{C} \otimes \cdots \otimes \overline{C} = \overline{C}^{\otimes k}$, and the differential is homogeneous in the sense that

$$d(\overline{C}^{\otimes k}) \subseteq \overline{C}^{\otimes (k+1)}.$$

Hence, the homology of the cobar construction splits into a direct sum of homogeneous pieces,

$$\begin{aligned}H\Omega C &= \text{Cotor}^C(\mathbb{Z}_{(p)}, \mathbb{Z}_{(p)}) \\ &= \bigoplus_{k \geq 0} \text{Cotor}^C_{-k}(\mathbb{Z}_{(p)}, \mathbb{Z}_{(p)}) \\ &= \bigoplus_{k \geq 0} z_k(C)/b_k(C)\end{aligned}$$

where

$$z_k(C) = \text{kernel } d : \overline{C}^{\otimes k} \to \overline{C}^{\otimes (k+1)} =$$

the cycles of tensor length k and

$$b_k(C) = \text{image } d : \overline{C}^{\otimes (k-1)} \to \overline{C}^{\otimes (k)} =$$

the boundaries of tensor length k.

11.4 The homology exponent of the loop space

Similarly,

$$H\Omega C' = \mathrm{Cotor}^{C'}(\mathbb{Z}_{(p)}, \mathbb{Z}_{(p)})$$
$$= \bigoplus_{k\geq 0} \mathrm{Cotor}^{C'}_{-k}(\mathbb{Z}_{(p)}, \mathbb{Z}_{(p)})$$
$$= \bigoplus_{k\geq 0} z_k(C')/b_k(C').$$

The fact that

$$\overline{H}(F^{2n+1}\{p^r\}) = \overline{C'} = \frac{1}{p^r}\overline{H}(\Omega S^{2n+1}) = \frac{1}{p^r}\overline{C}$$

and the fact that this is compatible with the reduced coproduct implies that

$$\overline{C'}^{\otimes k} = \frac{1}{p^{rk}}\overline{C}^{\otimes k},$$

$$z_k(C') = \frac{1}{p^{rk}} z_k(C),$$

$$b_k(C') = \frac{1}{p^{r(k-1)}} z_k(C).$$

There is a short exact sequence

$$0 \to \frac{\frac{1}{p^{rk}} b_k(C)}{\frac{1}{p^{r(k-1)}} b_k(C)} \to \frac{z_k(C')}{b_k(C')} \to \frac{\frac{1}{p^{rk}} z_k(C)}{\frac{1}{p^{rk}} b_k(C)} \to 0.$$

Since

$$\frac{\frac{1}{p^{rk}} z_k(C)}{\frac{1}{p^{rk}} b_k(C)}$$

is isomorphic to

$$\frac{z_k(C)}{b_k(C)} = \mathrm{Cotor}^{C}_{-k}(\mathbb{Z}_{(p)}, \mathbb{Z}_{(p)})$$

with p-torsion of exponent p and since

$$\frac{\frac{1}{p^{rk}} b_k(C)}{\frac{1}{p^{r(k-1)}} b_k(C)}$$

is isomorphic to

$$\frac{b_k(C)}{p^r b_k(C)}$$

of exponent p^r, it follows that the p-torsion in

$$\bigoplus_{k\geq 0} \frac{z_k(C')}{b_k(C')} = H\Omega C' = H\Omega F^{2n+1}\{p^r\}$$

has exponent p^{r+1}. □

11.5 The Bockstein spectral sequence of the loop space

Throughout this chapter, we assume that p is an odd prime and that all spaces are localized at p. In this section, we determine the mod p homology Bockstein spectral sequence of the loop space $\Omega F^{2n+1}\{p^r\}$.

We begin by computing the homology $H(\Omega F^{2n+1}\{p^r\}; \mathbb{Z}/p^s\mathbb{Z})$ with coefficients $\mathbb{Z}/p^s\mathbb{Z}$ where $s \leq r$.

When $s \leq r$, reduced homology with $\mathbb{Z}/p^s\mathbb{Z}$ coefficients,

$$\overline{H}P^{2n}(p^r) = \langle u, v \rangle$$

is a free module on generators u of degree $2n$ and v of degree $2n+1$. If $s = r$, then the short exact sequence

$$0 \to \mathbb{Z}/p^r\mathbb{Z} \to \mathbb{Z}/p^{2r}\mathbb{Z} \to \mathbb{Z}/p^r\mathbb{Z} \to 0$$

has the Bockstein connecting homomorphism $\beta v = u$, $\beta u = 0$. If $s = 1$, then the Bockstein differentials are

$$\beta^i v = 0, \ \beta^i u = 0 \ \text{ if } i \leq r-1$$

$$\beta^r v = u, \ \beta^r = 0.$$

Since $\Omega P^{2n+1}(p^r) = \Omega\Sigma P^{2n+1}(p^r)$, the Bott–Samelson theorem shows that, when $s \leq r$, the Hopf algebra

$$H(\Omega P^{2n+1}(p^r); \mathbb{Z}/p^r\mathbb{Z}) = T(u,v) = UL(u,v)$$

where $T(u,v)$ is the tensor algebra and $UL(u,v)$ is the universal enveloping algebra on the free Lie algebra $L(u,v)$. In case $s=1$ or $s=r$, the Hopf algebras and Lie algebras are differential objects with the differentials given by the respective formulas $\beta v = u$ and $\beta^r v = u$ as above.

The loops on the pinch map $\Omega q : \Omega P^{2n+1}(p^r) \to \Omega S^{2n+1}$ induces the map of differential Hopf algebras

$$H(\Omega P^{2n+1}(p^r)) = T(u,v) = UL(u,v) \to H(\Omega S^{2n+1}(p^r)) = T(\iota) = UL(\iota)$$

given on generators by

$$v \mapsto \iota, \quad u \mapsto 0$$

11.5 The Bockstein spectral sequence of the loop space 457

where degree ι equals to $2n$ and where the respective differentials are $\beta\iota = 0$ if $s = r$ and $\beta^i \iota = 0$ for all i.

Let $L = $ the kernel of the map of Lie algebras $L(u,v) \to L(\iota)$. We claim that L is a free Lie algebra $L_{k \geq 0}(\mathrm{ad}^k(v)(u))$ on generators

$$u, \, [v,u], \, [v,[v,u]], \, \ldots, \, \mathrm{ad}^k(v)(u), \, \ldots$$

and that:

Proposition 11.5.1. *With coefficients $\mathbb{Z}/p^s\mathbb{Z}$, $s \leq r$, there is an isomorphism of Hopf algebras*

$$H(\Omega F^{2n+1}\{p^r\}) = UL = UL(u, [v,u], [v,[v,u]], \ldots, \mathrm{ad}^k(v)(u), \ldots)$$
$$= UL_{k \geq 0}(\mathrm{ad}^k(v)(u)).$$

Proof: Consider the fibration sequence of loop spaces

$$\Omega F^{2n+1}\{p^r\} \to \Omega P^{2n+1}(p^r) \to \Omega S^{2n+1}.$$

Since there is a short exact sequence of Lie algebras

$$0 \to L \to L(u,v) \to L(\iota) \to 0,$$

it follows from Proposition 10.24.8 that this fibration sequence is totally nonhomologous to zero, and that as a $H(\Omega S^{2n+1}) = UL(\iota)$ comodule and UL module,

$$H(\Omega P^{2n+1}(p^r)) = UL(u,v) = UL \otimes UL(\iota).$$

Proposition 10.24.8 asserts that

$$H(\Omega F^{2n+1}\{p^r\}) = \mathrm{Cotor}^{UL(\iota)}(UL(u,v), \mathbb{Z}/p^s\mathbb{Z})$$
$$= (UL \otimes UL(\iota)) \square_{UL(\iota)} \mathbb{Z}/p^s\mathbb{Z} = UL$$

as Hopf algebras.

Now refer to Proposition 8.7.3. In brief, that argument goes as follows. Since L is a $\mathbb{Z}/p^s\mathbb{Z}$ split subalgebra of the free Lie algebra $L(u,v)$, it follows that L is also free, that is, $UL = T(W)$ and that its module of generators W is a free module over the algebra $T(u)$ via the adjoint action. Since

$$T(W) \otimes T(v) = T(u,v)$$

as graded modules, the Euler–Poincare series satisfy

$$\chi(T(u,v)) = \chi(T(W)) \cdot \chi(T(v)),$$

that is,

$$\frac{1}{1 - t^{2n} - t^{2n+1}} = \frac{1}{\chi W} \cdot \frac{1}{1 - t^{2n+1}},$$

which can be solved to give
$$\chi(W) = \Sigma_{k=0}^{\infty} t^{2n+(2n+1)k}.$$

Since u is surely a generator of $UL = T(W)$ and since W is a free $T(v)$ module, the computation of $\chi(W)$ forces W to have a basis
$$u, [v,u], [v,[v,u]], \ldots, \text{ad}^k(v)(u), \ldots.$$

Thus,
$$L = L(u, [v,u], [v,[v,u]], \ldots, \text{ad}^k(v)(u), \ldots) = L_{k \geq 0}(\text{ad}^k(v)(u)).$$

$\square$

Now the mod p homology Bockstein spectral sequence of the loops on the fibre of the pinch map is given by:

Proposition 11.5.2. *The terms of the homology Bockstein spectral sequence of $\Omega F^{2n+1}\{p^r\}$ are given by:*

(a) *As primitively generated Hopf algebras*
$$E^1 = \cdots = E^r = UL = UL_{k \geq 1}(\text{ad}^{k-1}(v)(u))$$
and
$$\beta^r \text{ad}^{k-1}(v)(u) = \sum_{j=1}^{k-1} (j, k-j)[\text{ad}^{j-1}(v)(u), \text{ad}^{k-j-1}(v)(u)].$$

(b) *As primitively generated Hopf algebras*
$$E^{r+1} = H(UL, \beta^r) == E(\tau_0, \tau_1, \tau_2, \ldots) \otimes P(\sigma_1, \sigma_2, \ldots)$$
where
$$\tau_k = \text{ad}^{p^k-1}(v)(u)$$
$$\sigma_k = \frac{1}{2} \sum_{j=1}^{p^k-1} [\text{ad}^{j-1}(v)(u), \text{ad}^{p^k-j-1}(v)(u)]$$
have respective degrees $2p^k n - 1$ and $2p^k n - 2$ and
$$\beta^{r+1} \tau_0 = 0, \ \beta^{r+1} \sigma_k = 0, \ \beta^{r+1} \tau_k = \lambda_k \sigma_k$$
where λ_k is a nonzero scalar for all $k \geq 1$.

(c) *And*
$$E^{r+2} = \cdots = E^{\infty} = E(\tau_0).$$

11.5 The Bockstein spectral sequence of the loop space

Remark. Since the scalar λ_k is nonzero, we will often redefine σ_k to include this scalar in its definition so that $\beta^{r+1}\tau_k = \sigma_k$.

Proof: The mod p homology of the fibration sequence

$$\Omega F^{2n+1}\{p^r\} \to \Omega P^{2n+1}(p^r) \to \Omega S^{2n+1}$$

is the sequence of universal enveloping algebras

$$UL \to UL(u,v) \to UL(\iota)$$

induced by the short exact sequence of Lie algebras. Since $\beta^s = 0$, $s \leq r-1$ and $\beta^r v = u$ in $UL(u,v) = E^1(\Omega P^{2n+1}(p^r)) = E^r(\Omega P^{2n+1}(p^r))$ in the temporarily constant Bockstein spectral sequence, it follows that $UL = E^1(\Omega F^{2n+1}\{p^r\}) = E^r(\Omega F^{2n+1}\{p^r\})$ in the temporarily constant Bockstein spectral sequence, and that the sequence

$$E^r(\Omega F^{2n+1}\{p^r\}) \to E^r(\Omega P^{2n+1}(p^r)) \to E^r(\Omega S^{2n+1})$$

of r-th terms of Bockstein spectral sequences is the same as the above sequence of universal enveloping algebras at E^1. But it now has the nonzero differential β^r and we wish to compute the homology

$$H(E^r(\Omega F^{2n+1}\{p^r\}), \beta^r) = E^{r+1}(\Omega F^{2n+1}\{p^r\}).$$

Proposition 8.3.5 implies that $UL(u,v) = UL \otimes UL(\iota)$ as a differential UL module and as a differential $UL(\iota)$ comodule. Thus, Exercise 2 of Section 10.3 implies that there is a twisting morphism τ which makes $UL(u,v) = UL \otimes_\tau UL(\iota)$ into a twisted tensor product. Since this tensor product is acyclic,

$$HUL(u,v) = HT(u,v) = TH\langle u,v\rangle = T0 = \mathbb{Z}/p\mathbb{Z},$$

it follows that the computation of the homology of its fibre HUL is the same as the computation in Corollary 10.26.4 of the mod p homology $H(\Omega^2 S^{2n+1})$ of the double loop space of the odd-dimensional sphere, that is,

$$E^{r+1}(\Omega F^{2n+1}\{p^r\}) = HUL = H(\Omega UL(\iota))$$
$$= H(\Omega T(\iota)) = \mathrm{Cotor}^{T(\iota)}(\mathbb{Z}/p\mathbb{Z}, \mathbb{Z}/p\mathbb{Z})$$
$$= E(\tau_0, \tau_1, \tau_2, \dots) \otimes P(\sigma_1, \sigma_2, \dots)$$

as algebras where $\mathrm{degree}(\tau_k) = 2p^k n - 1$ and $\mathrm{degree}(\sigma_k) = 2p^k n - 2$.

Furthermore, for $k \geq 1$, τ_k is characterized by its homology suspension in the fibration sequence, that is, $\sigma(\tau_k) = \iota^k$. Since Lemma 9.5.2 gives the computation $\beta^r v^k = \mathrm{ad}^{p^k-1}(v)(u)$, it follows that $\tau_k = \mathrm{ad}^{p^k-1}(v)(u)$ which is a Lie bracket and therefore primitive.

We also know that the Bockstein differential β^r in the graded Lie algebra $L = L_{k \geq 1}(\mathrm{ad}^{k-1}(v)(u))$ is compatible with the the trigrading imposed by total degree, the number of occurences of u, and the number of occurences of v. By Lemma 9.5.4, the element $\bar{\sigma}_k = \frac{1}{2} \sum_{j=1}^{p^k-1} [\mathrm{ad}^{j-1}(v)(u), \mathrm{ad}^{p^k-j-1}(v)(u)]$ is a nonzero cycle in L with precisely two occurences of u. Since L is generated by elements with one occurence of u and since $\beta^r \tau_k = 0$, it is clear that $\bar{\sigma}_k$ is not a boundary in HUL. It is also clear from degree reasons that its homology class cannot be a product of two classes, each with one occurence of u. Thus, the generator $\sigma_k = \bar{\sigma}_k$, which is a Lie bracket and therefore primitive.

We know from Section 11.4 that the p-torsion in the integral homology of $\Omega F^{2n+1}\{p^r\}$ has exponent p^{r+1} and we know that this space is rationally equivalent to the sphere S^{2n-1}. Thus, $E^{r+2} = E^{r+3} = \cdots = E^\infty = \mathbb{Z}/p\mathbb{Z}$ and the only way to make E^{r+1} acyclic is via $\beta^r \tau_k = \lambda_k \sigma_k$ where $\lambda_k \neq 0$ for $k \geq 1$. $\square$

11.6 The decomposition of the homology of the loop space

The mod p homology Bockstein spectral sequence of $\Omega F^{2n+1}\{p^r\}$ inspires this section's construction of a tensor product decomposition of the mod p homology of this space.

We begin with an algebraic version of the Hurewicz theorem which relates the homology of a free differential algebra with that of its module of indecomposables.

Algebraic Hurewicz theorem 11.6.1. *Let A be a connected supplemented differential algebra with A and the module of indecomposables $Q(A)$ free as modules. Suppose that the homology of the indecomposables is n connected, that is, $H_k Q(A) = 0$ for $k < n$, then the natural map*

$$\overline{H}_k A \to H_k Q(A)$$

is an isomorphism if $k < 2n$ and an epimorphism if $k = 2n$.

Proof: Let $I(A)$ be the augmentation ideal. The powers of the augmentation ideal satisfy $I(A)^{k+!} \subseteq I(A)^k$ and therefore define an **increasing** filtration of A via:

$$F_k A = \begin{cases} A & \text{if } k \geq 0 \\ I(A) & \text{if } k = -1 \\ I(A)^{-k} & \text{if } k \leq -2. \end{cases}$$

Since A is connected, this filtration is finite in each fixed degree.

If $V_{-1,*+1} = Q(A)_*$, then the associated graded object of the above filtration is the tensor algebra

$$E^0(A) = T(V_{-1,*}).$$

11.6 The decomposition of the homology of the loop space

The map $A \to A/I(A)^2 = R \oplus Q(A)$ induces a map of associated graded objects

$$E^0(A) \to R \oplus Q(A).$$

$E^0(A)$ is the first term of a convergent second quadrant homology spectral sequence of algebras $E^s(A)$ with abutment the homology HA and there is a map of spectral sequences

$$E^s(A) \to E^s(Q(A)) = R \oplus Q(A)$$

where the range is the constant spectral sequence with abutment $R \oplus Q(A)$.

Since these are homology spectral sequences, the differentials are

$$d^s : E^s_{i,j} \to E^s i - s, j + s - 1.$$

It follows that

$$E^1(A) = H(T(V_{-1,*}), d^0) = T(H(V_{-1,*}, d^0)) = T(H_{*-1}Q(A))$$

with $H(V_{-1,*}, d^0) = 0$ if $* - 1 < n$, that is, if $* < n + 1$. The first possibly nonzero differential is $d^1 : E^1_{-1,2n+2} \to E^1 -2, 2n + 2$ which lands in the indecomposables in total degree $2n$. Hence, for all $s \geq 1$,

$$E^s(A) \to E^s(Q(A))$$

is an isomorphism in total degrees $< 2n$ and an epimorphism in total degree $= 2n$. Thus the same is true for E^∞ and the abutments. $\square$

The algebraic Hurewicz theorem enables the construction of a decreasing filtration of the differential Lie algebra

$$L = L(u, [v, u], \ldots, \mathrm{ad}^j(v)(u), \ldots) = L_{j \geq 0}(\mathrm{ad}^j(v)(u))$$

with $\mathrm{degree}(v) = 2n$, $\mathrm{degree}(u) = 2n - 1$, and differential $dv = \beta^r = u$.

Proposition 11.6.2. *There is a decreasing filtration L_k of the differential graded Lie algebra L defined by*

$$L_0 = L$$

and

$$L_1 = [L, L] = \mathrm{kernel} \quad L_0 \to \langle u \rangle$$

and, for $k \geq 1$, there are short exact sequences of differential graded Lie algebras

$$0 \to L_{k+1} \to L_k \to \langle \tau_k, \sigma_k \rangle \to 0$$

where $\langle \tau_k, \sigma_k \rangle$ is an abelian Lie algebra and each L_k is a free Lie algebra and

$$HUL_k = \bigotimes_{j=k}^{\infty} S(\tau_j, \sigma_j) \quad k \geq 1$$

as primitively generated Hopf algebras.

Proof: We observe by inspection that $HQL = H\langle u, [v,u], [v,[v,u]], \ldots\rangle$ contains $\langle u \rangle$ as a summand, in other words, QL contains $\langle u \rangle$ as a differential summand. Hence, there is a map of differential objects

$$L \to QL \to \langle u \rangle$$

and L_1 is the kernel of this map. Since L_1 is a subalgebra of a free Lie algebra, it is also a free Lie algebra.

Although inspection shows that the first two generators of L_1 are $[u,u]$ and $[v,u]$, the algebraic Hurewicz theorem will tell us everything we need to know about the generators.

The short exact sequence

$$0 \to L_1 \to L_0 \to \langle u \rangle \to 0$$

of differential Lie algebras yields a sequence of universal enveloping algebras

$$UL_1 \to UL_0 \to U\langle u \rangle$$

with $U\langle u \rangle = S(u) = E(u)$ and with the middle

$$UL_O = UL_1 \otimes_\tau U\langle u \rangle$$

being a twisted tensor product of the ends. Since

$$HUL_O = S(u) \otimes \bigotimes_{j=1}^{\infty} S(\tau_j, \sigma_j) \to HU\langle u \rangle = S(u)$$

is an epimorphism of homology, the algebraic version of the Eilenberg–Moore spectral sequence in Proposition 10.24.8 shows that

$$HUL_1 = \mathrm{Cotor}^{HU\langle u \rangle}(HUL_0, \mathbb{Z}/p\mathbb{Z})$$

$$= HUL_0 \square_{HU\langle u \rangle} \mathbb{Z}/p\mathbb{Z} = \bigotimes_{j=1}^{\infty} S(\tau_j, \sigma_j)$$

as Hopf algebras.

The algebraic Hurewicz theorem implies that $\langle \tau_1, \sigma_1 \rangle$ is a summand of the homology of the indecomposables $HQUL_1$, in other words, $\langle \tau_1, \sigma_1 \rangle$ is a differential summand of the module of indecomposables QUL_1.

It follows that the composition

$$L_1 \to QL_1 \to \langle \tau_1, \sigma_1 \rangle$$

is a map of differential Lie algebras and we define

$$L_2 = \quad \text{kernel} \quad L_1 \to \langle \tau_1, \sigma_1 \rangle.$$

11.6 The decomposition of the homology of the loop space

Again, since L_2 is a subalgebra of a free Lie algebra, it is also a free Lie algebra. Just as before, the short exact sequence of differential Lie algebras

$$0 \to L_2 \to L_1 \to \langle \tau_1, \sigma_1 \rangle \to 0$$

yields to a sequence of universal enveloping algebras

$$UL_2 \to UL_1 \to U\langle \tau_1, \sigma_1 \rangle$$

with

$$U\langle \tau_1, \sigma_1 \rangle = S(\tau_1, \sigma_1) = E(\tau_1) \otimes P(\sigma_1)$$

and with the middle equal to a twisted tensor product of the ends

$$UL_1 = UL_2 \otimes_\tau U\langle \tau_1, \sigma_1 \rangle$$

and the algebraic version of the Eilenberg–Moore spectral sequence in Proposition 10.24.8 shows that

$$HUL_2 = \mathrm{Cotor}^{HU\langle \tau_1, \sigma_1 \rangle}(HUL_1, \mathbb{Z}/p\mathbb{Z})$$
$$= HUL_1 \square_{HU\langle \tau_1, \sigma_1 \rangle} \mathbb{Z}/p\mathbb{Z} = \bigotimes_{j=2}^{\infty} S(\tau_j, \sigma_j)$$

as Hopf algebras.

Continuing in this way, we get for all $k \geq 1$ short exact sequences of differential Lie algebras

$$0 \to L_{k+1} \to L_k \to \langle \tau_k, \sigma_k \rangle \to 0$$

where each $\langle \tau_k, \sigma_k \rangle$ is an abelian Lie algebra and each L_k is a free Lie algebra and

$$HUL_{k+1} = HUL_k \square_{U\langle \tau_k, \sigma_k \rangle} \mathbb{Z}/p\mathbb{Z} = \bigotimes_{j=k+1}^{\infty} S(\tau_j, \sigma_j)$$

as Hopf algebras. $\square$

Let

$$L_\infty = \bigcap_{k \geq 1} L_k = \lim_{k \to \infty} L_k.$$

This is an intersection of decreasing Lie subalgebras of L and the filtration is finite (or eventually constant) in each degree. L_∞ is a subalgebra of L and therefore is also free Lie algebra. Furthermore,

$$HUL_\infty = \lim_{k \to \infty} HUL_k = \mathbb{Z}/p\mathbb{Z}.$$

The algebraic Hurewicz theorem implies that the module of indecomposables $QUL_\infty = QL_\infty = L_\infty/[L_\infty, L_\infty]$ is acyclic. That is,

$$HQL_\infty = 0$$

and therefore there is a set of generators $\{x_\alpha, dx_\alpha\}_\alpha$ of L_∞ which is finite in each degree. (Note that the basis begins with $[u,u]$ and $[v,u]$ in degrees $2n-2$ and $2n-1$. The acyclity of the generating set is illustrated by $d[v,u] = [u,u]$.)

Hence,

$$L_\infty = L(x_\alpha, dx_\alpha)_\alpha$$

and

$$UL_\infty = UL(x_\alpha, dx_\alpha)_\alpha = T(x_\alpha, dx_\alpha)_\alpha$$

is an acyclic tensor algebra.

We note that the isomorphisms in Proposition 11.6.2 are defined by multiplication

$$UL_{k+1} \otimes U\langle \tau_k, \sigma_k \rangle \to UL_k \otimes UL_k \xrightarrow{\text{mult}} UL_k$$

where

$$UL_{k+1} \to UL_k$$

is the inclusion and

$$U\langle \tau_k, \sigma_k \rangle = S(\tau_k, \sigma_k) \to UL_k$$

is the obvious $S(\sigma_k) = P(\sigma_k)$ right equivariant section determined by

$$1 \mapsto 1, \sigma_k \mapsto \sigma_k, \tau_k \mapsto \tau_k.$$

If we iterate the steps in the proof of Proposition 11.6.2, we get

$$UL = UL_1 \otimes S(u)$$
$$= UL_2 \otimes S(\tau_1, \sigma_2) \otimes S(u)$$
$$= UL_k \otimes \bigotimes_{j=1}^{k-1} S(\tau_j, \sigma_j) \otimes S(u)$$
$$= UL_\infty \otimes \bigotimes_{j=1}^{\infty} S(\tau_j, \sigma_j) \otimes S(u)$$
$$= T(x_\alpha, dx_\alpha)_\alpha \otimes \bigotimes_{j=1}^{\infty} S(\tau_j, \sigma_j) \otimes S(u)$$

as differential coalgebras and also as modules over the tensor algebra $T(x_\alpha, dx_\alpha)_\alpha$.

By construction, this isomorphism is defined by the limit of the maps

$$UL_{k+1} \otimes S(\tau_k, \sigma_k) \otimes S(\tau_{k-1}, \sigma_{k-1}) \otimes \cdots \otimes S(\tau_1, \sigma_1) \otimes S(u)$$

$$\to \bigotimes_{j=1}^{k+2} UL_k \xrightarrow{\text{mult}} UL_k.$$

It is important to note that the injection

$$S(\tau_k, \sigma_k) \otimes \cdots S(\tau_1, \sigma_1) \to UL$$

extends to an injection

$$S(\tau_{k+1}, \sigma_{k+1}) \otimes \cdots S(\tau_1, \sigma_1) \to UL.$$

Thus, the maps

$$S(\tau_k, \sigma_k) \otimes \cdots \otimes S(\tau_1, \sigma_1) \to UL \otimes \cdots \otimes UL \to UL$$

give a limit map

$$\Psi : \bigotimes_{k=1}^{\infty} S(\tau_k, \sigma_k) \to UL.$$

We then multiply this with other maps

$$\Phi : UL_\infty \otimes \bigotimes_{k=1}^{\infty} S(\tau_k, \sigma_k) \otimes S(u) \xrightarrow{\text{incl} \otimes \Psi \otimes \iota} UL \otimes UL \otimes UL \xrightarrow{\text{mult}} UL$$

to get an isomorphism of differential coalgebras.

Exercise

(1) Use the techniques of chapter 9 to show that L_1 is the free Lie algebra generated by

$$[u, u], [v, u], , [u, [v, u]], [v, [v, u]], [u, [v, [v, u]]], [v, [v, [v, u]]], \ldots.$$

11.7 The weak product decomposition of the loop space

Recall that the weak product of a countable set X_j of pointed spaces is the direct limit of the finite products

$$\prod_{j=1}^{\infty} X_j = \lim_{k \to \infty} \prod_{j=1}^{k} X_j.$$

Thus, with any field coefficients, the homology of a weak product is the tensor product of the homologies,

$$H\left(\prod_{j=1}^{\infty} X_j\right) = \bigotimes_{j=1}^{\infty} H(X_j).$$

The purpose of this section is to prove a weak product decomposition of the loop space $\Omega F^{2n+1}\{p^r\}$. This weak product decomposition is the geometric realization of the tensor product decomposition of the homology

$$T(x_\alpha, \beta^r x_\alpha)_\alpha \otimes \bigotimes_{k=1}^{\infty} S(\tau_k, \sigma_k) \otimes S(u) \simeq H(\Omega F^{2n+1}\{p^r\}; \mathbb{Z}/p\mathbb{Z})$$

which is given in the previous section.

Proposition 11.7.1. *Localized at an odd prime p, there is a homotopy equivalence*

$$\Phi: \Omega\Sigma \bigvee_\alpha P^{n_\alpha}(p^r) \times \prod_{j=1}^{\infty} S^{2p^j n-1}\{p^{r+1}\} \times S^{2n-1} \to \Omega F^{2n+1}\{p^r\}$$

where

(a)

$$\bigvee_\alpha P^{n_\alpha}(p^r)$$

is an infinite bouquet of mod p^r Moore spaces with only finitely many in each dimension and where the least value of n_α is $4n-1$,

(b)

$$S^{2p^j n-1}\{p^{r+1}\}$$

is the fibre of the degree p^{r+1} map

$$p^{r+1}: S^{2pjn-1} \to S^{2pjn-1},$$

(c) S^{2n-1} is the localized sphere.

Remark. Proposition 11.7.1 can be thought of as illustrating the mod p homology Bockstein spectral sequence of $\Omega F^{2n+1}\{p^r\}$. In this spectral sequence, the homology of the first factor

$$H\left(\Omega\Sigma \bigvee_\alpha P^{n_\alpha}(p^r)\right) = T(x_\alpha, \beta x_\alpha)_\alpha$$

11.7 The weak product decomposition of the loop space

is acyclic with respect to the differential β^r, the homology of the second factor

$$H\left(\prod_{j=1}^{\infty} S^{2p^j n - 1}\{p^r\}\right) = \bigotimes_{j=1}^{\infty} S(\tau_j, \sigma_j), \quad \beta^{r+1}\tau_j = \sigma_j$$

is acyclic with respect to the differential β^{r+1}, and the homology of the third factor

$$H(S^{2n-1}) = S(u) = E(u)$$

survives to E^{∞}.

Remark. Although Proposition 11.7.1 is true for all odd primes, the failure of the Jacobi identity for mod 3 homotopy groups makes it necessary for the proof to become very, very much more complicated if $p = 3$ and $r = 1$. Hence, for the purpose of this proof, we will assume that $p > 3$ or that $r \geq 2$.

Proof: We need to construct three maps

$$\chi : \Omega\Sigma \bigvee_{\alpha} P^{n_\alpha}(p^r) \to \Omega F^{2n+1}\{p^r\},$$

$$\theta : \prod_{j=1}^{\infty} S^{2p^j n - 1}\{p^r\} \to \Omega F^{2n+1}\{p^r\},$$

$$\mu : S^{2n-1} \to \Omega F^{2n+1}\{p^r\}$$

with respective images in mod p homology isomorphic to the tensor algebra

$$T(x_\alpha, \beta^r x_\alpha)_\alpha,$$

the tensor product

$$\bigotimes_{j=1}^{\infty} S(\tau_j, \sigma_j),$$

and the exterior algebra

$$S(u) = E(u).$$

Then multiplying these maps together via

$$\Omega\Sigma \bigvee_{\alpha} P^{n_\alpha}(p^r) \times \prod_{j=1}^{\infty} S^{2p^j n - 1}\{p^r\} \times S^{2n-1} \xrightarrow{\chi \times \theta \times \mu}$$

$$\Omega F^{2n+1}\{p^r\} \times \Omega F^{2n+1}\{p^r\} \times \Omega F^{2n+1}\{p^r\} \xrightarrow{\text{mult}} \Omega F^{2n+1}\{p^r\}$$

gives a mod p homology isomorphism.

Since these spaces are all finite type localized at p, this is a homology isomorphism with local coefficients $\mathbb{Z}_{(p)}$ and hence a homotopy equivalence of spaces localized at p. (Since these are H-spaces, it is a homotopy equivalence even in the nonsimply connected case when $n = 1$.)

Before beginning the construction of these three maps, we need some preliminaries.

First, note that we have maps
$$\nu = \Sigma : P^{2n}(p^r) \to \Omega P^{2n+1}(p^r)$$
and that the Bockstein (= restriction to the bottom cell) $\beta \nu = \mu$ factors through $\Omega F^{2n+1}\{p^r\}$, that is,
$$S^{2n-1} \xrightarrow{\mu} \Omega F^{2n+1}\{p^r\} \xrightarrow{\iota} \Omega P^{2n+1}(p^r).$$

If we are careful, no confusion will result by identifying the map
$$\nu : P^{2n}(p^r) \to \Omega P^{2n+1}(p^r)$$
with its mod p reduction
$$\rho \nu = \nu \cdot \rho : P^{2n}(p) \xrightarrow{\rho} P^{2n}(p^r) \xrightarrow{\nu} \Omega P^{2n+1}(p^r).$$

Similarly, we will identify the map μ with its mod p^r and mod p reductions $\rho \mu = \beta^r \nu$
$$P^{2n-1}(p^r) \to S^{2n-1} \xrightarrow{\mu} \Omega F^{2n+1}\{p^r\}$$
$$P^{2n-1}(p) \xrightarrow{\rho} P^{2n-1}(p^r) \to S^{2n-1} \xrightarrow{\mu} \Omega F^{2n+1}\{p^r\}.$$

We note that the Bockstein formula $\beta \nu = \mu$ is valid in integral homotopy or in mod p^r homotopy and the Bockstein differential formula $\beta^r \nu = \mu$ is valid in the mod p homotopy Bockstein spectral sequence.

The Hurewicz maps satisfy
$$\phi(\nu) = v, \quad \phi(\mu) = u,$$
mod p^r, mod p, or integrally, whatever is appropriate.

Second, since the composition is null homotopic in the cofibration sequence
$$S^{2n_1} \xrightarrow{p^r} S^{2n-1} \to P^{2n+1}(p^r)$$

11.7 The weak product decomposition of the loop space

it follows that we have a lift of this to the path space $P(P^{2n+1}(p^r))$ and hence a map of fibration sequences

$$\begin{array}{ccc} S^{2n-1}\{p^r\} & \xrightarrow{\psi} & \Omega P^{2n+1}(p^r) \\ \downarrow & & \downarrow \\ S^{2n-1} & \to & P(P^{2n+1}(p^r)) \\ \downarrow p^r & & \downarrow \\ S^{2n-1} & \to & P^{2n+1}(p^r) \end{array}$$

and that the map $\psi : S^{2n-1}\{p^r\} \to \Omega P^{2n+1}(p^r)$ is equivariant with respect to the right actions of ΩS^{2n+1} on the domain and range.

Thus, the map in mod p homology

$$H(S^{2n-1}\{p^r\}) = S(u,v) \to H(\Omega P^{2n+1}(p^r)) = T(u,v)$$

is the standard $S(v) = P(u) = T(u)$ right equivariant injection determined by $v \mapsto v$, $u \mapsto u$.

We now construct the map

$$\chi : \Omega\Sigma \bigvee_\alpha P^{n_\alpha}(p^r) \to \Omega F^{2n+1}\{p^r\}.$$

Since the elements x_α which are generators of the tensor algebra $T(x_\alpha, \beta^r x_\alpha)_\alpha$ are Lie brackets of the elements v and u, they can be interpreted as relative Samelson products

$$\chi_\alpha : P^{n_\alpha}(p^r) \to \Omega F^{2n+1}\{p^r\}$$

of the homotopy classes ν and μ. We can also reduce mod p and, whether we do or not, the Hurewicz map satisfies

$$\phi(\chi_\alpha) = x_\alpha.$$

We also have the relative Samelson products

$$\beta^r \chi_\alpha : P^{n_\alpha - 1}(p^r) \to \Omega F^{2n+1}\{p^r\}$$

and the Hurewicz map satisfies

$$\phi(\beta^r \chi_\alpha) = \beta^r \phi(\chi_\alpha) = \beta^r x_\alpha.$$

For each α we have a map $\chi_\alpha : P^{n_\alpha}(p^r) \to \Omega F^{2n+1}\{p^r\}$ with the reduced homology image spanned by $\phi(\chi_\alpha) = x_\alpha$ and $\phi(\beta^r \chi_\alpha) = \beta^r x_\alpha$. We add these up to get a map

$$\bigvee_\alpha P^{n_\alpha}(p^r) \to \Omega F^{2n+1}\{p^r\}.$$

Denote the multiplicative extension of this map by

$$\chi : \Omega\Sigma \bigvee_\alpha P^{n_\alpha}(p^r) \to \Omega F^{2n+1}\{p^r\}.$$

We note that χ induces an isomorphism in mod p homology

$$\chi : H\left(\Omega\Sigma \bigvee_\alpha P^{n_\alpha}(p^r)\right) \to T(x_\alpha, \beta^r) \subseteq H(\Omega F^{2n+1}\{p^r\})$$

onto the tensor subalgebra.

We now construct the map

$$\theta : \prod_{j=1}^{\infty} S^{2p^j n - 1}\{p^{r+1}\} \to \Omega F^{2n+1}\{p^r\}.$$

It is sufficient to construct the maps

$$\theta_k : S^{2p^j n - 1}\{p^r\} \to \Omega F^{2n+1}\{p^r\}$$

which are mod p homology injections onto

$$S(\tau_k, \sigma_k) \subseteq UL$$

and then use the loop multiplication on $\Omega F^{2n+1}\{p^r\}$ to multiply them together. It is convenient to use the Moore loops with strict associativity and strict unit. In this way, the map of the finite product

$$\prod_{j=1}^{k} S^{2p^j n - 1}\{p^{r+1}\} \to \Omega F^{2n+1}\{p^r\}$$

extends to a map of one more factor

$$\prod_{j=1}^{k} S^{2p^j n - 1}\{p^{r+1}\} \to \Omega F^{2n+1}\{p^r\}$$

and one can take the direct limit to get a map of the infinite weak product.

Consider the relative Samelson product

$$\tau_k(\nu) = \mathrm{ad}^{p^k - 1}(\nu)(\mu) : P^{2p^k n - 1}(p^r) \to \Omega F^{2n+1}\{p^r\}.$$

In the presence of the Jacobi identity (which is valid if either $p > 3$ or if $r \geq 2$), Lemma 9.5.2 implies that the Bockstein $\beta\tau_k(\nu)$ is divisible by p ($=$ zero mod p), that is, with respect to the Bockstein cofibration sequence

$$P^{2p^k n - 2}(p) \xrightarrow{\beta} P^{2p^k n - 1}(p^r) \xrightarrow{\rho} P^{2p^k n - 1}(p^{r+1})$$

11.7 The weak product decomposition of the loop space

the composition

$$\beta \tau_k(\nu) = \tau_k(\nu) \cdot \underline{\beta} : P^{2p^k n-2}(p) \xrightarrow{\underline{\beta}} P^{2p^k n-1}(p^r) \xrightarrow{\tau_k(\nu)} \Omega F^{2n+1}\{p^r\}$$

is null homotopic. Hence, there exists an extension of $\tau_k(\nu)$ to a map

$$\overline{\tau}_k(\nu) : P^{2p^k n-1}(p^{r+1}) \to \Omega F^{2n+1}\{p^r\}$$

such that the mod p^r reduction satisfies

$$\rho \overline{\tau}_k(\nu) = \overline{\tau}_k \cdot \underline{\rho} = \tau_k(\nu).$$

Let e generate the top nonzero mod p homology group of $P^{2p^k n-1}(p^r)$ and let $\overline{e}$ generate the top nonzero mod p homology group of $P^{2p^k n-1}(p^{r+1})$. Then e maps to $\overline{e}$. Since the Hurewicz map in mod p homology satisfies

$$\phi(\tau_k(\nu)) = \tau_k(\nu)_*(e) = \tau_k(v) = \tau_k,$$

it follows that

$$(\overline{\tau}_k(\nu))_*(\overline{e}) = \tau_k(v) = \tau_k.$$

In addition,

$$(\overline{\tau}_k(\nu))_*(\beta^r \overline{e}) = \beta^r (\overline{\tau}_k(\nu))_*(\overline{e}) = \beta^r \tau_k(v) = \sigma_k(v)$$

up to a nonzero scalar.

Let

$$\theta_k : \Omega \Sigma P^{2p^k n-1}(p^{r+1}) \to \Omega F^{2n+1}\{p^r\}$$

be the multiplicative extension of $\overline{\tau}_k(\nu)$ and let

$$\theta_k : S^{2p^k n-1}\{p^{r+1}\} \xrightarrow{\psi} \Omega P^{2p^k n+1}(p^{r+1}) \to \Omega P^{2n+1}(p^r)$$

be the composition with the equivariant map ψ constructed above.

Then θ_k maps the mod p homology $H(S^{2p^k n-1}\{p^{r+1}\})$ isomorphically onto $S(\tau_k, \sigma_k) \subseteq UL$, as desired.

Finally, the fact that

$$\mu : S^{2n_1} \to \Omega F^{2n+1}\{p^r\}$$

maps the mod p homology isomorphically onto $S(u) = E(u) \subseteq UL$ completes the proof of the weak product decomposition Theorem 11.7.1. □

We adopt the shorthand notations

$$\Omega_r = \Omega\Sigma \bigvee_\alpha P^{n_\alpha}(p^r)$$

$$\prod_r = \prod_{j=1}^{\infty} S^{2p^j n-1}\{p^r\}$$

so that the equivalence in 11.7.1 becomes

$$\Phi : \Omega_r \times \prod_{r+1} \times S^{2n-1} \xrightarrow{\simeq} \Omega F^{2n+1}\{p^r\}.$$

We use the weak product decomposition of Proposition 11.7.1 to define:

Definition 11.7.2. For all $r \geq 1$, the map of localized spaces

$$\pi_r : \Omega S^{2n+1} \to S^{2n-1}$$

is defined to be the composition

$$proj \cdot \Phi^{-1} \cdot \Omega\partial : \quad \Omega^2 S^{2n+1} \to \Omega F^{2n+1}\{p^r\} \to$$

$$\Omega_r \times \prod_{r+1} \times S^{2n-1} \to S^{2n-1}$$

where $\Omega\partial$ is the loop of the connecting map in the fibration sequence, Φ^{-1} is the homotopy inverse to the equivalence in Proposition 11.7.1, and $proj$ is the projection onto the sphere factor of the product decomposition.

One easily checks that the map π_r has degree p^r on the bottom cell, that is, that the composition with the double suspension map

$$\pi_r \cdot \Sigma^2 : S^{2n-1} \to \Omega S^{2n+1} \to S^{2n-1}$$

has degree p^r. The case $r = 1$ is particularly important and will be denoted by $\pi = \pi_1$.

In the next section, we will prove another weak product decomposition which gives further information about the maps π_r.

Exercises

(1) Verify that the map π_r in Definition 11.7.2 has degree p^r on the bottom cell.

(2) Use Toda's odd primary exponent result

$$p^2 \pi * (\Omega S^{2n+1}) \subseteq \Sigma^2 \pi_*(S^{2n-1})$$

and the map π in Definition 11.7.2 to prove that

$$p^3 \pi_*(\Omega S^{2n+1}) \subseteq p^2 \Sigma^2 \pi_*(S^{2n-1}).$$

(3) If p is an odd prime, prove that p^{n+1} annihilates the p-primary component of $\pi_*(S^{2n+1})$ for all $n \geq 1$.

11.8 The odd primary exponent theorem for spheres

Let p be an odd prime. In this section, we prove the odd primary exponent theorem for odd dimensional spheres. This is a consequence of a weak product decomposition for the loops on the homotopy theoretic fibre $E^{2n+1}\{p^r\}$ of a map $P^{2n+1}(p^r) \to S^{2n+1}\{p^r\}$.

Since the composition in the cofibration sequence

$$P^{2n+1}(p^r) \xrightarrow{q} S^{2n+1} \xrightarrow{p^r} S^{2n+1}$$

is null homotopic, we can replace p^r by a fibration and (assuming that this has been done but not changing the notation) we get as in Proposition 3.2.3 the fibre extension of a totally fibred square, that is, up to homotopy a commutative diagram

$$\begin{array}{ccccc}
E^{2n+1}\{p^r\} & \to & P^{2n+1}(p^r) & \xrightarrow{t} & S^{2n+1}\{p^r\} \\
\downarrow \lambda & & \downarrow = & & \downarrow j \\
F^{2n+1}\{p^r\} & \to & P^{2n+1}(p^r) & \xrightarrow{q} & S^{2n+1} \\
\downarrow \kappa & & \downarrow & & \downarrow p^r \\
\Omega S^{2n+1} & \to & * & \to & S^{2n+1}
\end{array}$$

in which the rows and columns are all fibration sequences up to homotopy.

The subject of this section is the loop of this square and its continuation one step to the left as a 3×4 diagram in which all the rows and columns are fibration sequences up to homotopy. That is, in the homotopy commutative diagram below, all rows and columns are fibration sequences up to homotopy. (Recall that the loop of a k-th power map is also a k-th power map.)

$$\begin{array}{ccccccc}
\Omega^2 S^{2n+1}\{p^r\} & \xrightarrow{\delta} & \Omega E^{2n+1}\{p^r\} & \to & \Omega P^{2n+1}(p^r) & \xrightarrow{\Omega t} & \Omega S^{2n+1}\{p^r\} \\
\downarrow \Omega^2 j & & \downarrow \Omega \lambda & & \downarrow = & & \downarrow \Omega j \\
\Omega^2 S^{2n+1} & \xrightarrow{\Omega \partial} & \Omega F^{2n+1}\{p^r\} & \to & \Omega P^{2n+1}(p^r) & \xrightarrow{\Omega q} & \Omega S^{2n+1} \\
\downarrow \Omega^2 p^r & & \downarrow \Omega \kappa & & \downarrow & & \downarrow \Omega p^r \\
\Omega^2 S^{2n+1} & \xrightarrow{=} & \Omega^2 S^{2n+1} & \to & * & \to & \Omega S^{2n+1}
\end{array}$$

Let $C(n)$ be the homotopy theoretic fibre of the double suspension $\Sigma^2 : S^{2n-1} \to \Omega^2 S^{2n+1}$. Since the map

$$\Omega \kappa : \Omega F^{2n+1}\{p^r\} \to \Omega^2 S^{2n+1}$$

is degree 1 on the bottom factor S^{2n-1} of the weak product decomposition of $\Omega F^{2n+1}\{p^r\}$, there is a map of fibration sequences

$$\begin{array}{ccccc} C(n) & \to & S^{2n-1} & \xrightarrow{\Sigma^2} & \Omega^2 S^{2n+1} \\ \downarrow & & \downarrow \mu & & \downarrow = \\ \Omega E^{2n+1}\{p^r\} & \to & \Omega F^{2n+1}\{p^r\} & \xrightarrow{\Omega \kappa} & \Omega^2 S^{2n+1} \end{array}.$$

We note that multiplication of the double suspension sequence by the factor $\Omega_r \times \prod_{r+1}$ yields the fibration sequence

$$\Omega_r \times \prod_{r+1} \times C(n) \to \Omega_r \times \prod_{r+1} \times S^{2n-1} \xrightarrow{\tau} \Omega^2 S^{2n+1},$$

where the fibre space projection τ is the composition of the projection on the sphere factor with the double suspension, $\tau = \Sigma^2 \cdot proj_1$.

The main result of this section is that the above fibration sequence is equivalent to one which occurs in the above fibre extension of the totally fibred square.

Proposition 11.8.1. *There is a homotopy equivalence from the top fibration sequence to the bottom fibration sequence in*

$$\begin{array}{ccccc} \Omega_r \times \prod_{r+1} \times C(n) & \to & \Omega_r \times \prod_{r+1} \times S^{2n-1} & \to & \Omega^2 S^{2n+1} \\ \downarrow \Psi & & \downarrow \Phi & & \downarrow = \\ \Omega E^{2n+1}\{p^r\} & \to & \Omega F^{2n+1}\{p^r\} & \xrightarrow{\Omega \kappa} & \Omega^2 S^{2n+1} \end{array}.$$

Furthermore, the vertical maps are constructed by multiplying the above map of the double suspension sequence by lifts of the maps χ and θ used in Proposition 11.7.1.

One gets the immediate corollary:

Corollary 11.8.2. *If p is an odd prime, then the compositions*

$$\pi_r \cdot \Sigma^2 : S^{2n-1} \to \Omega^2 S^{2n+1} \to S^{2n-1}$$

and

$$\Sigma^2 \cdot \pi_r : \Omega^2 S^{2n+1} \to S^{2n-1} \to \Omega^2 S^{2n+1}$$

are both p^r power maps, that is,

$$\pi_r \cdot \Sigma^2 = p^r, \quad \Sigma^2 \cdot \pi_r = \Omega^2 p^r = p^r.$$

Proof of Corollary 11.8.2: The first equation $\pi_r \cdot \Sigma^2 = p^r$ is a consequence of the definition of π_r as the projection on the bottom cell of $\Omega F^{2n+1}\{p^r\}$. In detail, the fact that the integral homotopy exact sequence of the fibration

$$\pi_{2n-1}(\Omega^2 S^{2n+1}) \xrightarrow{\Omega \partial} \pi_{2n-1}(\Omega F^{2n+1}\{p^r\}) \to \pi_{2n-1}(\Omega P^{2n+1}(p^r))$$

11.8 The odd primary exponent theorem for spheres

is equal to

$$\mathbb{Z} \xrightarrow{p^r} \mathbb{Z} \to \mathbb{Z}/p^r\mathbb{Z}$$

implies the first equation.

If we use the equivalence of Proposition 11.8.1 to identify

$$\Omega_r \times \prod_{r+1} \times S^{2n-1} \cong \Omega F^{2n+1}\{p^r\},$$

then the map $\Omega\kappa$ annihilates the first two factors and therefore the composition $\Omega p^r = \Omega\kappa \cdot \Omega\partial$ factors through the bottom cell S^{2n-1}.

In detail, the equation $\Sigma^2 \cdot \pi_r = \Omega^2 p^r$ follows from the fact that

$$\Omega\kappa \cdot \Phi \left(\Omega_r \times \prod_{r+1} \right) = *$$

and thus

$$\Omega\kappa \cdot \Phi = \Omega\kappa \cdot \mu \cdot proj.$$

Hence,

$$\Omega^2 p^r = \Omega\kappa \cdot \Omega\partial = \Omega\kappa \cdot \Phi \cdot \Phi^{-1} \cdot \Omega\partial =$$
$$\Omega\kappa \cdot \mu \cdot proj \cdot \Phi^{-1} \cdot \Omega\partial = (\Omega\kappa \cdot \mu) \cdot (proj \cdot \Phi^{-1} \cdot \Omega\partial) = \Sigma^2 \cdot \pi_r.$$

□

By setting $r = 1$ in Corollary 11.8.2, we immediately get the odd primary exponent theorem for odd dimensional spheres.

Corollary 11.8.3. *If p is an odd prime, then the localized groups satisfy*

$$p\,\pi_*(\Omega^2 S^{2n+1}) \subseteq \Sigma^2 \pi_*(S^{2n-1})$$

and induction starting with $n = 1$ gives

$$p^n \pi_*(S^{2n+1}) = 0 \quad \text{for all } n \geq 1.$$

Proof of Proposition 11.8.1: We begin with a discussion of Samelson products in the loop space $\Omega P^{2n+1}(p^r)$, relative Samelson products in the fibration sequence

$$\Omega F^{2n+1}\{p^r\} \to \Omega P^{2n+1}(p^r) \to \Omega S^{2n+1},$$

and Samelson products over the loops on an H-space in the fibration sequence

$$\Omega E^{2n+1}\{p^r\} \to \Omega P^{2n+1}(p^r) \to \Omega S^{2n+1}\{p^r\}.$$

Samelson products of interest are formed from the even class $\nu : P^{2n}(p^r) \to \Omega P^{2n+1}(p^r)$ and the odd class $\mu : S^{2n-1} \to \Omega F^{2n+1}\{p^r\} \to \Omega P^{2n+1}(p^r)$.

It is obvious that the only nontrivial length 2 products are $[\nu, \mu]$, $[\mu, \mu]$ and, even if we know nothing about relative products or products over the loops on an H-space, it is obvious that these lift to $\Omega F^{2n+1}\{p^r\}$ since at least one of the factors vanishes in the base ΩS^{2n+1}. Moreover, these lifts will have proper Hurewicz images as commutators since there is a mod p homology monomorphism $\Omega F^{2n+1}\{p^r\} \to \Omega P^{2n+1}(p^r)$.

But we need to know about the theory of relative products in order to call these lifts Samelson products and to use the Lie identities. For example, the Lie identity $[\mu, [\nu, \mu]] = [[\mu, \nu], \mu] + [\nu, [\mu, \mu]]$ needs to be justified in $\Omega F^{2n+1}\{p^r\}$ by the theory of relative Samelson products. Lie identities are necessary in order to prove the vanishing of the r-th Bockstein differential, $\beta^r \tau_k(\nu) = 0$. This identity is vital in the construction of the map

$$\prod_{r+1} \to \Omega F^{2n+1}\{p^r\}.$$

Similarly, it is obvious that the nontrivial length 2 products $[\nu, \mu]$, $[\mu, \mu]$ lift to $\Omega E^{2n+1}\{p^r\}$ even though neither of the two classes ν, μ vanish in the base $\Omega S^{2n+1}\{p^r\}$. This base is the loops on an H-space and hence Samelson products vanish in it and thus lift to the fibre $\Omega E^{2n+1}\{p^r\}$. But we cannot call these lifts Samelson products without a theory to include them, and hence we cannot use the Lie identities. Fortunately, we have the theory of Samelson products over the loops on an H-space. (Historical note: When Proposition 11.8.1 was first proved, there was a theory of relative products but no theory of products over the loops on an H-space. This was overcome then by an ad hoc argument)

Since S^{2n+1} is an H-space, the relative Samelson products

$$\chi_\alpha : P^{n_\alpha}(p^r) \to \Omega F^{2n+1}\{p^r\}$$

which occur in the proof of Proposition 11.7.1 can be regarded as Samelson products defined over the loops on an H-space. Since the multiplication on $S^{2n+1}\{p^r\}$ is defined so that the map $S^{2n+1}\{p^r\} \to S^{2n+1}$ is an H-map, these Samelson products are natural with respect to the maps of fibration sequences

$$\begin{array}{ccccc} \Omega E^{2n+1}\{p^r\} & \to & \Omega P^{2n+1}(p^r) & \xrightarrow{\Omega t} & \Omega S^{2n+1}\{p^r\} \\ \downarrow \Omega \lambda & & \downarrow = & & \downarrow \Omega j \\ \Omega F^{2n+1}\{p^r\} & \to & \Omega P^{2n+1}(p^r) & \xrightarrow{\Omega q} & \Omega S^{2n+1} \end{array}$$

and we can regard the Samelson products χ_α as being in $\Omega E^{2n+1}\{p^r\}$. If we add them up and take the multiplicative extensions, we get a map

$$\bar\chi : \Omega_r \to \Omega E^{2n+1}\{p^r\}$$

which is a lift of the map

$$\chi : \Omega_r \to \Omega F^{2n+1}\{p^r\}.$$

11.8 The odd primary exponent theorem for spheres

Since we are localized at an odd prime we can assume that the H-space structures on S^{2n+1} and $S^{2n+1}\{p^r\}$ are homotopy commutative and hence that the Jacobi identities hold for Samelson products defined over the loops on an H-space. It follows that, not only do the elements

$$\tau_k(\nu) = \mathrm{ad}^{p^k-1}(\nu)(\mu)$$

lift from $\Omega F^{2n+1}\{p^r\}$ to $\Omega E^{2n+1}\{p^r\}$, the Bocksteins $\beta \tau_k(\nu)$ are zero mod p in both spaces. Hence there exist mod p^{r+1} extensions $\bar{\tau}_k(\nu)$ of the elements $\tau_k(\nu)$ in both spaces. That is, we have factorizations

$$P^{2p^k n-1}(p^r) \to P^{2p^k n-1}(p^{r+1}) \xrightarrow{\bar{\tau}_k(\nu)} \Omega E^{2n+1}\{p^r\} \to \Omega F^{2n+1}\{p^r\}.$$

Just as in the proof of Proposition 11.7.1, we get lifts to maps

$$\theta : S^{2p^k n+1}\{p^{r+1}\} \to \Omega E^{2n+1}\{p^r\}$$

and can multiply them together to get a map of the weak infinite product

$$\theta : \prod_{r+1} \to \Omega E^{2n+1}\{p^r\}$$

which is a lift of the previous map θ into $\Omega F^{2n+1}\{p^r\}$.

We start with a map of fibration sequences

$$\begin{array}{ccccc}
C(n) & \to & S^{2n-1} & \to & \Omega^2 S^{2n+1} \\
\downarrow & & \downarrow \Phi & & \downarrow = \\
\Omega E^{2n+1}\{p^r\} & \to & \Omega F^{2n+1}\{p^r\} & \xrightarrow{\Omega \kappa} & \Omega^2 S^{2n+1}
\end{array}$$

and multiply the left-hand vertical map by the maps

$$\Omega_r \to \Omega E^{2n+1}\{p^r\}, \quad \prod_{r+1} \to \Omega E^{2n+1}\{p^r\}$$

and the middle vertical map by

$$\Omega_r \to \Omega F^{2n+1}\{p^r\}, \quad \prod_{r+1} \to \Omega F^{2n+1}\{p^r\}$$

to get the map of fibration sequences

$$\begin{array}{ccccc}
\Omega_r \times \prod_{r+1} \times C(n) & \to & \Omega_r \times \prod_{r+1} \times S^{2n-1} & \to & \Omega^2 S^{2n+1} \\
\downarrow \Psi & & \downarrow \Phi & & \downarrow = \\
\Omega E^{2n+1}\{p^r\} & \to & \Omega F^{2n+1}\{p^r\} & \xrightarrow{\Omega \kappa} & \Omega^2 S^{2n+1}
\end{array}.$$

of Since the maps of the bases and the total spaces are homotopy equivalences of spaces localized at p, the map Ψ of the fibre spaces is also a homotopy equivalence of spaces localized at p. This completes the proof of Proposition 11.8.1. □

11.9 H-space exponents

The factorization of the p-th power on the double loop space

$$p = \Sigma^2 \cdot \pi : \Omega^2 S^{2n+1} \xrightarrow{\pi} S^{2n-1} \xrightarrow{\Sigma^2} \Omega^2 S^{2n+1}$$

lifts to a map of $2n-1$ connected covers via the maps of fibration sequences

$$\begin{array}{ccccc}
(\Omega^2 S^{2n+1})\langle 2n-1\rangle & \xrightarrow{\bar{\pi}} & S^{2n-1}\langle 2n-1\rangle & \xrightarrow{\overline{\Sigma^2}} & (\Omega^2 S^{2n+1})\langle 2n-1\rangle \\
\downarrow & & \downarrow & & \downarrow \\
\Omega^2 S^{2n+1} & \xrightarrow{\pi} & S^{2n-1} & \xrightarrow{\Sigma^2} & \Omega^2 S^{2n+1} \\
\downarrow & & \downarrow & & \downarrow \\
K(\mathbb{Z}_{(p)}, 2n-1) & \xrightarrow{p} & K(\mathbb{Z}_{(p)}, 2n-1) & = & K(\mathbb{Z}_{(p)}, 2n-1).
\end{array}$$

Thus we have a factorization of the p-th power on the connected covers

$$p : (\Omega^2 S^{2n+1})\langle 2n-1\rangle \xrightarrow{\bar{\pi}} S^{2n-1}\langle 2n-1\rangle \xrightarrow{\overline{\Sigma^2}} (\Omega^2 S^{2n+1})\langle 2n-1\rangle.$$

If we iterate this we get a factorization of the p^n-th power on universal covers of $2n$-fold loop spaces

$$p^n : (\Omega^{2n} S^{2n+1})\langle 1\rangle \to S^1\langle 1\rangle \xrightarrow{\overline{\Sigma^{2n}}} (\Omega^{2n} S^{2n+1})\langle 1\rangle.$$

Since the universal cover $S^1\langle 1\rangle = R$ is contractible, we get the following H-space exponent theorem.

Proposition 11.9.1. *If p is an odd prime, the p^n-th power map is null homotopic on the universal cover of the localized $2n$-fold loop space*

$$(\Omega^{2n} S^{2n+1})\langle 1\rangle = \Omega^{2n}(S^{2n+1}\langle 2n+1\rangle).$$

Remark. The above proposition was first proved for $n = 1$ by Selick, for $p > 3$ by Cohen, Moore, and Neisendorfer, and for $p = 3$ by Neisendorfer. Earlier results due to James for the prime 2 and due to Toda for odd primes had shown that the p^{2n}-th power map is null homotopic on the component of the identity in the localized iterated loop spaces

$$\Omega^{2n+1} S^{2n+1}.$$

Remark. Exercise 1 of Section 2.10 shows that, for any prime p, there is no H-space exponent for the localized loop space

$$\Omega^{2n-2}(S^{2n+1}\langle 2n+1\rangle),$$

that is, no matter how large k is, the p^k-th power map is not null homotopic on this space. The best possible results on number of loops [115] are known since Selick's result for the three-sphere combines with the above factorization of the

p-th power map to improve Proposition 11.9.1 to show that only $2n - 1$ loops are required, that is, p^n is null homotopic on the localized loop space

$$\Omega^{2n-1}(S^{2n+1}\langle 2n+1\rangle).$$

We conclude this section with:

Proposition 11.9.2. *If p is an odd prime, then the p-th power map is null homotopic on the localized fibre $C(n)$ of the double suspension $\Sigma^2 : S^{2n-1} \to \Omega^2 S^{2n+1}$.*

Proof: Recall that, if we replace the double suspension by a homotopy equivalent fibration

$$\tau : \tilde{S}^{2n-1} \to \Omega^2 S^{2n+1},$$

then the multiplication on the total space is homotopic to one for which τ is a strict H-map and the multiplication on the fibre $C(n)$ is defined so that the inclusion $C(n) \to \tilde{S}^{2n_1}$ is a strict H-map. It follows that the p-th power maps induce maps of fibration sequences

$$\begin{array}{ccc} C(n) & \xrightarrow{p} & C(n) \\ \downarrow & & \downarrow \\ \tilde{S}^{2n-1} & \xrightarrow{p} & \tilde{S}^{2n-1} \\ \downarrow & & \downarrow \\ \Omega^2 S^{2n+1} & \xrightarrow{p} & \Omega^2 S^{2n+1}. \end{array}$$

We claim that the above map of fibration sequences factors up to fibre homotopy as

$$\begin{array}{ccccc} C(n) & \to & * & \to & C(n) \\ \downarrow & & \downarrow & & \downarrow \\ \tilde{S}^{2n-1} & \xrightarrow{p} & S^{2n-1} & \xrightarrow{\simeq} & \tilde{S}^{2n-1} \\ \downarrow & & \downarrow = & & \downarrow \\ \Omega^2 S^{2n+1} & \xrightarrow{\pi} & S^{2n-1} & \xrightarrow{\Sigma^2} & \Omega^2 S^{2n+1}. \end{array}$$

Observe that the lower left-hand square is homotopy commutative but, since the identity map $S^{2n_1} \to S^{2n-1}$ is a fibration, it can be assumed to be strictly commutative. That the lower right-hand square is strictly commutative is a property of the standard construction which replaces a map by a fibration.

The maps on the bases of the two diagrams are homotopic. After a fibre homotopy on the first diagram we can assume that the maps on the bases are equal. Given that, the only obstruction to the existence of a fibre homotopy lies in

$$H^{2n-1}(\tilde{S}^{2n-1}; \pi_{2n-1}C(n)) = 0.$$

$\square$

11.10 Homotopy exponents of odd primary Moore spaces

In this section we prove exponent theorems for odd primary Moore spaces. For p an odd prime, we give complete details for the proof that $p^{2r+1}\pi_*(P^m(p^r)) = 0$ for all $m \geq 3$. We also prove the best possible result that $p^{r+1}\pi_*(P^m(p^r)) = 0$ for all $m \geq 3$, but for the latter we need a hard lemma, the proof of which is too complicated to give here. In both cases, we prove that there are H-space exponents for the double loop spaces $\Omega^2 P^m(p^r)$.

Remark. If we recall that the universal cover of $P^2(s)$ has the same homotopy type as a bouquet of $s - 1$ copies of S^2, then we see that the restriction to Moore spaces of dimension $m \geq 3$ is necessary in order for the homotopy groups to have an exponent.

We begin with the remarkable

Lemma 11.10.1. *If the suspension Σf of a map $f : \Omega\mathbb{Z} \to X$ out of a loop space admits a retraction $\theta : \Sigma X \to \Sigma\Omega\mathbb{Z}$, $\theta \cdot \Sigma f = 1_{\Sigma\Omega\mathbb{Z}}$, then f admits a retraction $\kappa : X \to \Omega\mathbb{Z}$, $\kappa \cdot f = 1_\mathbb{Z}$*

Proof: Consider the diagram

$$\begin{array}{ccccc} \Omega\mathbb{Z} & \xrightarrow{f} & X & & \\ \downarrow \Sigma & & \downarrow \Sigma & & \\ \Sigma\Omega\mathbb{Z} & \xrightarrow{\Omega\Sigma f} & \Omega\Sigma X & \xrightarrow{\Omega g} & \Omega\Sigma\Omega\mathbb{Z} \\ \downarrow \Omega e & & & & \downarrow \Omega e \\ \Omega\mathbb{Z} & & \xrightarrow{1} & & \Omega\mathbb{Z} \end{array}$$

where $e : \Sigma\Omega\mathbb{Z} \to \mathbb{Z}$ is the evaluation.

Then $\kappa = \Omega e \cdot \Omega g \cdot \Sigma$ is the retraction. $\square$

Lemma 11.10.2. *If $g : W \to Y$ is a mod p homology monomorphism and W and Y are both bouquets of mod p^r Moore spaces with r fixed, then there is a retraction $h : Y \to W$, $h \cdot g = 1_W$.*

We need a lemma to prove a lemma.

Lemma 11.10.3. *If A is an acyclic differential module over a field, then A has a so-called acyclic basis of the form $\{x_\alpha, dx_\alpha\}_\alpha$.*

Proof: We have $d : A/\ker d \xrightarrow{\cong} \ker d$. Pick elements x_α in A which project to a basis $\overline{x}_\alpha$ of $A/\ker d$. Then dx_α is a basis of $\ker d$. The exact sequence

$$0 \to \ker d \to A \to A/\ker d \to 0$$

shows that $\{x_\alpha, dx_\alpha\}_\alpha$ is a basis for A. $\square$

11.10 Homotopy exponents of odd primary Moore spaces

Proof of 11.10.2: With mod p coefficients and the r-th Bockstein differential β^r, the reduced homology groups $\overline{H}(W)$ and $\overline{H}(Y)$ are both acyclic. Hence the quotient $\overline{H}(Y)/\overline{H}(W)$ is also acyclic. Choose a subset $\{y_\beta, \beta^r y_\beta\}_\beta \subseteq \overline{H}(Y)$ which projects to an acyclic basis of $\overline{H}(Y)/\overline{H}(W)$.

If X is a bouquet of mod p^r Moore spaces and $x \in \overline{H}_k(X)$, there is a map $P^k(p^r) \to X$ which sends the k-dimensional generator in the mod p homology of $P^k(p^r)$ to x. Thus the image in reduced mod p homology is generated by x, $\beta^r x$.

Let

$$\mathbb{Z} = \bigvee_\beta P^{|y_\beta|}(p^r) \to Y$$

be a map of a bouquet of mod p^r Moore spaces whose span is $\langle y_\beta, \beta^r y_\beta \rangle_\beta$.

Then the map $W \vee \mathbb{Z} \to Y$ is a mod p homology isomorphism, therefore a homotopy equivalence. Using this equivalence, we can project onto the first summand, $Y \to W$, to get a retraction. $\square$

Lemmas 11.10.1 and 11.10.2 imply:

Corollary 11.10.4. *The map $\chi : \Omega_r \to \Omega P^{2n+1}(p^r)$ has a retraction $\kappa : \Omega P^{2n+1}(p^r) \to \Omega_r$, $\kappa \cdot \chi = 1_{\Omega_r}$.*

The classifying maps

$$\Sigma \bigvee_\alpha P^{n_\alpha} \to E^{2n+1}\{p^r\} \to F^{2n+1}\{p^r\} \to P^{2n+1}(p^r)$$

define a commutative diagram of horizontal fibration sequences

$$\begin{array}{ccccccc}
\Omega_r & \to & \Omega E^{2n+1}\{p^r\} & \to & V^{2n+1}\{p^r\} & \to & \Sigma \bigvee_\alpha P^{n_\alpha} & \to & E^{2n+1}\{p^r\} \\
\downarrow= & & \downarrow & & \downarrow & & \downarrow= & & \downarrow \\
\Omega_r & \to & \Omega F^{2n+1}\{p^r\} & \to & W^{2n+1}\{p^r\} & \to & \Sigma \bigvee_\alpha P^{n_\alpha} & \to & F^{2n+1}\{p^r\} \\
\downarrow= & & \downarrow & & \downarrow & & \downarrow= & & \downarrow \\
\Omega_r & \to & \Omega P^{2n+1}(p^r) & \to & T^{2n+1}\{p^r\} & \to & \Sigma \bigvee_\alpha P^{n_\alpha} & \to & P^{2n+1}(p^r).
\end{array}$$

The retraction in Corollary 11.10.4 gives compatible retractions

$$\begin{array}{ccc}
\Omega E^{2n+1}\{p^r\} & \to & \Omega_r \\
\downarrow & & \downarrow= \\
\Omega F^{2n+1}\{p^r\} & \to & \Omega_r \\
\downarrow & & \downarrow= \\
\Omega P^{2n+1}\{p^r\} & \to & \Omega_r
\end{array}$$

482 Odd primary exponent theorems

and hence multiplying the maps in the above fibration sequences with the retractions gives a commutative diagram of homotopy equivalences

$$\begin{array}{ccc} \Omega E^{2n+1}\{p^r\} & \xrightarrow{\simeq} & \Omega_r \times V^{2n+1}\{p^r\} \\ \downarrow & & \downarrow \\ \Omega F^{2n+1}\{p^r\} & \xrightarrow{\simeq} & \Omega_r \times W^{2n+1}\{p^r\}, \\ \downarrow & & \downarrow \\ P^{2n+1}(p^r) & \xrightarrow{\simeq} & \Omega_r \times T^{2n+1}\{p^r\}. \end{array}$$

Recall that there are homotopy equivalences

$$\Omega P^{2n+2}(p^r) \simeq S^{2n+1}\{p^r\} \times \Omega \bigvee_{k=0}^{\infty} P^{4n+2kn+3}(p^r)$$

and

$$P^m(p^r) \wedge P^n(p^r) \simeq P^{m+n}(p^r) \vee P^{m+n-1}(p^r).$$

Hence, the Hilton–Milnor theorem implies that all loops on odd primary Moore spaces are weak infinite products of two types of spaces.

Proposition 11.10.5. *Let p be an odd prime. For all integers $m \geq 3$, the spaces $\Omega P^m(p^r)$ are weak infinite products of spaces $T^{2k+1}\{p^r\}$ and $S^{2\ell+1}\{p^r\}$.*

The next two results say that the factor Ω_r can be excised from the fibration sequences in the second diagram which appears at the beginning of Section 11.8.

Lemma 11.10.6. *There is a homotopy equivalence of fibration sequences*

$$\begin{array}{ccc} \prod_{r+1} \times C(n) & \xrightarrow{\simeq} & V^{2n+1}\{p^r\} \\ \downarrow & & \downarrow \\ \prod_{r+1} \times S^{2n-1} & \xrightarrow{\simeq} & W^{2n+1}\{p^r\} \\ \downarrow & & \downarrow \\ \Omega^2 S^{2n+1} & \xrightarrow{=} & \Omega^2 S^{2n+1}. \end{array}$$

Lemma 11.10.7. *There is a homotopy commutative diagram*

$$\begin{array}{ccccc} V^{2n+1}\{p^r\} & \to & T^{2n+1}\{p^r\} & \to & \Omega S^{2n+1}\{p^r\} \\ \downarrow & & \downarrow= & & \downarrow \\ W^{2n+1}\{p^r\} & \to & T^{2n+1}\{p^r\} & \to & \Omega S^{2n+1} \\ \downarrow & & \downarrow & & \downarrow \\ \Omega^2 S^{2n+1} & \to & * & \to & \Omega S^{2n+1} \end{array}$$

in which the rows and columns are all fibration sequences up to homotopy.

11.10 Homotopy exponents of odd primary Moore spaces

Proof: Consider the fibre extension of the totally fibred square which begins Section 11.8, that is, the strictly commutative diagram

$$\begin{array}{ccccc}
E^{2n+1}\{p^r\} & \to & \hat{P}^{2n+1}(p^r) & \xrightarrow{t} & S^{2n+1}\{p^r\} \\
\downarrow \lambda & & \downarrow = & & \downarrow j \\
F^{2n+1}\{p^r\} & \to & \hat{P}^{2n+1}(p^r) & \xrightarrow{q} & \tilde{S}^{2n+1} \\
\downarrow \kappa & & \downarrow & & \downarrow p^r \\
\Omega S^{2n+1} & \to & C & \to & S^{2n+1},
\end{array}$$

where C is contractible and all the rows and columns are actual fibration sequences. Since the adjoint

$$\Sigma \bigvee_\alpha P^{n_\alpha}(p^r) \to P^{2n+1}(p^r)$$

factors through $E^{2n+1}\{p^r\}$, there is a strictly commutative diagram

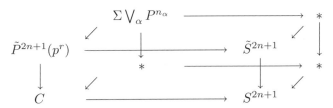

which we can enlarge into the fibre extension of a totally fibred cube, that is, extend the fibrations to fibre sequences so that all the rows and columns are fibration sequences up to homotopy.

The back face of this fibre extension is Lemma 11.10.6.

If we consider the left-hand corner of the back face and take one more step back, then we get a map of fibration sequences

$$\begin{array}{ccc}
\Omega E^{2n+1}\{p^r\} & \to & V^{2n+1}\{p^r\} \\
\downarrow & & \downarrow \\
\Omega F^{2n+1}\{p^r\} & \to & W^{2n+1}\{p^r\} \\
\downarrow & & \downarrow \\
\Omega^2 S^{2n+1} & \xrightarrow{=} & \Omega^2 S^{2n+1}.
\end{array}$$

Omitting the Ω_r factor in the decompositions of $\Omega E^{2n+1}\{p^r\}$ and $\Omega F^{2n+1}\{p^r\}$ gives Proposition 11.10.5. $\square$

Now the fact that $S^{2n+1}\{p^r\}$ has a null homotopic p^r-th power map, that $C(n)$ has a null homotopic p-th power map, the homotopy equivalences in Lemma 11.10.6, and the fact that the looped fibration sequence

$$\Omega \prod_{r+1} \times \Omega C(n) \to \Omega T^{2n+1}\{p^r\} \to \Omega S^{2n+1}\{p^r\}$$

is a multiplicative sequence of H-maps yields

Proposition 11.10.8. *If p is an odd prime, then the p^{2r+1}-st power map is null homotopic on $\Omega T^{2n+1}\{p^r\}$.*

Proof: The power map $p^r : \Omega T^{2n+1}\{p^r\} \to \Omega T^{2n+1}\{p^r\}$ projects to a null homotopic map in $\Omega S^{2n+1}\{p^r\}$ since this latter space has a null homotopic p^r-th power. Hence, this power map factors through the fibre $\Omega \prod_{r+1} \times \Omega C(n)$ which has a null homotopic p^{r+1}-st power. Hence, the power $p^{2r+1} = p^{r+1} \cdot p^r$ is null homotopic on $\Omega T^{2n+1}\{p^r\}$. □

The multiplicative decomposition of $\Omega^2 P^m(p^r)$ into a weak infinite product of spaces $\Omega T^{2k+1}\{p^r\}$ and $\Omega S^{2\ell+1}\{p^r\}$ yields:

Proposition 11.10.9. *If p is an odd prime, then the p^{2r+1}-st power map is null homotopic on $\Omega^2 P^{2n+1}(p^r)$.*

In order to prove a better result for the H-space exponent of a mod p^r Moore space we need the following theorem whose proof will not be given here.

Lemma 11.10.10. *There exists a map*

$$\theta : T^{2n+1}\{p^r\} \to \prod_r$$

such that the composition

$$\prod_{r+1} \to T^{2n+1}\{p^r\} \to \prod_r$$

has fibre $\prod_1$.

Lemma 11.10.10 has the following corollary:

Proposition 11.10.11. *There is a fibration sequence*

$$\prod_1 \times C(n) \to T^{2n+1}\{p^r\} \to \prod_r \times \Omega S^{2n+1}\{p^r\}.$$

Proof: Consider the commutative square

$$\begin{array}{ccc} T^{2n+1}\{p^r\} & \to & \Omega S^{2n+1}\{p^r\} \\ \downarrow \theta & & \downarrow \\ \prod_r & \to & * \end{array}$$

and enlarge it to the fibre extension in which the rows and columns are all fibration sequences up to homotopy

$$\begin{array}{ccc} Y & \to X & \to \Omega S^{2n+1}\{p^r\} \\ \downarrow & \downarrow & \downarrow = \\ \prod_{r+1} \times C(n) & \to T^{2n+1}\{p^r\} & \to \Omega S^{2n+1}\{p^r\} \\ \downarrow \psi & \downarrow \theta & \downarrow \\ \prod_r & = \prod_r & \to * \end{array}.$$

The composition

$$C(n) \to S^{2n-1} \to T^{2n+1}\{p^r\} \xrightarrow{\theta} \prod_r$$

is null homotopic for dimensional reasons. Since the map

$$\prod_{r+1} \times C(n) \to T^{2n+1}\{p^r\}$$

is defined by multiplying maps and since θ is null on the inclusion of $C(n)$, it follows that ψ factors through the projection on the first factor

$$\psi = \theta \cdot proj : \prod_{r+1} \times C(n) \to \prod_{r+1} \xrightarrow{\theta} \prod_r$$

and hence that the fibre Y of ψ is $\prod_1 \times C(n)$.

The fact that we have a fibre extension implies that there is a fibration sequence up to homotopy

$$Y \to T^{2n+1}\{p^r\} \to \prod_r \times \Omega S^{2n+1}\{p^r\}$$

where the base is the (homotopy) pullback of the original commutative square. □

Just as before, the loop of the fibration sequence in Proposition 11.10.11 is multiplicative, which implies the next two results.

Proposition 11.10.12. *If p is an odd prime, then the p^{r+1}-st power map is null homotopic on $\Omega T^{2n+1}\{p^r\}$.*

Proposition 11.10.13. *If p is an odd prime, then the p^{r+1}-st power map is null homotopic on $\Omega^2 P^{2n+1}(p^r)$.*

Recall that Proposition 9.6.3 shows that there exist infinitely many elements of order p^{r+1} in the integral homotopy groups of $P^{2n+1}(p^r)$ for all $n \geq 1$ and therefore also in the integral homotopy groups of $P^m(p^r)$ for all $m \geq 3$. Hence the result in Proposition 11.10.8 is the best possible H-space exponent.

11.11 Nonexistence of H-space exponents

The double loops in Proposition 11.10.8 is required since there is no H-space exponent for the single loop space $\Omega P^m(p^r)$. It is the object of this section to prove this fact. It is remarkable that this result is a consequence of the mod p Pontrjagin ring $H(\Omega P^m(p^r); \mathbb{Z}/p\mathbb{Z})$.

First we note that the localized homotopy groups imply:

Lemma 11.11.1. *Let X be a homotopy associative H-space localized at a prime p. If X is not contractible and k is an integer relatively prime to p, then the power map $k : X \to X$ is not null homotopic.*

Hence, for homotopy associative H-spaces localized at a prime p to have a null homotopic power map it must have one of the form $p^s : X \to X$.

Let A be a connected Hopf algebra and assume that A has a commutative and associative diagonal and an associative multiplication. This is called a homology Hopf algebra. If X be a connected homotopy associative H-space, then the homology $H(X)$ with field coefficients is a homology Hopf algebra.

Let k be a positive integer. Let

$$\Delta(k) : A \to \bigotimes^{k} A$$

be the k-fold diagonal of A. Since A has a commutative diagonal, $\Delta(k)$ is a map of Hopf algebras. Let

$$\mu(k) : \bigotimes^{k} A \to A$$

be the k-fold multiplication of A. Since A is a Hopf algebra, $\mu(k)$ is a map of coalgebras.

Definition 11.11.2. *The k-th power map of A is the map of coalgebras*

$$\rho(k) = \mu(k) \cdot \Delta(k) : A \to A.$$

We note the obvious

Lemma 11.11.3. *If X is a connected homotopy associative H-space, k is a positive integer, and F is a field, then the power map $k : X \to X$ induces the k-th power map $k_* = \rho(k) : H(X; F) \to H(X; F)$ on the homology Hopf algebra.*

The basic computation proving the nontriviality of power maps is

Proposition 11.11.4. *Let A be a connected homology Hopf algebra over a field of characteristic $p \neq 0$ and let $k = p^r$. If $x_1, \ldots, x_k$ are k primitive elements of A, then*

$$\rho(k)(x_1 \ldots x_k) = \sum_{\sigma \in \Sigma_n} \rho(k)\sigma(x_1 \otimes \cdots \otimes x_k).$$

Proof: The k-fold diagonal $\Delta(k)(x_1 \ldots x_k) = \Delta(k)(x_1) \ldots \Delta(k)(x_k)$ is the product of the elements

$$\Delta(k)(x_i) = \sum (1 \otimes \cdots \otimes 1 \otimes x_i \otimes 1 \otimes \cdots \otimes 1).$$

The distributive law yields

$$\Delta(k)(x_1 \ldots x_k) = \sum_y \mathrm{sgn}(y)\, y,$$

this sum being taken over all tensor products

$$y = y_1 \otimes \ldots y_k$$

where each y_j is either 1 or a nontrivial ordered product of some subset of the x_i with each x_i occurring precisely once in the set of products $y_1, \ldots, y_k$ and where $\mathrm{sgn}(y)$ is the sign introduced by shuffling the x_i.

Hence, the k-th power map is

$$\rho(k)(x_1 \ldots x_k) = \sum_y \mathrm{sgn}(y)\, y_1 \ldots y_k.$$

For each y we can associate the subset $\pi \subseteq \{1, \ldots, k\}$ which consists of those j such that $y_j = 1$. The j's which are not in π index the y_j which are a nontrivial ordered product of x_i s. Define $S_\pi = \sum \mathrm{sgn}(y)\mu(k)y$, the sum being taken over all y associated to π. Then

$$\rho(k)(x_1 \ldots x_k) = \sum_\pi S_\pi.$$

We observe that $S_\pi = 0$ if $\pi = \{1, \ldots, k\}$ and

$$S_\pi = \sum_{\sigma \in \Sigma_n} \rho(k)\sigma(x_1 \otimes \cdots \otimes x_k)$$

if $\pi = \phi$ is the empty set.

We also observe that $S_\pi = S_\tau$ if the cardinalities of the subsets π and τ are equal. The number of such subsets of cardinality ℓ is

$$\binom{k}{\ell} = \frac{(p^r)!}{\ell!(p^r - \ell)!}$$

which is divisible by p if $1 \leq \ell \leq k-1$.

Hence

$$\rho(k)(x_1 \ldots x_k) = \sum_{\sigma \in \Sigma_n} \rho(k)\sigma(x_1 \otimes \cdots \otimes x_k).$$

□

Lemma 11.11.5. *Suppose that A is a connected homology Hopf algebra over a field of characteristic $p \neq 0$. If A is isomorphic as an algebra to a tensor algebra on more than one generator, then, for all $r \geq 1$, the p^r-th power map $\rho(p^r)$ is nontrivial.*

Proof: Suppose A is the power map $\rho(p^r) : A \to A$ is trivial. Then it is also trivial for the associated graded Hopf algebra $E^0(A)$ which arises from the filtration of A by the powers of the augmentation ideal. But $E^0(A)$ is the primitively generated tensor algebra with the same module of indecomposables. Let u and v be two primitive generators of $E^0(A)$. Since the elements $\text{ad}^j(v)(u)$ are infinitely many algebraically independent primitive elements which freely generate a tensor subalgebra of $E^0(A)$ it follows from Proposition 11.11.4 that $\rho(p^r)$ is nontrivial. □

Proposition 11.11.6. *Let X be a connected space localized at p and suppose that the reduced mod p homology $\overline{H}(X)$ is nonzero. Then $\Omega\Sigma X$ has no null homotopic power maps.*

Proof: Suppose that $\Omega\Sigma X$ has a null homotopic power map. We may suppose that this is a power of p.

There is a least q such that the integral homology $H_q(X) \neq 0$. The first nonvanishing integral homotopy group $\pi_q(\Omega\Sigma X) = H_q(\Omega\Sigma X) = H_q(X)$ is a nonzero group localized at p which has an exponent. A theorem of Kaplansky [72] implies that $H_q(X)$ is a nontrivial direct sum of cyclic groups $\mathbb{Z}/p^s\mathbb{Z}$. Hence, the mod p homology $H(X; \mathbb{Z}/p\mathbb{Z}) = QH(\Omega\Sigma X; \mathbb{Z}/p\mathbb{Z})$ has more than two generators and no power map $\rho(p^r)$ can be trivial on the Hopf algebra $H(\Omega\Sigma X; \mathbb{Z}/p\mathbb{Z})$. □

Exercise

(1) Let A be a connected homology Hopf algebra over a field.

　(a) If k and ℓ are two positive integers, show that $\rho(k\ell) = \rho(k) \cdot \rho(\ell)$.

　(b) If the ground field of A has finite characteristic p and k is the least positive integer such that the power map $\rho(k) : A \to A$ is trivial, then k is a power of p.

12 Differential homological algebra of classifying spaces

This chapter completes the presentation of differential homological algebra which was started in Chapter 10. Following the presentation of Moore in the Cartan seminar [94], we show how the homology of a topological group determines the homology of its classifying space. This leads to the first quadrant Eilenberg–Moore spectral sequence which has also gone by a variety of other names, for example, the Milnor–Moore spectral sequence or the Rothenberg–Steenrod spectral sequence.

We try to emphasize the fact that the spectral sequence is a secondary object which comes from a chain model of the classifying space. Unfortunately, this approach is not as successful here as it was with the applications in Chapter 11 of the second quadrant Eilenberg–Moore spectral sequence. For example, the Borel transgression theorem requires the use of the spectral sequence.

The algebra of this first quadrant Eilenberg–Moore spectral sequence is based on the fact that the normalized chains of a topological group form a differential Hopf algebra. The second quadrant Eilenberg–Moore spectral sequence was based on the fact that the normalized chains on a topological space form an associative differential coalgebra.

In the case of the second quadrant Eilenberg–Moore models, the dominant algebraic structure is the structure of an associative algebra in differential Cotor. In the case of the first quadrant Eilenberg–Moore models to be studied here, the dominant algebraic structure is that of an associative coalgebra in differential Tor. There can be other structures but they are not always present. When present, they are consistent with the dominant structure. For example, if we get an algebra structure in differential Tor, it combines with the coalgebra structure to give a Hopf algebra structure.

We apply the first quadrant Eilenberg–Moore spectral sequence to the computation of the cohomology of the classifying spaces of the orthogonal and unitary groups. We use this to verify the axioms for Chern and Stiefel–Whitney classes. Then we close the book by following Milnor and Stasheff [91] in presenting the lovely applications of Stiefel–Whitney classes to nonimmersion and nonparallelizability results for real projective spaces and to the nonexistence of real division algebras. Their presentation of these applications cannot be improved upon. The major difference in our treatment from theirs is in our use of the Eilenberg–Moore spectral sequence to provide a uniform computation of the cohomology of the relevant classifying spaces.

In all honesty, it must be admitted that there is a computation of the cohomology of the classifying spaces of the orthogonal and unitary groups which uses only the Gysin sequence and some facts about Euler classes. It is given as an exercise in this chapter.

12.1 Projective classes

Let $\mathcal{A}$ be an abelian category. The definition of a projective class in $\mathcal{A}$ is the strict dual of the definition of injective class which was given in Chapter 10. It is of course due to Eilenberg and Moore [43].

Let $\mathcal{P}$ be a class of objects in $\mathcal{A}$. Define the associated class of relative epimorphisms as

$$\mathcal{P}^* = \{f : M \to N \ | \ f^* : \mathrm{map}_*(P, N) \to \mathrm{map}_*(P, M) \text{ is a surjection}$$
$$\forall P \in \mathcal{P}\}.$$

Let $\mathcal{E}$ be a class of morphisms in $\mathcal{A}$. Define the associated class of relative projective objects as

$$\mathcal{E}^* = \{P \ | \ f^* : \mathrm{map}_*(P, N) \to \mathrm{map}_*(P, M) \text{ is a surjection}$$
$$\forall f : M \to N \in \mathcal{E}\}.$$

Definition 12.1.1. The pair $(\mathcal{P}, \mathcal{E})$ is a projective class in $\mathcal{A}$ if

(a) $\mathcal{P}^* = \mathcal{E}$,

(b) $\mathcal{E}^* = \mathcal{P}$, and

(c) there are enough projective objects in the sense that

$$\forall M \in \mathcal{A}, \ \exists f : P \to M \ \text{with} \ P \in \mathcal{P} \ \text{and} \ f \in \mathcal{E}.$$

We note that

Lemma 12.1.2.

(a) $\mathcal{P}^*$ is always closed under right factorization, that is, $f \cdot g \in \mathcal{P}^*$ implies that $f \in \mathcal{P}^*$.

(b) $\mathcal{E}^*$ is always closed under retracts, that is, $P \in \mathcal{E}^*$ and Q a retract of P implies that $Q \in \mathcal{P}^*$.

12.1 Projective classes 491

The following lemma is often used to define projective classes.

Lemma 12.1.3. *Suppose that $\mathcal{E}$ is a class of morphisms in $\mathcal{A}$ which is closed under right factorization and define $\mathcal{P} = \mathcal{E}^*$. If there are enough projective objects, that is,*

$$\forall M \in \mathcal{A}, \quad \exists f : P \to M \quad \text{with} \quad P \in \mathcal{P} \quad \text{and} \quad f \in \mathcal{E},$$

then the pair $(\mathcal{P}, \mathcal{E})$ is a projective class in $\mathcal{A}$.

Definition 12.1.4. We say that a sequence $M_1 \xrightarrow{g} M_2 \xrightarrow{f} M_3$ is relative short exact if $f \in \mathcal{E}$ and $g = \text{kernel}(f)$.

We say that a complex

$$\ldots \xrightarrow{d_{n+1}} P_n \xrightarrow{d_n} P_{n-1} \xrightarrow{d_{n-1}} \ldots \xrightarrow{d_1} P_0 \xrightarrow{\epsilon} M$$

is relative exact (or relative acyclic) if all the

$$K_0 \to P_0 \xrightarrow{\epsilon} M$$

$$K_n \to P_n \xrightarrow{d_n} K_{n-1}, \quad n \geq 1$$

are relative short exact where $K_0 = \text{kernel}(\epsilon)$, $K_n = \text{kernel}(d_n)$, $n \geq 1$.

If all the P_n are relative projective objects, we say that $P_* \xrightarrow{\epsilon} M$ is a relative projective complex. A complex which is both a relative projective complex and is relative acyclic is a relative projective resolution. The usual proofs show that relative projective resolutions exist and are functorial up to chain homotopy, that is,

Proposition 12.1.5. *Given an object M in an abelian category with a projective class, there exists a relative projective resolution $P_* \xrightarrow{\epsilon} M$.*

Proposition 12.1.6.

(a) *Given a morphism $f : M \to N$, a relative projective complex $P_* \xrightarrow{\epsilon} M$, and a relative acyclic complex $Q_* \xrightarrow{\epsilon} N$, there exists a map of complexes $F : P_* \to Q_*$ which covers f, that is,*

$$\begin{array}{ccccccccc}
\ldots \xrightarrow{d_{n+1}} & P_n & \xrightarrow{d_n} & P_{n-1} & \xrightarrow{d_{n-1}} \ldots & \ldots \xrightarrow{d_1} & P_0 & \xrightarrow{\epsilon} & M \\
& \downarrow F_n & & \downarrow F_{n-1} & & & \downarrow F_0 & & \downarrow f \\
\ldots \xrightarrow{d_{n+1}} & Q_n & \xrightarrow{d_n} & Q_{n-1} & \xrightarrow{d_{n-1}} \ldots & \ldots \xrightarrow{d_1} & Q_0 & \xrightarrow{\epsilon} & N
\end{array}$$

commutes.

(b) *Furthermore, any two maps of complexes $F, G : P_* \to Q_*$ which cover the same f are chain homotopic, that is, there exists a chain homotopy $H : P_* \to Q_{*+1}$ such that*

$$d_1 \cdot H_0 = F_0 - G_0, \quad d_{n+1} \cdot H_n + H_{n-1} \cdot d_n = F_n - G_n \quad \forall n \geq 1.$$

We now specialize to the projective classes which are called proper and are the basis for the differential homological algebra in this section.

First, let R be a principal ideal domain and let $D\mathrm{mod}_R$ be the category of non-negatively graded differential modules over R. Similarly, let A be a differential graded algebra over R and let $D\mathrm{mod}_A$ be the category of negatively graded differential A modules. We have a forgetful functor $(\ )\#: D\mathrm{mod}_A \to D\mathrm{Mod}_R$. The projective classes which form the basis for this chapter are the so-called proper projective classes in these categories.

Definition 12.1.7.

(a) The proper projective class in $D\mathrm{mod}_R$ consists of all morphisms $f: M \to N$ with $f: M \to N$ and the map of cycles $\mathbb{Z}(f): \mathbb{Z}M \to \mathbb{Z}N$ surjections. These are called the proper epimorphisms. The relative projectives are all objects M with M and HM projective over R.

(b) The proper projective class in $D\mathrm{mod}_A$ consists of all morphisms $f: M \to N$ such that $f\#: M\# \to N\#$ is a proper epimorphism in $D\mathrm{mod}_R$. The relative projectives are all objects which are retracts of $A \otimes_R M$ with M a proper projective in $D\mathrm{mod}_R$,

It is clear that, if $f: M \to N$ is a proper epimorphism, then the induced map in homology $Hf: HM \to HN$ is an epimorphism. It is also clear that, if P is a proper projective, then HP is a projective HA module.

Proper projective classes were invented for the Eilenberg–Moore spectral sequence. The key properties are:

(1) If P is a proper projective and N is any differential A module, then

$$H(P \otimes_A N) = HP \otimes_{HA} HN.$$

(2) If $P_* \to M$ is a proper projective resolution, then $HP_* \to HM$ is a projective resolution of HA modules.

We leave it as an exercise to check that the definitions above constitute projective classes.

In Chapter 10, a notion of a proper injective class is defined for the category of differential graded comodules over a differential graded coalgebra. This notion is the dual of the following notion of very proper projective class of differential graded modules over a differential graded algebra.

Definition 12.1.8. A morphism $f: M \to N$ in $D\mathrm{mod}_A$ is a very proper epimorphism if $f\#: M\# \to N\#$ is a split epimorphism in $D\mathrm{mod}_R$. A very proper projective object is any retract of $A \otimes_R M$ where M is any object in $D\mathrm{mod}_R$.

The dual of this definition is used in the definition of differential Cotor. (But there is the difference that dualization here would produce differentials of degree $+1$ and not -1 as in Chapter 10!)

We shall leave further discussion of this definition and the verification that it is a projective class to the exercises.

Exercises

(1) Prove Lemma 12.1.2.

(2) Prove Lemma 12.1.3.

(3) Let $\mathcal{E}$, $\mathcal{F}$ be classes of morphisms and let $\mathcal{P}$, $\mathcal{Q}$ be classes of objects in $\mathcal{A}$. Show that

  (a) $\mathcal{E} \subseteq \mathcal{F}$ implies $\mathcal{F}^* \subseteq \mathcal{E}^*$

  (b) $\mathcal{P} \subseteq \mathcal{Q}$ implies $\mathcal{Q}^* \subseteq \mathcal{P}^*$

  (c) $\mathcal{E} \subseteq \mathcal{E}^{**}$, $\quad \mathcal{P} \subseteq \mathcal{P}^{**}$

  (d) $\mathcal{E}^* = \mathcal{E}^{***}$, $\quad \mathcal{P}^* = \mathcal{P}^{***}$

(4) Let $\mathcal{A}$, $\mathcal{B}$ be abelian categories and let
$$S : \mathcal{A} \to \mathcal{B}, \quad T : \mathcal{B} \to \mathcal{A}$$
be a pair of adjoint functors with a natural bijection
$$\mathcal{A}(TB, A) \simeq \mathcal{B}(B, SA) \quad \forall \quad A \in \mathcal{A}, \quad B \in \mathcal{B}.$$
If $(\mathcal{P}, \mathcal{E})$ is a projective class in $\mathcal{B}$, show that $(\overline{T\mathcal{P}}, S^{-1}\mathcal{E})$ is a projective class in $\mathcal{A}$ where
$$\overline{T\mathcal{P}} = \text{the set of all retracts of} \quad TP \quad \text{where} \quad P \in \mathcal{P}$$
$$S^{-1}\mathcal{E} = \text{the set of all} \quad f \in \mathcal{A} \quad \text{such that} \quad Sf \in \mathcal{E}.$$

(5) Verify that the proper projective classes in Definition 12.1.7 are projective classes.

(6) Verify that the very proper projective class in Definition 12.1.8 is a projective class.

(7) Suppose that $f : M \to N$ is a proper epimorphism in $D\text{mod}_A$ and that HN is projective over the ground ring R which is a principal ideal domain. Show that f is a very proper projective epimorphism.

(8) Suppose that HA and HM are projective over a principal ideal domain R. Show that there exists a proper projective resolution $P_* \to M$, which is also a very proper projective resolution.

12.2 Differential graded Hopf algebras

For any space X, the coalgebra structure on the normalized chains is given by the Alexander–Whitney map $\Delta: C(X) \to C(X) \otimes C(X)$. One of the remarkable technical facts discovered by Eilenberg and Moore is that the Eilenberg–Zilber map $\nabla: C(X) \otimes C(Y) \to C(X \times Y)$ is a map of differential coalgebras [42]. Furthermore, it is associative and commutative in the sense that the following diagrams are strictly commutative

$$\begin{array}{ccc}
C(X) \otimes C(Y) \otimes C(Z) & \xrightarrow{\nabla \otimes 1} & C(X \times Y) \otimes C(Z) \\
\downarrow 1 \otimes \nabla & & \downarrow \nabla \\
C(X) \otimes C(Y \times Z) & \xrightarrow{\nabla} & C(X \times Y \times Z)
\end{array}$$

$$\begin{array}{ccc}
C(X) \times C(Y) & \xrightarrow{\nabla} & C(X \times Y) \\
\downarrow T & & \downarrow T_* \\
C(Y) \otimes C(X) & \xrightarrow{\nabla} & C(Y \times X).
\end{array}$$

If we define the cross product of chains by $\alpha \times \beta = \nabla(\alpha \otimes \beta) \in C(X \times Y)$ for $\alpha \in C(X)$, $\beta \in C(Y)$, then the associativity and commutativity of the Eilenberg–Zilber map is:

$$(\alpha \times \beta) \times \gamma = \alpha \times (\beta \times \gamma), \quad T_*(\alpha \times \beta) = (-1)^{\deg(\alpha) \cdot \deg(\beta)} \beta \times \alpha.$$

Let G be a topological monoid with multiplication $\mu: G \times G \to G$. The Pontrjagin product of chains is defined by

$$\mu: C(G) \times C(G) \xrightarrow{\nabla} C(G \times G) \xrightarrow{\mu} C(G), \quad \mu(\alpha \otimes \beta) = \alpha \cdot \beta.$$

The good properties of the Eilenberg–Zilber map yield that $C(G)$ is a differential graded Hopf algebra and, if G is commutative, it has a graded commutative multiplication.

Furthermore

Lemma 12.2.1. *If G and H are topological monoids, then*

$$\nabla: C(G) \otimes C(H) \to C(G \times H)$$

is a map of differential Hopf algebras.

Proof: The only question is whether ∇ is a map of algebras. Consider the diagram

$$\begin{array}{ccc}
C(G) \otimes C(H) \otimes C(G) \otimes C(H) & \xrightarrow{\nabla \otimes \nabla} & C(G \times H) \otimes C(G \times H) \\
\downarrow T & & \downarrow \nabla \\
C(G) \otimes C(G) \otimes C(H) \otimes C(H) & & C(G \times H \times G \times H) \\
\downarrow \nabla \otimes \nabla & & \downarrow 1 \times T \times 1 \\
C(G \times G) \otimes C(H \times H) & \xrightarrow{\nabla} & C(G \times G \times H \times H) \\
\downarrow \mu \otimes \mu & & \downarrow \mu \times \mu \\
C(G) \otimes C(H) & \xrightarrow{\nabla} & C(G \times H)
\end{array}$$

Since the Eilenberg–Zilber map is a natural transformation, the bottom square commutes. Since

$$(-1)^{\deg(\beta)\cdot\deg(\gamma)}(\alpha \times \gamma \times \beta \times \delta) = (\alpha \times T(\beta \times \gamma) \times \delta),$$

the top square commutes. Hence, ∇ is a map of algebras. □

Let G be a topological monoid and let X be a left G space with action $\mu: G \times X \to X$. Just as above, we have that

$$\mu: C(G) \otimes C(X) \xrightarrow{\nabla} C(G \times X) \xrightarrow{\mu} C(X)$$

is a map of differential coalgebras which gives $C(X)$ the structure of a differential $C(G)$ module coalgebra.

Let G_i be a topological monoid and let X_i be a left G_i space for $i = 1, 2$. Just as above,

$$C(X_1) \otimes C(X_2) \xrightarrow{\nabla} C(X_1 \times X_2)$$

is a map of differential module coalgebras with respect to the map of differential Hopf algebras

$$C(G_1) \otimes C(G_2) \xrightarrow{\nabla} C(G_1 \times G_2).$$

Exercise

(1) Prove the two statements on differential module coalgebras which conclude this section.

12.3 Differential Tor

The algebraic analog of a space with an action of a topological group is a differential module M over a differential algebra A. If we start with an arbitrary differential module M over a differential algebra A, we can find a homologically equivalent differential module by a process of assembling a proper projective resolution of M into a total complex. We describe this process in detail now.

Let $P_* \xrightarrow{\epsilon} M$ be a proper projective resolution. Then P_* has two differentials, an internal differential and an external differential. The internal differential $d_I : P_n \to P_n$ lowers degree by 1, satisfies $d_I \cdot d_I = 0$ and is a derivation over A, that is, $d_I(ax) = (da)x + (-1)^{\deg(a)}a(d_I x)$. The external differential $d_E : P_n \to P_{n-1}$ preserves degree, satisfies $d_E \cdot d_E = 0$, and is linear over A, that is, $d_E(ax) = a(d_E x)$. These internal and external differentials commute with each other, that is, $d_E \cdot d_I = d_I \cdot d_E$. We assemble a resolution P_* into a total complex $T(P_*)$ as follows:

Definition and Lemma 12.3.1.

(a)
$$T(P_*) = \bigoplus_{n \geq 0} s^n P_n$$

where $(sM)_n = M_{n-1}$ is the suspension operator.

(b) d_I and d_E are defined on $T(P_*)$ by

$$d_I(s^n x) = (-1)^n s^n (d_I x), \quad d_E(s^n x) = s^{n-1} d_E(x)$$

where

$$d_I \cdot d_I \cdot s^n = 0, \quad d_E \cdot d_E = 0, \quad d_I \cdot d_E \cdot s^n = -d_E \cdot d_I \cdot s^n.$$

(c) If A acts on $T(P_*)$ by $as^n(x) = (-1)^n s^n(ax)$, then d_E and d_I on $T(P_*)$ satisfy

$$d_E(as^n x) = (-1)^{\deg(a)} a(d_E s^n x),$$
$$d_I(as^n x) = da(s^n x) + (-1)^{\deg(a)} a(d_I s^n x).$$

(d) The internal differential makes $s^n P_n$ into a proper projective module over A.

We leave the verification of the above lemma to the exercises.

Definition and Lemma 12.3.2.

(a) The total differential on $T(P_*)$ is $d_T = d_I + d_E$. The total differential makes $T(P_*)$ into a differential module over A, that is,

$$d_T \cdot d_T = 0, \quad d_T(ax) = (da)x + (-1)^{\deg(a)} a(d_T x) \quad \forall x \in T(P_*).$$

(b) The map $\epsilon : T(P_*) \to M$ defined by

$$\epsilon : P_0 \to M, \quad \epsilon = 0 : s^n P_n \to M \quad \forall n \geq 1$$

is a chain equivalence (= homology isomorphism).

Proof: Part (a) is an immediate consequence of Lemma 12.3.1. To prove part (b), filter $T(P_*)$ by internal degree so that

$$E^0_{p,q} T(P_*) = P_{q,p}, \quad d^0 = d_E : E^0_{p,q} = P_{q,p} \to P_{q-1,p} = E^0_{p,q-1}.$$

Then

$$E^1_{p,q} = \begin{cases} M_p, & q = 0 \\ 0, & q \neq 0. \end{cases}$$

Hence $T(P_*) \xrightarrow{\epsilon} M$ is a homology isomorphism. $\square$

12.3 Differential Tor

Lemma 12.3.3. *Let* $H : P_* \to Q_{*+1}$ *be a chain homotopy between chain maps* $F, G : P_* \to Q_*$ *of complexes over* A. *That is, each* $H : P_n \to Q_{n+1}$ *is linear over* A *and*

$$H \cdot d_I = d_I \cdot H, \quad d_E \cdot H + H \cdot d_E = F - G.$$

Then defining

$$F \cdot s^n = s^n \cdot F, \quad G \cdot s^n = s^n \cdot G, \quad H \cdot s^n = s^{n+1} \cdot H$$

creates A *linear maps* $F, G : T(P_*) \to T(Q_*)$ *and a chain homotopy* $H : T(P_*) \to T(Q_*)$ *which satisfies*

$$H(as^n x) = (-1)^{\deg(a)} a H(s^n x), \quad H \cdot d_I \cdot s^n = -d_I \cdot H \cdot s^n,$$
$$(d_E \cdot H + H \cdot d_I) \cdot s_E^n = (F - G) \cdot s^n, \quad (d_T \cdot H + H \cdot d_T) \cdot s_E^n = (F - G) \cdot s^n.$$

This lemma is also an exercise.

We are now ready to define differential Tor.

Definition 12.3.4. Let M be a right differential module and let N be a left differential module over a differential algebra A. Choose proper projective resolutions $P_* \to M$ and $Q_* \to N$ and define

$$\mathrm{Tor}^A(M, N) = H\{T(P_*) \otimes_A T(Q_*)\}.$$

The lemmas above insure that $\mathrm{Tor}^A(M, N)$ is well defined and functorial up to natural isomorphism.

The following notion of a construction is a generalization of the total complex of a proper projective resolution and is sufficient to define $\mathrm{Tor}^A(M, N)$. A construction is the appropriate algebraic analog of a principal bundle.

Definition 12.3.5. A (left) construction over a differential algebra A is a differential module D over A such that there is an increasing filtration $F_n(D)$ of D by differential submodules with associated graded

$$E_n^0(D) = A \otimes \overline{D_n}, \quad d^0 = d \otimes 1 : A \otimes \overline{D} \to A \otimes \overline{D}$$

where $\overline{D}$ is projective over R.

Since the $\overline{D_j}$ are projective, there is no problem in lifting them up to D so that

$$D = A \otimes \bigoplus_j \overline{D_j}$$

and the filtration is then

$$F_n(D) = A \otimes \bigoplus_{j \leq n} \overline{D_j}.$$

The differential is
$$d(a \otimes x) = da \otimes x + \sum b \otimes y, \quad a, b \in A, \ x \in \overline{D_n}, \ y \in \overline{D_j}, \ j < n.$$
Examples of left constructions over A are twisted tensor products $A \otimes_\tau C$ with
$$F_n(A \otimes_\tau C) = A \otimes C_{j \le n}$$
and total complexes of resolutions $T(P_*)$ with $T(P_*) = A \otimes \overline{T}(P_*)$ and
$$F_n(T(P_*)) = A \otimes \{\overline{T}(P_*)\}_{\le n}.$$

The next proposition and its corollary show that constructions can replace resolutions and also that $\operatorname{Tor}^A(M, N)$ is a balanced functor in the sense that we can get differential Tor by resolving only one of the variables.

Proposition 12.3.6. *Let $f : M_1 \to M_2$ be a homology isomorphism of right differential modules over a differential algebra A. Let D be a left construction over A. Then we have a homology isomorphism*
$$M_1 \otimes_A D \xrightarrow{f \otimes 1} M_2 \otimes_A D.$$

Proof: Consider the increasing filtrations of $M_i \otimes_A D$ defined by
$$F_n(M_i \otimes_A D) = M_i \otimes_A F_n(D).$$
There is a map of filtered objects and we have the following maps of associated graded objects
$$E_n^0(M_1 \otimes_A D) = M_1 \otimes \overline{D_n} \to E_n^0(M_2 \otimes_A D) = M_2 \otimes \overline{D_n}$$
and, in the domain and the range, we have the formula for the differential $d^0(x \otimes y) = dx \otimes y$. Thus there is an isomorphism of E^1 terms and also a homology isomorphism. □

Of course, we can interchange left and right in the above proposition and this gives:

Corollary 12.3.7. *Choose proper projective resolutions $P_* \to M$ and $Q_* \to N$ and suppose there is a homology isomorphism $D \xrightarrow{\simeq} M$ where D is a construction. Then there are homology isomorphisms*
$$D \otimes_A N \xleftarrow{\simeq} D \otimes_A T(Q_*) \xrightarrow{\simeq} M \otimes_A T(Q_*) \xleftarrow{\simeq} T(P_*) \otimes_A T(Q_*)$$

Remark. The above corollary has the consequence that the induced maps on $\operatorname{Tor}^A(M, N)$ depend only on the maps of the variable entries A, M, N. They do not depend on the covering maps of resolutions or constructions since we can, one at a time, omit the resolution or construction. We already knew this when differential Tor was defined by proper projective resolutions. Now we know it even if we define it by constructions.

12.3 Differential Tor

We record the following Künneth formula.

Künneth formula 12.3.8.
$$\mathrm{Tor}^{A \otimes B}(M_1 \otimes M_2, N_1 \otimes N_2) \simeq \mathrm{Tor}^A(M_1, N_1) \otimes \mathrm{Tor}^B(M_2, N_2)$$
provided that either $\mathrm{Tor}^A(M_1, N_1)$ *or* $\mathrm{Tor}^A(M_1, N_1)$ *is flat over* R.

Proof: If $P_{1*} \to M_1$, $P_{2*} \to M_2$ are proper projective right resolutions over A and B respectively and $Q_{1*} \to N_1$, $Q_{2*} \to N_2$ are proper projective left resolutions over A and B respectively, then
$$P_{1*} \otimes P_{2*} \to M_1 \otimes M_2$$
and
$$Q_{1*} \otimes Q_{2*} \to N_1 \otimes N_2$$
are right and left proper projective resolutions over $A \otimes B$.

Thus,
$$\{T(P_{1*}) \otimes T(P_{2*})\} \otimes_{A \otimes B} \{T(Q_{1*}) \otimes T(Q_{2*})\}$$
$$\xrightarrow{1 \otimes T \otimes 1} \{T(P_{1*}) \otimes_A T(Q_{1*})\} \otimes_R \{T(P_{2*}) \otimes_B T(Q_{2*})\}$$
is an isomorphism. □

We now come to the important (first quadrant) algebraic Eilenberg–Moore spectral sequence. Let A be a differential algebra and let M and N be right and left differential modules over A.

Algebraic Eilenberg–Moore spectral sequence 12.3.9. *There exists a functorial first quadrant homology spectral sequence with*
$$E^2_{p,q} = \{\mathrm{Tor}^{HA}_p(HM, HN)\}_q$$
$$d^r : E^r_{p,q} \to E^r_{p-r, q+r-1}$$
and converging strongly to its filtered abutment
$$\{HM \otimes_{HA} HN\}_{p+q} = F_0 \subseteq F_1 \subseteq \cdots \subseteq F_p \subseteq \cdots \subseteq F_\infty = \mathrm{Tor}^A(M, N)_{p+q}.$$

Proof: Let $Q_* \to N$ be a proper projective resolution of N so that
$$\mathrm{Tor}^A(M, N) = H(M \otimes_A T(Q_*))$$
and
$$\mathrm{Tor}^{HA}(HM, HN) = H(HM \otimes_{HA} T(HQ_*)).$$
Without loss of generality, we may assume that each $Q_p = A \otimes \overline{Q_p}$ is a free A module with tensor product differential $d_I = d_A \otimes 1 + 1 \otimes \overline{d_I}$, so that $HQ_p = HA \otimes H\overline{Q_p}$ with respect to the internal differential d_I.

Filter $M \otimes_A T(Q_*)$ by the increasing filtration of resolution degree, that is,

$$F_n(M \otimes_A T(Q_*)) = \bigoplus_{p \leq n} M \otimes_A s^p Q_p.$$

Then the associated graded object is

$$E_{p,q}^0 = \{M \otimes_A s^p Q_p\}_{p+q} = \{M \otimes s^p \overline{Q_p}\}_{p+q}$$
$$d^0 = d_M \otimes_A 1 + 1 \otimes_A d_I = d_M \otimes 1 + 1 \otimes \overline{d_I}$$

and hence

$$E_{p,q}^1 = \{HM \otimes s^p \overline{HQ_p}\}_{p+q} = \{HM \otimes_{HA} s^p HQ_p\}_{p+q}$$
$$= \{HM \otimes_{HA} HQ_p\}_q.$$

The differential is $d^1 = 1 \otimes_{HA} d_E$ and hence

$$E_{p,q}^2 = \{\mathrm{Tor}_p^{HA}(HM, HN)\}_q.$$

The convergence of the spectral sequence is strong since the filtrations are finite in each fixed degree. $\square$

The convergence of the algebraic Eilenberg–Moore spectral sequence immediately yields the homological invariance of differential Tor.

Corollary 12.3.10. *Homology isomorphisms* $A_1 \to A_2$, $M_1 \to M_2$ *and* $N_1 \to N_2$ *induce an isomorphism*

$$\mathrm{Tor}^{HA_1}(HM_1, HN_1) \to \mathrm{Tor}^{HA_2}(HM_2, HN_2).$$

The fact that the Eilenberg–Zilber map induces homology isomorphisms implies the geometric Künneth theorem:

Corollary 12.3.11. *Let G_i be topological monoids with right and left G_i spaces X_i and Y_i for $i = 1, 2$. Then there are isomorphisms*

$$\mathrm{Tor}^{C(G_1) \otimes C(G_2)}(C(X_1) \otimes C(X_2), C(Y_1) \otimes C(Y_2)) \xrightarrow{\mathrm{Tor}^\nabla(\nabla, \nabla)}$$
$$\mathrm{Tor}^{C(G_1 \times G_2)}(C(X_1 \times X_2), C(Y_1 \times Y_2)).$$

The algebraic Eilenberg–Moore spectral sequence has the Künneth spectral sequence as an interesting special case.

Künneth spectral sequence 12.3.12. *Let M and N be right and left chain complexes over an algebra with zero differential A. Suppose that M is flat over A. Then there is a first quadrant homology spectral sequence with*

$$E_{p,q}^2 = \{\mathrm{Tor}_p^A(HM, HN)\}_q$$

and strongly converging to $H_{p+q}(M \otimes_A N)$.

Proof: The algebraic Eilenberg–Moore spectral sequence has the correct E^2 term. All we have to prove is that $\operatorname{Tor}^A(M, N) = H(M \otimes_A N)$.

Filter M by total degree,
$$F_p M = \bigoplus_{r \leq p} M_r.$$

Filter $T(Q_*)$ by complementary degree,
$$F_q T(Q_*) = \bigoplus_{p \geq 0, s \leq q} s^p Q_{p,q}.$$

Filter $M \otimes_A T(Q_*)$ by the product filtration,
$$F_n(M \otimes_A T(Q_*)) = \bigoplus_{p+q=n} F_p(M) \otimes_A F_q(T(Q_*))$$

and then the associated graded object is
$$E^0 = M \otimes_A T(Q_*), \quad d^0 = 1 \otimes d_E.$$

Since M is flat over A,
$$E^1 = M \otimes_A H(T(Q_*), d_E) = M \otimes_A N.$$

Hence, $E^2 = H(M \otimes_A N)$. Since the spectral sequence is confined to a single line, it collapses, $E^2 = E^\infty$, and there are no extension problems. Therefore,
$$\operatorname{Tor}^A(M, N) = H(M \otimes_A T(Q_*)) = H(M \otimes_A N).$$
$\square$

Exercises

(1) Prove Lemma 12.3.1.

(2) Prove Lemma 12.3.2.

(3) Refer to Chapter 10 and to Section 12.1 in order to do this exercise on duality.

   (a) If A, M, N are all finite type and projective over R, show that there is an isomorphism
   $$(M \otimes_A N)^* \simeq M^* \square_{A^*} N^*.$$

   (b) If A, , M, N are all finite type and projective over a principal ideal domain R, show that there is an isomorphism
   $$\{\operatorname{Tor}^A(M, N)\}^* \simeq \operatorname{Cotor}_{A^*}(M^*, N^*).$$

(4) Let R be a principal ideal domain, let K be an R module, and let M and N be chain complexes over R where M is flat over R. Use the Künneth spectral sequence to prove the Künneth and universal coefficient exact sequences:

$$0 \to \{HM \otimes_R HN\}_n \to H(M \otimes_R N) \to \{\operatorname{Tor}_1^R(M,N)\}_{n-1} \to 0$$

$$0 \to H_n M \otimes_R K \to H(M \otimes_R K) \to \operatorname{Tor}_1^R(HM_{n-1}, K) \to 0.$$

12.4 Classifying spaces

According to Dold, a cover $\mathcal{V} = \{V_\lambda\}_\lambda$ of a topological space X is called numerable if it has a partition of unity subordinate to it and a bundle $\pi : E \to X$ is called a numerable bundle if it has a numerable cover $\mathcal{V}$ such that the restrictions $\pi : \pi^{-1} V_\lambda \to V_\lambda$ are trivial bundles for all λ.

The definition of a numerable bundle can be applied to any class of bundles, for example, if G is a topological group, there are numerable principal G bundles. And there are numerable real vector bundles and numerable complex vector bundles. We shall denote by $k_G(X)$, $k_\mathcal{R}$, and $k_\mathcal{C}$ the respective isomorphism classes of numerable bundles of these three types.

Definition 12.4.1. A principal G bundle $\pi : E \to B$ is called a universal bundle if it is numerable and the natural transformation

$$[X, B] \to k_G(X), \quad [f] \mapsto f^* E$$

gives a bijection for all X between homotopy classes and isomorphism classes. The base B of a universal bundle is called a classifying space. Universal real and complex vector bundles are defined in a similar fashion.

Dold has proved the definitive theorem in this direction [32].

Dold's theorem 12.4.2. *If $\pi : E \to B$ is a numerable principal G bundle and E is contractible, then it is a universal principal G bundle.*

Proof: We will be content to sketch the demonstration of the bijection in the simple case when the base is a CW complex.

Let $A \subseteq Y$ be a subspace. A map of a trivial bundle into E, $f : A \times G \to E$, is completely determined by the restriction to a section $h = f(\ ,e) : A \to E$ with $f(a, g) = f(a, e)g$. So an extension $H : X \to E$ of $h : A \to E$ defines an extension of f to a bundle map $F : X \times G \to E$.

We know that bundles are trivial over contractible spaces. If $\pi : D \to X$ is any principal G bundle over a CW complex X, then D is trivial when pulled back to any cell. Hence, induction over the cells of a CW complex X shows that there is a bundle map $D \to E$. Hence, the natural transformation is surjective for X.

There is a principal G bundle $\tau = 1 \times \pi : I \times D \to I \times X$ and the same argument as above shows that any bundle map of $\tau^{-1}(1 \times X \cup 0 \times X) \to E$ can be extended to a bundle map of $I \times D \to E$. Hence, the natural transformation is injective for X. □

Milnor has given an elegant and functorial construction [86] of a classifying space for a topological group G. Let

$$G^{*n} = \{t_0 g_0 + t_1 g_1 + \cdots + t_n g_n \mid t_i \geq 0, \sum t_i = 1, g_i \in G\}$$

be the n-fold join where we identify

$$t_0 g_0 + t_1 g_1 + \cdots + t_n g_n = t_0 g_0 + t_1 g_1 + \cdots + t_i \hat{g}_i + \cdots + t_n g_n, \quad t_i = 0.$$

Thus, there is an embedding $G^{*n} \subseteq G^{*(n+1)}$.

Clearly G acts on the right (or the left) of G^{*n} compatible with the embeddings, on the right by

$$\left(\sum t_i g_i\right) g = \sum t_i (g_i g).$$

The direct limit

$$EG = \lim_{n \to \infty} G^{*n}$$

is a numerable principal G bundle for which EG is contractible. The base space $BG = EG/G$ is called the classifying space of G. This construction was introduced by Milnor who showed that it was a principal G bundle with $\pi_i EG = 0$ for all $i \geq 0$. He based this on the simple fact that the iterated joins become more and more connected. The fact that it is numerable and contractible was shown by Dold [32].

If $U(n)$ is the unitary group, then a principal $U(n)$ bundle is the same thing as a bundle of orthonormal complex n-frames. Let $\pi : E \to X$ be a principal $U(n)$ bundle with fibres $\pi^{-1} x = \{(f_1, \ldots, f_n)\} \simeq U(n)$ where $(f_1, \ldots, f_n)$ is a complex n-frame. There is an associated complex n-plane bundle $\tau : CE \to X$ with fibres $\tau^{-1} x = \langle f_1, \ldots, f_n \rangle$.

Conversely, if $\tau : D \to X$ is any numerable complex n-plane bundle with fibres $\tau^{-1} x$, it has a choice of a Hermitian metric and an associated n-frame bundle $\pi : \mathcal{F} D \to X$ with fibres $\pi^{-1} x = \{(f_1, \ldots, f_n)\}$ where $(f_1, \ldots, f_n)$ is an n-frame in the complex vector space $\tau^{-1} x$.

Given two choices of Hermitian metrics on D, they are homotopic through Hermitian metrics, that is, there is a Hermitian metric on $I \times D \to I \times B$ which restricts to the two choices on the ends. The bundle $\mathcal{F}(I \times D) \to I \times B$ restricts to the two choices of $\mathcal{F} D$ on the ends. Hence, classification of n-frame bundles via the universal $U(n)$ bundle $EU(n) \to BU(n)$ shows that the construction $\mathcal{F}D$ is independent up to isomorphism of the choice of the Hermitian metric.

Hence, we have natural isomorphisms

$$\mathcal{F}CE \simeq E, \quad \mathcal{C}\mathcal{F}D \simeq D.$$

It follows that: $\mathcal{C}EU(n) \to BU(n)$ is a universal complex n-plane bundle. In this case, the total space has the same homotopy type as the base space. The universal bundle is contractible only in the case of principal bundles.

Similarly, $\mathcal{R}EO(n) \to BO(n)$ is a universal real n-plane bundle.

Exercise

(1) Show that any two classifying spaces for principal G bundles are homotopy equivalent.

12.5 The Serre filtration

Let G be a topological group. If G_0 is the component of the identity, it is a normal subgroup and the group of components G/G_0 acts on HG_0 by conjugation. Note that $H_0G = HG/G_0 = R[G/G_0]$. In homology, the conjugation is

$$H_0 G \otimes HG_0 \to HG_0, \quad \sum_{g \in G/G_0} r_g g \otimes \beta \mapsto \sum_{g \in G/G_0} r_g g^{-1} \beta g.$$

If $\alpha = \sum_{g \in G/G_0} r_g g$, we write this action as $\alpha \otimes \beta \mapsto c(\alpha)\beta$.

The algebra HG is the semi-direct tensor product

$$HG = H_0 G \otimes HG_0, \quad (\alpha \otimes \beta) \cdot (\gamma \otimes \delta) = \alpha \cdot \gamma \otimes (c(\gamma)\beta) \cdot \delta.$$

Suppose $F \to E \xrightarrow{\pi} B$ is a fibration sequence. Recall that the Serre filtration on the chains $C(E)$ is the inverse image of the skeletal filtration on the chains $C(B)$. When the base and fibre are connected, then Serre [116] has computed that the E^1 term of the associated Serre spectral sequence satisfies

$$E^1 = HF \otimes C(B).$$

Consider a left principal G bundle of a topological group,

$$G \to E \to B$$

where

$$B = * \times_G E = \{(*, e) \mid e \in E, \ (*, e) \sim (*, ge), \ g \in G\}.$$

Since E is also a principal G_0 bundle, that is, there is a principal bundle sequence $G_0 \to E \to B_0$ with $B_0 = * \times_{G_0} E$.

There is a factorization $E \to B_0 \to B$ where the second map is a covering space with discrete fibre G/G_0. Hence, $C(B_0)$ is a free $H_0 G = R[G/G_0]$ module and, as a nondifferential module, is a tensor product

$$C(B_0) = H_0 G \otimes C(B).$$

The differential has a "twisting" coming from the action of the covering group G/G_0.

The Serre filtration on $G \to E \to B$ is identical to the Serre filtration on $G_0 \to E \to B_0$. Serre's computation of the E^1 term applies to the latter, so we have

$$E^1 = HG_0 \otimes C(B_0) = HG_0 \otimes H_0 G \otimes C(B) = HG \otimes C(B)$$

but the last identification does not respect the differential and therefore it is better to write

$$E^1 = HG \otimes_{H_0 G} C(B_0).$$

Consider the Serre filtration on a Borel construction

$$X \times_G E = \{(x, e) \mid x \in X, \ e \in E, \ (xg, e) \sim (x, ge), \ g \in G\}$$

where X is a right G space and E is a left principal G bundle. If X is connected, we can apply Serre's computation of E^1 directly to the fibration sequence

$$X \to X \times_G E \to * \times_G E = B$$

with fibre X and we get

$$E^1 = HX \otimes C(B) = HX \otimes_{H_0 G} C(B_0).$$

12.6 Eilenberg–Moore models for Borel constructions

In this section we construct the chain models for Borel constructions $X \times_G E$ where X is a right G space and E is a left principal G bundle.

Let $P_* \to C(E)$ be a proper projective resolution of $C(E)$ over the differential algebra $C(G)$. The total complex of the resolution gives a homology isomorphism $T(P_*) \to C(E)$ and the main result of this section is the following theorem.

Geometric approximation theorem 12.6.1 [94]. *There is a homology isomorphism*

$$C(X) \otimes_{C(G)} T(P_*) \to C(X \times_G E),$$

that is, there is an isomorphism

$$\operatorname{Tor}^{C(G)}(C(X), C(E)) \cong H(X \times_G E).$$

Proof: First of all, we construct the map. The associativity of the Eilenberg–Zilber map gives the commutative diagram

$$\begin{array}{ccc} C(X) \otimes C(G) \otimes C(E) & \xrightarrow{\nabla} & C(X \times G) \times E) \\ \mu \otimes 1 \downarrow\downarrow 1 \otimes \mu & & \mu \times 1 \downarrow\downarrow 1 \times \mu \\ C(X) \otimes C(E) & \xrightarrow{\nabla} & C(X \times E) \\ \downarrow & & \downarrow \\ C(X) \otimes_{C(G)} C(E) & \dashrightarrow & C(X \times_G E) \end{array}$$

which defines a map $C(X) \otimes_{C(G)} C(E) \to C(X \times_G E)$ and composition gives the required map

$$C(X) \otimes_{C(G)} T(P_*) \to C(X) \otimes_{C(G)} C(E) \to C(X \times_G E).$$

Filter $T(P_*)$ by the $C(G)$ module filtration generated by skeleta, that is,

$$F_p T(P_*) = \text{image } C(G) \otimes \{T(P_*)\}_{\leq p} \xrightarrow{\mu} T(P_*).$$

The p-th filtration is generated by elements of the form ax with $a \in C(G)$ and $\deg(x) \leq p$. On such an element, the total differential is $d_T(ax) = (da)x + \pm a(d_T x) \equiv (da)x$ mod filtration $p - 1$.

In addition, if $x = s^n y$, $y \in P_n$, and $d_E y = \sum bw + \sum ct$ with $b \in C_0(G)$, $c \in C_{>0}(G)$, then

$$d_T x \equiv d_I x \pm \sum b s^{n-1} w$$

mod filtration $p - 2$.

If $C(E)$ is given the Serre filtration, then the map $T(P_*) \to C(E)$ is filtration preserving.

The map induces a map of spectral sequences, converging to an isomorphism of HE, with the map of E^1 terms being

$$HG \otimes (R \otimes_{C(G)} T(P_*)) \to HG \otimes C(B).$$

The comparison theorem of Moore, as quoted in [22] and proved in the exercises below, shows that

$$H_0 G \otimes (R \otimes_{C(G)} T(P_*)) \to H_0 G \otimes C(B)$$

is a homology isomorphism on the edge, that is,

$$E^1_{*,0} \to E^1_{*,0}$$

is an isomorphism.

Remark. In the domain of this map on the edge, the above computation of the differential mod filtration $p - 2$ shows that it is important that the $H_0 G$ factor be included in $E^1_{*,0}$ in order to keep the edge closed under the d^1 differential. Similarly, in order for $E^1_{*,0}$ in the range to be closed under the d^1 differential, $E^1_{*,0}$ must be $C(B_0) = H_0 G \otimes C(B)$ where $B_0 = E/G_0$ is the covering space of B with covering group G/G_0.

The Künneth spectral sequence implies that
$$R \otimes_{CG} T(P_*) = R \otimes_{H_0 G} H_0 G \otimes_{CG} T(P_*) \to C(B)$$
$$= R \otimes_{H_0 G} H_0 G \otimes C(B)$$

is a homology isomorphism.

In other words, we have proved the following special case of Theorem 12.6.1.

Proposition 12.6.2.
$$\operatorname{Tor}^{C(G)}(R, C(E)) \cong HB$$

Filter $C(X) \otimes_{C(G)} T(P_*)$ by
$$F_p\{C(X) \otimes_{C(G)} T(P_*)\} = C(X) \otimes_{C(G)} F_p T(P_*).$$

If $C(X \times_G E)$ is given the Serre filtration, then the map
$$C(X) \otimes_{C(G)} T(P_*) \to C(X \times_G E)$$

is filtration preserving and the resulting map of the E^1 terms of the spectral sequences is
$$HX \otimes_{H_0 G} H_0 G \otimes (R \otimes_{C(G)} T(P_*)) \to HX \otimes_{H_0 G} H_0 G \otimes C(B)$$

which is a homology isomorphism since
$$H_0 G \otimes R \otimes_{C(G)} T(P_*) \to H_0 G \otimes C(B)$$

is a homology isomorphism. Thus, the map of spectral sequences is an isomorphism at E^2 and therefore also at E^∞. Hence,
$$C(X) \otimes_{C(G)} T(P_*) \to C(X \times_G E)$$

is a homology isomorphism. $\square$

Exercises

(1) Let $f : M \to N$ be a map of differential graded modules over a differential graded algebra A. The mapping cylinder of f is the differential graded module defined by
$$\mathbb{Z}(f) = M \oplus sM \oplus N$$

with differential d defined by
$$d_{|M} = d_M, \quad d_{|N} = d_N, \quad , d(sx) = -s(dx) + x - fx \quad \forall x \in M.$$

(a) Show that $\mathbb{Z}(f)$ is a differential graded module over A, that is, show
$$d^2 = 0, \quad d(a\, y) = (da)\, y + (-1)^{\deg(a)} a\, (dy), \quad a \in A,\ y \in \mathbb{Z}(f).$$

(b) There is a factorization of f into a composite of maps of differential graded modules $M \to \mathbb{Z}(f) \to N$.

(c) The map $\mathbb{Z}(f) \to N$ is a homology isomorphism.

(d) The map $Hf : HM \to HN$ is an isomorphism if and only if $H(\mathbb{Z}(f)/M) = 0$.

(2) Suppose M is a filtered differential graded module over a differential graded algebra A and assume that
$$E^1_{p,q} = E^1_{p,0} \otimes_{H_0 A} H_q A$$
and that $HA = H_0 A \otimes F$ is free over $H_0 A$. Show that $HM = 0$ if and only if $E^1_{p,0} = 0$, $\forall p \geq 0$.

(3) Suppose that $f : M \to N$ is a map of filtered differential graded modules, both satisfying the hypotheses of Exercise 2. Suppose that the induced map $E^1_{p,q}(M) \to E^1_{p,q}(N)$ is isomorphic to a tensor product map
$$E^1_{p,0}(M) \otimes_{H_0 A} H_q A \to E^1_{p,0}(N) \otimes_{H_0 A} H_q A.$$
Show that $Hf : HM \to HN$ is an isomorphism if and only if $E^1_{p,0}(M) \to E^1_{p,0}(N)$ is a homology isomorphism, that is, if and only if $E^2_{p,0}(M) \to E^2_{p,0}(N)$ is an isomorphism.

12.7 Differential Tor of several variables

The several variable version of differential Tor is the way to establish the connection between the algebraic coproduct and the geometric diagonal. For example, the diagonal is represented by
$$E \times_G G \times_G E \to E \times_G * \times_G E$$
which corresponds to
$$\operatorname{Tor}^A(R, R) = \operatorname{Tor}^A(R, A, R) \to \operatorname{Tor}(R, R, R)$$
$$= \operatorname{Tor}^A(R, R) \otimes \operatorname{Tor}^A(R, R).$$

This idea was introduced in Moore's expose 7 in the Cartan seminar [95]. One could get away without it by just using the geometric diagonal since $C(G)$ is a differential Hopf algebra. But you cannot do this evasive sort of thing in the cobar world. Hence, differential Cotor is used in Chapter 10. This is here in Chapter 12

12.7 Differential Tor of several variables

to be symmetric with that, and it is the neatest way to establish the connection between the geometric diagonal and the coproduct in the bar construction.

Let A and B be differential graded algebras. A differential graded R module M is an $A - B$ differential bimodule if it is a differential left A module and a differential right B module together with an extra associativity which makes $\mu : A \otimes M \otimes B \to M$ well defined, that is, the diagram below commutes

$$\begin{array}{ccc} A \otimes M \otimes B & \xrightarrow{\mu \otimes 1} & M \otimes B \\ \downarrow 1 \otimes \mu & & \downarrow \mu \\ A \otimes M & \xrightarrow{\mu} & M \end{array}.$$

The opposite algebra B^{op} is: $B^{op} = B$ as an R module and with multiplication

$$\mu^{op} = \mu \cdot T : B \otimes B \to B \otimes B \to B.$$

That is,

$$b_1 * b_2 = \mu^{op}(b_1 \otimes b_2) = (-1)^{\deg(b_1)\deg(b_2)} b_2 b_1$$

where $b_1 * b_2$ is the multiplication in the opposite algebra.

Observe that, if M a left differential B^{op} module, then M is a right differential B module via the action

$$M \otimes B \xrightarrow{T} B^{op} \otimes M \xrightarrow{\mu} M, \quad mb = (-1)^{\deg(m) \cdot \deg(b)} bm, \quad b \in B, \, m \in M.$$

The notion that M is an $A - B$ differential bimodule is equivalent to the notion that M is a left $A \otimes B^{op}$ differential module.

Explicitly, given an $A - B$ differential module M, it is left $A \otimes B^{op}$ differential bimodule via the left action

$$A \otimes B^{op} \otimes M \xrightarrow{1 \otimes T} A \otimes M \otimes B \xrightarrow{\mu} M$$

$$(a \otimes b)m = (-1)^{\deg(m) \cdot \deg(b)} amb, \quad a \in A, \, m \in M, \, b \in B.$$

Conversely, given a left $A \otimes B^{op}$ differential module, it is by restriction both a left A and a left B module and this makes it an $A - B$ differential bimodule via:

$$amb = (-1)^{\deg(m) \cdot \deg(b)} (a \otimes b)m$$

Hence, if M is an $A - B$ differential bimodule, we know that there exists a proper projective resolution of bimodules $P_* \to M$ since this is just a proper projective resolution of differential left $A \otimes B^{op}$ modules. And these proper projective resolutions are functorial up to chain homotopy.

We now define the notion of differential Tor of several variables:

Let M_1 be a right differential A module, M_n be a left differential A module, and M_i a differential $A - A$ bimodule for $i = 2, \ldots, n - 1$.

Let $P_{i,*} \to M_i$ be proper projective resolutions and $T_i = T(P_i, *)$ their respective total complexes.

Definition 12.7.1. Differential Tor of n variables is:
$$\operatorname{Tor}^A(M_1, \ldots, M_n) = H\{T_1 \otimes_A T_2 \otimes_A \cdots \otimes_A T_n\}$$

Since proper projective resolutions are functorial up to chain homotopy, the definition of differential Tor in Definition 12.7.1 is well defined and functorial up to natural isomorphism.

Just as in the case of 2 variables, we can use constructions instead of resolutions to compute differential Tor of several variables. Even though the notion of a biconstruction is already defined as a left construction over the differential algebra $A \otimes B^{op}$, it is worthline to make it explicit:

Definition 12.7.2. A biconstruction over a pair of differential algebras A and B is a differential $A - B$ bimodule D over A such that there is an increasing filtration $F_n(D)$ of D by differential subbimodules with associated graded
$$E_n^0(D) = A \otimes \overline{D_n} \otimes B,$$
$$d^0 = d \otimes 1 \otimes 1 + 1 \otimes 1 \otimes d : A \otimes \overline{D} \otimes B \to A \otimes \overline{D} \otimes B$$
where $\overline{D}$ is projective over R.

Thus the filtration of a biconstruction is
$$F_n(D) = A \otimes \bigoplus_{j \leq n} \overline{D_j} \otimes B.$$

Observe that a total complex $T(P_*)$ of a $A - B$ proper projective resolution is also a left construction with the filtration
$$G_n(T(P_*)) = A \otimes \left\{ \bigoplus_{p+q \leq n} \{\overline{T}(P_*)\}_{\leq p} \otimes B_{\leq q} \right\}.$$

The several variable version of differential Tor has three useful properties. It is balanced, it splits, and it collapses. We proceed to explain these ideas.

Tor is balanced in the sense that, in the computation of Tor, we may omit the total complex T_i in any one variable (in the middle or on the ends) and replace it by M_i, that is:
$$T_1 \otimes_A T_2 \otimes_A \cdots \otimes_A T_i \otimes_A \cdots \otimes_A T_n$$
$$\to T_1 \otimes_A T_2 \otimes_A \cdots \otimes_A M_i \otimes_A \cdots \otimes_A T_n$$

12.7 Differential Tor of several variables

is a homology isomorphism. Thus

$$\text{Tor}^A(M_1, \ldots, M_n) = H\{T_1 \otimes_A T_2 \otimes_A \cdots \otimes_A M_i \otimes_A \cdots \otimes_A T_n\}.$$

More generally, let $D_j \xrightarrow{\simeq} M_j$ be homology isomorphisms for $j \neq i$ and suppose that D_j is a right construction for $j < i$ and a left construction for $j > i$. The interior D_j are also required to be bimodules. Then Proposition 12.3.6 implies that there are homology isomorphisms

$$D_1 \otimes_A \ldots D_{i-1} \otimes_A T_i \otimes_A \ldots D_n \to D_1 \otimes_A \ldots D_{i-1} \otimes_A M_i \otimes_A \ldots D_n.$$

With a little work, this implies that:

Proposition 12.7.3. *For any choice of $1 \leq i \leq n$, differential Tor of n variables can be computed with $n - 1$ constructions as follows:*

$$\text{Tor}^A(M_1, \ldots, M_n) = H\{D_1 \otimes_A \cdots \otimes_A M_j \otimes_A \cdots \otimes D_n\}.$$

Proof: We give the proof that $\text{Tor}^A(M_1, M_2, M_3) = H\{D_1 \otimes_A M_2 \otimes_A D_3\}$. There are homology isomorphisms

$$T_1 \otimes_A T_2 \otimes_A T_3 \xrightarrow{\simeq} M_1 \otimes_A T_2 \otimes_A T_3 \xleftarrow{\simeq} D_1 \otimes_A T_2 \otimes_A T_3$$
$$\xrightarrow{\simeq} D_1 \otimes_A T_2 \otimes_A M_3 \xleftarrow{\simeq} D_1 \otimes_A T_2 \otimes_A D_3$$
$$\xrightarrow{\simeq} D_1 \otimes_A M_2 \otimes_A D_3.$$

$\square$

Remark. This is a repeat of the remark about differential Tor for 2 variables. The above proposition has the consequence that the induced maps on differential Tor depend only on the maps of the variable entries. They do not depend on the covering maps of resolutions or constructions since we can, one at a time, omit the resolution or construction. We already knew this when differential Tor was defined by proper projective resolutions. Now we know it even if we define it by constructions.

Tor is split in the interior variables in the sense that the isomorphism $R \equiv R \otimes_R R$ implies the isomorphism

$$T_1 \otimes_A T_2 \otimes_A \cdots \otimes_A T_{i-1} \otimes_A R \otimes_A T_{i+1} \otimes_A \cdots \otimes_A T_n$$
$$\equiv \{T_1 \otimes_A T_2 \otimes_A \cdots \otimes_A T_{i-1} \otimes_A R\} \otimes_R \{R \otimes_A T_{i+1} \otimes_A \cdots \otimes_A T_n\}$$

and hence

Proposition 12.7.4.

$$\text{Tor}^A(M_1, \ldots, M_{i-1}, R, M_{i+1}, \ldots, M_n)$$
$$\equiv \text{Tor}^A(M_1, \ldots, M_{i-1}, R) \otimes \text{Tor}^A(R, M_{i+1}, \ldots, M_n)$$

whenever one or both of the tensor factors is projective over R.

Tor collapses in the sense that

$$T_1 \otimes_A T_2 \otimes_A \cdots \otimes_A T_{i-1} \otimes_A A \otimes_A T_{i+1} \otimes_A \cdots \otimes_A T_n$$
$$\equiv T_1 \otimes_A T_2 \otimes_A \cdots \otimes_A T_{i-1} \otimes_A T_{i+1} \otimes_A \cdots \otimes_A T_n$$

and hence

Proposition 12.7.5.

$$\operatorname{Tor}^A(M_1, \ldots, M_{i-1}, A, M_{i+1}, \ldots, M_n)$$
$$\equiv \operatorname{Tor}^A(M_1, \ldots, M_{i-1}, M_{i+1}, \ldots, M_n).$$

Exercise

(1) If M is a right B module and $b * c$ is the multiplication in the opposite algebra, check that M is a left B^{op} algebra by verifying

$$(b * c)m = b(cm)$$

where the definition of the left action is $bm = (-1)^{\deg(b) \cdot \deg(m)} mb$.

12.8 Eilenberg–Moore models for several variables

Let G be a topological group. With the exception of one fixed $1 \leq i \leq n$, suppose that E_1 is a free right G space (= right principal G bundle), E_n is a free left G space, and, for $i = 2, \ldots n-1$, E_i is a free $G-G$ space (= free left $G^{op} \times G$ space). In the case of the one $E_i = X$ we drop all free conditions on it.

Let $P_{i,*} \to C(E_i)$ be proper projective resolutions and let $T_i = T(P_*) \to C(E_i)$ be the maps of total complexes. As with two variables, the Eilenberg–Zilber map

$$\nabla : C(E_1) \otimes \cdots \otimes C(E_n) \to C(E_1) \times \ldots C(E_n)$$

induces a map

$$C(E_1) \otimes_{C(G)} \cdots \otimes_{C(G)} C(E_n) \to C(E_1 \times_G \cdots \times_G E_n).$$

The Eilenberg–Moore geometric approximation theorem for several variables is

Proposition 12.8.1. *The composite map*

$$T_1 \otimes_{C(G)} \cdots \otimes_{C(G)} C(X) \otimes_{C(G)} \cdots \otimes_{C(G)} T_n \xrightarrow{\simeq} C(E_1 \times_G \cdots \times_G E_n)$$

is a homology isomorphism. That is, there is a natural isomorphism

$$\operatorname{Tor}^{C(G)}(C(E_1), \ldots, C(E_n)) \simeq H(E_1 \times_G \cdots \times_G E_n).$$

12.8 Eilenberg–Moore models for several variables

Proof: The proof is a simple induction using the two-variable theorem. We do it in detail for the case

$$T_1 \otimes_{C(G)} C(X) \otimes_{C(G)} T_3 \xrightarrow{\simeq} C(E_1 \times_G X \times_G E_3).$$

The two-variable theorem says that there is a homology isomorphism

$$T_1 \otimes_{C(G)} C(X) \to C(E_1 \times_G X).$$

It follows from the two-variable theorem and the fact that T_3 is a construction that there are homology isomorphisms

$$T_1 \otimes_{C(G)} C(X) \otimes_{C(G)} T_3 \xrightarrow{\simeq} C(E_1 \times_G X) \otimes_{C(G)} T_3 \xrightarrow{\simeq} C(E_1 \times_G X \times_G E_3).$$

$\square$

The identity $* \equiv * \times *$ gives the geometric splitting

$$E_1 \times_G * \times_G E_3 \equiv (E_1 \times_G * *) \times_G (* \times_G \times_G E_3).$$

The fact that the Eilenberg–Zilber map

$$\nabla : C(E_1) \times C(X) \times C(Y) \times C(E_3) \to C(E_1 \times X \times Y \times E_3)$$

factors as $\nabla \cdot (\nabla \otimes \nabla)$ and the compatibility with the diagonal shows that

$$\begin{array}{ccc} C(E_1) \otimes_{C(G)} R \otimes_{C(G)} C(E_3) & \xrightarrow{\simeq} & C(E_1) \otimes_{C(G)} R \otimes R \otimes_{C(G)} C(E_3) \\ \downarrow \nabla & & \downarrow \nabla \\ C(E_1 \times_G * \times_G E_3) & \xrightarrow{\equiv} & C(E_1 \times_G * \times * \times_G E_3) \end{array}$$

commutes.

Hence,

$$\begin{array}{ccc} T_1 \otimes_{C(G)} R \otimes_{C(G)} T_3 & \xrightarrow{\simeq} & T_1 \otimes_{C(G)} R \otimes R \otimes_{C(G)} T_3 \\ \downarrow \nabla & & \downarrow \nabla \\ C(E_1 \times_G * \times_G E_3) & \xrightarrow{\equiv} & C(E_1 \times_G * \times * \times_G E_3) \end{array}$$

commutes. If at least one of $H(E_1 \times_G *)$ or $H(* \times_G E)$ is projective over R, we can combine this with the Künneth isomorphisms to get that

Proposition 12.8.2.

$$\begin{array}{ccc} \operatorname{Tor}^{C(G)}(C(E_1), R, C(E_3)) & \xrightarrow{\simeq} & \operatorname{Tor}^{C(G)}(C(E_1), R) \otimes \operatorname{Tor}^{C(G)}(R, C(E_3)) \\ \downarrow \nabla & & \downarrow \nabla \\ H(E_1 \times_G * \times_G E_3) & \xrightarrow{\equiv} & H(E_1 \times_G * \times * \times_G E_3) \end{array}$$

is a commutative diagram of isomorphisms.

We have shown that the splitting of differential Tor is compatible with the geometric splitting.

Similarly, we have the compatibility of the collapse of the differential Tor in the geometric approximation with geometric collapse. It follows from the associativity and naturality of the Eilenberg–Zilber maps.

Proposition 12.8.3.
$$\begin{array}{ccc}
\operatorname{Tor}^{C(G)}(C(E_1), C(G), C(E_3)) & \stackrel{\equiv}{\to} & \operatorname{Tor}^{C(G)}(C(E_1), C(E_3)) \\
\downarrow \nabla & & \downarrow \nabla \\
H(E_1 \times_G G \times_G E_3) & \stackrel{\equiv}{\to} & H(E_1 \times_G E_3)
\end{array}$$
is a commutative diagram of isomorphisms.

Finally, we have the following geometric form of the Künneth theorem: Let G and H be topological groups, let X be a right G space and E a free left G space, and let Y be a right H space and F a free left H space.

Geometric Künneth theorem 12.8.4. *There is a commutative diagram:*
$$\begin{array}{ccc}
\operatorname{Tor}^{C(G)}(C(X), C(E)) \otimes \operatorname{Tor}^{C(H)}(C(Y), C(F)) & \to & \operatorname{Tor}^{C(G \times H)}(C(X \times Y), C(E \times F)) \\
\downarrow \equiv & & \downarrow \equiv \\
H(X \times_G E) \otimes H(Y \times_H F) & \to & H((X \times Y) \times_{G \times H} (E \times F))
\end{array}$$
where the vertical maps are all isomorphisms and the top horizontal map is the composition
$$\operatorname{Tor}^{C(G)}(C(X), C(E)) \otimes \operatorname{Tor}^{C(H)}(C(Y), C(F))$$
$$\to \operatorname{Tor}^{C(G) \otimes C(H)}(C(X) \otimes C(Y), C(E) \otimes C(F))$$
$$\xrightarrow{\operatorname{Tor}^{\nabla}(\nabla, \nabla)} \operatorname{Tor}^{C(G \times H)}(C(X \times Y), C(E \times F)).$$

If $H(X \times_G E)$ and $H(Y \times_H F)$ are R projective, then the horizontal maps are isomorphisms.

Proof: The associativity and commutativity of the Eilenberg–Zilber maps give a commutative diagram
$$\begin{array}{ccc}
C(X) \otimes C(Y) \otimes_{C(G) \otimes C(H)} C(E) \otimes C(F) & \xrightarrow{\nabla \otimes \nabla} & C(X \times Y) \otimes_{C(G \times H)} C(E \times F) \\
\downarrow \nabla \cdot (1 \otimes T \otimes 1) & & \downarrow \nabla \\
C(X \times_G E \times Y \times_H F) & \stackrel{\equiv}{\to} & C(X \times Y \times_{G \times H} E \times F)
\end{array}$$
Denote the total complexes of the appropriate proper projective resolutions by $T_X, T_Y, T_E, T_F, T_{X \times Y}, T_{E \times Y}$. Then we get a map
$$T_X \otimes T_Y \otimes_{C(G) \otimes C(H)} T_E \otimes T_F \to T_{X \times Y} \otimes_{C(G \times H)} T_{E \times F}$$
which covers the top map and induces the map of differential Tors in the top horizontal map of the proposition. $\square$

Exercise

(1) Check the details in the proofs of 12.8.2 and 12.8.3.

12.9 Coproducts in differential Tor

The collapse and splitting of differential Tor enable us to define a coproduct in an R projective differential $\mathrm{Tor}^A(R, R)$ via the fact that the augmentation $\epsilon: A \to R$ is a map of $A - A$ bimodules:

$$\mathrm{Tor}^A(R, R) \equiv \mathrm{Tor}^A(R, A, R) \xrightarrow{\mathrm{Tor}^A(1, \epsilon, 1)} \mathrm{Tor}^A(R, R, R)$$
$$\equiv \mathrm{Tor}^A(R, R) \otimes \mathrm{Tor}^A(R, R).$$

Up to the isomorphisms of collapse and splitting, the coproduct is the map

$$\Delta: \mathrm{Tor}^A(R, \epsilon, R): \mathrm{Tor}^A(R, A, R) \to \mathrm{Tor}^A(R, R, R)$$

which can be described in terms of resolutions as follows:

Let T_1 be the total complex of a right resolution of R and let T_3 be the total complex of a left resolution of R. Let $T_2 = S \otimes T$ where $S = T_3$ is the total complex of a left resolution of R and $T = T_1$ is the total complex of a right resolution of R. Let $\hat{T}_2$ be the total complex of a biresolution of A. Then the coproduct Δ is defined by the map of complexes

$$T_1 \otimes_A \hat{T}_2 \otimes T_3 \to T_1 \otimes_A T_2 \otimes T_3.$$

The iterated coproducts

$$(\Delta \otimes 1) \cdot \Delta, \quad (\Delta \otimes 1) \cdot \Delta: \mathrm{Tor}^A(R, A, A, R) \to \mathrm{Tor}^A(R, R, R, R)$$

are defined by the two natural composite maps of the complexes, depending on whether the second or third variable is mapped first,

$$T_1 \otimes_A \hat{T}_2 \otimes_A \hat{T}_2 \otimes_A T_3 \to T_1 \otimes_A T_2 \otimes_A T_2 \otimes_A T_3.$$

Since these are equal maps on the level of complexes, the coproduct is associative. A counit $\epsilon: \mathrm{Tor}^A(R, R) \to R$ is defined by the map of complexes

$$T_1 \otimes_A T_3 \to R \otimes_R R.$$

Since the composition

$$T_1 \otimes_A \hat{T}_2 \otimes_A T_3 \to T_1 \otimes_A T_2 \otimes_A T_3 \to R \otimes_R \otimes_A T_3$$

is the natural equivalence, it follows that ϵ is a left (and right) counit.

We have shown

Proposition 12.9.1. *For any differential algebra A, there is a natural coalgebra structure on any R projective $\mathrm{Tor}^A(R, R)$. This is called the canonical coalgebra structure.*

We claim that

Proposition 12.9.2. *There is a natural isomorphism*
$$HBA = \mathrm{Tor}^A(R,R)$$
where BA is the classifying or bar construction. This is an isomorphism of coalgebras if HBA is R projective.

Proof: There are acyclic constructions $BA \otimes_\tau A$ and $A \otimes_\tau BA$. In terms of these constructions, the coproduct is defined by the map of complexes
$$BA \otimes_\tau A \otimes_A A \otimes_A A \otimes_\tau BA \to BA \otimes A \otimes_A R \otimes_A A \otimes_\tau BA.$$
(See Section 10.4 for the definition of the differentials of these twisted tensor products.)

Up to natural isomorphism, the coproduct is defined by the map of complexes
$$BA \otimes_\tau A \otimes_\tau BA \to BA \otimes_\tau R \otimes_\tau BA, \quad x \otimes a \otimes y \mapsto x \otimes \epsilon(a) \otimes y$$
where the differential on the range is given by
$$d(x \otimes a \otimes y)$$
$$= dx \otimes a \otimes y + (-1)^{\deg(x)} x \otimes da \otimes y + (-1)^{\deg(x)+\deg(a)} x \otimes a \otimes dy$$
$$+ \sum x' \otimes \tau(x'')a \otimes y + (-1)^{\deg(x)+\deg(a)}(x \otimes a\tau(y') \otimes y''$$
and where $\Delta(x) = \sum x' \otimes x''$ and $\Delta(y) = \sum y' \otimes y''$ are the diagonals in the bar construction.

Lemma 12.9.3. *The map defined by the diagonal in the bar construction*
$$\delta: BA \to BA \otimes_\tau A \otimes_\tau BA, \quad , \delta(x) = \sum x' \otimes 1 \otimes x''$$
is a chain map which is naturally isomorphic to the identity in homology.

Proof: It is easy to check that this map is a chain map.

The composition
$$BA \to BA \otimes_\tau A \otimes_\tau BA \to R \otimes R \otimes BA$$
is the identity and, reversing the collapse of the tensor product, the second map
$$BA \otimes_\tau A \otimes_A A \otimes_\tau BA \to R \otimes R \otimes_A A \otimes_\tau BA$$
is the natural homology isomorphism which is the tensor product over A with the construction in the second factor. □

We have the general result:

Proposition 12.9.4. *Let $A \otimes_\sigma C$ be an acyclic twisted tensor product with twisting morphism $\sigma: C \to A$. There is an isomorphism*
$$HC \to \mathrm{Tor}^A(R,R)$$
which is an isomorphism of coalgebras if HC is R projective.

Proof: Refer to Section 10.5 on universal twisting morphisms.

The universal property of the twisting morphism of the bar construction yields a factorization

$$\sigma : C \xrightarrow{\overline{\sigma}} BA \xrightarrow{\tau} A$$

where $\overline{\sigma} : C \to BA$ is a morphism of differential coalgebras which is a homology isomorphism when $A \otimes_\sigma C$ is acyclic. $\square$

Exercise

(1) Prove directly from the splitting of differential Tor of several variables that the coproduct in $\operatorname{Tor}^A(R, R)$ is associative. Can you think of another (easier) proof?

12.10 Künneth theorem

The purpose of this section is to prove the Künneth theorem.

Künneth theorem 12.10.1. *There is a map*

$$\operatorname{Tor}^A(R, R) \otimes \operatorname{Tor}^B(R, R) \to \operatorname{Tor}^{A \otimes B}(R, R)$$

which is an isomorphism of coalgebras if both factors are projective over R.

Let C be a differential coalgebra and let A be a differential algebra.

Definition 12.10.2. Given a twisting cochain $\sigma : C \to A$, there is a chain map

$$\delta : C \to C \otimes_\sigma A \otimes_\sigma C \equiv C \otimes_\sigma A \otimes_A A \otimes_\sigma C$$

given by $\delta(c) = \sum c' \otimes 1 \otimes c''$ where $\Delta(c) = \sum c' \otimes c''$ is the diagonal.

The composition

$$C \xrightarrow{\delta} C \otimes_\sigma A \otimes_\sigma C \to C \otimes R \otimes C \equiv C \otimes C$$

is the diagonal.

Definition 12.10.3. Given twisted tensor products $A \otimes_\sigma C$ and $B \otimes_\gamma D$, the tensor product

$$(A \otimes_\sigma C) \otimes (B \otimes_\gamma D) \equiv (A \otimes B) \otimes_\xi (C \otimes D)$$

is a twisted tensor product via the twisting morphism

$$\xi = (\sigma \otimes 1)(\eta \otimes \epsilon) + (1 \otimes \gamma)(\epsilon \otimes \eta) : C \otimes D \to C \otimes R \oplus R \otimes D \to A \otimes B.$$

Apply Definition 12.10.3 to the tensor product of the bar constructions $A \otimes_\sigma BA$ and $B \otimes_\gamma BB$ to get the twisting morphism $\xi : BA \otimes BB \to A \otimes B$. The universal property of the twisting morphism of the bar construction $\tau : B(A \otimes B) \to A \otimes B$ gives a lift to a map of differential coalgebras $\bar\xi : BA \otimes BB \to B(A \otimes B)$, that is, this diagram below commutes

$$\begin{array}{ccc} BA \otimes BB & \xrightarrow{\bar\xi} & B(A \otimes B) \\ & \searrow \xi & \downarrow \tau \\ & & A \otimes B \end{array}$$

Lemma 12.10.4. *The map $\bar\xi : BA \otimes BB \to B(A \otimes B)$ is a homology isomorphism of differential coalgebras.*

Proof: This is a consequence of the Zeeman comparison theorem and the fact that the twisted tensor products $(A \otimes B) \otimes_\xi (BA \otimes BB)$ and $(A \otimes B) \otimes_\tau B(A \otimes B)$ are both acyclic. $\square$

Clearly, the above proves the Künneth Theorem 12.10.1.

Exercises

(1) Verify that the map δ in Definition 12.10.2 is a chain map.

(2) Verify that the tensor product of twisted tensor products is given by the twisting morphism in Definition 12.10.3.

12.11 Products in differential Tor

The Künneth theorem in the previous section shows that other algebraic structures can sometimes be introduced into differential Tor.

For example, if A is a commutative differential graded algebra, then the multiplication $A \otimes A \to A$ is a map of differential algebras and induces a map

$$\mathrm{Tor}^A(R,R) \otimes \mathrm{Tor}^A(R,R) \to \mathrm{Tor}^{A \otimes A}(R,R) \to \mathrm{Tor}^A(R,R).$$

Whether or not $\mathrm{Tor}^A(R,R)$ is projective over R, this makes it into a (graded) commutative algebra over R. If it is projective over R, then the above is a map of coalgebras and hence $\mathrm{Tor}^A(R,R)$ is a Hopf algebra.

Suppose that A is a differential Hopf algebra, for example, A could be $C(G)$ where G is a topological group. Suppose that M and N are right and left differential module coalgebras over A. Then the diagonals $\Delta : A \to A \otimes A, M \to M \otimes M, N \to N \otimes N$ are map of differential algebras and module coalgebras and induce a map

$$\Delta_* : \mathrm{Tor}^A(M,N) \to \mathrm{Tor}^{A \otimes A}(M \otimes M, N \otimes N)$$

12.11 Products in differential Tor

which is a map of coalgebras if $\operatorname{Tor}^A(M,N)$ is projective over R. Hence, if $\operatorname{Tor}^A(M,N)$ is projective over R, this gives a coalgebra structure

$$\delta : \operatorname{Tor}^A(M,N) \to \operatorname{Tor}^A(M,N) \otimes \operatorname{Tor}^A(M,N)$$

which is also a morphism of the usual coalgebra structure discussed in Section 9 and and which is denoted by $\Delta : \operatorname{Tor}^A(M,N) \to \operatorname{Tor}^A(M,N) \otimes \operatorname{Tor}^A(M,N)$. Since the structure δ is the one which is often computable, it is fortunate that we have

Proposition 12.11.1. *If C is a coalgebra with diagonal $\Delta : C \to C \otimes C$ and $\delta : C \to C \otimes C$ is a map of coalgebras with a counit in common with Δ, then they are equal, $\Delta = \delta$, and both are (graded) commutative, $T \cdot \delta = \delta$.*

Proof: The common counit hypothesis is

$$\Delta(c) = c \otimes 1 + 1 \otimes c + \sum c^\alpha \otimes c^\beta, \quad \delta(c) = c \otimes 1 + 1 \otimes c + \sum c_\alpha \otimes c_\beta$$

and the following diagram commutes

$$\begin{array}{ccc}
C & \xrightarrow{\Delta} & C \otimes C \\
\downarrow \delta & & \downarrow \delta \otimes \delta \\
C \otimes C & \xrightarrow{\Delta \otimes \Delta} C \otimes C \otimes C \otimes C \xrightarrow{1 \otimes T \otimes 1} & C \otimes C \otimes C \otimes C.
\end{array}$$

Start in the upper left-hand corner. Go right and down and see the terms $\sum 1 \otimes c^\alpha \otimes c^\beta \otimes 1$. Go down and right and see the terms $\sum (-1)^{\deg(c_\alpha) \cdot \deg(c_\beta)} 1 \otimes c_\beta \otimes c_\alpha \otimes 1$. Thus, $T \cdot \delta = \Delta$.

Go right and down and see the terms $\sum c^\alpha \otimes 1 \otimes 1 \otimes c^\beta$. Go down and right and see the terms $\sum c_\alpha \otimes 1 \otimes 1 \otimes c_\beta$. Therefore, $\Delta = \delta$. $\square$

Here is the algebraic form of the Borel transgression theorem.

Borel transgression Theorem 12.11.2. *Let $E(V)$ be an exterior algebra with V concentrated in odd degrees (but with no such degree restrictions if R is a field of characteristic 2). Then*

$$\operatorname{Tor}^{E(V)}(k,k) = \Gamma(sV)$$

is a divided power algebra as Hopf algebras.

Proof: Since E and Γ both commute with direct limits and since

$$E(V \oplus W) = E(V) \otimes E(W), \quad \Gamma(sV \oplus sW) = \Gamma(sV) \otimes \Gamma(sW),$$

it is sufficient to demonstrate the one generator case

$$\operatorname{Tor}^{E(x)}(R,R) = \Gamma(sx).$$

Recall that $\Gamma(sx)$ has a basis $1 = \gamma_0, \gamma_1 = sx, \gamma_2 = \gamma_2(sx), \ldots, \gamma_i = \gamma_i(sx), \ldots$ with multiplication

$$\gamma_i \cdot \gamma_j = (i,j)\gamma_{i+j}$$

and diagonal

$$\Delta(\gamma_n) = \sum_{p+q=n} \gamma_p \otimes \gamma_q.$$

The differential Hopf algebra

$$\mathcal{R} = E(x) \otimes \Gamma(sx), \quad dx = 0, \quad d(\gamma_i) = x\gamma_{i-1}, \ i \geq 1$$

is an acyclic construction over $E(V)$. Let

$$\mathcal{A} = \mathcal{R} \otimes_{E(V)} \mathcal{R} = \Gamma(sx).$$

The multiplication $\mathcal{R} \otimes \mathcal{R} \to \mathcal{R}$ is a map of constructions with respect to the multiplication map $E(V) \otimes E(V) \to E(V)$ and the diagonal map $\mathcal{R} \to \mathcal{R} \otimes \mathcal{R}$ is a map of constructions with respect to the diagonal map $E(V) \to E(V) \otimes E(V)$.

Hence, these induce a Hopf algebra structure on $\mathcal{A}$ which has zero differential and hence is $\text{Tor}^{E(x)}(R, R)$ as a Hopf algebra. $\square$

Exercise

(1) Assuming that the differential Tors are all projective over R, verify that

$$\text{Tor}^{A \otimes B}(R, R) \equiv \text{Tor}^A(R, R) \otimes \text{Tor}^B(R, R)$$

as Hopf algebras if A and B are commutative differential graded algebras.

12.12 Coproducts and the geometric diagonal

Let G be a topological group with a right G space X and a free left G space E.

Proposition 12.12.1. *If $H(X \times_G E)$ is R projective, then*

$$\text{Tor}^{C(G)}(C(X), C(E)) = H(X \times_G E)$$

as coalgebras.

Proof: Since the geometric diagonal defines a differential Hopf algebra structure on $C(G)$ and differential module coalgebra structures on $C(X)$ and $C(E)$, it defines a coalgebra structure on $\text{Tor}^{C(G)}(C(X), C(E))$ which is identical to the canonical one when both are defined.

We have a commutative diagram

$$\begin{array}{ccc} \operatorname{Tor}^{C(G)}(C(X), C(E)) & \xrightarrow{\operatorname{Tor}^{\Delta}(\Delta,\Delta)} & \operatorname{Tor}^{C(G \times G)}(C(X \times X), C(E \times E)) \\ \downarrow & & \downarrow \\ H(X \times_G E) & \xrightarrow{\Delta} & H((X \times X) \times_{G \times G} (E \times E)) \end{array}.$$

When combined with the Künneth isomorphisms, this gives the desired result. □

Suppose that G and H are topological groups. Using Milnor's functorial construction of the classifying space, the projection maps define a homotopy equivalence $B(G \times H) \xrightarrow{\simeq} BG \times BH$.

If G is a commutative topological group, the multiplication $\mu : G \times G \to G$ is a homomorphism and the commutative diagram below shows that BG is an H-space:

$$\begin{array}{ccc} & B(* \times G) & \\ & \downarrow & \searrow = \\ BG \times BG \xleftarrow{\simeq} & B(G \times G) & \xrightarrow{B\mu} BG \\ & \uparrow & \nearrow = \\ & B(G \times *) & \end{array}$$

The details of the following proposition are similar to those of Proposition 12.12.1:

Proposition 12.12.2. *If G is a commutative topological group, then there is an isomorphism of algebras*

$$\operatorname{Tor}^{C(G)}(R, R) \to HBG.$$

If, in addition, HBG is R projective, then this is an isomorphism of Hopf algebras.

We illustrate Proposition 12.12.2 with the computation of the homology Hopf algebra structures of HBG when G is a cyclic group.

Discrete groups

In the case of a discrete group π, the normalized chains $C\pi$ are nothing but the integral group ring $R[\pi]$. The multiplication of the group defines the algebra structure on generators of $R[\pi]$ via $g \cdot h = gh$ and the coalgebra structure $\delta : R[\pi] \to R[\pi] \otimes R[\pi]$ is given on generators by $\delta(g) = g \otimes g$. The element 1 is a unit and the augmentation $\epsilon : R[\pi] \to R, \epsilon(g) = 1$ is the counit.

In this case, the differential Hopf algebra $C\pi$ has zero differential and hence the proper projective class over A is the same as the usual projective class. In other words, a proper projective resolution is the same as a projective resolution. Hence, $HB\pi = HK(\pi, 1) = \operatorname{Tor}^{R[\pi]}(R, R)$.

The infinite cyclic group

The integral group ring $\mathbb{Z}[\mathbb{Z}] = \mathbb{Z}[t, t^{-1}] = A =$ the ring of finite Laurent polynomials. There is a projective resolution of $\mathbb{Z}$ over A given by

$$0 \to Ae_1 \xrightarrow{d} Ae_0 \xrightarrow{\epsilon} \mathbb{Z} \to 0$$

where

$$de_1 = (t-1)e_0, \quad \epsilon\left(\sum a_i t^i\right) e_0 = \sum a_i.$$

We denote this resolution by $\mathcal{S} \to \mathbb{Z}$ and note that

$$\mathbb{Z} \otimes_A \mathcal{S} = \langle e_0 = 1,\ e_1 \rangle.$$

Hence,

$$HB\mathbb{Z} = \langle 1, e_1 \rangle = E(e_1)$$

and it is clear that this is even an isomorphism of primitively generated Hopf algebras. We have computed the homology of a circle!

Finite cyclic groups

Let $\pi = \{1, t, , t^{n-1}\}$, $t^n = 1$. The integral group ring is

$$A = \mathbb{Z}[t]/\langle t^n - 1 \rangle.$$

Let $\Delta = t - 1$ and $N = 1 + t + \cdots + t^{n-1}$. There is a projective resolution $\mathcal{S} \xrightarrow{\epsilon} R$, that is,

$$\cdots \xrightarrow{d} Ae_i \xrightarrow{d} Ae_{i-1} \to \cdots \xrightarrow{d} Ae_0 \xrightarrow{\epsilon} R \to 0$$

where

$$\epsilon(e_0) = 1, \quad de_{2i+1} = \Delta e_{2i}, \quad de_{2i+2} = N e_{2i+1}.$$

Let $R = \mathbb{Z}/n\mathbb{Z}$. With R coefficients,

$$HB\pi = R \otimes_A \mathcal{S} = \langle e_0 = 1, e_1, \ldots, e_i, \ldots \rangle =$$

a free R module with one generator in each positive degree.

It is easy to compute the algebra structure in $HB\pi$ but the coalgebra structure is somewhat more difficult.

Define an algebra structure on $\mathcal{S}$ by a map $\mu : \mathcal{S} \otimes \mathcal{S} \to \mathcal{S}$ extending the multiplication $A \otimes A \to A$. On generators the map is

$$\mu(e_{2r} \otimes e_{2s}) = (r, s) e_{2r+2s}$$
$$\mu(e_{2r+1} \otimes e_{2s}) = \mu(e_{2r} \otimes e_{2s+1}) = (r, s) e_{2r+2s+1}$$
$$\mu(e_{2r+1} \otimes e_{2s+1}) = 0.$$

12.12 Coproducts and the geometric diagonal

It is easy to see that this multiplication commutes with the differential and hence induces the multiplication on $HB\pi$ given by the above formulas on generators. Hence, as an algebra it is the tensor product of an exterior algebra with a divided power algebra,

$$HB\pi = E(e_1) \otimes \Gamma(e_2).$$

In order to compute the coalgebra structure in $HB\pi$, we follow Steenrod and introduce the following element

$$\Omega = \sum_{0 \le i < j < n} t^i \otimes t^j \in R[\pi] \otimes R[\pi].$$

Define a coproduct

$$\delta : S \to S \otimes S$$

on generators by

$$\delta e_{2i} = \sum_{j=0}^{i} e_{2j} \otimes e_{2i-2j} + \sum_{j=0}^{i-1} \Omega(e_{2j+1} \otimes e_{2i-2j-1})$$

$$\delta e_{2i+1} = \sum_{j=0}^{i} (e_{2j} \otimes e_{2i-2j+1} + e_{2j+1} \otimes te_{2i-2j}).$$

Make this a map of modules with respect to the diagonal $\delta : A \to A \otimes A$.

Lemma 12.12.3. *The following identities hold:*

$$t \otimes t - 1 \otimes 1 = 1 \otimes \Delta + \Delta \otimes t$$

$$(t \otimes t)\Omega - \Omega = N \otimes 1 - 1 \otimes N$$

$$1 \otimes 1 + t \otimes t + \cdots + t^{n-1} \otimes t^{n-1} = 1 \otimes N + \Omega(\Delta \otimes 1)$$

$$1 \otimes t + t \otimes t^2 + \cdots + t^{n-1} \otimes 1 = N \otimes 1 - \Omega(1 \otimes \Delta).$$

These identities show that the above map $\delta : S \to S \otimes S$ is a map of chain complexes. Hence, with coefficients $R = \mathbb{Z}$, the coalgebra structure on $HB\pi = R \otimes_A S$ is given by the formulas

$$\delta e_{2i} = \sum_{j=0}^{i} e_{2j} \otimes e_{2i-2j} + \sum_{j=0}^{i-1} \frac{n(n-1)}{2}(e_{2j+1} \otimes e_{2i-2j-1})$$

$$\delta e_{2i+1} = \sum_{j=0}^{i} (e_{2j} \otimes e_{2i-2j+1} + e_{2j+1} \otimes e_{2i-2j}).$$

We break this into three cases.

If n is odd, take coefficients $R = \mathbb{Z}/n\mathbb{Z}$, and note that the second term of the coproduct on an even dimensional class vanishes. Hence, the coalgebra is exterior tensor divided power, that is,

$$HB\pi = HK(\mathbb{Z}/n\mathbb{Z}, 1); \mathbb{Z}/n\mathbb{Z}) = E(e_1) \otimes \Gamma(e_2)$$

as coalgebras. If we compare this with the algebra computation, we see that this is an isomorphism of Hopf algebras.

If $n = 2^r$ with $r \geq 2$, then take coefficients $R = \mathbb{Z}/2\mathbb{Z}$, and note that the second term of the coproduct on an even dimensional class vanishes. Hence, as above,

$$HB\pi = HK(\mathbb{Z}/2^r\mathbb{Z}, 1); \mathbb{Z}/2\mathbb{Z}) = E(e_1) \otimes \Gamma(e_2)$$

as Hopf algebras.

But, if $n = 2$ and we take coefficients $R = \mathbb{Z}/2\mathbb{Z}$, the second term of the coproduct on an even dimensional class does not vanish and we see that

$$HB\pi = HK(\mathbb{Z}/2\mathbb{Z}, 1); \mathbb{Z}/2\mathbb{Z}) = H(RP^\infty; \mathbb{Z}/2\mathbb{Z}) = \Gamma(e_1)$$

is a divided power algebra as a Hopf algebra.

Exercises

(1) Verify the details of Proposition 12.12.2.

(2) Verify Lemma 12.12.3 and the fact that the map $\delta : S \to S \otimes S$ is a map of chain complexes.

(3) Use the cohomology Serre spectral sequence of the fibration sequence

$$S^1 \to S^\infty \to CP^\infty$$

to show that the Hopf algebra

$$H^*CP^\infty = P(u) = $$

a primitively generated polynomial algebra generated by an element u of dimension 2.

(4) The fibration sequence $S^1 \to K(\mathbb{Z}/n\mathbb{Z}, 1) \to K(\mathbb{Z}, 2) \xrightarrow{n} K(\mathbb{Z}, n)$ begins with a principal bundle $S^1 \to K(\mathbb{Z}/n\mathbb{Z}, 1) \to K(\mathbb{Z}, 2)$. This principal bundle allows us to compute the cohomology Hopf algebra of $K(\mathbb{Z}/n\mathbb{Z}, 1)$ without the use of resolutions.

(a) Use the cohomology Serre spectral sequence to show that

$$H^*(K(\mathbb{Z}/n\mathbb{Z}, 1); \mathbb{Z}/n\mathbb{Z}) = E(e) \otimes H^*CP^\infty = E(e) \otimes P(u)$$

as modules over the polynomial algebra $P(u)$ where the degree of e is 1 and the degree of u is 2.

(b) If $n = 2$, show that $Sq^1(e) = e \cup e = u$ and conclude that
$$H^*(K(\mathbb{Z}/2\mathbb{Z}, 1), \mathbb{Z}/2\mathbb{Z}) = P(e) =$$
a polynomial algebra on a generator of dimension 1.

(c) If $n = 2^r$ with $r \geq 2$, show that, as algebras,
$$H^*(K(\mathbb{Z}/2^r\mathbb{Z}, 1), \mathbb{Z}/2\mathbb{Z}) = E(e) \otimes P(u) =$$
an exterior algebra tensor a polynomial algebra.

(d) If n is odd, show that
$$H^*(K(\mathbb{Z}/n\mathbb{Z}, 1); \mathbb{Z}/n\mathbb{Z}) = E(e) \otimes P(u)$$
as algebras.

(e) Show that the cohomology generator u is primitive in all cases (a) through (d) and hence that the Hopf algebra structure of $H^*(K(\mathbb{Z}/n\mathbb{Z}, 1))$ is determined with the above coefficients.

12.13 Suspension and transgression

In Section 10.25 the homology suspension is treated in the context of the second quadrant Eilenberg–Moore models such as the cobar construction. In this section we treat it in the context of the first quadrant models such as the bar construction.

Let $F \xrightarrow{\iota} E \xrightarrow{\pi} B$ be a fibration sequence. In general, the homology suspension is the relation
$$\sigma : \overline{H_*F} \xleftarrow{\partial} \overline{H_{*+1}(E, F)} \xrightarrow{\pi} \overline{H_{*+1}B}$$
and its inverse relation is called the transgression
$$\tau : \overline{H_{*+1}B} \xleftarrow{\pi} \overline{H_{*+1}(E, F)} \xrightarrow{\partial} \overline{H_*F}.$$
Both of these concepts are useful in general, but they are most useful when some ambiguity is removed and one or both of the homology suspension and the transgression become well defined maps.

In particular, the homology suspension is a well defined map if the total space E is contractible and hence $\partial : H(E, F) \to \overline{H}F$ is an isomorphism. For example, let $E = PB$ be the path space and then the fibre is the loop space $F = \Omega B$. We have the fundamental proposition originally due to G. Whitehead:

Proposition 12.13.1. *If B is simply connected with HB projective over R, the homology suspension factors through the indecomposables of the loop space and*

the primitives of the base as follows
$$\sigma : \overline{H}_*\Omega B \to Q_*H\Omega B \to PH_{*+1}B \subseteq \overline{H}_{*+1}B.$$

Proof: Since the base is simply connected the E^2 term of the Serre spectral sequence has the simple form $E^2_{p,q} = H_p(B) \otimes H_q F$. It is a fundamental property of the homology Serre spectral sequence of the fibration that the suspension can be defined as follows

$$H_{n-1}\Omega B = E^2_{0,n-1} \to E^n_{0,n-1} \xleftarrow{d^n \cong} E^n_{n,0} \subseteq E^2_{n,0} = HB.$$

The acyclicity of the total space guarantees that the last "transgressive" differential is an isomorphism. (It is for this reason that the inverse to the homology suspension is called the transgression.)

It is sufficient to prove two things:

(1) the kernel of $E^2_{0,n-1} \to E^n_{0,n-1}$ contains all decomposables.

(2) the module of transgressive elements $E^n_{n,0}$ is contained in the module of primitives.

Let $z = xy$ be a Pontrjagin product of two positive degree classes x $y \in H\Omega B$. Since the spectral sequence is "acyclic", we know that $x = d^r w$ for some $r < n$. Since the spectral sequence is a module over the algebra $H\Omega B$, we have

$$d^r(wy) = (d^r w)y = xy.$$

Hence, the decomposable xy is zero in $E^n_{0,n-1}$.

Now, let $z \in E^n_{n,0} \subseteq \overline{H}B$. We claim that z is primitive. Since the spectral sequence is acyclic, we have that $E^n_{r,0} = 0$ for all $0 < r < n$. But the diagonal satisfies

$$\Delta(z) \in \sum_{r+s=n} E^n_{r,0} \otimes E^n_{s,0}.$$

Hence, the transgressive element z is primitive. □

Exactly the same proof applies to universal principal G bundles. However, with the use of a little more geometry and algebra, we can eliminate the hypothesis that the base be simply connected with projective homology over R.

Proposition 12.13.2. *If G is a topological group, then the homology suspension factors as*

$$\sigma : \overline{H}_*G \to Q_*HG \to PH_{*+1}BG \subseteq \overline{H}_{*+1}BG.$$

Proof: Consider the embedding $G * G \subseteq E_G$ of the two-fold join into the infinite join. Under the action of G, the orbit spaces are $\Sigma G \subseteq B_G$ where the suspension

space
$$\Sigma G = I \times G/(1.g) \sim *_1, \ (0,g) \sim *_2$$
is the unreduced suspension which is the quotient of the cone
$$G * e \to \Sigma G \quad tg + (1-t)e \mapsto [t,g].$$
The commutative diagram

$$\begin{array}{ccccc} G & \to & G*e & \to & \Sigma G \\ \downarrow = & & \downarrow \subseteq & & \downarrow \subseteq \\ G & \to & E_G & \to & B_G \end{array}$$

shows that the homology suspension factors as
$$\sigma : \overline{H}_*G \xleftarrow{\partial \simeq} \overline{H}_{*+1}(G*e, G) \to \overline{H}_{*+1}\Sigma G \to \overline{H}_{*+1}B_G.$$

But with any coefficients, the positive degree elements in the homology of a suspension are primitive. Hence, the image of the homology suspension is primitive.

In order to show that the homology suspension annihilates the decomposables in HG, we interpret the homology suspension in terms of the algebraic bar construction. Let $A = CG$ be the chains on the topological group and consider the acyclic twisted tensor product $A \otimes_\tau BA$ where the twisting morphism $\tau : BA \to A$ is nontrivial only on length one tensors and is given on those by $\tau[a] = a$. There is a chain equivalence $BA \to C(B_G)$. In fact, the sequence $A \to A \otimes_\tau BA \to BA$ is chain equivalent to the sequence $CG \to CE_G \to CB_G$.

Let a be a cycle in A. Since $d[a] = a$ in the twisted tensor product $A \otimes_\tau BA$, it follows that the homology suspension is given by $\sigma a = [a]$. By the way, this shows again that σa is primitive, even as a chain.

Let $a = bc$ be a product of two cycles in A. Since $d(b[c]) = \pm bc = \pm a$, it follows that $\sigma a = 0 = $ the image of $b[c]$ in BA. Hence, the homology suspension annihilates decomposable elements. □

Exercises

(1) If M and N are differential modules over a differential Hopf algebra A, show that $M \otimes N$ is a differential module over A via the diagonal action

$$A \otimes M \otimes N \xrightarrow{\Delta \otimes 1 \otimes 1} A \otimes A \otimes M \otimes N$$
$$\xrightarrow{1 \otimes T \otimes 1} A \otimes M \otimes A \otimes N \xrightarrow{\mu \otimes \mu} M \otimes N.$$

(2) If A is a differential Hopf algebra and E and F are acyclic constructions over A, show that there are homology equivalences of differential modules over A,

$$F \xleftarrow{\simeq} E \otimes F \xrightarrow{\simeq} E$$

and hence a there are homology equivalences

$$R \otimes_A F \xleftarrow{\simeq} R \otimes_A (E \otimes F) \xrightarrow{\simeq} R \otimes_A E.$$

(3) If G is a topological group, show that there is a sequence of homology equivalences leading from the bar construction on the chains BCG to the chains on the classifying space CBG. Conclude that the description given in the proof of Proposition 12.13.2 of the homology suspension given in terms of the bar construction is correct.

12.14 Eilenberg–Moore spectral sequence

The algebraic Eilenberg–Moore spectral sequence 12.3.9 specializes to the geometric version below.

Proposition 12.4.1. *Let G be a topological group with right G-space X and left free G-space E. Then there is a first quadrant homology spectral sequence with*

$$E^2_{p,q} = \{\operatorname{Tor}^{HG}_p(HX, HE)\}_q$$
$$d^r : E^r_{p,q} \to E^r_{p-r, q+r-1}$$

and converging strongly to its filtered abutment

$$\{HX \otimes_{HA} HE\}_{p+q} = F_0 \subseteq F_1 \subseteq \cdots \subseteq F_p \subseteq \cdots \subseteq F_\infty = H_{p+q}(X \times_G E).$$

If all E^r are projective over R, then this is a spectral sequence of commutative coalgebras.

The classifying space situation is important enough to be mentioned separately:

Proposition 12.14.2. *If G is a topological group, then there is a first quadrant homology spectral sequence with*

$$E^2_{p,q} = \{\operatorname{Tor}^p(R, R)\}_q$$

converging strongly to its filtered abutment

$$R = F_0 \subseteq F_1 \subseteq \cdots \subseteq F_p \subseteq \cdots \subseteq F_\infty = HBG$$

and the homology suspension $\sigma : \overline{H}_*G \to \overline{H}_{*+1}BG$ factors as follows

$$\overline{H}_n G \to Q_n HG \equiv \operatorname{Tor}^{HG}_1(R, R) \to F_1 \to F_\infty = HBG.$$

When the terms of the spectral sequence are all projective over R, then it is a spectral sequence of commutative coalgebras.

If all the terms E^r are projective over R and of finite type, taking duals implies that the above propositions can be given in cohomology versions which we record here.

Proposition 12.14.3. *Let G be a topological group with right G-space X and left free G-space E. Suppose all homologies HG, HX, HE are all projective and finite type over R and, in addition, all E_r are projective over R, then there is a first quadrant cohomology spectral sequence of commutative algebras with*

$$E_2^{p,q} = \{\mathrm{Cotor}_{HG}^p(HX, HE)\}^q$$
$$d_r : E_r^{p,q} \to E_r^{p+r,q-r+1}$$

and converging strongly to its filtered abutment

$$\{H^*X \square_{H^*A} H^*E\}^{p+q} \leftarrow F^0 \leftarrow F^1 \leftarrow \cdots \leftarrow F^p \leftarrow \cdots \leftarrow F^\infty$$
$$= H^{p+q}(X \times_G E).$$

The classifying space situation is:

Proposition 12.14.4. *Let G be a topological group with R projective finite type homology. If all E_r are projective over R, then there is a first quadrant cohomology spectral sequence of commutative algebras with*

$$E_2^{p,q} = \{\mathrm{Cotor}_p^{H^*G}(R,R)\}^q$$

converging strongly to its filtered abutment

$$R^{p+q} = F^0 \leftarrow F_1 \leftarrow \cdots \leftarrow F_\infty = H^{p+q}BG$$

and the cohomology suspension $\sigma^* : \overline{H}^{*+1}BG \to \overline{H}^*G$ factors as follows

$$\overline{H}^{n+1}BG \twoheadrightarrow Q^{n+1}H^*BG \equiv F^\infty \twoheadrightarrow F^1 \equiv E_\infty^{1,n} \to E_2^{1,n}$$
$$\equiv \{\mathrm{Cotor}_1^{H^*G}(R,R)\}^n \equiv \{PH^*G\}^n.$$

We close with the following statement of the Borel transgression theorem [10] in terms of cohomology.

Borel trangression theorem 12.14.5. *Suppose*

$$H_*G = E(V)$$

as algebras with V concentrated in odd degrees, projective and finite type over R. Then the cohomology of the classifying space is the polynomial algebra

$$H^*BG = P(sV^*)$$

where the the transgression $\tau : PH^*G = V^* \to sV^* = QH^*BG$ is an isomorphism from the exterior primitives to the polynomial generators.

Proof: In the cohomology version of the Eilenberg–Moore spectral sequence,

$$E_2 = \mathrm{Cotor}_{H^*G}(R,R) = \mathrm{Cotor}^{E(V^*)}(R,R) = P(sV^*).$$

Since the spectral sequence is concentrated in even degrees, it must collapse and $E_2 = E_\infty = P(sV^*)$. Since the abutment H^*BG is a commutative algebra and E_∞ is a free commutative algebra, there are no extension problems. That is, lifting the polynomial generators up to $F^\infty = H^*BG$ gives an isomorphism $H^*BG \simeq P(sV^*)$. □

12.15 Euler class of a vector bundle

A real vector bundle ξ of dimension n (a real n-plane bundle) consists of a total space $E = E(\xi)$, a base space $B = B(\xi)$ and a continuous map $\pi : E \to B$ such that:

(1) for all $x \in B$, each $E_x = \pi^{-1}(x)$ is a real vector space of dimension n and

(2) there exists an open cover U of B and homeomorphisms $\psi_U : U \times R^n \to \pi^{-1}(U)$ covering the identity map on U such that the restriction, for all $x \in B$ to each fibre $\{x\} \times R^n \to E_x$ is a vector space isomorphism.

We adopt the following notation: $E_0 = E(\xi)_0 = E - \{0_x \mid x \in B\}$ = the complement of the 0-section, $R_0^n = R^n - \{0\}$ = the complement of the origin, if $A \subseteq B$, then $E_A = \pi^{-1}(A)$ is the restriction of the bundle over A, and

$$\psi_{U,V} = \psi_V^{-1} \cdot \psi_U : (U \cap V) \times R^n \to (V \cap U) \times R^n,$$
$$\psi_{U,V}(x,v) = (x, g_{U,V}(x)(v))$$

for the transition functions. Thus, each $g_{U,V}(x)$ is a linear isomorphism which depends continuously on $x \in U \cap V$.

There are three definitions of orientability for a real vector bundle.

Definition 12.15.1. A real vector bundle is geometrically orientable if all of the $g_{U,V}(x)$ can be chosen to have positive determinant.

Definition 12.15.2. A real vector bundle is locally orientable with coefficients R if, for each $x \in B$, there is a continuous choice of a generator $\mu_x \in H^n(E_x, E_{x0}; R)$. The choice is continous in the sense that each x has a neighborhood $U \subseteq B$ and a class $\mu_U \in H^n(E_U, E_{U,0}; R)$ which restricts to μ_y for all $y \in U$.

Definition 12.15.3. A real vector bundle is orientable with coefficients R if there is a class $\mu \in H^n(E, E_0; R)$ which restricts to a generator $\mu_x \in H^n(E_x, E_{x0}; R)$ for all $x \in B$. The class $\mu = \mu_\xi$ is called the Thom class of the bundle.

The following proposition describes the sense in which these three definitions are equivalent.

Proposition 12.15.4.

(a) *Local orientability with coefficients R is equivalent to orientability with coefficients R.*

(b) *Orientability with integral coefficients $\mathbb{Z}$ implies orientability with all coefficients R.*

(c) *Geometric orientability is equivalent to orientability with integral coefficients $\mathbb{Z}$.*

Before proving the above, we prove [91]

Thom isomorphism Theorem 12.15.5. *Let ξ be a real n-plane bundle which is orientable over R, There is an isomorphism*

$$\Phi : H^k(B) \otimes H^n(R^n, R_0^n) \to H^{n+k}(E, E_0)$$

given by $\Phi(\alpha \otimes e_n) = \pi^*\alpha \cup \mu$ with $\pi^*\alpha \in H^k(E)$ and $\mu \in H^n(E, E_0)$ is the Thom class.

Proof: In the cohomology Serre spectral sequence of the bundle pair $(E, E_0) \to (B, B)$,

$$E_2^{p,q} = H^p B \otimes H^q(R^n, R_0^n)$$

and the orientability implies that the generator $e \in H^n(R^n, R_0^n)$ survives to represent the Thom class in E_∞. Hence, $E_2 = E_\infty$ and the map $\Phi : H^*(B) \otimes H^*(R^n, R_0^n) \to H^*(E, E_0)$ is an isomorphism. $\square$

Remark. The Thom isomorphism theorem may be restated as the fact that cupping with the Thom class is an isomorphism,

$$(\ ) \cup \mu : H^k E \xrightarrow{\cong} H^{k+n}(E, E_0), \quad \alpha \mapsto \alpha \cup \mu.$$

Proof of Proposition 12.15.4

Part (a1): Orientability with coefficients R implies local orientability with coefficients R. This follows since we can define $\mu_x \in H^n(E_x, E_{x0})$ to be the restriction of the Thom class.

Part (a2): Local orientability with coefficients R implies orientability with coefficients R. Since we have orientability over the neighborhoods U, we have the Thom isomorphism neighborhood over these neighborhoods and, in particular, we have $H^k(E_U, E_{U0}) = 0$ for $k < n$.

Now suppose that A is a maximal open subset of B for which E_U is orientable with coeffiencts R. If there exists $x \in B - A$, then the Meyer–Vietoris exact sequence formed from the restriction maps

$$0 \to H^n(E_{U_x \cup A}, E_{U_x \cup A\, 0}) \xrightarrow{\rho \oplus \rho} H^n(E_{U_x}, E_{U_x\, 0}) \oplus H^n(E_A, E_{A\, 0})$$
$$\xrightarrow{(\rho, -\rho)} H^n(E_{U_x \cap A}, E_{U_x \cap A\, 0})$$

shows that there exists a unique Thom class for the bundle over $U_x \cup A$, which contradicts the maximality of A.

Part (b): If the bundle is orientable over the integers, then $H^n(E, E_0; R) = H^n(E, E_0; \mathbb{Z}) \otimes R$ shows that it is orientable with any coefficients.

Part (c1): Geometric orientability implies local orientability over the integers: For each set U in an open cover of B, the maps

$$\phi_U^* : H^n(E_U, E_{U0}) \to H^0(U) \otimes H^n(R^n, R_0^n)$$

are isomorphisms. Choose $\mu_U \in H^n(E_U, E_{U0})$ which map to the preferred generator $1 \otimes e_n \in H^0(U) \otimes H^n(R^n, R_0^n)$. Geometric orientability implies that this is a coherent choice of generators.

Part (c2): Orientability over the integers implies geometric orientability. For each connected open set U choose the maps $\psi_U : U \times R^n \to E_U$ so that $\psi_U^* : H^n(E, E_0) \to H^0(U) \otimes H^n(R^n, R_0^n)$ sends the restriction of the Thom class to the preferred generator. $\square$

Remark. It is clear that all vector bundles are orientable with mod 2 coefficients. It is also clear that complex n-plane bundles are integrably orientable when regarded as real $2n$-plane bundles since elements of the complex general linear group all have positive determinant.

Definition 12.15.6. Let ξ be a real n-plane bundle which is orientable with R coefficients, with maps $\pi : E \to B$ and $j : E \to (E, E_0)$. The Euler class $\chi_\xi \in H^n(B; R)$ is the unique class related to the Thom class by

$$\pi^* \chi_\xi = j^* \mu_\xi.$$

Suppose that ξ_1 and ξ_2 are real vector bundles of respective dimensions m and n. Then the product bundle $\xi_1 \times \xi_2$ is the bundle with dimension $m + n$ and with product total space and product base space, that is,

$$E_1 \times E_2 \to B_1 \times B_2.$$

If $B_1 = B_2$, the Whitney sum $\xi_1 \oplus \xi_2$ is the bundle with dimension $m + n$ which is the pullback of the product bundle over the diagonal

$$\xi_1 \oplus \xi_2 = \Delta^*(\xi_1 \times \xi_2), \quad E_1 \oplus E_2 \to B.$$

Since

$$(R^{m+n}, R_0^{m+n}) = (R^n, R_0^n) \times (R^m, R_0^m)$$
$$(E_1 \times E_2, (E_1 \times E_2)_0) = (E_1, E_{10}) \times (E_2, E_{20})$$

it is clear that the product bundle $\xi_1 \times \xi_2$ is orientable if ξ_1 and ξ_2 are and that it has Thom class

$$\mu_{\xi_1 \times \xi_2} = \mu_{\xi_1} \times \mu_{\xi_2} \in H^{m+n}(E_1 \times E_2, (E_1 \times E_2)_0).$$

The Euler class is given by the product formula

$$\chi_{\xi_1 \times \xi_2} = \chi_{\xi_1} \times \chi_{\xi_2} \in H^{m+n}(B_1 \times B_2).$$

12.15 Euler class of a vector bundle

Similarly, the Whitney sum of two orientable bundles is orientable and has Thom class

$$\mu_{\xi_1 \oplus \xi_2} = \Delta^*(\mu_{\xi_1} \times \mu_{\xi_2}) = \mu_{\xi_1} \cup \mu_{\xi_2} \in H^{m+n}(E_1 \oplus E_2, (E_1 \oplus E_2)_0).$$

The Euler class is given by the product formula

$$\chi_{\xi_1 \oplus \xi_2} = \chi_{\xi_1} \cup \chi_{\xi_2} \in H^{m+n}(B).$$

The Euler class of an orientable bundle can also be defined in terms of the transgression in the cohomology Serre spectral sequence.

The cohomology exact sequence

$$\cdots \to H^i(E, E_0) \to H^i(E) \to H^i(E_0) \xrightarrow{\delta} H^{i+1}(E, E_0) \to \cdots$$

of the pair (E, E_0) transforms via the Thom isomorphism $(\ )\cup\mu : H^{i-n}(E) \to H^i(E, E_0)$ into the exact sequence

$$\cdots \to H^{i-n}(E) \xrightarrow{(\ )\cup\mu} H^i(E) \to H^i(E_0) \xrightarrow{\delta} H^{i-n+1}E \to \cdots.$$

The isomorphism $\pi^* : H^*B \xrightarrow{\cong} H^*E$ transforms the exact sequence into the exact Gysin sequence

$$\cdots \to H^{i-n}(B) \xrightarrow{(\ )\cup\chi_\xi} H^i(B) \xrightarrow{\pi^*} H^i(E_0) \xrightarrow{\delta} H^{i-n+1}B \to \cdots.$$

Hence, the Euler class χ_ξ is a generator of the kernel of the map $H^n(B) \xrightarrow{\pi^*} H^n(E_0)$. But, up to a unit multiple, this is clearly the transgression of the preferred generator in the cohomology of the fibre of the fibration sequence

$$R_0^n \to E_0 \xrightarrow{\pi} B.$$

Proposition 12.15.7. *If ξ is an orientable bundle, then, up to a unit multiple, the transgression in the cohomology Serre spectral sequence of the fibration $R_0^n \to E_0 \xrightarrow{\pi} B$ is given by the Euler class,*

$$\tau e_{n-1} = \chi_\xi.$$

Exercises

(1) If ξ and ρ are n-plane bundles, a bundle map $f = (\overline{f}, \underline{f}) : (E(\xi), B(\xi)) \to (E(\rho), B(\rho))$ is a commutative diagram of maps

$$\begin{array}{ccc} E(\xi) & \xrightarrow{\overline{f}} & E(\rho) \\ \downarrow & & \downarrow \\ B(\xi) & \xrightarrow{\underline{f}} & B(\rho) \end{array}$$

where $\overline{f}$ restricts to a linear isomorphism on each fibre. Show that ξ is isomorphic to the pullback bundle $\underline{f}^*\rho$.

(2) If the bundles are orientable and the base $B(\xi)$ is connected, then show that $\overline{f}$ either preserves orientation on all fibres or reverses orientation on all fibres.

(3) If the map in Exercise 1 is a bundle map of complex n-plane bundles, then show that it preserves orientation on all fibres.

(4) If the map in Exercise 1 preserves orientation on all fibres, then show that the Euler class is natural in the sense that $\underline{f}^*\chi_\rho = \chi_\xi$. If it reverses orientation on all fibres, then show that $\underline{f}^*\chi_\rho = -\chi_\xi$.

(5) If $E \to B$ is an orientable vector bundle of dimension n, then the map $\pi^* : H^i B \to H^i E_0$ is an isomorphism for $i < n-1$ and a monomorphism for $i = n-1$.

(6) Let ξ be a bundle over B and let $B_2 \xrightarrow{f} B_1 \xrightarrow{g,h} B$ be maps. Show that

(a) There is an isomorphism of pullback bundles $f^*(g^*\xi) \simeq (gf)^*\xi$.

(b) If g and h are homotopic, there is a bundle isomorphism $g^*\xi \simeq h^*\xi$. (Hint: You may do this the easy way by using classification.)

(c) If g is null homotopic, then $g^*\xi$ is isomorphic to a trivial (product) bundle.

12.16 Grassmann models for classifying spaces

It is convenient to introduce the classical Grassmann models for the classifying spaces of the orthogonal and unitary groups.

Let $G_k(R^{n+k})$ denote the set of real k-planes V through the origin in R^{n+k}. Every such k-plane V and its orthogonal complement $V^\perp$ are spanned by orthonormal bases $\mathcal{B}$ and $\mathcal{C}$, the assignment $(\mathcal{B}, \mathcal{C}) \mapsto V$ defines a surjective map

$$\pi : O(n+k) \to G_k(R^{n+k})$$

and we give the so-called Grassmann manifold $G_k(R^{n+k})$ the quotient topology. There is a homeomorphism

$$O(n+k)/O(k) \times O(n) \xrightarrow{\simeq} G_k(R^{n+k})$$

Similarly, let $V_k(R^{n+k})$ denote the set of orthonormal k-frames in R^{n+k} and topologize it via the quotient map

$$O(n+k) \to V_k(R^{n+k})$$

so that there is a homeomorphism

$$O(n+k)/1_k \times O(n) \xrightarrow{\simeq} V_k(R^{n+k}).$$

12.16 Grassmann models for classifying spaces

The frame bundle $\pi : V_k(R^{n+k}) \to G_k(R^{n+k})$ is a principal bundle with fibre $O(k)$.

The universal $O(k)$ bundle $\pi : V_k^R \to G_k^R$ is gotten by letting n go to infinitity. That is,
$$V_k^R = \lim_{n \to \infty} V_k(R^{n+k}), \quad G_k^R = \lim_{n \to \infty} G_k(R^{n+k}).$$

Since $V_k(R^{n+k})$ is n-connected, it follows that all the homotopy groups of the limit V_k^R are zero. Hence, according to Steenrod [125], it is a universal bundle for all principal $O(k)$ bundles where the base is a CW complex.

Remark. In fact, it is a universal bundle for all principal bundles over a paracompact base space. This follows from a result of Milnor and Stasheff [91]. See the remark below.

Using the correspondence between frame bundles and vector bundles, we see that the so-called canonical k-plane bundle γ_k is a universal vector bundle. In detail, this vector bundle has the total space
$$E_k^R = E(\gamma_k) = \{(V, v) \mid V \in B_k^R, \ v \in V\}$$
and projection map
$$\pi : E(\gamma_k) \to B_k^R, \quad \pi(V, v) = V.$$

Remark. In Milnor and Stasheff [91], it is verified that this vector bundle has a paracompact base and is a universal vector bundle for all vector bundles over paracompact bases. Hence, the corresponding frame bundle V_k^R is a universal principal $O(k)$ bundle for all principle bundles over paracompact base spaces.

Let
$$\tilde{G}_{k-1}^R = \{(V, v) \mid V \in G_k^R, \ 0 \neq v \in V\}$$
and define a canonical $(k-1)$-plane bundle $\tilde{\gamma}_{k-1}$ over $\tilde{G}_k^R$ by
$$E(\tilde{\gamma}_{k-1}) = \tilde{E}_{k-1} = \{(V, v, w) \mid (V, v) \in \tilde{G}_{k-1}^R,$$
$$w \in v^\perp = \{w \in V \mid \langle w, v \rangle = 0\}$$
with the obvious projection map.

Recall that bundle maps
$$f = (\overline{f}, \underline{f}) : (E(\xi), B(\xi)) \to (E(\rho), B(\rho))$$
are defined to be vector space isomorphisms on each vector space fibre so that $\underline{f}^* \rho \simeq \xi$.

Since the vectors v span a one-dimensional trivial bundle ϵ over $\tilde{G}_{k-1}^R$, we have a bundle map of k-plane bundles

$$\tilde{\gamma}_{k-1} \oplus \epsilon \to \gamma_k, \quad (V, v, w, tv) \mapsto (V, w + tv)$$

which covers the obvious map

$$\pi : \tilde{G}_{k-1}^R \to G_k^R, \quad (V, v) \mapsto V.$$

We also have a bundle map of $(k-1)$-plane bundles

$$\tilde{\gamma}_{k-1} \to \gamma_{k-1}, \quad (V, v, w) \mapsto (v^\perp, w)$$

which covers the obvious map

$$\tau : \tilde{G}_{k-1}^R \to G_{k-1}^R, \quad (V, v) \mapsto v^\perp.$$

In the diagram below, the horizontal and vertical rows are bundle sequences

$$\begin{array}{ccccc} & & R_0^\infty & & \\ & & \downarrow & & \\ R_0^k & \to & \tilde{G}_{k-1}^R & \xrightarrow{\pi} & G_k^R \\ & & \downarrow \tau & & \\ & & G_{k-1}^R & & \end{array}$$

where the fibres are punctured Euclidean spaces.

Since R_0^∞ is contractible, the map τ is a homotopy equivalence, and up to homotopy equivalence, the map π produces a bundle sequence

$$R_0^n \to G_{k-1}^R \to G_k^R.$$

Or, in the language of classifying spaces, up to homotopy equivalence, we have a commutative diagram of bundle sequences

$$\begin{array}{ccccc} O(k) & \to & EO(k) & \to & BO(k) \\ \downarrow & & \downarrow & & \downarrow = \\ S^{k-1} & \to & BO(k-1) & \to & BO(k) \end{array}.$$

The first vertical map selects the last element of a k-frame, the second vertical map uses the last element of a k-frame to determine an orthogonal complement in the vector space spanned by the frame, the third vertical map is the identity.

Of course, we could have replaced the real numbers R with the complex numbers C and repeated all of the above, using Hermitian metrics, complex orthonormal frames, the unitary groups, and ending up with a bundle sequence

$$C_0^k \to \tilde{G}_{k-1}^C \to G_k^C$$

and, up to homotopy equivalence, a commutative diagram of bundle sequences

$$\begin{array}{ccccc} U(k) & \to & EU(k) & \to & BU(k) \\ \downarrow & & \downarrow & & \downarrow= \\ S^{2k-1} & \to & BU(k-1) & \to & BU(k) \end{array}.$$

Exercise

(1) (a) Show that the map $\tilde{G}^R_{k-1} \to G^R_k$ can be covered by a map from a $(k-1)$-frame bundle to a k-frame bundle which is equivariant with respect to the inclusion $O(k-1) \to O(k)$.

(b) Show that the map $\tilde{G}^R_{k-1} \to G^R_{k-1}$ can be covered by a map from a $(k-1)$-frame bundle to a $(k-1)$ frame bundle.

(c) Show that there is a map $G^R_{k-1} \to G^R_k$ which can be covered by a map of universal frame bundles which is equivariant with respect to the inclusion $O(k-1) \to O(k)$.

(d) Conclude that the above map $BO(k-1) \to BO(k)$ is the map induced by the inclusion of topological groups.

12.17 Homology and cohomology of classifying spaces

Since the homology of a unitary group is an exterior algebra on generators of odd degree, it follows directly from the homology version of the Borel transgression theorem that:

Proposition 12.7.1. *With any coefficients*

$$HBU(n) = \Gamma(\sigma QHU(n))$$

as a coalgebra.

But we prefer the cohomology version.

Proposition 12.7.2. *With any coefficients*

$$H^*(BU(n)) = H^*(G^C_n) = P(c_1, \ldots, c_n) =$$

a polynomial algebra where the generators c_i have dimension $2i$ and c_n is the Euler class of the universal complex n-plane bundle.

The classes $c_i = c_i(\gamma_n)$ are called the universal Chern classes of the universal complex n-plane bundle γ_n over $BU(n) = G^C_n$.

Included in the Borel transgression theorem is that the statement that the generators c_i are the transgressions of the generators of the module of primitives

$$PH^*U(n) = \langle e_1, e_2, \ldots, e_n \rangle, \quad \deg(e_i) = 2i - 1$$

There is a map of bundle sequences

$$\begin{array}{ccccc} U(n) & \to & V_n^C & \to & G_n^C \\ \downarrow & & \downarrow & & \downarrow \\ C_0^n & \to & \tilde{G}_{n-1}^C & \to & G_n^C \end{array}$$

where the left-hand map sends the generator of the cohomology of the punctured complex n-plane to the primitive generator e_n. The naturality of the transgression and Proposition 12.15.7 show that the transgression $\tau e_n = c_n$ is the Euler class χ of the universal complex n-plane bundle.

Furthermore, Proposition 12.15.7 shows that the Chern class $c_n = c_n(\gamma_n)$ maps to zero via

$$H^* G_n^C \to H^* \tilde{G}_{n-1}^C.$$

Since the canonical $(n-1)$-plane bundle $\tilde{\gamma}_{k-1}$ is a universal vector bundle, the associated $(n-1)$-frame bundle $\mathcal{F}E(\tilde{\gamma}_{k-1}) = \tilde{V}_{n-1}^C$ is a universal $(n-1)$-frame bundle.

The map of bundle sequences

$$\begin{array}{ccccc} U(n-1) & \to & \tilde{V}_{n-1}^C & \to & \tilde{G}_{n-1}^C \\ \downarrow & & \downarrow & & \downarrow \\ U(n) & \to & V_n^C & \to & G_n^C \end{array}$$

and the naturality of the transgression shows that, for all $i < n$, the Chern class $c_i = c_i(\gamma_n)$ maps to $c_i = c_i(\tilde{\gamma}_{n-1})$ via

$$H^* G_n^C \to H^* \tilde{G}_{n-1}^C.$$

$\square$

The computation of the cohomology ring of $BO(n)$ requires mod 2 coefficients and slightly more care. But the result is similar.

Proposition 12.17.3. *With mod 2 coefficients*

$$H^*(BO(n)) = H^*(G_n^R) = P(w_1, \ldots, w_n) =$$

a polynomial algebra where the generators w_i have dimension i and, if $n > 1$, w_n is the mod 2 Euler class of the universal real n-plane bundle.

The classes $w_i = w_i(\gamma_n)$ are called the universal Stiefel–Whitney classes of the universal real n-plane bundle γ_n over $BO(n) = G_n^R$.

We prove the result by induction on n. The case $n = 1$ is simply the cohomology of the infinite real projective space,

$$H^* BO(1) = H^*(G_1^R) = H^* RP^\infty = P(w_1).$$

Assume that we know the result for $n-1$. Consider the map $O(n-1) \to O(n)$ and the corresponding map of cohomology Eilenberg–Moore spectral sequences. At the E_2 level we have the surjective map of algebras

$$P(w_1,\ldots,w_n) \to P(w_1,\ldots,w_{n-1}),$$

where the w_i are the transgressions of the primitive generators of the cohomology of the groups. We know that the generators w_i of bidegree $(1, i-1)$ survive to E_∞ in the range. Since the differentials d_r go from $(1, i-1)$ to $(1+r, i-r)$, the differentials on w_i for $i < n$ cannot involve w_n and must be zero in the domain also. In the domain, the differentials on w_n can be nonzero only if some polynomial in the w_i for $i < n$ is annihilated. That cannot happen since these w_i generate a free commutative algebra in the range. Hence, all of the generators $w_1, \ldots, w_n$ are infinite cycles in the domain. Thus, $E_2 = E_\infty = P(w_1,\ldots,w_n)$ in the domain. Lifting up the w_n to $H^*BO(n)$ shows that $H^*BO(n) = P(w_1,\ldots, w_n)$.

In order to ensure that w_n goes to zero in $H^*BO(n-1)$, it may have to be adjusted by the addition of a polynomial in the lower degree Stiefel–Whitney classes. But this being done, Proposition 12.15.7 says that w_n must be the mod 2 Euler class of the universal n-plane bundle. $\square$

Exercises

(1) Use the Gysin sequences of the sphere bundles

$$S^{n-1} \to BO(n-1) \to BO(n)$$

and

$$S^{2n-1} \to BU(n-1) \to BU(n)$$

to give inductive proofs of the computations in Propositions 12.17.2 and 12.17.3. Hint: You may assume that $w_n \neq 0 \in H^*(B0(n); \mathbb{Z}/2\mathbb{Z})$. Use the inductive hypotheses that

$$0 \to H^i B0(n) \xrightarrow{(\ )\cup w_n} H^{i+n} BO(n) \to H^{i+n} BO(n-1) \to 0$$

is exact. Get a similar inductive hypothesis for the unitary groups without assuming anything about c_n.

12.18 Axioms for Stiefel–Whitney and Chern classes

We define Stiefel–Whitney classes for all real vector bundles by means of the computation of the mod 2 cohomology ring of the classifying spaces G_n^R.

Definition 12.18.1. If ξ is a real n-plane bundle with a classifying map $f : B \to G_n^R$ such that $f^*\gamma_n \simeq \xi$, the Stiefel–Whitney classes $w_i(\xi) \in H^i(B; \mathbb{Z}/2\mathbb{Z})$ are

$$w_i(\xi) = \begin{cases} 1 & i = 0 \\ f^*w_i(\gamma_n) & 1 \leq i \leq n \\ 0 & i > n. \end{cases}$$

These Stiefel–Whitney classes satisfy the following axioms.

Axioms for Stiefel–Whitney classes 12.18.2.

(a) Dimension axiom: If dimension ξ equals n, then $w_0(\xi) = 1$ and $w_i(\xi) = 0$ for $i > n$.

(b) Naturality under bundle maps: If $\xi \to \rho$ is a bundle map of n-plane bundles covering a map $f : B(\xi) \to B(\rho)$, then $f^*w_i(\rho) = w_i(\xi)$ for all $i \geq 0$.

(c) Stability: If ϵ and ξ are bundles over the same base space B and ϵ is a trivial (product) bundle, then $w_i(\xi \oplus \epsilon) = w_i(\xi)$ for all $i \geq 0$.

(d) Nontriviality: $w_1(\gamma_1)$ is a generator of $H^1(RP^\infty; \mathbb{Z}/2\mathbb{Z})$.

(e) Whitney product formula: If η and ξ are bundles over bases B_ξ and B_ρ, then

$$w_k(\eta \times \xi) = \sum_{i+j=k} w_i(\eta) \times w_j(\xi) \in H^*(B_\xi \times B_\rho; \mathbb{Z}/2\mathbb{Z}).$$

Proof: The dimension axiom, the naturality axiom, and the nontriviality axiom are true by definition.

In order to check the stability axiom, it is sufficient to consider the case where ϵ is a trivial line bundle and ξ is a bundle of dimension $n-1$. Consider the diagram

$$\begin{array}{ccc} & \tilde{G}_{n-1}^R & \xrightarrow{h} G_n^R \\ {}^g\nearrow & \downarrow k & \\ B & \xrightarrow{f} & G_{n-1}^R \end{array}$$

where k and h are the natural maps and $f^*\gamma_{n-1} \simeq \xi$.

Since $\tilde{\gamma}_{n-1}$ is a universal $(n-1)$-plane bundle, there is a map $g : B \to \tilde{G}_{n-1}^R$ such that $g^*\tilde{\gamma}_{n-1} \simeq \xi$. Since $k^*\gamma_{n-1} \simeq \tilde{\gamma}_{n-1}$, it follows that $k \cdot g \sim f$. Since $h^*\gamma_n \simeq \tilde{\gamma}_{n-1} \oplus \epsilon$, it follows that $g^*h^*\gamma_n \simeq \xi \oplus \epsilon$ and hence

$$w_i(\xi \oplus \epsilon) = g^*h^*w_i(\gamma_n) = g^*w_i(\tilde{\gamma}_{n-1})$$
$$= g^*k^*w_i(\gamma_{n-1}) = f^*w_i(\gamma_{n-1}) = w_i(\xi).$$

12.18 Axioms for Stiefel–Whitney and Chern classes

We now verify the Whitney product formula. Since $\gamma_m \times \gamma_n$ is an $m + n$-plane bundle over $G_m^R \times G_n^R$, it is classified by a map $\phi_{m,n} : G_m^R \times G_n^R \to G_{m+n}^R$, that is,

$$\phi_{m,n}^* \gamma_{m+n} \simeq \gamma_m \times \gamma_n.$$

Since Stiefel–Whitney classes are natural with respect to maps, the general Whitney product formula is equivalent to the universal product formula

$$\phi_{m,n}^* w_k(\gamma_{m+n}) = \sum_{i+j=k} w_i(\gamma_m) \times w_j(\gamma_n).$$

The mod 2 cohomology ring $H^*(G_m^R \times G_n^R) = H^*(G_m^R) \otimes H^*(G_n^R)$ is the polynomial ring freely generated by the Stiefel–Whitney classes

$$w_1(\gamma_m), \ldots, w_m(\gamma_m), w_1(\gamma_n), \ldots, w_n(\gamma_n).$$

Hence, $\psi_{m,n}^* w_k(\gamma_{m+n})$ must be equal to a polynomial in these classes.

We claim that this polynomial is

$$\phi_{m,n}^* w_k(\gamma_{m+n}) = \sum_{i+j=k} w_i(\gamma_m) w_i(\gamma_n).$$

Assume that this formula is true for all $m' \leq m$ and $n' \leq m$ with $m' + n' < m + n$ and hence that the Whitney product formula is true for all bundles of these dimensions.

Replacing γ_m by $\epsilon \oplus \gamma_{m-1}$ and γ_{m+n} by $\epsilon \oplus \gamma_{m+n+1}$ has the effect of setting $w_m(\gamma_m) = 0$ and $w_{m+n}(\gamma_{m+n}) = 0$ and leaving all the other Stiefel–Whitney classes unchanged. Hence, our inductive hypotheses shows that the formula is true modulo the ideal generated by $w_m(\gamma_m)$.

Similarly, the formula is true modulo the ideal generated by $w_n(\gamma_n)$. Since we are in a unique factorization domain, the formula is true modulo the ideal generated by the product $w_m(\gamma_m) w_n(\gamma_n)$. Thus the formula is true for all $k < m+n$ and

$$\phi_{m,n}^* w_{m+n}(\gamma_{m+n}) = w_m(\gamma_m) w_n(\gamma_n)$$

modulo the ideal generated by $w_m(\gamma_m) w_n(\gamma_n)$. But this is the top Stiefel–Whitney class and is therefore equal to the mod 2 Euler class. The product formula for mod 2 Euler classes says that this formula is exactly true. $\square$

Since the Whitney sum bundle $\xi \oplus \rho$ is the pullback of the product bundle via the diagonal $\Delta^*(\xi \times \rho)$, we can also write the Whitney product formula in cup product form:

Whitney product formula 12.18.2.

$$w_k(\xi \oplus \rho) = \sum_{i+j=k} w_i(\xi) \cup w_j(\rho).$$

If ξ is any complex n-plane bundle over B with classifying map $f: B \to G_n^C$, that is, there is an isomorphism with the pullback of the universal complex vector bundle, $f^*\gamma_n \simeq \xi$, then the Chern classes can be defined in a similar fashion $f^* c_i(\gamma_i) = c_i(\xi)$ and the obvious analogs of Definition 12.18.1 and Axioms 12.18.2 can be verified by essentially the same proofs.

12.19 Applications of Stiefel–Whitney classes

Let ξ be a real vector bundle over a base space B. The total Stiefel–Whitney class of ξ is the formal sum

$$w(\xi) = 1 + w_1(\xi) + w_2(\xi) + \cdots + w_k(\xi) + \cdots$$

in the mod 2 cohomology ring

$$H^\Pi(B) = H^0 B \oplus H^1 B \oplus H^2 B \oplus \cdots \oplus H^k B \oplus \cdots.$$

Since the leading coefficient is 1, $w(\xi)$ is a unit in this ring with multiplicative inverse

$$\overline{w}(\xi) = 1 + \overline{w}_1(\xi) + \overline{w}_2(\xi) + \cdots + \overline{w}_k(\xi) + \cdots.$$

The terms $\overline{w}_k(\xi)$ are uniquely determined by the relations

$$\overline{w}_k(\xi) + \overline{w}_{k-1}(\xi) w_1(\xi) + \overline{w}_{k-2}(\xi) w_2(\xi) + \cdots + \overline{w}_1(\xi) w_{k-1}(\xi) + w_k(\xi) = 0.$$

In terms of the total Stiefel–Whitney class, the Whitney product formula for Whitney sums becomes

$$w(\xi \oplus \eta) = w(\xi) w(\eta).$$

Suppose that M is a differentiable manifold of dimension n and that there is an immersion $f: M \to R^{n+k}$ into some Euclidean space. We will denote the n-dimensional tangent bundle of M by τ_m and the k-dimensional normal bundle to the immersion by ν_M. Since $\tau_M \oplus \nu_M \simeq \epsilon^{n+k}$ is an $(n+k)$-dimensional trivial bundle, we have

Whitney duality theorem 12.19.1. *The total Stiefel–Whitney class of the normal bundle is the multiplicative inverse of the total Stiefel–Whitney class of the tangent bundle,*

$$w(\nu_M) = \overline{w}(\tau_M) = w^{-1}(\tau_M).$$

Hence, in order to study immersions of manifolds, we need to identify the tangent bundle.

12.19 Applications of Stiefel–Whitney classes

Let $\gamma_{1,n}$ be the canonical line bundle over the real projective space RP^n. The total space is

$$E(\gamma_{1,n}) = \{(L,v) \mid L \in RP^n \text{ is a line through the origin in } R^{n+1}, v \in L\}.$$

Since this is the pullback (= restriction) of the universal line bundle γ_1 over RP^∞, we have the total Stiefel–Whitney class

$$w(\gamma_{1,n}) = 1 + u$$

where u is the generator of the mod 2 cohomology group $H^1(RP^n)$.

Proposition 12.9.2. *If ϵ^1 is a trivial line bundle, then*

$$\tau_{RP^n} \oplus \epsilon^1 \simeq \gamma_{1,n} \oplus \cdots \oplus \gamma_{1,n} = (n+1)\gamma_{1,n}$$

is a Whitney sum of $n+1$ copies of the canonical line bundle.

Proof: The total space of the canonical line bundle has two descriptions as identification spaces

$$E(\gamma_{1,n}) = \begin{cases} \{(x, tx) \mid x \in S^n, (x, tx) \sim (-x, tx)\} \text{ or, sending } (x, tx) \mapsto (x, t), \\ \{(x, t) \mid x \in S^n, \quad (x, t) \sim (-x, -t)\}. \end{cases}$$

The total space of the tangent bundle is the identification space

$$E(\tau_{RP^n}) = \{(x, v) \mid x \in S^n, v \in R^{n+1}, v \perp x, \quad (x, v) \sim (-x, -v)\}.$$

Hence, the total space

$$E(\tau_{RP^n} \oplus \epsilon^1) = \{(x, v, t) \mid x \in S^n, v \in R^{n+1}, v \perp x,$$
$$(x, v, t) \sim (-x, -v, t)\} = \{(x, v, tx) \mid x \in S^n, v \in R^{n+1}, v \perp x,$$
$$(x, v, tx) \sim (-x, -v, -tx)\} = \{(x, w) \mid x \in S^n, w \in R^{n+1},$$
$$(x, w) \sim (-x, -w)\}.$$

Writing $w = t_1 e_1 + t_2 e_2 + \cdots + t_{n+1} e_{n+1}$ shows that $\tau_{RP^n} \oplus \epsilon^1 \simeq (n+1)\gamma_{1,n}$. $\square$

The Whitney product formula shows

Corollary 12.9.3. *The total Stiefel–Whitney class of the tangent bundle of the n-dimensional real projective space is*

$$w(\tau_{RP^n}) = (1+u)^{n+1} = 1 + (n+1)u + \binom{n+1}{2} u^2 + \cdots + \binom{n+1}{n} u^n.$$

We can prove the following nonimmersion theorem.

Proposition 12.19.4. *If RP^n immerses in the Euclidean space R^{n+k} and $n = 2^r$ is a power of 2, then $k \geq n-1$.*

Proof: Since
$$w(\tau_{RP^n}) = (1+u)^{n+1} = (1+u)^{2^r}(1+u) = (1+u^{2^r})(1+u) = 1+u+u^{2^r}$$
mod 2, it follows that the total Stiefel–Whitney class of the normal bundle is
$$w(\nu_{RP^N}) = w^{-1}(\tau_{RP^n})$$
$$= 1 + (u + u^{2^r}) + (u + u^{2^r})^2 + \cdots = 1 + u + u^2 + \cdots + u^{2^r - 1}.$$
Hence, the dimension of the normal bundle is $\geq n - 1$. $\square$

Remark. In fact the Whitney immersion theorem says that every n-dimensional manifold can be immersed in the Euclidean space R^{2n-1} if $n > 1$. Hence, in this one case, the above nonimmersion result is the best possible.

We note that the total Stiefel–Whitney class $w(\tau_{RP^n}) = (1+u)^{n+1} = 1$ if and only if $n + 1 = 2^r$ is a power of 2. Hence

Proposition 12.19.5. *If the projective space RP^n is parallelizable, that is, has a trivial tangent bundle, then $n = 2^r - 1$ must be one less than a power of 2.*

This is related to the nonexistence of real divison algebras by the following result of Stiefel.

Proposition 12.19.6. *If there exists a bilinear product operation*
$$R^n \times R^n \to R^n, \quad (x, y) \mapsto x \cdot y$$
with no zero divisors, then the projective space RP^{n-1} is parallelizable, hence, n must be a power of 2.

Proof: Let $e_1, \ldots, e_n$ be a fixed basis of R^n. Given an x the equation $a \cdot e_1 = x$ can be solved uniquely for $a = f(x)$ where $f(x)$ is a one to one onto linear function of x. Suppose that x and hence a are nonzero.

The elements $a \cdot e_1, \ldots, a \cdot e_2, \ldots, a \cdot e_n$ are linearly independent and hence so are the elements $x, x_2, \ldots, x_n$ where x_i is the projection of $a \cdot e_i$ on the orthogonal complement of x. Since changing the sign of x changes the sign of the x_i, the vectors $x_2, \ldots, x_n$ provide a framing of the tangent bundle of RP^{n-1}. $\square$

Remark. The real numbers, the complex numbers, the quaternions, and the Cayley numbers provide examples of real division algebras in the only dimensions 1,2,4, and 8 where they actually exist. [2, 74, 12]

Bibliography

[1] J. F. Adams. On the cobar construction. *Proc. Natl. Acad. Sci. USA*, **42**:409–412, 1956.
[2] J. F. Adams. On the nonexistence of elements of hopf invariant one. *Ann. Math.*, **72**:20–104, 1960.
[3] J. F. Adams. *Algebraic Topology – A Student's Guide, London Math. Soc. Lecture Notes Series 4*. Cambridge University Press, 1972.
[4] J. F. Adams. *Stable Homotopy and Generalized Homology*. Chicago University Press, 1974.
[5] J. F. Adams. *Infinite Loop Spaces, Annals of Math. Studies 90*. Princeton University Press, 1978.
[6] D. Anderson. Localizing cw complexes. *Ill. Jour. Math.*, **16**:519–525, 1972.
[7] M. F. Atiyah and I. G. Macdonald. *Introduction to Commutative Algebra*. Addison-Wesley, 1969.
[8] M. G. Barratt. Track groups i. *Proc. London Math. Soc.*, **s3–5**:71–106, 1955.
[9] M. G. Barratt. Spaces of finite characteristic. *Quart. J. Math. Oxford*, **11**:124–136, 1960.
[10] A. Borel. *Topics in the Homology Theory of Fibre Bundles*. Springer-Verlag, 1967.
[11] R. Bott. The stable homotopy of the classical groups. *Ann. Math.*, **70**:313–337, 1959.
[12] R. Bott and J. W. Milnor. On the parallelizability of spheres. *Bull. Amer. Math. Soc.*, **64**:87–89, 1958.
[13] R. Bott and H. Samelson. On the pontrjagin product in spaces of paths. *Comm. Math. Helv.*, **27**:320–337, 1953.
[14] A. K. Bousfield. The localization of spaces with respect to homology. *Topology*, **14**:133–150, 1975.
[15] A. K. Bousfield. Localization and periodicity in unstable homotopy theory. *J. Amer. Math. Soc.*, **7**:831–874, 1994.
[16] A. K. Bousfield. Unstable localizations and periodicity. In C. Brota, C. Casacuberta, and G. Mislin (editors), *Algebraic Topology: New Trends in Localization and Periodicity*. Birkhäuser, 1996.
[17] A. K. Bousfield and D. M. Kan. *Homotopy Limits, Completions, and Localization, Lecture Notes in Math 304*. Springer-Verlag, 1972.
[18] A. K. Bousfield and D. M. Kan. The homotopy spectral sequence of a space with coefficients in a ring. *Topology*, **11**:79–106, 1972.
[19] W. Browder. Torsion in H-spaces. *Ann. Math.*, **74**, 1961.
[20] E. H. Brown. Twisted tensor products i. *Ann. Math.*, **59**:223–246, 1960.
[21] K. S. Brown. *Cohomology of Groups*. Springer-Verlag, 1982.
[22] H. Cartan. *Algebres d'Eilenberg-MacLane, Seminaire Henri Cartan 1954/55, exposes 2-11*. Ecole Normal Supérieure, 1955.
[23] H. Cartan and S. Eilenberg. *Homological Algebra*. Princeton University Press, 1956.
[24] W. Chachólski and J. Sherer. *Homotopy Theory of Diagrams*. Amer. Math. Soc., 2002.
[25] F. R. Cohen. Splitting certain suspensions via self-maps. *Illinois J. Math.*, **20**:336–347, 1976.
[26] F. R. Cohen, J. C. Moore, and J. A. Neisendorfer. The double suspension and exponents of the homotopy groups of spheres. *Ann. Math.*, **110**:549–565, 1979.
[27] F. R. Cohen, J. C. Moore, and J. A. Neisendorfer. Torsion in homotopy groups. *Ann. Math.*, **109**:121–168, 1979.

[28] F. R. Cohen, J. C. Moore, and J. A. Neisendorfer. Exponents in homotopy theory. In W. Browder (editor), *Algebraic Topology and Algebraic K-Theory*, pp. 3–34. Princeton University Press, 1987.
[29] F. R. Cohen and J. A. Neisendorfer. Note on desuspending the Adams map. *Proc. Camb. Phil. Soc.*, **99**:59–64, 1986.
[30] J. Cohen. The homotopy groups of inverse limits. *Proc. London Math Soc.*, **27**:159–177, 1973.
[31] E. B. Curtis. Simplicial homotopy theory. *Adv. Math*, **6**:107–209, 1971.
[32] A. Dold. Partitions of unity in the theory of fibrations. *Ann. Math.*, **78**:223–255, 1963.
[33] A. Dold. *Lectures on Algebraic Topology*. Springer-Verlag, 1972.
[34] A. Dold and R. Thom. Quasifaserungen und unendliche symmetriche produkte. *Ann. Math.*, **67**:239–281, 1958.
[35] E. Dror, W. G. Dwyer, and D. M. Kan. An arithmetic square for virtually nilpotent spaces. *Illinois J. Math.*, **21**:242–254, 1977.
[36] E. Dror Farjoun. *Cellular Spaces, Null Spaces, and Homotopy Localization, Lecture Notes in Math. 1622*. Springer-Verlag, 1995.
[37] W. G. Dwyer, P. S. Hirschhorn, D. M. Kan, and J. H. Smith. *Homotopy Limit Functors on Model Categories and Homotopical Categories*. Amer. Math. Soc., 2004.
[38] W. G. Dwyer and C. W. Wilkerson. Homotopy fixed point methods for lie groups and classifying spaces. *Ann. Math.*, **139**:395–442, 1984.
[39] E. Dyer and J. Roitberg. Note on sequences of Mayer–Vietoris type. *Proc. Amer. Math. Soc.*, **80**:660–662, 1980.
[40] S. Eilenberg. Singular homology theory. *Ann. Math.*, **45**:63–89, 1944.
[41] S. Eilenberg and S. MacLane. On the groups $H(\pi, n)$, *I*. *Ann. Math.*, **58**:55–106, 1953.
[42] S. Eilenberg and J. C. Moore. Homology and fibrations I, coalgebras, cotensor product and its derived functors. *Comm. Math. Helv.*, **40**:199–236, 1966.
[43] S. Eilenberg and J. C. Moore. *Foundation of Relative Homological Algebra*. Memoirs Amer. Math. Soc. 55, 1968.
[44] S. Eilenberg and N. Steenrod. *Foundations of Algebraic Topology*. Princeton University Press, 1952.
[45] H. Freudenthal. Uber die klassen der spharenabbildungen. *Comp. Math.*, **5**:299–314, 1937.
[46] B. I. Gray. On the sphere of origin of infinite families in the homotopy groups of spheres. *Topology*, **8**:219–232, 1969.
[47] B. I. Gray. *Homotopy Theory*. Academic Press, 1975.
[48] B. I. Gray. Associativity in two-cell complexes. *Contemp. Math.*, **258**:185–196, 2000.
[49] M. J. Greenburg and J. R. Harper. *Algebraic Topology, A First Course*. Benjamin-Cummings, 1981.
[50] J. R. Harper. *Secondary Cohomology Operations, Graduate Studies in Mathematics 49*. Amer. Math. Soc., 2002.
[51] A. Hatcher. *Algebraic Topology*. Cambridge University Press, 2001.
[52] K. Hess and R. Levi. An algebraic model for the loop space homology of a homotopy fiber. *Algebr. Geom. Topol.*, **7**:1699–1765, 2007.
[53] P. Hilton, G. Mislin, and J. Roitberg. *Localization of Nilpotent Groups and Spaces, North Holland Math. Studies 15*. Elsevier, 1975.
[54] P. J. Hilton. On the homotopy groups of a union of spheres. *J. London Math. Soc.*, **30**:154–172, 1955.
[55] P. J. Hilton and J. Roitberg. On principal S^3 bundles over spheres. *Ann. Math.*, **90**:91–107, 1969.
[56] P. J. Hilton and J. Roitberg. On the classification problem for H-spaces of rank two. *Comm. Math. Helv.*, **46**:506–516, 1971.
[57] P. Hirschhorn. *Model Categories and Their Localizations*. Amer. Math. Soc., 2003.

[58] H. Hopf. Uber die abbildungen von spharen niedriger dimensionen. *Fund. Math.*, **25**:427–440, 1935.
[59] H. Hopf. Uber die topologie der gruppen-mannigfaltigkeiten und ihrer verallgemeinerungen. *Ann. Math.*, **42**:22–52, 1941.
[60] M. Hovey. *Model Categories*. Amer. Math. Soc., 1991.
[61] J. R. Hubbuck. Two lemmas on primary cohomology operations. *Camb. Phil. Soc.*, **68**:631–636, 1970.
[62] W. Hurewicz. Beitrage zur topologie der deformationen. *Nedrl. Akad. Wetensch. Proc. Ser. A*, **38**, **39**:521–528,117–126,215–224, 1935,1936.
[63] W. Hurewicz. On the concept of fiber space. *Proc. Nat. Acad. USA*, **41**:956–961, 1953.
[64] D. Husemoller, J. C. Moore, and J. D. Stasheff. Differential homological algebra and homogeneous spaces. *J. Pure Appl. Alg.*, **5**:113–185, 1974.
[65] N. Jacobson. *Lie Algebras*. Dover, 1962.
[66] I. M. James. Reduced product spaces. *Ann. Math.*, **62**:170–197, 1955.
[67] I. M. James. On the suspension sequence. *Ann. Math.*, **65**:74–107, 1957.
[68] B. W. Jordan. A lower central series for split hopf algebras with involution. *Trans. Amer. Math. Soc.*, **257**:427–454, 1980.
[69] D. M. Kan. A combinatorial definition of homotopy groups. *Ann. Math.*, **67**:282–312, 1958.
[70] R. K. Kane. *The Homology of Hopf Spaces, North Holland Math. Studies 40*. North Holland, 1988.
[71] I. Kaplansky. Projective modules. *Ann. Math.*, **68**:372–377, 1958.
[72] I. Kaplansky. *Infinite Abelian Groups*. University of Michigan Press, 1971.
[73] I. Kaplansky. *Set Theory and Metric Spaces*. Amer. Math. Soc. Chelsea, 1972.
[74] M. Kervaire. Non-parallelizability of the n-sphere for $n > 7$. *Proc. Natl. Acad. Sci. USA*, **44**:280–283, 1958.
[75] J. Lannes. Sur la cohomologie modulo p des p-groups abeliens elementaires. In E. Ress and J. D. S. Jones (editors), *Homotopy Theory, Proc. Durham Symp*. Cambridge University Press, 1985.
[76] A. Liulevicius. *The factorization of cyclic reduced powers by secondary cohomology operations, Memoirs. Amer. Math. Soc.*, **42**, 1962.
[77] S. MacLane. *Homology*. Springer-Verlag, 1963.
[78] W. Massey. Exact couples in algebraic topology I. *Ann. Math.*, **56**:363–396, 1952.
[79] W. Massey. Exact couples in algebraic topology II. *Ann. Math.*, **57**:248–286, 1953.
[80] W. Massey. Products in exact couples. *Ann. Math.*, **59**:558–569, 1954.
[81] J. P. May. *Simplicial Objects in Algebraic Topology*. Van Nostrand, 1967.
[82] J. McCleary. *A User's Guide to Spectral Sequences*. Cambridge University Press, 2nd edition, 2001.
[83] C. A. McGibbon and J. A. Neisendorfer. On the homotopy groups of a finite dimensional space. *Comm. Math. Helv.*, **59**:253–257, 1984.
[84] H. R. Miller. The Sullivan conjecture on maps from classifying spaces. *Ann. Math.*, **120**:39–87, 1984.
[85] J. W. Milnor. Construction of universal bundles I. *Ann. Math.*, **63**:272–284, 1956.
[86] J. W. Milnor. Construction of universal bundles II. *Ann. Math.*, **63**:430–436, 1956.
[87] J. W. Milnor. On spaces having the homotopy type of a cw-complex. *Trans. Amer. Math. Soc.*, **90**:272–280, 1959.
[88] J. W. Milnor. *Morse Theory, Annals of Math. Studies 51*. Princeton University Press, 1963.
[89] J. W. Milnor. On the construction fk, 1956 Princeton notes. In Adams J. F. (editor), *Algebraic Topology–A Student's Guide*. Cambridge University Press, 1972.
[90] J. W. Milnor and J. C. Moore. On the structure of Hopf algebras. *Ann. Math.*, **81**:211–264, 1965.
[91] J. W. Milnor and J. E. Stasheff. *Characteristic Classes, Annals of Math. Studies 76*. Princeton University Press, 1974.

[92] J. C. Moore. On the homotopy groups of spaces with a single non-vanishing homology group. *Ann. Math.*, **59**:549–557, 1954.

[93] J. C. Moore. The double suspension and p-primary components of the homotopy groups of spheres. *Boll. Soc. Mat. Mexicana*, **1**:28–37, 1956.

[94] J. C. Moore. *Algebre homologique et homologie des espace classificants, Seminaire Henri Cartan 1959/60, expose 7*. Ecole Normal Supérieure, 1960.

[95] J. C. Moore. *La suspension, Seminaire Henri Cartan 1959/60, expose 6*. Ecole Normal Supérieure, 1960.

[96] J. C. Moore and J. A. Neisendorfer. A view of some aspects of unstable homotopy theory since 1950. In J. Jones (editor), *Homotopy Theory*. Cambridge University Press, 1987.

[97] J. C. Moore and J. A. Neisendorfer. Equivalence of Toda Hopf inavariants. *Israel J. Math.*, **66**:300–318, 1989.

[98] R. Mosher and M. Tangora. *Cohomology Operations and Applications in Homotopy Theory*. Harper and Row, 1968.

[99] J. A. Neisendorfer. *Primary Homotopy Theory, Memoirs A.M.S. 232*. Amer. Math. Soc., 1980.

[100] J. A. Neisendorfer. 3-primary exponents. *Math. Proc. Camb. Phil. Soc.*, **90**:63–83, 1981.

[101] J. A. Neisendorfer. Properties of certain H-spaces. *Quart. J. Math. Oxford*, **34**:201–209, 1981.

[102] J. A. Neisendorfer. The exponent of a Moore space. In W. Browder (editor), *Algebraic Topology and Algebraic K-Theory*, pages 35–71. Princeton University Press, 1987.

[103] J. A. Neisendorfer. Localization and connected covers of finite complexes. *Contemp. Math.*, **181**:385–390, 1995.

[104] F. P. Peterson. Generalized cohomotopy groups. *Amer. J. Math.*, **78**:259–281, 1956.

[105] F. P. Peterson and N. Stein. Secondary cohomology operations: two formulas. *Amer. J. Math.*, **81**:281–305, 1959.

[106] H. Poincare. Analysis situs. *Jour. Ecole Polytech.*, **1**:1–123, 1895.

[107] D. Puppe. Homotopiemengen und ihre induzierten abbildungen. *Math. Zeit.*, **69**:299–344, 1958.

[108] D. Quillen. *Homotopical algebra*. Springer-Verlag, 1967.

[109] D. Quillen. The geometric realization of a Kan fibration is a Serre fibration. *Proc. Amer. Math. Soc.*, **19**:1499–1500, 1968.

[110] D. Quillen. Rational homotopy theory. *Ann. Math.*, **90**:295–295, 1969.

[111] H. Samelson. A connection between the Whitehead and the Pontrjagin product. *Amer. J. Math.*, **75**:744–752, 1953.

[112] P. S. Selick. Odd primary torsion in $\pi_k(S^3)$. *Topology*, **17**:407–412, 1978.

[113] P. S. Selick. 2-primary exponents for the homotopy groups of spheres. *Topology*, **23**:97–98, 1984.

[114] P. S. Selick. *Introduction to Homotopy Theory, Fields Inst. Monographs 9*. Amer. Math. Soc., 1991.

[115] P. S. Selick. Space exponents for loop spaces of spheres. In W. Dwyer *et al.* (editor), *Stable and Unstable Homotopy*. Amer. Math. Soc., 1998.

[116] J.-P. Serre. Homologie singuliere des espaces fibre. *Ann. Math.*, **54**:425–505, 1951.

[117] J.-P. Serre. Cohomologie modulo 2 des complexes d'Eilenberg–MacLane. *Comm. Math. Helv.*, **27**:198–231, 1953.

[118] J.-P. Serre. Groupes d'homotopie et classes de groupes abeliens. *Ann. Math.*, **58**:258–294, 1953.

[119] J.-P. Serre. *Lie Algebras and Lie Groups*. Benjamin, 1965.

[120] S. J. Shiffman. *Ext p-completion in the homotopy category*. PhD thesis, Dartmouth College, 1974.

[121] N. Shimada and T. Yamanoshita. On triviality of the mod p Hopf invariant. *Japan J. Math.*, **31**:1–25, 1961.

[122] L. Smith. Homological algebra and the Eilenberg-Moore spectral sequence. *Trans. Amer. Math. Soc.*, **129**:58–93, 1967.

[123] E. H. Spanier. *Algebraic Topology*. McGraw-Hill, 1966.

[124] J. D. Stasheff. *H-spaces from the Homotopy Point of View*. Springer-Verlag, 1970.
[125] N. E. Steenrod. *The Topology of Fibre Bundles*. Princeton University Press, 1951.
[126] N. E. Steenrod. A convenient category of topological spaces. *Michigan Math. J.*, **14**:133–152, 1967.
[127] N. E. Steenrod and D. B. A. Epstein. *Cohomology Operations, Annals of Math. Studies 50*. Princeton University Press, 1962.
[128] D. Sullivan. *Geometric Topology*. MIT, 1970.
[129] D. Sullivan. Genetics of homotopy theory and the Adams conjecture. *Ann. Math.*, **100**:1–79, 1974.
[130] H. Toda. On the double suspension E^2. *J. Inst. Polytech. Osaka City Univ. Ser. A*, **7**:103–145, 1956.
[131] H. Toda. p-primary components of homotopy groups II, mod p Hopf invariant. *Mem. Coll. Sci. Univ. of Kyoto Ser. A*, **31**:143–160, 1958.
[132] H. Toda. *Composition Methods in the Homotopy Groups of Spheres, Princeton Math Series 49*. Princeton University Press, 1962.
[133] G. W. Whitehead. On mappings into group-like spaces. *Comm. Math. Helv.*, **28**:320–328, 1954.
[134] G. W. Whitehead. *Elements of Homotopy Theory*. Springer-Verlag, 1978.
[135] J. H. C. Whitehead. Combinatorial homotopy I. *Bull. Amer. Math. Soc.*, **55**:213–245, 1949.
[136] J. H. C. Whitehead. On simply connected 4-dimensional polyhedra. *Comm. Math. Helv.*, **22**:48–92, 1949.
[137] A. Zabrodsky. Homotopy associativity and finite CW complexes. *Topology*, **9**:121–128, 1970.
[138] A. Zabrodsky. The classification of simply connected H-spaces with three cells I,II. *Math. Scand.*, **30**:193–210,211–222, 1972.
[139] A. Zabrodsky. On the construction of new finite CW H-spaces. *Invent. Math.*, **16**:260–266, 1972.
[140] A. Zabrodsky. On the genus of new finite CW H-spaces. *Comm. Math. Helv.*, **49**:48–64, 1972.
[141] A. Zabrodsky. On the homotopy type of principal classical group bundles over spheres. *Israel J. Math.*, **11**:315–325, 1972.
[142] A. Zabrodsky. *Hopf Spaces*. North-Holland, 1976.
[143] A. Zabrodsky. Phantom maps and a theorem of H. Miller. *Israel J. Math.*, **58**:129–143, 1987.
[144] E. C. Zeeman. A proof of the comparison theorem for spectral sequences. *Proc. Camb. Phil. Soc.*, **53**:57–62, 1957.

Index

adjoint equivalences, 335
adjoint functors between algebras and coalgebras B and Ω, 333
adjunction maps, 334
Alexander–Whitney maps, 380
   associative, 380
   define coalgebra structure, 381
algebraic Hurewicz theorem, 460
augmented algebras, 315
   derivations, 317
   differential, 316
   generating modules, 316
   indecomposables, 316

balanced functor, 495
bar construction, 326
   suspension, 525
bar resolution, 275
biconstructions, 371
bimodules, 509
Bockstein exact homotopy sequence, 22
Bockstein spectral sequence, 5
   homology of the loop space, 456
   homotopy, 5
Bockstein spectral sequences, 221
   cohomology, 238
   convergence, 230
   detecting the order of torsion, 227
   extensions of maps, 230
   homology, 235
   homotopy, 225
   Samelson products, 232
      Lie identities at the prime 3, 234
   universal coefficient theorem, 227
Borel construction, 505
Borel transgression theorem, 519, 529
Bott–Samelson, 6, 107, 110, 417
Browder's implication theorem, 243

Cartan's construction characterization., 370
Cartan's constructions, 369

differential Cotor, 376
   homological invariant of cotensor product, 375
chain models
   fibre of the pinch map, 450
   loops on spheres, 450
change of rings isomorphism, 274
   Lie bracket action, 277
Chern classes, 537
   axioms, 542
   definition, 537, 542
classifying construction on a differential algebra, 326
classifying spaces, 502
   cohomology, 537
   commutative topological groups, 521
   Dold's theorem, 502
   Grassmann models, 534
   homology, 537
   Milnor's construction, 503
coalgebra kernels, 410
cobar construction, 313, 326
comodules, 345
   differential, 346
   extended, 347
   injective, 347
   injective and bounded below implies extended, 347
   primitives, 346
completion, 58, 63
   abelian groups, 58
   spaces, 63
connected cover of S^3, 124
cotensor product, 356
   associative, distributive, unit, 358
   derived functors, 359
      balanced, 360
   primitives, 358
   proper left exact, 357
Cotor, 359

derived functors of tensor product, 343
   balanced, 344

descending central series, 178
    associated Lie algebra, 178
differential Cotor, 314, 368
    acyclic twisted tensor product
        gives algebra structure, 407
    balanced, 377
    coalgebra structure, 393
    coalgebra structure not intrinsic, 394
    cobar construction
        gives algebra structure, 406
    commutative algebra structures, 408
    equal algebra structures, 408
    homological invariance, 390, 400
    Hopf algebra structure, 409
    several variables, 397
        algebra structure, 404
        balanced, 397
        coalgebra structure, 402
        collapse, 398
        tensor splitting, 398
    well defined, 400
differential graded Hopf algebras, 494
differential homological algebra, 313
differential module coalgebras, 495
differential Tor, 314, 495, 497
    commutative coalgebra structures, 518
    coproducts, 515
    geometric diagonal, 520
    homological invariance, 500
    Hopf algebra structures, 518
    several variables, 510
        balanced, 510
        collapse, 512
        splitting, 511
differentials
    internal, external, 363, 368
divided power algebras, 135, 136
division algebras, 544
double loop spaces of spheres
    homology, 418
double suspension
    factorization, 148

EHP sequence, 118, 132
Eilenberg subcomplexes, 195
Eilenberg–Moore
    Cotor approximation, 315, 385
        algebra structure, 405
        coalgebra structure, 403
        Hopf algebra structure, 408
        several variables, 401

spectral sequence, 315, 387
    algebra and coalgebra structure, 389
    coalgebra structure, 394
    edge homomorphism, 389
Eilenberg–Moore models
    Borel constructions, 505
    several variables, 512
Eilenberg–Moore spectral sequence, 528
    algebraic , 499
    suspension, 528
Eilenberg–Zilber maps, 380, 494
    associative, 381
    maps of differential coalgebras, 381
Euler class, 532
    definition via sphere bundles, 533
    definition via vector bundles, 532
    product formula, 532
exact couples, 221, 222
exponent theorems, 2
    Cohen–Moore–Neisendorfer, 3, 136, 475
    Gray, 3, 136, 438
    H-space exponents, 2, 3
    homology of the loop space, 453
    James 2-primary, 2, 107, 121
    odd primary Moore spaces, 10, 438, 480
    Selick, 3, 135
    Toda's odd primary, 3, 135, 155
exponents of H-spaces, 81
    nonexistence, 81
extended ideals in graded Lie algebras, 200
    differential, 200

factorization of p-th powers, 474
factorization of degree p maps, 472
fibre extensions, 94
    totally fibred cubes, 101
    totally fibred squares, 96
fibre of the pinch map, 131, 160, 447
    homology, 447
fibres of degree k maps, 171
frame bundles, 534

generating complexes, 425
geometric collapse, 513
geometric splitting, 513
graded Lie algebras, 189, 251, 252
    differential, 190
        cycles, 299
        homology, 306
        embeds in universal enveloping algebra, 265

graded Lie algebras (*cont.*)
  free, 270
    generators of commutator subalgebra, 287
    subalgebras, 278
    universal enveloping algebras, 255
  free subalgebras, 251
  generators, 251
  generators of subalgebras of free, 279
  indecomposables, 256
  quasi, 253
  short exact sequences, 266
  universal enveloping algebras, 251, 253
    free modules over subalgebras, 265
    primitives, 265
group models for loop spaces, 193
  fibrations, 196
  Kan, 195
  Milnor, 195

H-space exponents, 123, 155, 438
  Cohen–Moore–Neisendorfer, 478
  double loops on Moore spaces, 484, 485
  fibre of the double suspension, 479
  fibres of power maps, 446
  James, 478
  nonexistence, 81, 478
  nonexistence on loop suspensions, 485
  Selick, 478
  Toda, 478
  universal covers of iterated loops on spheres, 478
H-spaces
  fibre of the double suspension, 443
  fibres of power maps, 443
  keeping basepoints stationary, 440
  localized spheres, 443
  Moore loop space, 442
Hilton–Hopf invariants, 116, 145
Hilton–Roitberg examples, 86
homology invariant functors, 374
homology suspension, 169, 413
  Eilenberg–Moore spectral sequence, 415
  inverse to transgression, 414
  primitives and indecomposables, 413
homotopy groups, 1
  elements of order 4, 128, 132
  finite generation, 2
  higher order torsion, 9, 284, 303
  Hopf, 1
  Hurewicz, 1

infinitely many nonzero, 2, 78
  Poincaré, 1
  Serre's α_1, 1, 126
homotopy groups with coefficients, 4, 11
  definition, 13
  exponents, 20, 164
  group structure, 14, 34
  nonfinitely generated coefficients, 23
  Peterson, 4
  universal coefficient theorem, 11, 16
homotopy invariant functors, 374
homotopy pullbacks
  several variables, 395
  balanced, 396
  collapse, 395
  loop multiplication, 403
  product splitting, 395
homotopy theoretic fibres, 95
Hopf algebra kernels, 411
Hopf algebras, 241
  Borel, 248
  Hopf–Borel, 248
  primitive-indecomposable exact sequence, 241, 422
  when primitives are indecomposable, 308
Hopf invariant one, 90, 284
  equivalent forms, 294
Hurewicz homomorphism, 4, 25
  factored, 305
Hurewicz isomorphism theorem, 27, 68

immersions, 542
injective class
  proper
    comodules, 352
    differential comodules, 353
  split, 351
injective classes, 314, 349
injective resolutions, 348
  existence, 348
  uniqueness up to chain homotopy, 348

James construction, 107, 111
James fibrations, 118, 135, 138, 143
James–Hopf invariants, 141
  combinatorial definition, 141
  decomposition definition, 144

Künneth formula, 392
Künneth spectral sequence, 500

Künneth theorem, 495, 517, 518
  algebraic, 499
  geometric, 500, 514

Lie algebras, 7
  differential graded, 7
  free, 7
  homology
    free differential graded, 9
    kernels, 7
  limits and derived functors, 53
localization, 35
  abelian groups, 46
  classical localization of spaces, 47
  fracture lemmas, 70
localization of connected covers, 76
loop construction on a differential coalgebra, 326
loop structures
  completions of spheres, 88
loops on the fibre of the pinch map
  Bockstein spectral sequence, 458
  mod p homology, 457

mapping cylinder, 507
Mayer–Vietoris homotopy sequence, 41
Miller's theorem, 78
Milnor–Moore, 5
mod 2 Euler class, 532
modules, 337
  differential, 337
  free, 338
  indecomposables, 337, 343
  projective, 338
  projective and bounded below implies free, 339
Moore comparison theorem, 506

Nakayama's lemma, 267, 338
  dual, 346
normal bundles, 542
normalized chains, 379
numerable bundles, 502

opposite algebras, 509

Peterson–Stein formula, 99
Poincaré–Birkhoff–Witt, 251, 257
  basis free, 264
product decomposition theorems, 6
  Cohen–Moore–Neisendorfer, 8, 9

geometric realization, 285, 287, 290, 437
  fibres of power maps, 289
Hilton–Milnor, 2, 7, 107, 113, 280, 290
  homology of the loop space, 464
  lift to loop space, 474
  loops on even dimensional Moore spaces, 283, 286
  loops on the fibre of the pinch map, 466
  Serre, 1, 136, 281, 283, 284
  universal enveloping algebras, 6
projective class, 490
projective classes, 314
projective resolutions
  existence, 339
  uniqueness up to chain homotopy, 339
proper injective comodule, 354
proper injective resolutions
  functorial, 354
proper monomorphism, 353
proper projective class, 492

real projective space, 542
  canonical line bundle, 542
  immersions in Euclidean space, 543
  parallelizability, 544
  tangent bundle, 543
reflection via adjoint functors, 351, 493
relative epimorphisms, 490
relative injective resolutions, 354
  existence, 354
  uniqueness up to chain homotopy, 355
relative injectives, 351
relative monomorphisms, 351
relative projective resolutions, 491
  existence, 491
  uniqueness up to chain homotopy, 491
relative projectives, 490
relative short exact sequences, 354, 491

Samelson products, 4, 158
  Bockstein derivations, 189
  external, 158, 179
    Lie identities, 184
    failure of Jacobi identity, 188
  internal, 158, 186
    Lie identities, 187
  Jacobi identity, 4

Samelson products (cont.)
  over loops on an H-space, 210, 475
    universal models, 210
  over loops on H-spaces, 159
  relative, 159, 198, 202, 475
    universal models, 202
  vanishing, 168
Serre filtration, 331, 383, 504
Serre's conjecture, 79
simplicial sets, 378
  skeletal filtration, 382, 383
smash products of Moore spaces, 158
  existence, 166
  uniqueness, 171
special orthogonal groups
  homology away from 2, 432
  mod 2 generating complexes, 431
  mod 2 homology, 431
special unitary groups
  generating complexes, 426
  homology, 427
  homology of loop space, 430
spectral systems, 221
Spin(n)
  homology of loop space away from 2, 435
standard proper embedding, 354
Stiefel–Whitney classes, 537
  axioms, 540
  definition, 538, 539
supplemented coalgebras, 317
  coderivations, 321
  differential, 318
  primitives, 319
  reduced coalgebra filtration, 319
  retraction onto primitives, 320
suspension, 525

tapered resolutions, 363
tensor algebras
  generators of subalgebras, 278
  homological characterization, 273
tensor products of graded modules, 341
  right exactness, 341
Toda fibrations, 135, 148, 154
Toda–Hopf invariants, 135, 151
torsion in H-spaces, 241
total complexes, 363, 495
total Stiefel–Whitney class, 542
totally nonhomologous to zero, 410
  short exact sequence of graded Lie algebras, 412
transgression, 525
  geometric suspension, 525
twisted tensor products, 313, 329, 373
  acyclic, 334
  homology of fibre equals differential Cotor, 412
  universal acyclic, 334, 335
twisting morphisms, 329

universal associative algebras, 323
universal associative coalgebra, 324
universal bundles, 502
universal twisting morphisms, 332
  universal property, 333

vector bundles, 530, 535
  orientable, 530

Whitehead products, 168
  vanishing, 169
Whitney duality theorem, 542

Zabrodsky mixing, 86
Zeeman comparison theorem, 50, 332